Springer Series in Optical Sciences

Volume 198

Springer Series in Optical Sciences

The Springer Series in Optical Sciences, under the leadership of Editor-in-Chief William T. Rhodes, Georgia Institute of Technology, USA, provides an expanding selection of research monographs in all major areas of optics: lasers and quantum optics, ultrafast phenomena, optical spectroscopy techniques, optoelectronics, quantum information, information optics, applied laser technology, industrial applications, and other topics of contemporary interest.
With this broad coverage of topics, the series is of use to all research scientists and engineers who need up-to-date reference books.
The editors encourage prospective authors to correspond with them in advance of submitting a manuscript. Submission of manuscripts should be made to the Editor-in-Chief or one of the Editors. See also www.springer.com/series/624

More information about this series at http://www.springer.com/series/624

John J. Healy • M. Alper Kutay
Haldun M. Ozaktas • John T. Sheridan
Editors

Linear Canonical Transforms

Theory and Applications

Springer

Editors
John J. Healy
School of Electrical and Electronic Engineering
University College Dublin
Belfield, Dublin, Ireland

Haldun M. Ozaktas
Department of Electrical Engineering
Bilkent University
Bilkent, Ankara, Turkey

M. Alper Kutay
The Scientific and Technological Research Council of Turkey
Kavaklıdere, Ankara, Turkey

John T. Sheridan
School of Electrical and Electronic Engineering
University College Dublin
Belfield, Dublin, Ireland

ISSN 0342-4111 ISSN 1556-1534 (electronic)
Springer Series in Optical Sciences
ISBN 978-1-4939-3027-2 ISBN 978-1-4939-3028-9 (eBook)
DOI 10.1007/978-1-4939-3028-9

Library of Congress Control Number: 2015946861

Springer New York Heidelberg Dordrecht London

Printed on acid-free paper

Springer International Publishing AG Switzerland is part of Springer Science+Business Media (www.springer.com)

Foreword

Waves underlie a wealth of natural phenomena, ranging from seismic activity to elementary particles, and encompassing light and sound. Mathematical tools that are useful for modeling and understanding the behavior of waves are therefore of central importance in science and engineering. The insights of Huygens and Fresnel led to the description of wave propagation in terms of secondary waves: if a wave field is known at a given initial plane, its propagation away from it can be modeled by expressing the field as a continuous superposition of secondary waves emanating from all points over the plane. This beautiful interpretation provides a connection between wave propagation and linear integral transformations. As underlined by Feynman's path-integral formalism of quantum mechanics, this interpretation also holds for the description of the temporal evolution of quantum-mechanical wave functions, where time plays the role of the propagation direction, and instead of an initial plane one must consider all space at an initial time.

The mathematical similarity between different wave phenomena becomes more accentuated when regimes that allow certain approximations are considered. For example, in the description of optical waves of a given temporal frequency, one is often interested in highly collimated beams that propagate mainly around a specific direction. In this case, the electric field distribution satisfies approximately what is known as the paraxial wave equation. This equation is mathematically similar to the Schrödinger equation ruling the evolution of quantum wave functions in the nonrelativistic regime. Further, the effect of some refractive index distributions on the propagation of an optical field can be formally analogous to that of some potentials over the evolution of a particle's wave function. It is then natural that the same propagation models be employed in the description of these systems.

This book gives a thorough overview of a class of integral transformations, known as linear canonical transformations, which are remarkable both for their mathematical elegance and for their range of physical applications. Mathematically, linear canonical transformations are defined by their simple properties: (1) each of these transformations is associated with, and fully determined by, a 2×2 matrix (or $2N \times 2N$ matrix, when applied to functions that depend on N variables); (2) a concatenation of a series of linear canonical transformations can be reduced to a single linear canonical transformation whose matrix is the product of the matrices for the original independent transformations. Physically, linear canonical transformations describe wave propagation in cases where the Hamiltonian is at most quadratic in both position (e.g., thin lenses and quadratic gradient index media in optics, or harmonic-oscillator potentials in quantum mechanics) and momentum (i.e., within the paraxial approximation in optics or the non-relativistic approximation in quantum mechanics). In these contexts, the matrix associated with the transformation turns out to be the transfer matrix that maps the initial position and momentum of a classical particle or ray to the final ones for the system in question. Linear canonical transformations include as special cases the Fourier transformation, the fractional Fourier transformation (which describes the paraxial propagation of optical fields in quadratic gradient index fibers, as well as the evolution of quantum states in a harmonic oscillator potential), the Fresnel transformation (which describes free propagation of paraxial wave beams), and even simple multiplication by quadratic phase factors.

This book is, to my knowledge, the first devoted fully to providing a comprehensive study of linear canonical transformations and their applications. Some previous publications have included some discussions on these transformations, while others have focused on specific special cases like the Fourier or even the fractional Fourier transformations. While some of these special cases are standard items in the toolbox of most physicists and engineers, the more general class of transformations discussed here is not as widely known. The present book is therefore a very timely and welcome addition to the scientific literature. Further, its chapters are authored by some of the most influential researchers in the subject. The first part of the book concentrates on the origins, definition, and properties of linear canonical transformations. Chapter 1, by Kurt Bernardo Wolf, gives a historical perspective on the independent development of linear canonical transformations in optics and nuclear physics, from the point of view of someone at the intersection of these two communities. In Chap. 2, Martin J. Bastiaans and Tatiana Alieva provide a detailed treatment of the definition and properties of linear canonical transformations, paying careful attention to cases of special interest. The eigenfunctions of linear canonical transformations, i.e. those functions that retain their functional form following transformation, are discussed by Soo-Chang Pei and Jian-Jiun Ding in Chap. 3. The different types of uncertainty relations between functions and their linear canonical transforms are the subject of Chap. 4, by R. Tao. In Chap. 5, Tatiana Alieva, José A. Rodrigo, Alejandro Cámara, and Martin J. Bastiaans discuss the application of linear canonical transformations to the modeling of light propagation through paraxial optical systems. Complementarily, M. Alper Kutay, Haldun M. Ozaktas, and José A. Rodrigo consider the use of simple optical systems for implementing linear canonical transformations, both in one and two variables in Chap. 6. The second part of the book focuses on practical aspects of the numerical implementation of linear canonical transformations. In Chap. 7, Figen S. Oktem and Haldun M. Ozaktas discuss the degrees of freedom involved in the implementation of a linear canonical transformation. The effects of sampling and discretization of linear canonical transformations are presented by John J.

Healy and Haldun M. Ozaktas in Chap. 8. Markus Testorf and Brian Hennelly investigate in Chap. 9 the effect known as self-imaging in systems described by linear canonical transformations. This part concludes with a discussion by Aykut Koç and Haldun M. Ozaktas in Chap. 10 about fast computational implementations of linear canonical transformations. The third and final part of the book is devoted to applications. This part opens with a study in Chap. 11 by Unnikrishnan Gopinathan, John Healy, Damien P. Kelly, and John T. Sheridan of the connection between linear canonical transformations and the retrieval of the phase of a field from the knowledge of its intensity. In Chap. 12, Damien P. Kelly and John T. Sheridan discuss the application of these transformations in digital holography. Applications to signal encryption are presented in Chap. 13 by Pramod Kumar, Joby Joseph and Kehar Singh. Steen G. Hanson, Michael L. Jakobsen and Harold T. Jura explore the use of these transformations for speckle metrology in Chap. 14. Lastly, the use of linear canonical transformations in quantum optics is presented by Gabriel F. Calvo and Antonio Picón.

This volume will be a very useful reference for specialists working in the fields of optical system design and modeling, image and signal processing, and quantum optics, to name a few. It will also be a great resource for graduate students in physics and engineering, as well as for scientists in other areas seeking to learn more about this important yet relatively unfamiliar class of integral transformations.

The Institute of Optics
University of Rochester
Rochester, New York

Miguel A. Alonso

Preface

Linear canonical transforms (LCTs) are a three-parameter family of linear integral transformations, which have a quadratic-phase kernel. For this reason, they have also been called quadratic-phase transforms or quadratic-phase systems (as well as other names). They are unitary transforms that correspond to linear, area-preserving distortions in phase space, a fact which underlies certain invariance properties. Combinations of LCTs are again LCTs. The family includes important operations or transforms such as chirp multiplication, chirp convolution (Fresnel transforms), fractional Fourier transforms, and of course the ordinary Fourier transform, as special cases. Arbitrary LCTs can be written as combinations of these simpler transforms. This leads to fast algorithms for approximately calculating LCTs, much as the ordinary Fourier transform can be calculated with fast algorithms.

LCTs have been rediscovered many times in different contexts, a fact we consider evidence of their ubiquity. Their significance in optics was recognized at least as early as the 1970s. Later, interest in the fractional Fourier transform during the 1990s led to renewed interest in LCTs from new perspectives.

This book deals with LCTs primarily from the perspective of signal and image processing, and optical information processing. Part I presents the mathematical theory of LCTs in the style of signal theory and analysis, as well as the foundations of how LCTs are related to optical systems. Part II deals with issues of degrees of freedom, sampling, numerical implementation, and fast algorithms. Part III is a survey of various applications. No attempt is made here to discuss canonical transformations as they appear in classical Hamiltonian mechanics and symplectomorphisms. These are well-established subjects in physics. However, we note that it is quite possible that a crossover of concepts and techniques between the different approaches to these transforms may be quite fruitful, and we hope this book may contribute to that end, in addition to being useful for its primary audience in the areas of signal processing and optics.

Overview

The opening chapters cover a range of fundamental topics. We start with a discussion of the twin discovery of LCTs in two different areas: paraxial optics and nuclear physics. This provides a fascinating window into more than 40 years of parallel scientific progress. This chapter also contrasts two parallel efforts to define a discrete counterpart to the LCTs—one based on group theory, the other on sampling theory. Chapter 2 provides a self-contained introduction to LCTs and their properties, so the reader who just wishes to dip into the subject may be advised to start here. Chapter 3 discusses the eigenfunctions of the LCTs. These functions are important for analyzing the characteristics of the transforms. Since the LCT can be used to describe wave propagation, they also play important roles in the analysis of self-imaging and resonance phenomena. Chapter 4 continues the theme of key properties of the transform with a discussion of the uncertainty principle. Heisenberg's principle provides a lower bound on the spread of signal energy in the time and frequency domains, and there has been a good deal of work on extending this work to LCTs. The first part of the book is rounded out by Chaps. 5 and 6 that discuss the relationship of LCTs to optics. These chapters deal with both how LCTs can be used to model and analyze optical systems and how LCTs can be optically implemented.

The modern age is digital, whether we are working with spatial light modulators and digital cameras or processing the resulting signals with a computer. In the second part of the book, we have a number of chapters on topics relevant to discrete signals and their processing. Chapter 7 discusses a modern interpretation of the relationship between sampling and information content of signals. Chapter 8 discusses sampling theory and builds up to a discrete transform. Periodic gratings have long been known to produce discrete signals at certain distances, and in Chap. 9 this Talbot effect and hence the relationship between discrete and periodic signals are examined. Just as the fast Fourier transform is key to the utility of conventional spectral analysis, corresponding fast algorithms are critical to our ability to use LCTs in a range of applications. Chapter 10 examines how to calculate the LCT numerically in a fast and accurate fashion.

In the final part of the book, we turn to a series of chapters in which linear canonical transforms are used in a variety of optical applications. One of the fundamental problems in optics is that our detectors are insensitive to phase. Chapter 11 discusses phase retrieval from the field intensity captured in planes separated by systems that can be described using LCTs, focusing particularly on non-iterative techniques. Another way to find the full wave field (amplitude and phase) is to record a hologram, a topic which experienced a revival in the past 20 years due to the rapid improvement in digital cameras. Digital holography is the focus of Chap. 12. Chapter 13 examines optical encryption by means of random phase encoding in multiple planes separated by systems that may be described using LCTs. Coherent light reflected from a rough surface develops laser speckle, a characteristic of the wave field, which may be beneficial in metrology or a nuisance in display technologies. Chapter 14 examines complex-parametered LCTs as a means of modelling speckle fields propagating through apertured optical systems. With Chap. 15, the book is rounded off with a discussion of the use of LCTs in quantum optics.

Dublin, Ireland	John J. Healy
Ankara, Turkey	M. Alper Kutay
Ankara, Turkey	Haldun M. Ozaktas
Dublin, Ireland	John T. Sheridan

Acknowledgements

Healy acknowledges partial support of Science Foundation Ireland, the Consejo Nacional de Ciencia y Tecnologia (Mexico), the Irish Research Council, and the National University of Ireland.

Ozaktas acknowledges partial support of the Turkish Academy of Sciences.

Sheridan acknowledges partial support of Enterprise Ireland, The Irish Research Council and Science Foundation Ireland.

We would like to thank series editor William Rhodes and our editor at Springer, Ho Ying Fan. We thank Miguel Alonso for writing the foreword and of course all the authors who contributed. Finally, we sincerely thank all our families, collaborators, and students for their invaluable inputs and assistance.

Contents

Part I
Fundamentals

Chapter 1
Development of Linear Canonical Transforms: A Historical Sketch

Kurt Bernardo Wolf

Abstract Linear canonical transformations (LCTs) were introduced almost simultaneously during the early 1970s by Stuart A. Collins Jr. in paraxial optics, and independently by Marcos Moshinsky and Christiane Quesne in quantum mechanics, to understand the conservation of information and of uncertainty under linear maps of phase space. Only in the 1990s did both sources begin to be referred jointly in the growing literature, which has expanded into a field common to applied optics, mathematical physics, and analogic and digital signal analysis. In this introductory chapter we recapitulate the construction of the LCT integral transforms, detailing their Lie-algebraic relation with second-order differential operators, which is the origin of the metaplectic phase. Radial and hyperbolic LCTs are reviewed as unitary integral representations of the two-dimensional symplectic group, with complex extension to a semigroup for systems with loss or gain. Some of the more recent developments on discrete and finite analogues of LCTs are commented with their concomitant problems, whose solutions and alternatives are contained the body of this book.

1.1 Introduction

The discovery and development of the theory of linear canonical transforms (LCTs) during the early seventies was motivated by the work on two rather different physical models: paraxial optics and nuclear physics. The integral LCT kernel was written as a descriptor for light propagation in the paraxial régime by Stuart A. Collins Jr., working in the ElectroScience Laboratory of Electrical Engineering at Ohio State University. On the other hand, Marcos Moshinsky and his postdoctoral associate Christiane Quesne, theoretical physicists at the Institute of Physics of the Universidad Nacional Autónoma de México, while working among other problems on the alpha clustering and decay of radioactive nuclei, saw LCTs as

K.B. Wolf (✉)
Instituto de Ciencias Físicas, Universidad Nacional Autónoma de México,
Av. Universidad s/n, Cuernavaca, Morelos 62251, México
e-mail: bwolf@fis.unam.mx

J.J. Healy et al. (eds.), *Linear Canonical Transforms*, Springer Series in Optical Sciences 198, DOI 10.1007/978-1-4939-3028-9_1

the key to understand the conservation of uncertainty as a matter of intrinsic mathematical interest. Some two decades elapsed before the two currents of research acknowledged each other. For this reason alone, the 45-year history of LCTs could provide an interesting case study on the intertwining of basic and applied endeavors. The more recent trend towards the analysis of discrete and finite data sets such as computers can handle also evinces a bifurcation between the search for efficient algorithms and the quest for subtler constructions based on symmetry. Usually mathematics yields more results than can be useful for applications. Applications have generated admirable technology, while symmetry catches the eye and pleases the mind.

The two seminal papers on LCTs, of Collins [1], and of Moshinsky and Quesne [2–4], are highly referenced (>657 and >390 joint citations, respectively, 11/10/15). Yet closer analysis shows that the authors who cited each of them have been mostly disjoint up to recent years: there was an optics community and a theoretical physics community, each with its own preferred journals, interests, and working styles [5]. The author's [6] grievously omits Collins' work—and any reference to optics as well. Fortunately, during the early eighties a mathematician colleague brought to my attention a series of papers by Nazarathy, Shamir, and Hardy on linear systems with loss or gain [7–11], and the work of Alex J. Dragt (University of Maryland) and several of his collaborators [12, 13] who had been developing techniques to control charged particle beams for the Superconducting Supercollider project [14, 15], which started a learning process on optical systems seen as a group-theoretical construct.

It should not be a matter of apology to focus this introductory chapter toward a review of LCTs seen from a more mathematical perspective. Section 1.2 contains the Collins and Moshinsky–Quesne approaches to LCTs, and the context in which our local research continued to develop. Thus, Sect. 1.3 reviews the salient properties of LCTs as integral transform realizations of the *double cover* of the group $\mathsf{Sp}(2,\mathsf{R})$ of 2×2 real matrices of unit determinant, and as generated by an algebra of second-order differential operators in Sect. 1.4. Section 1.5 recapitulates the radial and the (lesser-known) hyperbolic LCTs, geared to answer the question "*what are LCTs?*" In that section we propose what seems to be the proper context to accommodate all realizations ("faces") of integral and (infinite) matrix LCTs. Section 1.6 recalls complex extensions of LCTs that can be made unitary, such as heat diffusion, and a hint of applications to special function theory. Realizations of LCTs as finite matrices are addressed in Sect. 1.7 because there is a growing interest in fast algorithms to digitally treat LCTs for finite signals or pixelated images, where several tactics have been proposed to handle them, and on which I add a few words in the concluding Sect. 1.8. Here too it seems that at least two schools of thought contend, one strives for æsthetics and the other for efficacy.

1.2 Diffraction Integrals, Uncertainty Relations

Geometric and wave optics, as well as classical and quantum mechanics, agree with each other in the linear approximation—except for complex phases. It should be evident therefore that the paraxial régime of optics and quadratic systems in mechanics are closely related in their mathematical structure. They are both Hamiltonian systems whose waveforms, or states in any number of dimensions, can be displayed on a flat phase space. There, evolution is canonical (keeping the symplectic structure invariant) and linear (consisting only of translations, rotations, and shears). In paraxial wave optics, shears of phase space result from thin lenses and empty spaces, which, respectively, multiply the input functions by quadratic phases, and subject them to an isotropic Fresnel integral transform. In quantum mechanics on the other hand, beside the shear of free propagation, the harmonic oscillator is the most privileged actor; it generates a fractional Fourier transform on the initial state—times a phase.

1.2.1 Matrix Representation of Paraxial Optical Systems

The evolution in linear systems can be represented mathematically in three ways: by linear operators, by integral kernels, and by finite or infinite matrices. These will act on the states of the system, which in turn are realized, respectively, as differentiable and/or integrable functions of position (or momentum, or other observables), and as finite- or infinite-dimensional vectors. Since LCTs form a *group*, there will be locally a 1:1 correspondence between the three realizations, so one can use the algebraically simpler finite matrix realization to compute products and actions. Many authors point to the books by Willem Brouwer [16] and by Gerrard and Burch [17] for introducing the use of matrix algebra to paraxial optical design for resonators and the evolution of Gaussian beams therein. In two-dimensional (2*D*) optics, free propagation by z is represented by the 2×2 matrix $\begin{pmatrix} 1 & z \\ 0 & 1 \end{pmatrix}$, and a thin lens of focal distance f by $\begin{pmatrix} 1 & 0 \\ -1/f & 1 \end{pmatrix}$; these act on rays represented by a two-vector $\begin{pmatrix} x \\ p \end{pmatrix}$, where x is the *position* of the ray on the $z = 0$ *screen*, and $p = n \sin\theta \approx n\theta$ is the *momentum* of the ray that crosses the screen with the "small" angle θ to its normal, in a transparent optical medium of refractive index n. In the paraxial régime one lets the phase space coordinates (x, p) roam over the full plane R^2. Products of these matrices correspond with the concatenation of the optical elements, and every paraxial 2*D* optical system is thus represented by a 2×2 matrix $\begin{pmatrix} a & b \\ c & d \end{pmatrix}$, with $ad - bc = 1$ because the two generator matrices have unit determinant.

The paper by Stuart A. Collins, Jr. [1] considered the generic 3*D* paraxial, generally nonsymmetric but centered and aligned system.[1] These systems are represented by a 4×4 matrix $\mathbf{M} = \begin{pmatrix} \mathbf{a} & \mathbf{b} \\ \mathbf{c} & \mathbf{d} \end{pmatrix}$,

$$\begin{pmatrix} \mathbf{x}' \\ \mathbf{p}' \end{pmatrix} = \begin{pmatrix} \mathbf{a} & \mathbf{b} \\ \mathbf{c} & \mathbf{d} \end{pmatrix} \begin{pmatrix} \mathbf{x} \\ \mathbf{p} \end{pmatrix}, \quad \text{i.e.,} \quad \mathbf{w}' = \mathbf{M}\mathbf{w}, \tag{1.1}$$

where $\mathbf{w} := \begin{pmatrix} \mathbf{x} \\ \mathbf{p} \end{pmatrix}$, with $\mathbf{x}, \mathbf{x}'$, $\mathbf{p}, \mathbf{p}'$ being 2-vectors, and $\mathbf{a}$, $\mathbf{b}$, $\mathbf{c}$, $\mathbf{d}$ are the 2×2 submatrices of $\mathbf{M}$. Since free propagation of an input function $f(\mathbf{x})$ is described by the Fresnel transform, whose integral kernel has a quadratic phase, and thin lenses multiply the function by a quadratic phase also, one should guess that the output $f_{\mathrm{M}}(\mathbf{x})$ of an $\mathbf{M} = \begin{pmatrix} a & b \\ c & d \end{pmatrix}$-transform should be an integral transform which, for the generic N-dimensional case is

$$f_{\mathrm{M}}(\mathbf{x}) \equiv (\mathcal{C}_{\mathrm{M}} f)(\mathbf{x}) := \int_{\mathsf{R}^N} \mathrm{d}^N\mathbf{x}' \, C_{\mathrm{M}}(\mathbf{x}, \mathbf{x}') f(\mathbf{x}'), \tag{1.2}$$

with a quadratic phase kernel $C_{\mathrm{M}}(\mathbf{x}, \mathbf{x}')$ in the components of $\mathbf{x}$ and $\mathbf{x}'$, and the matrix parameters of $\mathbf{M}$. The Collins paper considers transverse scalar fields $E_i = A_i \exp(\mathrm{i}kL_i)$ in each element of the optical setup, using the Fermat principle to show how the eikonal (optical distance) can be expressed in terms of the initial and final ray positions and slopes.[2] The resulting linear relations between these two 4-vectors with the parameters of the optical system turn out to be equivalent to the definition of *symplectic* matrices, whose generic form is

$$\mathbf{M}\,\mathbf{\Omega}\,\mathbf{M}^\top = \mathbf{\Omega}, \qquad \mathbf{\Omega}^\top = -\mathbf{\Omega}, \quad \mathbf{\Omega}^2 = -\mathbf{1}, \tag{1.3}$$

where the skew-symmetric *metric* matrix $\mathbf{\Omega}$ is usually written as $\mathbf{\Omega} = \begin{pmatrix} \mathbf{0} & \mathbf{1} \\ -\mathbf{1} & \mathbf{0} \end{pmatrix}$. In the 2×2 submatrix form (1.1), this is

$$\begin{pmatrix} \mathbf{a} & \mathbf{b} \\ \mathbf{c} & \mathbf{d} \end{pmatrix} \begin{pmatrix} \mathbf{0} & \mathbf{1} \\ -\mathbf{1} & \mathbf{0} \end{pmatrix} \begin{pmatrix} \mathbf{a}^\top & \mathbf{c}^\top \\ \mathbf{b}^\top & \mathbf{d}^\top \end{pmatrix} = \begin{pmatrix} \mathbf{0} & \mathbf{1} \\ -\mathbf{1} & \mathbf{0} \end{pmatrix}, \tag{1.4}$$

which implies that the following submatrix products are symmetric,

$$\mathbf{a}\,\mathbf{b}^\top = (\mathbf{a}\,\mathbf{b}^\top)^\top, \quad \mathbf{c}\,\mathbf{d}^\top = (\mathbf{c}\,\mathbf{d}^\top)^\top, \quad \mathbf{a}\,\mathbf{d}^\top - \mathbf{b}\,\mathbf{c}^\top = \mathbf{1}. \tag{1.5}$$

[1] All lens centers are assumed to be on a common straight optical axis with their planes orthogonal to it; the "center" of cylindrical lenses is a line that should also intersect this axis. The consideration of displacement and (paraxial) tilt can be made using $2 + 2$ more parameters for *inhomogeneous* LCTs, which are not explicitly considered here. See [18].

[2] The paper by Collins uses momenta in the form $n_i\mathbf{p}_i$ with $|\mathbf{p}_i| = \sin\theta_i$, and orders the 4-vector components as $(x_1, p_1, x_2, p_2)^\top$.

These conditions (for $N = 2$) were found [1, Appendix B] and thereby the optical distance between initial and final ray positions $L := L_0 + L_M$, consisting of the distance L_0 along the axis plus that gained for rays between positions off this axis, $L_M(\mathbf{x}, \mathbf{x}')$, which is a quadratic function of its arguments and contains the parameters of the transfer matrix $\mathbf{M}$. The integral kernel (1.2) is thus determined to be of the form $A_M \exp(\mathrm{i}kL_M)$. The normalization factor A_M is computed by demanding the conservation of energy, and its phase is taken from the Fresnel diffraction kernel [1, Eq. (28)]. The paper by Collins applies this result for the analysis of Hermite–Gaussian beams in resonators and for the reconstruction of holographic images.

1.2.2 Evolution in Quadratic Quantum Systems

Marcos Moshinsky had been studying the harmonic motion of Gaussian wavepackets that represent alpha bondings in various oscillator models of the nucleus. This is the context in which he seems to have been motivated to touch upon canonical transformations in quantum mechanics. His paper was presented at the XV Solvay Conference in Physics of 1970 [2], whose Proceedings were delayed 4 years. Upon returning to Mexico with the Belgian postdoctoral associate Dr. Christiane Quesne, they stated the problem in the following terms [3, 4]: What are the transformations of phase space that leave the structure of quantum mechanics invariant? This included the important uncertainty relation $\Delta_f \Delta_{\tilde{f}} \geq \frac{1}{4}$ ($\hbar \equiv 1$) that is a mathematical property of the Fourier integral transform. The question remitted them to the basic Heisenberg commutators

$$[\hat{x}_i, \hat{p}_j] := \hat{x}_i\hat{p}_j - \hat{p}_j\hat{x}_i = \mathrm{i}\,\delta_{i,j}, \tag{1.6}$$

between the Schrödinger position operators $\hat{x}_i = x_i\,\cdot$ and the momentum operators $\hat{p}_j = -\mathrm{i}\partial_j$ (where $\partial_j \equiv \partial/\partial x_j$), for $i, j = 1, 2, \ldots, N$ in N dimensional systems. Such transformations can be linear or nonlinear; some of the latter were examined a few years later, but the more immediate ones were the linear, for N-vector operators $\hat{\mathbf{x}}$ and $\hat{\mathbf{p}}$ forming a $2N$-vector $\hat{\mathbf{w}} \equiv \begin{pmatrix}\hat{\mathbf{x}}\\ \hat{\mathbf{p}}\end{pmatrix}$ as before, acted upon by a transformation $\mathcal{C}_M$ depending on the elements of a $2N \times 2N$ matrix $\mathbf{M}$. For operators, these are written somewhat differently from (1.1),

$$\mathcal{C}_M\,\hat{\mathbf{w}}\,\mathcal{C}_M^{-1} = \mathbf{M}^{-1}\hat{\mathbf{w}}. \tag{1.7}$$

The reason for having the inverse matrix on the right-hand side is that this alone ensures that the composition of transforms follows that of the matrices: $\mathcal{C}_{M_1}\mathcal{C}_{M_2} = \varphi\,\mathcal{C}_{M_1M_2}$, with φ a constant undetectable in (1.7). Next, direct replacement into (1.6) yields the symplectic conditions (1.3)–(1.5) for $\mathbf{M}$. Symplectic matrices are invertible,

$$\begin{pmatrix} \mathbf{a} & \mathbf{b} \\ \mathbf{c} & \mathbf{d} \end{pmatrix}^{-1} = \boldsymbol{\Omega}\mathbf{M}^\top\boldsymbol{\Omega}^\top = \begin{pmatrix} \mathbf{d}^\top & -\mathbf{b}^\top \\ -\mathbf{c}^\top & \mathbf{a}^\top \end{pmatrix}, \tag{1.8}$$

the unit $\mathbf{1}$ is symplectic and associativity holds. Hence symplectic matrices that are real form the *real symplectic group* $\mathsf{Sp}(2N, \mathsf{R})$ with $N(2N+1)$ independent parameters. When $N = 1$, $\mathsf{Sp}(2, \mathsf{R})$ is identical with the group of all 2×2 real matrices of unit determinant. (The complex case will be considered in Sect. 1.6.)

The action of the linear operators $\mathcal{C}_\mathrm{M}$ on the usual Hilbert space $\mathcal{L}^2(\mathsf{R}^N)$ of quantum mechanical Lebesgue square-integrable functions, $f \mapsto f_\mathrm{M} \equiv \mathcal{C}_\mathrm{M} f$, is expected to be integral in R^N as (1.2), and unitary, because such is quantum evolution. The integral kernel can be found applying $\mathcal{C}_\mathrm{M}$ to $\hat{x}_i f$ and to $\hat{p}_j f$ using (1.7) and (1.8),

$$\mathcal{C}_\mathrm{M}(\hat{x}_i f) = (\mathcal{C}_\mathrm{M}\hat{x}_i\mathcal{C}_\mathrm{M}^{-1}) f_\mathrm{M} = \textstyle\sum_j (d_{j,i}\hat{x}_j - b_{j,i}\hat{p}_j) f_\mathrm{M}, \tag{1.9}$$

$$\mathcal{C}_\mathrm{M}(\hat{p}_i f) = (\mathcal{C}_\mathrm{M}\hat{p}_i\mathcal{C}_\mathrm{M}^{-1}) f_\mathrm{M} = \textstyle\sum_j (-c_{j,i}\hat{x}_j + a_{j,i}\hat{p}_j) f_\mathrm{M}. \tag{1.10}$$

On the right, $\hat{x}_i$ and $\hat{p}_i$ act outside of the integral, on the $\mathbf{x}$ argument of the kernel $C_\mathrm{M}(\mathbf{x}, \mathbf{x}')$, while those on the left act inside, on $f(\mathbf{x}')$; the derivatives of the latter can be integrated by parts to act on the $\mathbf{x}'$ argument of the kernel. Since f is arbitrary, one obtains the $2N$ simultaneous linear differential equations satisfied by the LCT kernel,

$$x_i' C_\mathrm{M}(\mathbf{x}, \mathbf{x}') = \textstyle\sum_j \Big(d_{j,i}x_i + \mathrm{i}b_{j,i}\partial_j\Big) C_\mathrm{M}(\mathbf{x}, \mathbf{x}'), \tag{1.11}$$

$$\partial_i' C_\mathrm{M}(\mathbf{x}, \mathbf{x}') = \textstyle\sum_j \Big(\mathrm{i}c_{j,i}x_i - a_{j,i}\partial_j\Big) C_\mathrm{M}(\mathbf{x}, \mathbf{x}'). \tag{1.12}$$

The solution, up to a multiplicative constant K_M, is

$$C_\mathrm{M}(\mathbf{x}, \mathbf{x}') := K_\mathrm{M} \exp \mathrm{i}\Big(\tfrac{1}{2}\mathbf{x}^\top\mathbf{b}^{-1}\mathbf{d}\mathbf{x} - \mathbf{x}^\top\mathbf{b}^{-1}\mathbf{x}' + \tfrac{1}{2}\mathbf{x}'^\top\mathbf{a}\mathbf{b}^{-1}\mathbf{x}'\Big). \tag{1.13}$$

The constant K_M is found from the limit to the $2N \times 2N$ unit matrix, $\mathbf{M} \to \mathbf{1}$ (with $\det \mathbf{b}$ in the lower-half complex plane), so that $C_\mathrm{M}(\mathbf{x}, \mathbf{x}') \to \delta^N(\mathbf{x} - \mathbf{x}')$, regaining the unit transform $\mathcal{C}_\mathbf{1} = \mathit{1}$. The result is

$$K_\mathrm{M} = \frac{1}{\sqrt{(2\pi\mathrm{i})^N \det \mathbf{b}}} \equiv \frac{e^{-\mathrm{i}\pi N/4} \exp \mathrm{i}(-\frac{1}{2} \arg \det \mathbf{b})}{\sqrt{(2\pi)^N |\det \mathbf{b}|}}. \tag{1.14}$$

Finally, when only $\mathbf{b} \to \mathbf{0}$ from the lower complex half-plane, the matrix is $\mathbf{M}_0 := \begin{pmatrix} \mathbf{a} & \mathbf{0} \\ \mathbf{c} & \mathbf{a}^{\top-1} \end{pmatrix}$, the Gaussian kernel converges weakly to a Dirac δ, and the integral operator action becomes a change of scale of the function multiplied by a quadratic phase,

$$(\mathcal{C}_{\mathrm{M}_0} f)(\mathbf{x}) = \frac{\exp \mathrm{i}(\frac{1}{2}\mathbf{x}^\top \mathbf{c}\mathbf{a}^{-1}\mathbf{x})}{\sqrt{\det \mathbf{a}}} f(\mathbf{a}^{-1}\mathbf{x}). \tag{1.15}$$

In the case of $N = 1$-dimensions, Eqs. (1.2) and (1.13)–(1.14) simplify to the best-known form of LCTs,

$$f_{\mathrm{M}}(x) \equiv (\mathcal{C}_{\mathrm{M}} f)(x) = \int_{\mathsf{R}} \mathrm{d}x' \, C_{\mathrm{M}}(x, x') f(x'), \tag{1.16}$$

$$C_{\mathrm{M}}(x, x') := \frac{1}{\sqrt{2\pi \mathrm{i}\, b}} \exp\left(\frac{\mathrm{i}}{2b}(dx^2 - 2xx' + ax'^2)\right), \tag{1.17}$$

where it should be understood that $1/\sqrt{ib} = \exp(-\mathrm{i}\frac{1}{2}\pi(\operatorname{sign} b + \frac{1}{2}))/\sqrt{|b|}$. The generalization of the Fourier–Heisenberg uncertainty relation to LCTs is of the form $\Delta_f \Delta_{f_{\mathrm{M}}} \geq \frac{1}{4}|b|$. The last two chapters of [6] were written based on the works of Marcos Moshinsky and his associates on LCTs, complemented with results by the author on translations of phase space, complex extensions, and applications to the evolution of Gaussians and other wavefunctions of quantum quadratic systems (oscillator wavefunctions, parabolic cylinder and Airy functions) under diffusion.

1.2.3 LCTs in a Broader Context

Optical models are richer than mechanical ones because they provide a wider view of canonical transformations beyond the linear regime. Mechanical Hamiltonians are mostly of the form $\frac{1}{2}p^2 + V(x)$, where the potential $V(x)$ with a smooth minimum may be expanded using perturbation series in powers of x around the harmonic oscillator; in geometric and magnetic metaxial optics on the other hand, the presence of aberrations generally requires evolution Hamiltonians expressible in series of terms $p^n x^m$. As Alex J. Dragt applied for accelerators [12–15, 19], Hamiltonian and Lie-theoretic tools served to calculate carefully one turn in the accelerator, and then one raises that transformation to the power of any number of turns, while canonicity ensures the conservation of the beam area in phase space. The usefulness of these techniques for optical design was facilitated by a neat theorem on the canonical transformations produced by refraction between two media separated by a surface of smooth but arbitrary shape [20]: they can be factored into the product of two canonical transformations, each depending on the surface and only one of the media. This allowed the computation of the aberration coefficients for polynomial surfaces of revolution, and the group structure translated the concatenation of optical elements along the optical axis into matrix multiplication. Interest in these lines led us to convene two gatherings on *Lie optics* (a convenient neologism), in 1985 and 1988 [21, 22]. In fact, LCTs were taken for granted and served as basis for chapters on Fourier optics, coherent states, holography, computational aspects for aberrations, and optical models that take into account that the optical momentum vector ranges over a sphere, and not over a plane as the paraxial theory assumes. Yet, it is the linear regime (paraxial optics or quadratic mechanics) that displays naturally the cleanest symmetries.

Closely related with LCTs, a line of research on the Wigner distribution function applied to optical waveforms and their transformation in first-order optical systems was opened by Martin J. Bastiaans by the end of the 1970s [23, 24]. Both papers are highly cited (>330 and >400 citations), indicating that many authors have followed the analysis of non-imaging linear systems in phase space [25–27]. More recent work with Tatiana Alieva, María Luisa Calvo, and several coworkers addressed LCTs to obtain phase information out of intensity measurements [28–30], and the processing of two-dimensional images [31, 32] by means of optical setups of cylindrical lenses that can be rotated in fixed positions to synthesize any LCT transformation [33], in particular fractional Fourier transforms [34, 35] and gyrators [36, 37]. Both the Wigner function and the two-dimensional LCTs that form the group $\mathsf{Sp}(4, \mathsf{R})$ cannot be surveyed in this chapter for reasons of space even though they are now widely used for many applications in quantum optics. See, for example, [38] (>1120 citations).

Linear canonical transformations include fractional Fourier transforms in the subgroup $\mathcal{F}^\nu \equiv \mathcal{F}_\theta \in \mathsf{SO}(2) \subset \mathsf{Sp}(2, \mathsf{R})$ of matrices $\begin{pmatrix} \cos\theta & \sin\theta \\ -\sin\theta & \cos\theta \end{pmatrix}$, of power $\nu \in \mathsf{R}$ or angle $\theta = \frac{1}{2}\pi\nu$, times the metaplectic phase (to be seen below). This development also has a story behind: in 1937, Edward Condon thanks Profs. Bochner, von Neumann, and Bohnenblust for conversations leading to the article [39], where he clearly defines the fractional Fourier transform and finds its kernel following the reasoning in (1.11)–(1.12), recognizing the metaplectic problem. Condon's result seems to have been in suspended animation for decades, unnoticed by Victor Namias [40] who in 1980 rediscovered $\mathcal{F}^\nu$ proposing that it self-reproduces the harmonic oscillator wavefunctions with a phase $(-\mathrm{i})^\nu$ (to be taken as $e^{-\mathrm{i}\pi\nu/2}$), and the kernel found from the bilinear generating function of Hermite polynomials (inexplicably, [6] disregarded this specialization of LCTs). Interest of the optical community in fractional Fourier transforms grew in the early nineties around their optical implementation through the slicing of graded-index media and non-imaging lens systems, by Mendlovic and Ozaktas [41–44] (>780, >437, >250 and >254 citations). Their work was formalized in the 2001 book [45] by Ozaktas, Zalevsky, and Kutay, which spread the use of the fractional Fourier transform and LCTs in general. This book contains a bibliography of >500 references which hardly any of us can read entirely, and which I certainly cannot reproduce.

1.3 LCTs, Matrices, Signs and Covers

An important property of the LCTs (1.13)–(1.15) is that they conserve the norms [1] and overlaps [3], i.e., the transformations are unitary under the $\mathcal{L}^2(\mathsf{R}^N)$ inner product,

$$(f,g)_{\mathcal{L}^2(\mathsf{R}^N)} := \int_{\mathsf{R}^N} \mathrm{d}\mathbf{x} f(\mathbf{x})^* g(\mathbf{x}) = (f_{\mathrm{M}}, g_{\mathrm{M}})_{\mathcal{L}^2(\mathsf{R}^N)}, \tag{1.18}$$

because

$$C_{\mathrm{M}}(\mathbf{x},\mathbf{x}') = C_{\mathrm{M}^{-1}}(\mathbf{x}',\mathbf{x})^*. \tag{1.19}$$

However, the group composition property of LCTs is satisfied by the integral kernels only as

$$\int_{\mathsf{R}^N} \mathrm{d}\mathbf{x}'\, C_{\mathrm{M}_1}(\mathbf{x},\mathbf{x}')\, C_{\mathrm{M}_2}(\mathbf{x}',\mathbf{x}'') = \sigma\, C_{\mathrm{M}_1\mathrm{M}_2}(\mathbf{x},\mathbf{x}''), \tag{1.20}$$

where σ is a phase—the *metaplectic* phase (actually a sign). This problem is announced by the square root in the denominator of (1.14) and (1.15); it can be seen most clearly in the Fourier integral transform $\mathcal{F}$ for $N = 1$, which for dimensionless matrix elements corresponds to $\mathbf{F} = \boldsymbol{\Omega}$[3]; the integral kernel is then $C_{\mathrm{F}}(x,x') = e^{-\mathrm{i}\pi/4} e^{-\mathrm{i}xx'}/\sqrt{2\pi}$, so

$$\mathcal{C}_{\mathrm{F}} = e^{-\mathrm{i}\pi/4}\mathcal{F}, \qquad \mathbf{F} = \begin{pmatrix} 0 & 1 \\ -1 & 0 \end{pmatrix}. \tag{1.21}$$

Thus, while $\mathcal{F}^4 = \mathit{1}$ we have $\mathcal{C}_{\mathrm{F}}^4 = -\mathit{1}$; this is reminiscent of the behavior of spin under 2π rotations.

The metaplectic sign has bedeviled many papers, and it can be said that it was not really understood until the group theory behind brought to the fore the fact that the correspondence between integral LCTs and matrices is not 1:1, but 2:1. The problem is not crucial in optical setups because overall phases are commonly not registered, but in mathematics signs cannot be just ignored. Indeed, the structure of the symplectic groups (even that of 2×2 matrices) is unexpectedly imbricate [46]. The problem for $N = 1$ was clarified early by Valentin Bargmann in 1947 [47, Sects. 3, 4] using the *polar* decomposition of matrices. This is a generalization of the factorization of complex numbers $z = e^{\mathrm{i}\phi}|z|$ into a phase $e^{\mathrm{i}\phi}$ times a positive number $|z|$; multiple Riemann sheets of a function around its branch points need the phase ϕ to range beyond its basic interval modulo 2π. A real 2×2 matrix can be similarly decomposed into the product of a unitary and a symmetric positive definite matrix,

[3] I thank Dr. George Nemeş for the remark that when dimensions are respected, $\mathbf{F} \neq \boldsymbol{\Omega}$ because the parameters b and $1/c$ have units of momentum/position, while a and d have no units. In our presentation of the kernel (1.17) we assume that momentum p bears no units (as in optics), and that a unit of distance has been agreed for position so that x is its numerical multiple.

$$\begin{pmatrix} a & b \\ c & d \end{pmatrix} = \begin{pmatrix} \cos\phi & -\sin\phi \\ \sin\phi & \cos\phi \end{pmatrix} \begin{pmatrix} \lambda + \operatorname{Re}\mu & \operatorname{Im}\mu \\ \operatorname{Im}\mu & \lambda - \operatorname{Re}\mu \end{pmatrix}, \tag{1.22}$$

where μ is complex, and $\lambda := +\sqrt{|\mu|^2 + 1} \geq 1$. Under multiplication of two matrices, their Bargmann parameters (with subindices 1 and 2) compose through

$$\phi = \phi_1 + \phi_2 + \arg\nu, \qquad \mu = e^{-\mathrm{i}\arg\nu}(\lambda_1\mu_2 + e^{-2\mathrm{i}\phi_2}\mu_1\lambda_2), \tag{1.23}$$

where $\nu := 1 + e^{-2\mathrm{i}\phi_2}\mu_1\mu_2/\lambda_1\lambda_2$ is an auxiliary complex quantity whose phase is determined to range in $\arg\nu \in (-\frac{1}{2}\pi, \frac{1}{2}\pi)$, and $\lambda = \lambda_1|\nu|\lambda_2 \geq 1$. The composite ϕ can thus take values on the full real line R and hence parametrize all elements of $\overline{\mathsf{Sp}(2,\mathsf{R})}$, the *infinite cover* of the group $\mathsf{Sp}(2,\mathsf{R})$. Thus, while the unitary spin group $\mathsf{SU}(2)$ covers twice the orthogonal rotation group $\mathsf{SO}(3)$, the symplectic group is infinitely covered; the realization by LCTs is then a *two*fold cover of the group of 2×2 real matrices of unit determinant. Below we shall comment on this feature of the group of integral transforms, called the *metaplectic group* $\mathsf{Mp}(2,\mathsf{R})$. (See also [48, Sect. 9.4].)

The generic case of $\mathsf{Sp}(2N,\mathsf{R})$ follows suit, as proved by Bargmann some years later [49]. The polar decomposition is then into a real $2N \times 2N$ orthosymplectic matrix that represents the group $\mathsf{U}(N)$ of $N \times N$ unitary matrices, and again a symmetric positive definite matrix [48, p. 173]. This $\mathsf{U}(N)$ group is the maximal *compact* (i.e., of finite volume) subgroup of $\mathsf{Sp}(2N,\mathsf{R})$, and has been called the *Fourier* group [50]. In the $N = 2$-dimensional case, $\mathsf{U}(2)$ contains the isotropic and anisotropic fractional Fourier and gyration integral transforms [34, 36], as well as joint rotations of position and momentum around the optical center and axis. In turn, this $\mathsf{U}(2)$ is the direct product of a $\mathsf{U}(1)$ subgroup of isotropic fractional Fourier transforms (a circle), times the group $\mathsf{SU}(2)$ of 2×2 matrices of unit determinant; the latter is simply connected, so the onus of multivaluation falls on the former. For $N = 2$ and the 4×4 Fourier matrix $\mathbf{F} = \boldsymbol{\Omega} = \begin{pmatrix} \mathbf{0} & \mathbf{1} \\ -\mathbf{1} & \mathbf{0} \end{pmatrix}$ the relation between the LCT and the $2D$ Fourier integral transform is thus $\mathcal{C}_{\mathrm{F}} = e^{-\mathrm{i}\pi/2}\mathcal{F}$.

1.4 LCTs Are Generated by Second-Order Differential Operators

In retrospect it is obvious that unitary LCTs $\mathcal{C}_{\mathrm{M}}$ and self-adjoint second-order differential operators $\hat{J} = \alpha\hat{p}_i\hat{p}_j + \beta\frac{1}{2}(\hat{x}_i\hat{p}_j + \hat{p}_j\hat{x}_i) + \gamma\hat{x}_i\hat{x}_j$ should be closely related, the latter generating the former through $\mathcal{C}_{\mathrm{M}(\tau)} = \exp(\mathrm{i}\tau\hat{J})$. The LCT integral kernels $C_{\mathrm{M}(\tau)}(\mathbf{x},\mathbf{x}')$ are Green functions of quadratic Hamiltonians that can be found through

$$\hat{J}f(\mathbf{x}) = -\mathrm{i}\frac{\partial}{\partial\tau}\int_{\mathrm{R}^N} \mathrm{d}\mathbf{x}'\, C_{\mathrm{M}(\tau)}(\mathbf{x},\mathbf{x}')f(\mathbf{x}')\Big|_{\tau=0}, \tag{1.24}$$

and $C_{\mathrm{M}(0)}(\mathbf{x},\mathbf{x}') = \delta^N(\mathbf{x}-\mathbf{x}')$, as was done in [51]. Probably the reason for not having recognized this relation earlier was that since the time of Sophus Lie only first-order differential operators, $f(x)\partial_x + g(x)$, were used to generate Lie groups.

Writing $\mathcal{C}_{\mathrm{M}} \equiv \mathcal{C}(\mathbf{M})$, we have the following $N = 1$ paraxial optical elements generated by operators and their LCTs,

$$\text{thin lens:} \qquad \exp\left(\mathrm{i}\tfrac{1}{2}\tau\hat{x}^2\right) = \mathcal{C}\begin{pmatrix} 1 & 0 \\ \tau & 1 \end{pmatrix}, \tag{1.25}$$

$$\text{free flight:} \qquad \exp\left(\mathrm{i}\tfrac{1}{2}\tau\hat{p}^2\right) = \mathcal{C}\begin{pmatrix} 1 & -\tau \\ 0 & 1 \end{pmatrix}, \tag{1.26}$$

$$\text{magnifier:} \quad \exp\left(\mathrm{i}\tfrac{1}{2}\tau(\hat{p}\hat{x}+\hat{x}\hat{p})\right) = \mathcal{C}\begin{pmatrix} e^{-\tau} & 0 \\ 0 & e^{\tau} \end{pmatrix}, \tag{1.27}$$

$$\text{repulsive guide:} \quad \exp\left(\mathrm{i}\tfrac{1}{2}\tau(\hat{p}^2-\hat{x}^2)\right) = \mathcal{C}\begin{pmatrix} \cosh\tau & -\sinh\tau \\ -\sinh\tau & \cosh\tau \end{pmatrix}, \tag{1.28}$$

$$e^{\mathrm{i}\pi\tau/4}\times\text{Fourier}^{-\tau}: \quad \exp\left(\mathrm{i}\tfrac{1}{2}\tau(\hat{p}^2+\hat{x}^2)\right) = \mathcal{C}\begin{pmatrix} \cos\tau & -\sin\tau \\ \sin\tau & \cos\tau \end{pmatrix}. \tag{1.29}$$

For vanishing τ, $\mathbf{M}(\tau) \approx \mathbf{1} + \tau\mathbf{m}$, we can associate the generator operators with traceless 2×2 matrices $\mathbf{m}$: thin lens, $\begin{pmatrix} 0 & 0 \\ 1 & 0 \end{pmatrix}$; free flight, $\begin{pmatrix} 0 & -1 \\ 0 & 0 \end{pmatrix}$; magnifier, $\begin{pmatrix} -1 & 0 \\ 0 & 1 \end{pmatrix}$; repulsive guide, $\begin{pmatrix} 0 & -1 \\ -1 & 0 \end{pmatrix}$; and harmonic guide, $\begin{pmatrix} 0 & -1 \\ 1 & 0 \end{pmatrix}$. This infinitesimal "portion" of $\mathsf{Sp}(2,\mathsf{R})$ constitutes a linear space called its *Lie algebra*, denoted by the lowercase name $\mathsf{sp}(2,\mathsf{R})$, and whose structure is determined by the commutators of its elements. Under $\mathcal{C}_{\mathrm{M}}$ the "infinitesimal" matrices $\mathbf{m} \in \mathsf{sp}(2,\mathsf{R})$ will transform by similarity as $\mathbf{m} \mapsto \mathbf{m}' = \mathbf{M}\,\mathbf{m}\,\mathbf{M}^{-1}$, and with all $\mathbf{M} \in \mathsf{Sp}(2,\mathsf{R})$ we build the *orbit* of $\mathbf{m}$. Thus the generators of lenses and of free flights are in the same orbit related by the Fourier matrix $\mathbf{F} = \begin{pmatrix} 0 & 1 \\ -1 & 0 \end{pmatrix}$, and the generators of magnifiers are the same with those of repulsive guides, related by the square root $\mathbf{F}^{1/2} = \frac{1}{\sqrt{2}}\begin{pmatrix} 1 & 1 \\ -1 & 1 \end{pmatrix}$. Analysis shows that $\mathsf{sp}(2,\mathsf{R})$ has three orbits (excluding the orbit of $\mathbf{0}$): *elliptic* containing (1.29); *hyperbolic* (1.27)–(1.28); and *parabolic* (1.25)–(1.26). The last forms a cone in R^3, the first and second fill the inside and outside of that cone. This division into disjoint orbits in the R^3 linear space of the algebra extends to the group, but the group $\mathsf{Sp}(2,\mathsf{R})$ of matrices $\begin{pmatrix} a & b \\ c & d \end{pmatrix}$ has an extra *non-exponential* region identified by the range of the trace, $a + d \in (-\infty, -2)$, where the matrices have no real logarithm. For $N = 2$ dimensions, the identification of generating Hamiltonians in $\mathsf{sp}(4,\mathsf{R})$ with optical elements can be found in [48, Chap. 12]; there are 4 continua of orbits and 12 isolated points, few of which have been exploited.

The relations (1.25)–(1.29) also determine that the eigenfunctions of an operator $\hat{J}\,\psi_\nu = E_\nu\,\psi_\nu$ (whose eigenvalues E_ν are common to all elements in its orbit), will self-reproduce under the generated LCT as $\mathcal{C}_{\mathrm{M}(\tau)}\psi_\nu = e^{\mathrm{i}\tau E_\nu}\psi_\nu$. In particular, the harmonic oscillator Hermite–Gauss eigenfunctions $\Psi_n(x)$ correspond to energies $E_n = n + \frac{1}{2}$, $n \in \{0,1,2,\ldots\}$. Thus, the $\mathcal{C}_{\mathrm{F}}$ LCT of the eigenfunctions $\psi_n^{\mathrm{M}} = \mathcal{C}_{\mathrm{M}}\psi_n$ of all operators in the elliptic orbit is

$$\mathcal{C}_{\mathrm{F}}\psi_n^{\mathrm{M}} = \exp\left[-\mathrm{i}\tfrac{1}{2}\pi(n+\tfrac{1}{2})\right]\psi_n^{\mathrm{M}} = e^{-\mathrm{i}\pi/4}(-\mathrm{i})^n\psi_n^{\mathrm{M}}, \tag{1.30}$$

having set $\tau = -\frac{1}{2}\pi$ in (1.29) and in agreement with (1.21). Here again the phase evinces the double cover of $\mathcal{C}_{\mathrm{F}^\alpha} \in \mathsf{Mp}(2,\mathsf{R})$ over the circle of fractional Fourier matrices $\mathbf{F}^\alpha \in \mathsf{Sp}(2,\mathsf{R})$. We may also see the metaplectic phase as the energy of the vacuum, $E_0 = \frac{1}{2}$.

We have thus associated three classes of mathematical actors in the $\mathsf{Sp}(2,\mathsf{R})$ troupe: LCT integral transforms, hyperdifferential (exponentials of second order) operators, and matrices (modulo a sign). Product operations in one class correspond with products in the other two. Hence, we can easily write Baker–Campbell–Hausdorff relations between quadratic operators [6, Sect. 9.3.2], and the LCTs of the eigenfunction set of one under LCTs generated by another, including phase space translations [6, Chap. 10]. Certainly, other authors have considered various aspects of the above constructions (see, e.g., [52]), so it is as grievous not to mention one as it is to mention all.

1.5 Radial, Hyperbolic, and Other LCTs

Isotropic LCTs in $N = 2$ or more dimensions that are represented by matrices $\mathbf{M} = \begin{pmatrix} a\mathbf{1} & b\mathbf{1} \\ c\mathbf{1} & d\mathbf{1} \end{pmatrix}$ with diagonal submatrices can be reduced to $\begin{pmatrix} a & b \\ c & d \end{pmatrix}$ *radial* LCTs acting on eigenspaces of functions of the radius and with definite angular momentum. One may also ask for separation of variables in other sets of coordinates and select eigenspaces under other operators, to find, e.g., *hyperbolic* LCTs. Not surprisingly, it turns out that for $N = 1$ the theory of $\mathsf{Sp}(2,\mathsf{R})$ representations studied by Bargmann [47], and Gel'fand and Naĭmark [53]—also in the same year 1947, provides an appropriate framework to phrase these and other derivate LCTs.

1.5.1 Radial Canonical Transforms

Shortly after completing the initial two papers on LCTs based on the 2×2 $\mathsf{Sp}(2,\mathsf{R})$ matrices [3, 4], and Dr. Quesne having returned to Belgium, Marcos Moshinsky extended his inquiry to canonical transformations which he deemed to be nonlinear, but were closely related to the two-dimensional oscillator through the subgroup chain $\mathsf{Sp}(4,\mathsf{R}) \supset \mathsf{SO}(2) \otimes \mathsf{Sp}(2,\mathsf{R})$, where $\mathsf{SO}(2)$ is the group of rotations in the plane [54]. The representations of the two subgroups are *conjugate*, i.e., the representation $m \in \mathsf{Z}$ of $\mathsf{SO}(2)$ fixes the discrete-series representation $k = \frac{1}{2}(|m|+1)$ of $\mathsf{Sp}(2,\mathsf{R})$ (see below). This approach considered isotropic LCTs (1.2) in the *polar* coordinates of R^2,

$$x_1 = r\cos\theta, \quad x_2 = r\sin\theta, \qquad r \in \mathsf{R}_0^+ = [0,\infty), \quad \theta \in \mathsf{R} \bmod 2\pi. \tag{1.31}$$

Since angular momentum $\hat{L} = -\mathrm{i}(x_1\partial_2 - x_2\partial_1) = -\mathrm{i}\partial_\theta$ commutes with these LCTs, we can isolate an eigenspace of functions $f(\mathbf{x}) \sim f(r)\, e^{\mathrm{i}m\theta}/\sqrt{2\pi}$ with integer $m \in \mathsf{Z}$, to find the corresponding "radial" LCTs (RLCTs). There, $\nabla^2 = \partial_r^2 + r^{-1}\partial_r + r^{-2}\partial_\theta^2$ where with $\partial_\theta^2 \mapsto -m^2$ is self-adjoint under the measure $r\,\mathrm{d}r$. In order to have "m-radial" spaces where ∂_r^2 be self-adjoint, we need the inner product

$$(f, g)_{\mathcal{L}^2(\mathsf{R}+)} := \int_0^\infty \mathrm{d}r f(r)^*\, g(r) \tag{1.32}$$

with measure $\mathrm{d}r$, so previous operators should be transformed through $\hat{J} \mapsto \sqrt{r}\,\hat{J}/\sqrt{r}$ to keep self-adjointness.

To find the RLCT integral kernel under (1.32), we project out the Fourier series coefficient of the $e^{\mathrm{i}m\theta}$ component of the $N = 2$ isotropic LCT kernel (1.13),

$$C_{\mathrm{M}}^{(m)}(r, r') = \frac{1}{2\pi}\int_{-\pi}^{\pi} \mathrm{d}\theta\, C_{\mathrm{M}}(\mathbf{x}, \mathbf{x}') e^{-\mathrm{i}m\theta}. \tag{1.33}$$

Noting that only the factor $e^{-\mathrm{i}\mathbf{x}\cdot\mathbf{x}'/b}$ contains the mutual angle through $\mathbf{x}\cdot\mathbf{x}' = rr'\cos(\theta - \theta')$, we fix the reference axes by $\mathbf{x}$ to perform the integration. This is the angular momentum decomposition of the LCT, and defines the m-RLCT by

$$f_{\mathrm{M}}^{(m)}(r) \equiv (C_{\mathrm{M}}^{(m)} f)(r) = \int_{\mathsf{R}+} \mathrm{d}r'\, C_{\mathrm{M}}^{(m)}(r, r') f(r'), \tag{1.34}$$

$$C_{\mathrm{M}}^{(m)}(r, r') = \frac{e^{\mathrm{i}\pi(m+1)/2}}{b} \exp\Big(\frac{\mathrm{i}}{2b}(dr^2 + ar'^2)\Big) J_m\Big(\frac{rr'}{b}\Big), \tag{1.35}$$

where $J_m(z)$ is the Bessel function of the first kind. An alternative derivation of this kernel can be found in [55].

1.5.2 Hyperbolic Canonical Transforms

Hyperbolic canonical transforms are obtained when instead of the polar coordinates (1.31), one introduces the two-chart *hyperbolic* coordinates [56],

$$\left.\begin{array}{ll} \sigma = + : x_1 = \rho\cosh\zeta, & x_2 = \rho\sinh\zeta, \\ \sigma = - : x_1 = \rho\sinh\zeta, & x_2 = \rho\cosh\zeta, \end{array}\right\} \quad \begin{array}{l} \rho,\ \zeta \in \mathsf{R}, \\ \sigma := \operatorname{sign}(x_1^2 - x_2^2). \end{array} \tag{1.36}$$

Here the subgroup chain to be used is $\mathsf{Sp(4, R)} \supset \mathsf{O(1, 1)} \otimes \mathsf{Sp(2, R)}$, where now $\mathsf{O}(1, 1)$ consists of pseudo-orthogonal ("1+1 Lorentz") matrices, and inversions $\Pi : \mathbf{x} = -\mathbf{x}$ that also commute with $\mathsf{Sp(2, R)}$, reducing the range of the "hyperbolic radius" ρ to $[0, \infty)$. Instead of the isotropic LCTs used for RLCTs above, we now consider LCTs of the form $\mathbf{M} = \begin{pmatrix} a\mathbf{1} & b\overline{\mathbf{1}} \\ c\overline{\mathbf{1}} & d\mathbf{1} \end{pmatrix}$ with $\overline{\mathbf{1}} := \begin{pmatrix} 1 & 0 \\ 0 & e^{-\mathrm{i}\pi} \end{pmatrix}$, where the phase $e^{-\mathrm{i}\pi}$ is important. Then in (1.13) the first exponential term is $\mathbf{x}^\top \mathbf{b}^{-1}\mathbf{dx} = \sigma d\rho^2/b$, and only the term $\mathbf{x}^\top \mathbf{b}^{-1}\mathbf{x}' = \sigma\rho\rho' \cosh(\zeta - \zeta')$ contains the boost "angle" $\zeta \in \mathsf{R}$ that will be subject to integration.

Fourier integral decomposition of the LCT kernel (1.13) into plane waves and parity yield the "hyperbolic" LCTs (HLCTs), characterized now by the Fourier conjugate variable $s \in \mathsf{R}$ and the parity eigenvalue $\varpi \in \{+1, -1\}$. But note that now there are also two charts $\sigma \in \{+, -\}$, so that functions should be represented by two σ-component functions with definite parity ϖ, as $\mathbf{f}^{\varpi}(\rho) = \begin{pmatrix} f^{+,\varpi}(\rho) \\ f^{-,\varpi}(\rho) \end{pmatrix}$, with $f^{\sigma,\varpi}(\rho) = \varpi f^{\sigma,\varpi}(-\rho)$, and the inner product

$$(\mathbf{f}, \mathbf{g})_{\mathcal{L}^2(\varpi,\mathsf{R}+)} := \sum_{\sigma\in\{+,-\}} \int_0^\infty \mathrm{d}\rho\, f^{\sigma,\varpi}(\rho)^*\, g^{\sigma,\varpi}(\rho). \tag{1.37}$$

The HLCT of a function $f(\rho)$ is then

$$\mathbf{f}_{\mathrm{M}}^{\varpi,s}(\rho) \equiv (\mathcal{C}_{\mathrm{M}}^{\varpi,s}\,\mathbf{f})(\rho) = \int_{\mathsf{R}+} \mathrm{d}\rho'\, \mathbf{C}_{\mathrm{M}}^{(\varpi,s)}(\rho,\rho')\,\mathbf{f}(\rho'), \tag{1.38}$$

where the matrix integral kernel is

$$\mathbf{C}_{\mathrm{M}}^{(\varpi,s)}(\rho,\rho') = \begin{pmatrix} G_{\mathrm{M},+,+}(\rho,\rho')\,H_{+,+}^{(\varpi,s)}(\rho\rho'/b) & G_{\mathrm{M},+,-}(\rho,\rho')\,H_{+,-}^{(\varpi,s)}(\rho\rho'/b) \\ G_{\mathrm{M},-,+}(\rho,\rho')\,H_{-,+}^{(\varpi,s)}(\rho\rho'/b) & G_{\mathrm{M},-,-}(\rho,\rho')\,H_{-,-}^{(\varpi,s)}(\rho\rho'/b) \end{pmatrix}, \tag{1.39}$$

$$G_{\mathrm{M},\sigma,\sigma'}(\rho,\rho') = \frac{\sqrt{\rho\rho'}}{2\pi\,|b|} \exp\Big(\mathrm{i}\frac{\sigma d\rho^2 + \sigma' a\rho'^2}{2b}\Big), \tag{1.40}$$

$$H_{+,+}^{(\varpi,s)}(\xi) = \mathrm{i}\pi\,[\varpi e^{-\pi s} H_{2\mathrm{i}s}^{(1)}(\xi + \mathrm{i}0^+) - \varpi e^{\pi s} H_{2\mathrm{i}s}^{(2)}(\xi - \mathrm{i}0^+)] \tag{1.41}$$

$$= \varpi H_{-,-}^{(\varpi,s)}(\xi),$$

$$H_{+,-}^{(\varpi,s)}(\xi) = 4c_\xi^{\varpi,s}\, K_{2\mathrm{i}s}(|\xi|) = \varpi H_{-,+}^{(\varpi,s)}(\xi), \tag{1.42}$$

and where $H_\nu^{(1)}$ and $H_\mu^{(2)}$ are Hankel functions of the first and second kind valued above and below the branch cut, K_ν is the MacDonald function, $c_\xi^{+1,s} := \cosh \pi s$ and $c_\xi^{-1,s} := -\mathrm{sign}\,\xi\, \sinh \pi s$.

1.5.3 LCTs as Representations of Sp(2, R)

In Sect. 1.3 I warned that the theory of 2×2 real matrices is more imbricate than expected.[4] Yet I believe that the natural context to understand the foundations and

[4]Once I said in front of a large student audience that I had devoted much work to understand 2×2 matrices, the giggles in the hall were sobering.

see the possible incarnations of linear canonical transformations is in the theory of unitary irreducible representations of the Lorentz group $\mathsf{SO}(2,1)$ of "2+1" special relativity [46]. Let me now place LCTs in this context.

After relating paraxial optical elements to LCTs and second-order differential operators in (1.25)–(1.29), we note further that the following operators

$$\hat{J}_1 := \frac{1}{4}\Big(-\frac{\mathrm{d}^2}{\mathrm{d}r^2} + \frac{\gamma}{r^2} - r^2\Big), \tag{1.43}$$

$$\hat{J}_2 := \frac{-\mathrm{i}}{4}\Big(r\frac{\mathrm{d}}{\mathrm{d}r} + \frac{\mathrm{d}}{\mathrm{d}r}r\Big), \tag{1.44}$$

$$\hat{J}_3 := \frac{1}{4}\Big(-\frac{\mathrm{d}^2}{\mathrm{d}r^2} + \frac{\gamma}{r^2} + r^2\Big), \tag{1.45}$$

are essentially self-adjoint under the inner product (1.32) of $\mathcal{L}^2(\mathsf{R}^+)$, and that they close into an algebra with the commutation relations

$$[\hat{J}_1, \hat{J}_2] = -\mathrm{i}\hat{J}_3, \quad [\hat{J}_2, \hat{J}_3] = \mathrm{i}\hat{J}_1, \quad [\hat{J}_3, \hat{J}_1] = \mathrm{i}\hat{J}_2, \tag{1.46}$$

that characterize the isomorphic algebras $\mathsf{sp}(2,\mathsf{R}) = \mathsf{so}(2,1)$. Instead of starting with the preservation of the Heisenberg canonical commutation relations (1.6) between the Schrödinger quantum position and momentum operators, here we start from the preservation of the commutators (1.46) and their realization by the three operators (1.43)–(1.45). Their commutators are preserved under linear transformations with parameters taken from $\mathbf{M} = \left(\begin{smallmatrix} a & b \\ c & d \end{smallmatrix}\right) \in \mathsf{Sp}(2,\mathsf{R})$,

$$\begin{pmatrix} \hat{J}_1 \\ \hat{J}_2 \\ \hat{J}_3 \end{pmatrix} \overset{\left(\begin{smallmatrix} a & b \\ c & d \end{smallmatrix}\right)}{\longrightarrow} \begin{pmatrix} \frac{1}{2}(a^2-b^2-c^2+d^2) & bd-ac & \frac{1}{2}(a^2-b^2+c^2-d^2) \\ cd-ab & ad+bc & -cd-ab \\ \frac{1}{2}(a^2+b^2-c^2-d^2) & -bd-ac & \frac{1}{2}(a^2+b^2+c^2+d^2) \end{pmatrix} \begin{pmatrix} \hat{J}_1 \\ \hat{J}_2 \\ \hat{J}_3 \end{pmatrix}. \tag{1.47}$$

These 3×3 matrices form the "2+1" Lorentz group $\mathsf{SO}(2,1)$ with metric $(--+)$. Since both $\mathbf{M}$ and $-\mathbf{M}$ yield the same 3×3 matrix, this Lorentz group is covered 2:1 by $\mathsf{Sp}(2,\mathsf{R})$; however, their Lie *algebras*, defined by their commutation relations, are the same.

In Sect. 1.4 we came upon the three orbits of $\mathsf{sp}(2,\mathsf{R}) = \mathsf{so}(2,1)$, which can be also be characterized by the distinctive spectrum of the generator that we choose to be the operator of position, $\{\rho\}$, which can be discrete or continuous. We can use the realization (1.43)–(1.45) in $\mathcal{L}^2(\mathsf{R}^+)$ for $\gamma > 0$ to evince those spectra.[5] The are:

[5]For $\gamma < 0$ there is a doubling of the Hilbert space that requires some extra analytical finesse [57], which stems from a separation in hyperbolic coordinates such as that seen in the previous subsection.

- In the ω = **elliptic** orbit of the compact "harmonic oscillator $+\gamma/r^2$" operator, $\hat{J}_3$ in (1.45) has a discrete spectrum $\{\rho\}_3$ bounded from below, and equally spaced by 1.
- In the ω = **parabolic** orbit of the "thin lens" generator in (1.25), here $\hat{J}_- := \hat{J}_3 - \hat{J}_1 = \frac{1}{2}r^2 \geq 0$, the spectrum $\{\rho\}_-$ is continuous and non-negative. Its Fourier-Bessel transform is $\hat{J}_+ := \hat{J}_3 + \hat{J}_1$, which is the Hamiltonian of "free flight in a γ/r^2 potential," and has the same spectrum.
- In the ω = **hyperbolic** orbit of the "repulsive oscillator $+\gamma/r^2$" operator, $\hat{J}_1$ in (1.43), the spectrum $\{\rho\}_1$ is the real line.

Thus, while $\mathsf{su}(2) = \mathsf{so}(3)$ contains a single orbit and the spectrum $\{\mu\}$ of any generator J_z can provide the row and column labels—positions—for the spin j representation matrices and vectors, bound by integer-spaced $|\mu| \leq j$, in $\mathsf{sp}(2,\mathsf{R}) = \mathsf{so}(2,1)$ we have *three* orbits and *three* choices for the position $\{\rho\}$: discrete, continuous positive, or real. Moreover, while the representations of $\mathsf{so}(3)$ are simply labelled by the non-negative integers $j \in \mathsf{Z}_0^+$ in the eigenvalues $j(j+1)$ of the square angular momentum, the representation structure of $\mathsf{so}(2,1)$ and the bounds it imposes on $\{\rho\}$ are more complicated. The parameter γ in (1.43)–(1.45) is the strength of the centrifugal ($\gamma > 0$) or centripetal potential ($\gamma < 0$); the special case $\gamma = 0$ will remit us back to the original and best-known LCT face in (1.16)–(1.17). This parameter γ determines *almost* (see below) the representation of the algebra through the eigenvalues of the $\mathsf{so}(2,1)$ invariant *Casimir* operator,

$$\hat{C} := \hat{J}_1^2 + \hat{J}_2^2 - \hat{J}_3^2 = (-\tfrac{1}{4}\gamma + \tfrac{3}{16})\, 1 =: k(1-k)\, 1, \tag{1.48}$$

$$\gamma = (2k-1)^2 - \tfrac{1}{4}, \qquad k = \tfrac{1}{2}(1 \pm \sqrt{\tfrac{1}{4}+\gamma}). \tag{1.49}$$

Here, k is the all-important *Bargmann index*; it distinguishes the two main series of representations:

- Bargmann **discrete** $\mathcal{D}_k^\pm$ representations [47] (called *complementary* by Gel'fand and Naĭmark [53]). When the coefficient γ is of centrifugal origin in two dimensions, angular momentum $\mu \in \mathsf{Z}$ determines $\gamma = \mu^2 - \frac{1}{4} \geq -\frac{1}{4}$, which implies the range $k = \frac{1}{2}(|\mu| + 1) \in \{\frac{1}{2}, 1, \frac{3}{2}, \ldots\}$. This series can be extended to continuous $k \in \mathsf{R}^+$, representing multiple covers of $\mathsf{so}(2,1)$. In particular for k quarter-integers, they are faithful representations of $\mathsf{Mp}(2;\mathsf{R})$. The $\mathcal{D}_k^-$ representations are related to the $\mathcal{D}_k^+$ ones by an outer automorphism of the group that in geometric optics is reflection [48, Sect. 10.4].
- Bargmann **continuous** $\mathcal{C}_s^\varepsilon$ representations (called *principal* by Gel'fand and Naĭmark). When $\gamma < 0$, the potential is centripetal and we must further distinguish the *exceptional* range $-\frac{1}{4} \leq \gamma < 0$ where $\frac{1}{2} \leq k < 1$ is real, from the *principal* range $\gamma < -\frac{1}{4}$ where $k = \frac{1}{2} + \mathrm{i}s$, with $s = \pm\frac{1}{2}\sqrt{(|\gamma| - \frac{1}{4})} \in \mathsf{R}$, and $\varepsilon \in \{0, \frac{1}{2}\}$ is a multivaluation index. We shall exclude the exceptional range

$0 < k < 1$ from further detailed considerations.[6] We treat this interval as an extension of the $\mathcal{D}_k^{\pm}$ discrete series.

The best-known one-dimensional LCT in (1.16)–(1.17) occurs for $\gamma = 0$, namely the quarter-integers $k = \frac{1}{4}$ and $k = \frac{3}{4}$, for the subspaces of even and odd functions of ρ, respectively—recall that here we are on the "radial" half line for the inner product (1.32) of $\mathcal{L}^2(\mathsf{R}^+)$.

Using Dirac's shorthand notation, let $|k, \rho\rangle^{\omega}$ be a basis vector for the unitary irreducible representation k (in $\mathcal{D}_k^+$ or $\mathcal{C}_s^{\varepsilon}$), with row ρ (discrete or continuous) determined by the orbit ω of the chosen position operator. We may then understand LCTs as the unitary irreducible representations of $\mathbf{M} \in \mathsf{Sp}(2, \mathsf{R})$ acting on those Hilbert space bases and functions,

$$f_{\mathrm{M}}^{k,\omega}(\rho) = \underset{\rho'(k,\omega)}{\mathsf{S}} D_{\rho,\rho'}^{k,\omega}(\mathbf{M}) f(\rho'), \tag{1.50}$$

$$D_{\rho,\rho'}^{k,\omega}(\mathbf{M}) := {}^{\omega}\langle k, \rho | \mathcal{C}_{\mathrm{M}} | k, \rho' \rangle^{\omega}, \qquad \text{i.e.,} \tag{1.51}$$

$${}^{\omega}\langle k, \rho | f_{\mathrm{M}} \rangle = {}^{\omega}\langle k, \rho | \mathcal{C}_{\mathrm{M}} | f \rangle = {}^{\omega}\langle k, \rho | \mathcal{C}_{\mathrm{M}} | k, \rho' \rangle^{\omega}\, {}^{\omega}\langle k, \rho' | f \rangle, \tag{1.52}$$

where S_{ρ} is a sum or integral over the range of eigenvalues of position $\rho(k, \omega)$ allowed in the representation k, where the chosen position operator is in the orbit ω. The ranges of its "position coordinate" are:

$$\begin{array}{lll} & \mathcal{D}_k^+ & \mathcal{C}_s^{\varepsilon} \\ \omega \text{ elliptic:} & \rho = k+n,\ n \in \mathsf{Z}_0^+ & \rho - \varepsilon \in \mathsf{Z} \\ \omega \text{ parabolic:} & \rho \in \mathsf{R}^+ & \rho \in \mathsf{R}^+ \oplus \mathsf{R}^+ \\ \omega \text{ hyperbolic:} & \rho \in \mathsf{R} & \rho \in \mathsf{R} \oplus \mathsf{R} \end{array} \tag{1.53}$$

The orthogonality and completeness of the bases $|k, \rho'\rangle^{\omega}$ guarantees that the group composition property holds and that the transformation is unitary and hence invertible,

$$\underset{\rho'(k,\omega)}{\mathsf{S}} D_{\rho,\rho'}^{k,\omega}(\mathbf{M}_1)\, D_{\rho',\rho''}^{k,\omega}(\mathbf{M}_2) = D_{\rho,\rho''}^{k,\omega}(\mathbf{M}_1\mathbf{M}_2), \tag{1.54}$$

$$D_{\rho,\rho'}^{k,\omega}(\mathbf{M}^{-1}) = D_{\rho',\rho}^{k,\omega}(\mathbf{M})^*. \tag{1.55}$$

The matrices and integral kernels $D_{\rho,\rho'}^{k,\omega}(\mathbf{M})$ are known in the literature. They were written out for ω = elliptic by Bargmann [47]; for ω = hyperbolic by Mukunda and Radhakrishnan [60]; and for ω = parabolic they are the radial and hyperbolic LCT kernels of this section. In [57] all ${}^{\omega}\langle k, \rho | \mathcal{C}_{\mathrm{M}} | k, \rho' \rangle^{\omega'}$ are listed, including the mixed cases $\omega \neq \omega'$; these were later used to find the $\mathsf{so}(2, 1)$

[6] The generators present a one-parameter family of self-adjoint extensions with non-equally spaced spectra [58] and also harbor the $\mathcal{E}_k$ exceptional (or *supplementary*) representation series [59].

Clebsch–Gordan coefficients between all representation series [61]. Finally, while writing this chapter, I completed the work in [62], giving explicitly (in the present notation) the six distinct *faces* of LCTs, $D^{k,\omega}_{\rho,\rho'}(\mathbf{M})$ for the three orbits in the two nonexceptional representation series. I close this section reminding the readers that there is a theorem stating that noncompact groups (i.e., of infinite volume) do *not* have faithful finite-dimensional unitary representations; thus, $\mathsf{Sp}(2,\mathsf{R})$ only has finite representations that are *not* unitary—such as the 2×2 matrix $\mathbf{M} = \left(\begin{smallmatrix} a & b \\ c & d \end{smallmatrix}\right)$, the 3×3 matrix in (1.47), or others of "spin" k given in [48, Eq. (13.6)] that are used for Lie aberration optics.

1.6 Complex Extensions of LCTs

While LCTs allow a transparent formulation of the properties of resonators, where a paraxial wavefield is bounced repeatedly between two end-mirrors, it is natural to inquire about systems with loss or gain [10, 11]. On the other hand, applications to clustering in nuclei [63] required the description of Gaussian packets in terms of the raising and lowering operators of the harmonic oscillator [64], i.e.,

$$\begin{pmatrix} \hat{z}^{\uparrow} \\ -\mathrm{i}\hat{z}^{\downarrow} \end{pmatrix} = \frac{1}{\sqrt{2}}\begin{pmatrix} 1 & -\mathrm{i} \\ -\mathrm{i} & 1 \end{pmatrix}\begin{pmatrix} \hat{x} \\ \hat{p} \end{pmatrix} = \frac{1}{\sqrt{2}}\begin{pmatrix} \hat{x} - \mathrm{i}\hat{p} \\ -\mathrm{i}(\hat{x} + \mathrm{i}\hat{p}) \end{pmatrix}, \qquad -\mathrm{i} \equiv e^{-\mathrm{i}\pi/2}. \tag{1.56}$$

Issues related to the meshing between Bargmann and LCT transforms were discussed in a think-tank at the Centre de Recherches Mathématiques (Université de Montréal) during the closing months of 1973. It was also noted that the real heat diffusion kernel to time $t > 0$ is the $\left(\begin{smallmatrix} 1 & -2\mathrm{i}t \\ 0 & 1 \end{smallmatrix}\right)$ complex LCTs with kernel $\sim \exp(-(x-x')^2/4t)/\sqrt{t}$; for $t > 0$ these transforms form a *semi*-group (i.e., without inverses). Indeed, one can extend the $\left(\begin{smallmatrix} a & b \\ c & d \end{smallmatrix}\right)$ parameters as long as the LCT kernel (1.17) is a *decreasing* Gaussian in the argument x' subject to integration, namely $\mathrm{Re}\,(\mathrm{i}a/b) < 0$. If a is real, this means that the complex value of b must be in the lower complex half-plane, $-\pi < \arg b < 0$.

But unitarity is a cherished property among group theorists, so the question was posed to find appropriate Hilbert spaces to comply with this requirement. There was the precedent of Bargmann's space for analytic functions $f(z)^* = f(z^*)$ [49], so it was not difficult [51] to follow his construction in proposing a *measure* for the sesquilinear inner product that integrates over the whole complex plane $z \in \mathsf{C}$, of $\mathcal{L}^2(\mathsf{R})$ functions that have been transformed by a complex $\mathbf{M} = \left(\begin{smallmatrix} a & b \\ c & d \end{smallmatrix}\right)$,

$$(\mathbf{f}_{\mathrm{M}}, \mathbf{g}_{\mathrm{M}})_{\mathcal{B}_{\mathrm{M}}} := \int_{\mathsf{C}} \mathrm{d}^2\mu_{\mathrm{M}}(z, z^*) f_{\mathrm{M}}(z)^*\, g_{\mathrm{M}}(z) = (\mathbf{f}, \mathbf{g})_{\mathcal{L}^2(\mathsf{R})}, \tag{1.57}$$

$$\mathrm{d}^2\mu_{\mathrm{M}}(z, z^*) = \nu_{\mathrm{M}}(z, z^*)\, \mathrm{d}\,\mathrm{Re}\, z\, \mathrm{d}\,\mathrm{Im}\, z, \quad \text{where} \tag{1.58}$$

$$\nu_{\mathrm{M}}(z, z^*) = \sqrt{\frac{2}{\pi v}} \exp\left(\frac{uz^2 - 2zz^* + u^* z^{*2}}{2v}\right), \tag{1.59}$$

$$u := a^* d - b^* c, \quad v := 2\,\mathrm{Im}\,(a\, b^*) > 0. \tag{1.60}$$

This defines Bargmann-type Hilbert spaces $\mathcal{B}_{\mathrm{M}}$ such that the complex LCT between $\mathcal{L}^2(\mathsf{R}) = \mathcal{B}_1$ and $\mathcal{B}_{\mathrm{M}}$ is unitary, and can be inverted back to $\mathcal{L}^2(\mathsf{R})$ through

$$f(x) = \int_{\mathsf{C}} \mathrm{d}^2 \mu_{\mathrm{M}}(z', z'^*)\, C_{\mathrm{M}^{-1}}(x, z') f_{\mathrm{M}}(z'). \tag{1.61}$$

In the limit when **M** becomes real, the measure weight function $\nu_{\mathrm{M}}(z, z^*)$ in (1.59) is a Gaussian that collapses to a Dirac δ on the $\mathrm{Re}\, z'$ axis [6, Sect. 9.2.2]. Of interest to mathematicians is the use of the hyperdifferential operator realization of complex LCTs to find an expression for Hermite polynomials, such as [51, App. A]

$$H_n(x) = \exp\left(-\frac{1}{4}\frac{\mathrm{d}^2}{\mathrm{d}x^2}\right)(2x)^n, \qquad x = \exp\left(\frac{1}{4}\frac{\mathrm{d}^2}{\mathrm{d}x^2}\right) H_n(x). \tag{1.62}$$

Similar relations could be found for parabolic cylinder and other special functions, but have not been investigated.

Radial LCTs can also be extended to the complex domain [55] when the radial kernel (1.35) is a decreasing Gaussian, $\mathrm{Re}\,(\mathrm{i}a/b) < 0$ as before. But now, noting in (1.32) that the argument of the functions is $r \in [0, \infty)$, it turns out that the complex-transformed functions will be analytic only in the right half-plane $\varrho \in \mathsf{C}^+$ where $\mathrm{Re}\, \varrho > 0$. The Bargmann-type inner products that preserve the unitarity of the complex RLCTs that map $f(r) \in \mathcal{L}^2(\mathsf{R}^+)$ to $f_{\mathrm{M}}^{(m)}(r) \in \mathcal{B}_{\mathrm{M}}^{(m)}$ are

$$(\mathbf{f}_{\mathrm{M}}^{(m)}, \mathbf{g}_{\mathrm{M}}^{(m)})_{\mathcal{B}_{\mathrm{M}}^{(m)}} := \int_{\mathsf{C}+} \mathrm{d}^2 \mu_{\mathrm{M}}^{(m)}(\varrho, \varrho^*) f_{\mathrm{M}}^{(m)}(\varrho)^*\, g_{\mathrm{M}}^{(m)}(\varrho) = (\mathbf{f}, \mathbf{g})_{\mathcal{L}^2(\mathsf{R}+)}, \tag{1.63}$$

$$\mathrm{d}^2 \mu_{\mathrm{M}}^{(m)}(\varrho, \varrho^*) = \nu_{\mathrm{M}}^{(m)}(\varrho, \varrho^*)\, \mathrm{d}\,\mathrm{Re}\, \varrho\; \mathrm{d}\,\mathrm{Im}\, \varrho, \quad \text{where} \tag{1.64}$$

$$\nu_{\mathrm{M}}^{(m)}(\varrho, \varrho^*) = \frac{2}{\pi v} \exp\left(\frac{u\varrho^2 + u^* \varrho^{*2}}{2v}\right) K_m\left(\frac{\varrho \varrho^*}{v}\right), \tag{1.65}$$

where $K_m(z)$ is the MacDonald function, while u and v are given by (1.60). The inversion and real limit properties are similar to those of the complex LCTs seen above.

A specific case of interest is the treatment of the Barut–Girardello transform and coherent state [65]. Similar to (1.62), one obtains a hyperdifferential form for the Laguerre polynomials [55],

$$L_n^{(m)}(\tfrac{1}{2} r^2) = \frac{(-1)^n}{n!\, 2^n} r^{-m} \exp\left[\frac{-1}{2}\left(\frac{\mathrm{d}^2}{\mathrm{d}r^2} + \frac{1}{r}\frac{\mathrm{d}}{\mathrm{d}r} - \frac{m^2}{r^2}\right)\right] r^{2n+m}. \tag{1.66}$$

Finally, hyperbolic LCTs do not allow for any complex extension [56].

1.7 Finite Data Sets and LCTs

Most data sets in the real world are finite and can be represented as N-component vectors, $\mathbf{f} \equiv \{f_m\}_{m=1}^N$. If the numbers come from sensing a continuous wavefield $f(x)$ at N discrete points, how can we compute their propagation through an optical LCT setup? The most direct answer is sampling the assumed smooth wavefield $f_m := f(x_m)$ and the LCT kernel (1.17) at the same points $\{x_m\}_{m=1}^N$, and simply performing the product of the $N \times N$ matrix $\mathbf{C}_\mathrm{M}$, function of $\mathbf{M} = \begin{pmatrix} a & b \\ c & d \end{pmatrix}$, and the vector $\mathbf{f}$. Following the discretization adopted in [66] for $x_m = m\sqrt{(2\pi/N)}$, this is

$$(\mathbf{C}_\mathrm{M}\,\mathbf{f})_m := \sum_{m'=1}^{N} \frac{1}{\sqrt{N}} \exp\Big(\frac{\mathrm{i}\pi}{bN}(dm^2 - 2mm' + am'^2)\Big) f_{m'}, \tag{1.67}$$

$$(\mathbf{C}_\mathrm{M}\,\mathbf{f})_m := \exp(\mathrm{i}\pi\, cm^2/aN) f_m \qquad \text{when } b = 0, \tag{1.68}$$

where we leave out phases. Yet this transformation is only an approximation to the LCT, and it is generally not unitary.

Regarding the spacing of the sampling points, Ding [67] has given a sampling theorem that generalizes that of Shannon in terms of the desired extent of the LCT transform signal. The requirement of unitarity on the kernels (1.67) is that $\mathbf{C}_\mathrm{M}^\dagger \mathbf{C}_\mathrm{M} = \mathbf{1}$; this occurs only for values of the parameter b such that $1/b$ is an integer relatively prime to N [68]. Combining both results, in [69], the authors present sufficient conditions on the sampling rate of $\{x_m\}_{m=1}^N$ for any one LCT to ensure its unitarity. However, two such matrices will not concatenate as integral LCTs do because, as we mentioned at the end of Sect. 1.5, $\mathsf{Sp}(2,\mathsf{R})$ has no finite-dimensional unitary irreducible representations. Alternatively, if one discretizes the LCT kernel by using the LCT sampling theorem [67, 70], a unitary discrete LCT which provides a provably good approximation to the continuous LCT can be obtained [71, 72]. In principle, one would like to have a relation between the discrete and continuous LCTs that mirrors and generalizes the corresponding relation for ordinary Fourier transforms. Such a relation has been provided in [71], showing that the discrete LCT, as defined in [73], approximates the LCT in the same sense that the discrete Fourier transform approximates the continuous Fourier transform, provided that the number of samples and the sampling intervals are chosen according to the LCT sampling theorem [67, 70–72].

There are some other problems to define subsets of finite LCTs that form a group, which we can point out for finite analogues of fractional Fourier transforms. If we search for a one-parameter subgroup of unitary $N \times N$ matrices $\mathbf{F}^\nu \in \mathsf{U(N)}$ such that $\mathbf{F}^\nu \mathbf{F}^\mu = \mathbf{F}^{\nu+\mu}$, with $\mathsf{F}^1 = \mathbf{F}$ being the well-known finite Fourier transform matrix,

$$F_{m,m'} = \frac{1}{\sqrt{N}} \exp\Big(-\mathrm{i}\frac{2\pi\, m\, m'}{N}\Big), \tag{1.69}$$

which is unitary and idempotent, $\mathbf{F}^4 = \mathbf{1}$, what we find is a deluge of possibilities: we can draw the N^2 real parameters of unitary $N \times N$ matrices inside a sphere in R^{N^2} space; the four matrix powers of $\mathbf{F}$ are but four points—with $\mathbf{1}$ on the origin. Unitary matrices of unit determinant form the simply connected subgroup $\mathsf{SU}(N)$, whose $\mathsf{SO}(2)$ subgroups of possible fractional $\mathbf{F}^{\nu}$'s are closed lines (picture them as circles) that can be freely rotated keeping the origin fixed, and are only required to pass through $\mathbf{F}$ for $\nu = 1$, since automatically the circles will pass also through its integer powers. Clearly, for dimensions $N > 2$ there is a continuum of such circles that can be drawn through two points. From that perspective, we analyzed this freedom in [74] in terms of choosing "good" bases for the R^N manifold. Alternative approaches to define good bases to build finite fractional Fourier transform matrices have used sampled harmonic oscillator wavefunctions, as done by Pei et al. [73, 75, 76], or other candidates such as the Harper functions [77] by Ozaktas et al.

Additionally, there is a problem with the phase space interpretation of these finite fractional Fourier transforms, which we can see through the commonly used finite Wigner function [25]. The finite Fourier transform matrix $\mathbf{F}$ brings N-vectors of position to N-vectors of momentum. And, being cyclic, $F_{m,m'} = F_{m+N,m'} = F_{m,m'+N}$ leads to consider a phase space that is discrete and connected as a torus. The "front face" of this torus is the origin of phase space $m = 0 = m'$. So, while the *integral* fractional Fourier transform rotates the phase space plane around the origin, we *cannot* rotate the front face of a torus without tearing it. Yet to be applicable, finite LCTs must be computed efficiently for one- or two-dimensional signals and images in real time. This line of research has been developed by Sheridan et al. in [66, 78–81] with the strategy of separating the finite LCT in (1.67) into a Fourier transform factor, for which the FFT algorithm exists, and factors of (1.68). Another strategy for fast and accurate computation of LCTs has been developed by Ozaktas et al. using the Iwasawa decomposition [82, 83]. Alternatively, a chirp-Fourier-chirp factorization with a fast-convergent quadrature formula was proposed in [84].

Finally, we should mention another fast computation method that also involves chirp multiplication, fast Fourier transform, and a second chirp multiplication [71]: this method has the advantage of involving the least number of samples possible as determined from the LCT sampling theorem [67, 70]. This discrete LCT has a well-defined relation to the continuous LCT and can be made unitary by adding a factor in front [71, 72]. This approach is attractive because it combines a desirable analytic discrete LCT definition with a computational method that is nearly as fast and accurate as the fast Fourier transform algorithm to compute continuous Fourier transforms.

1.8 Conclusion

I was deeply honored by the invitation of the Editors to write some pages about the development of linear canonical transforms, in company with distinguished researchers who are applying them in encryption, metrology, holography, and optical implementations. Emeritus Professor Stuart Collins is still active and has registered six patents to his name between 1982 and 2008; his work in optoelectronics has been applied for space science. I was an apprentice of Marcos Moshinsky and developed his work on quantum mechanics as a fruitful model for LCTs and related transforms. Only later did I learn that LCTs were excellent tools for paraxial optics as used by a community with whom I could then establish dialogue.

Perhaps a similar bifurcation of viewpoints may occur concerning finite LCTs. The previous section contains problems which I regard as indicative that a different approach can be useful to understand finite signals on phase space and their canonical transformations. Based on the rotation algebra $\mathsf{so}(3)$, instead of the Heisenberg–Weyl algebra of quantum mechanics, we have proposed a model for discrete Hamiltonian systems where phase space is a sphere [85]. When the number of position points and their density increase, the model contracts to that of quantum mechanics, the sphere blowing up into the quantum phase space plane. Canonical transformations in $\mathsf{so}(3)$ are those that preserve the surface elements of the sphere. *Linear* transformations of N-point signals are the rigid rotations of that sphere. Among these, the fractional Fourier-*Kravchuk* transform [86] describes the time evolution of this finite harmonic oscillator. Moreover, *nonlinear* canonical transformations can be defined in correspondence with optical aberrations as matrices in the full $\mathsf{U}(N)$ group of linear transformations of N-vectors [87]. Based on the Euclidean and Lorentz algebras, other discrete models are available in one and two dimensions [88].

On the other hand, it is not clear that expansions in group-theoretic bases have any advantage over other bases for expansion [89], since they do not seem amenable to fast algorithms. Still, based on previous experiences, I harbor the hope that the mathematical landscape succinctly described here can be of use to broaden the perspective we have of canonical transformations of phase space. The founders of this field must have been quite unaware of the full panorama they opened for us to see.

Acknowledgements I must thank the inspiring interaction I had with Professor Marcos Moshinsky for many years, the continuing interest of my colleagues, and especially that of the Editors of this volume. I have also incorporated some remarks graciously offered by Professor Stuart A. Collins Jr. Support for this research has been provided by the *Óptica Matemática* projects of UNAM (PAPIIT IN101115) and by the National Council for Science and Technology (SEP-CONACYT 79899).

References

1. S.A. Collins Jr., Lens-system diffraction integral written in terms of matrix optics. J. Opt. Soc. Am. **60**, 1168–1177 (1970)
2. M. Moshinsky, C. Quesne, Oscillator systems, in *Proceedings of the 15th Solvay Conference in Physics (1970)* (Gordon and Breach, New York, 1974)
3. M. Moshinsky, C. Quesne, Linear canonical transformations and their unitary representation. J. Math. Phys. **12**, 1772–1780 (1971)

4. C. Quesne, M. Moshinsky, Linear canonical transformations and matrix elements. J. Math. Phys. **12**, 1780–1783 (1971)
5. S. Liberman, K.B. Wolf, Independent simultaneous discoveries visualized through network analysis: the case of Linear Canonical Transforms. Scientometrics (2015). doi:10.1007/s/11192-015-1602-x online 20/06
6. K.B. Wolf, *Integral Transforms in Science and Engineering* (Plenum Publishing Corporation, New York, 1979)
7. M. Nazarathy, J. Shamir, Fourier optics described by operator algebra. J. Opt. Soc. Am. **70**, 150–159 (1980)
8. M. Nazarathy, J. Shamir, Holography described by operator algebra. J. Opt. Soc. Am. **71**, 529–541 (1981)
9. M. Nazarathy, J. Shamir, First-order optics—a canonical operator representation: lossless systems. J. Opt. Soc. Am. **72**, 356–364 (1982)
10. M. Nazarathy, J. Shamir, First-order optics—operator representation for systems with loss or gain. J. Opt. Soc. Am. **72**, 1398–1408 (1982)
11. M. Nazarathy, A. Hardy, J. Shamir, Generalized mode propagation in first-order optical systems with loss or gain. J. Opt. Soc. Am. **72**, 1409–1420 (1982)
12. A.J. Dragt, Lie algebraic theory of geometrical optics and optical aberrations. J. Opt. Soc. Am. **72**, 372–379 (1982)
13. A.J. Dragt, *Lectures on Nonlinear Orbit Dynamics*. American Institute of Physics Conference Proceedings, vol. 87 (American Institute of Physics, New York, 1982)
14. A.J. Dragt, Elementary and advanced Lie algebraic methods with applications to accelerator design, electron microscopes, and light optics. Nucl. Instrum. Methods Phys. Res. A **258**, 339–354 (1967)
15. A.J. Dragt, *Lie Methods for Nonlinear Dynamics with Applications to Accelerator Physics*, University of Maryland (2015). http://www.physics.umd.edu/dsat/dsatliemethods.html
16. W. Brouwer, *Matrix Methods in Optical Instrument Design* (Benjamin, New York, 1964)
17. A. Gerrard, B. Burch, *Introduction to Matrix Methods in Optics* (Wiley, New York,1975)
18. M. Nazarathy, A. Hardy, J. Shamir, Misaligned first-order optics: canonical operator theory. J. Opt. Soc. Am. A **3**, 1360–1369 (1986)
19. A.J. Dragt, E. Forest, K.B. Wolf, Foundations of Lie algebraic theory of geometrical optics, in *Lie Methods in Optics*, ed. by J. Sánchez-Mondragón, K.B. Wolf. Lecture Notes in Physics, vol. 250 (Springer, Heidelberg, 1986)
20. M.N. Saad, K.B. Wolf, Factorization of the phase–space transformation produced by an arbitrary refracting surface. J. Opt. Soc. Am. A **3**, 340–346 (1986)
21. J. Sánchez-Mondragón, K.B. Wolf (eds.), *Lie Methods in Optics*. Lecture Notes in Physics, vol. 250 (Springer, Heidelberg, 1986)
22. K.B. Wolf (ed.), *Lie Methods in Optics, II*. Lecture Notes in Physics, vol. 352 (Springer, Heidelberg, 1989)
23. M.J. Bastiaans, Wigner distribution function applied to optical signals and systems. Opt. Commun. **25**, 26–30 (1978)
24. M.J. Bastiaans, Wigner distribution function and its applications to first-order optics. J. Opt. Soc. Am. **69**, 1710–1716 (1979)
25. G.W. Forbes, V.I. Man'ko, H.M. Ozaktas, R. Simon, K.B. Wolf (eds.), Feature issue on Wigner distributions and phase space in optics. J. Opt. Soc. Am. A **17**(12), 2274–2274 (2000)
26. M.J. Bastiaans, Wigner distribution function and its application to first-order optics, in *Selected Papers on Phase-Space Optics*, ed. by M.E. Testorf, J. Ojeda-Castañeda, A.W. Lohmann (SPIE, Bellingham, 2006), pp. 315–321
27. T. Alieva, M.J. Bastiaans, Properties of the linear canonical integral transformation. J. Opt. Soc. Am. A **24**, 3658–3665 (2007)
28. T. Alieva, M.J. Bastiaans, Finite-mode analysis by means of intensity information in fractional optical systems. J. Opt. Soc. Am. A **19**, 481–484 (2002)
29. M.J. Bastiaans, K.B. Wolf, Phase reconstruction from intensity measurements in linear systems. J. Opt. Soc. Am. A **20**, 1046–1049 (2003)

30. T. Alieva, M.J. Bastiaans, M.L. Calvo, Fractional transforms in optical information processing. EURASIP J. Appl. Signal Process. **2005**, 1498–1519 (2005)
31. T. Alieva, M.J. Bastiaans, Wigner distribution and fractional Fourier transform for two-dimensional symmetric optical beams. J. Opt. Soc. Am. A **17**, 2319–2323 (2000)
32. T. Alieva, M.J. Bastiaans, Alternative representation of the linear canonical integral transform. Opt. Lett. **30**, 3302–3304 (2005)
33. M.J. Bastiaans, T. Alieva, Synthesis of an arbitrary ABCD-system with fixed lens positions. Opt. Lett. **16**, 2414–2416 (2006)
34. T. Alieva, M.J. Bastiaans, Orthonormal mode sets for the two-dimensional fractional Fourier transformation. Opt. Lett. **33**, 1226–1228 (2007)
35. T. Alieva, M.J. Bastiaans, Classification of lossless first-order optical systems and the linear canonical transformation. J. Opt. Soc. Am. A **24**, 1053–1062 (2007)
36. J.A. Rodrigo, T. Alieva, M.L. Calvo, Gyrator transform: properties and applications. Opt. Express **15**, 2190–2203 (2007)
37. J.A. Rodrigo, T. Alieva, M.J. Bastiaans, Phase space rotators and their applications in optics, in *Optical and Digital Image Processing: Fundamentals and Applications*, ed. by G. Cristóbal, P. Schelkens, H. Thienpont (Wiley-VCH, Weinheim, 2011), pp. 251–271
38. R. Simon, Peres-Horodecki separability criterion for continuous variable systems. Phys. Rev. Lett. **84**, 2726–2729 (2000)
39. E.U. Condon, Immersion of the Fourier transform in a continuous group of functional transformations. Proc. Natl. Acad. Sci. **23**, 158–163 (1937)
40. V. Namias, The fractional order Fourier transform and its applications in quantum mechanics. IMA J. Appl. Math. **25**, 241–265 (1980)
41. D. Mendlovic, H.M. Ozaktas, Fractional Fourier transforms and their optical implementation: I. J. Opt. Soc. Am. A **10**, 1875–1881 (1993)
42. H.M. Ozaktas, D. Mendlovic, Fractional Fourier transforms and their optical implementation: II. J. Opt. Soc. Am. A **10**, 2522–2531 (1993)
43. H.M. Ozaktas, D. Mendlovic, Fourier trasforms of fractional order and their optical interpretation. Opt. Commun. **101**, 163–169 (1993)
44. H.M. Ozaktas, D. Mendlovic, Fractional Fourier optica. J. Opt. Soc. Am. A **12**, 743–751 (1995)
45. H.M. Ozaktas, Z. Zalevsky, M.A. Kutay, *The Fractional Fourier Transform with Applications in Optics and Signal Processing* (Wiley, Chichester, 2001)
46. M. Kauderer, *Symplectic Matrices. First Order Systems and Special Relativity* (World Scientific, Singapore, 1994)
47. V. Bargmann, Irreducible unitary representations of the Lorentz group. Ann. Math. **48**, 568–642 (1947)
48. K.B. Wolf, *Geometric Optics on Phase Space* (Springer, Heidelberg, 2004)
49. V. Bargmann, Group representation in Hilbert spaces of analytic functions, in *Analytical Methods in Mathematical Physics*, ed. by P. Gilbert, R.G. Newton (Gordon & Breach, New York, 1970), pp. 27–63
50. R. Simon, K.B. Wolf, Fractional Fourier transforms in two dimensions. J. Opt. Soc. Am. A **17**, 2368–2381 (2000)
51. K.B. Wolf, Canonical transforms, I. Complex linear transforms. J. Math. Phys. **15**, 1295–1301 (1974)
52. B. Mielnik, J. Plebański, Combinatorial approach to Baker-Campbell-Hausdorff exponents. Ann. Inst. H. Poincaré A **12**, 215–254 (1970)
53. I.M. Gel'fand, M.A. Naĭmark, Unitary representations of the Lorentz group. Izvestiya Akad. Nauk SSSR. Ser. Mat. **11**, 411–504 (1947)
54. M. Moshinsky, T.H. Seligman, K.B. Wolf, Canonical transformations and the radial oscillator and Coulomb problems. J. Math. Phys. **13**, 901–907 (1972)
55. K.B. Wolf, Canonical transforms. II. Complex radial transforms. J. Math. Phys. **15**, 2101–2111 (1974)
56. K.B. Wolf, Canonical transforms. IV. Hyperbolic transforms: continuous series of SL(2,R) representations. J. Math. Phys. **21**, 680–688 (1980)

57. D. Basu, K.B. Wolf, The unitary irreducible representations of SL(2,*R*) in all subgroup reductions. J. Math. Phys. **23**, 189–205 (1982)
58. K.B. Wolf, F. Aceves de la Cruz, Dependence of *s*-waves on continuous dimension: the quantum oscillator and free systems. Fortsch. Phys. **54**, 1083–1108 (2006)
59. D. Basu, T. Bhattacharya, The Gel'fand realization and the exceptional representations of SL(2,*R*). J. Math. Phys. **26**, 12–17 (1985)
60. N. Mukunda, B. Radhakrishnan, New forms for representations of 3-dimensional Lorentz group. J. Math. Phys. **14**, 254–258 (1973)
61. D. Basu, K.B. Wolf, The Clebsch–Gordan coefficients of the three-dimensional Lorentz algebra in the parabolic basis. J. Math. Phys. **24**, 478–500 (1983)
62. K.B. Wolf, A top-down account of linear canonical transforms. SIGMA **8**, art. 033, 13 p. (2012)
63. P. Kramer, M. Moshinsky, T.H. Seligman, Complex extensions of canonical transformations and quantum mechanics, in *Group Theory and Its Applications*, vol. III, ed. by E.M. Loebl (Academic, New York, 1975), pp. 250–332
64. V. Bargmann, On a Hilbert space of analytic functions and an associated integral transform, Part I. Commun. Pure Appl. Math. **20**, 187–214 (1961)
65. A.O. Barut, L. Girardello, New coherent states associated with noncompact groups. Commun. Math. Phys. **21**, 1–41 (1971)
66. J.J. Healy, J.T. Sheridan, Fast linear canonical transforms. J. Opt. Soc. Am. A **27**, 21–30 (2010)
67. J.-J. Ding, Research of fractional Fourier transform and linear canonical transform. Ph.D. Dissertation, National Taiwan University, 2001
68. J.J. Healy, K.B. Wolf, Discrete canonical transforms that are Hadamard matrices. J. Phys. A **44**, art. 265302, 10 p. (2011)
69. L. Zhao, J.J. Healy, J.T. Sheridan, A unitary discrete linear canonical transform: analysis and application. Appl. Optics Appl. Optics ID 177405. **52**, C30–C36 (2013)
70. A. Stern, Sampling of linear canonical transformed signals. Signal Process. **86**, 1421–1425 (2006)
71. F.S. Oktem, H.M. Ozaktas, Exact relation between continuous and discrete linear canonical transforms. IEEE Signal Process. Lett. **16**, 727–730 (2009)
72. A. Stern, Why is the linear canonical transform so little known? AIP Conf. Proc. **860**, 225–234 (2006)
73. S.-C. Pei, J.-J. Ding, Closed-form discrete fractional and affine transforms. IEEE Trans. Signal Process. **48**, 1338–1353 (2000)
74. K.B. Wolf, G. Krötzsch, Geometry and dynamics in the fractional discrete Fourier transform. J. Opt. Soc. Am. A **24**, 651–658 (2007)
75. S. C. Pei, M. H. Yeh, Improved discrete fractional transform. Opt. Lett. **22**, 1047–1049 (1997)
76. S.-C. Pei, M.-H. Yeh, C.-C. Tseng, Discrete fractional Fourier transform based on orthogonal projections. IEEE Trans. Signal Process. **47**, 1335–1348 (1999)
77. L. Barker, Ç. Çandan, T. Hakioğlu, M.A. Kutay, H.M. Ozaktas, The discrete harmonic oscillator, Harper's equation, and the discrete fractional Fourier transform. J. Phys. A **33**, 2209–2222 (2000)
78. B.M. Hennely, J.T. Sheridan, Fast algorithm for the determination of linear canonical transform and fractional Fourier transform. SPIE Proc. **5456**, 472–483 (2004)
79. B.M. Hennely, J.T. Sheridan, Efficient algorithms for linear canonical transforms. SPIE Proc. **5557**, 267–278 (2004)
80. B.M. Hennely, J.T. Sheridan, A recursive fast algorithm for the linear canonical transforms. SPIE Proc. **5823**, 1–12 (2004)
81. J.J. Healy, J.T. Sheridan, Sampling and discretization of the linear canonical transform. Signal Process. **89**, 641–648 (2009)
82. H.M. Ozaktas, A. Koç, I. Sari, M.A. Kutay, Efficient computation of quadratic-phase integrals in optics. Opt. Lett. **31**, 35–37 (2006)
83. A. Koç, H.M. Ozaktas, Ç. Çandan, M.A. Kutay, Digital computation of linear canonical transforms. IEEE Trans. Signal Process. **56**, 2383–2394 (2008)

84. R.G. Campos, J. Figueroa, A fast algorithm for the linear canonical transform. Signal Process. **91**, 1444–1447 (2011)
85. N.M. Atakishiyev, G.S. Pogosyan, K.B. Wolf, Finite models of the oscillator. Phys. Part. Nucl. Suppl. 3 **36**, 521–555 (2005)
86. N.M. Atakishiyev, K.B. Wolf, Fractional Fourier-Kravchuk transform. J. Opt. Soc. Am. A **14**, 1467–1477 (1997)
87. K.B. Wolf, Linear transformations and aberrations in continuous and in finite systems. J. Phys. A **41**, art. 304026, 19 p. (2008)
88. K.B. Wolf, Discrete systems and signals on phase space. Appl. Math. Inf. Sci. **4**, 141–181 (2010)
89. L.E. Vicent, K.B. Wolf, Analysis of digital images into energy-angular momentum modes. J. Opt. Soc. Am. A **28**, 808–814 (2011)

Chapter 2
The Linear Canonical Transformation: Definition and Properties

Martin J. Bastiaans and Tatiana Alieva

Abstract In this chapter we introduce the class of linear canonical transformations, which includes as particular cases the Fourier transformation (and its generalization: the fractional Fourier transformation), the Fresnel transformation, and magnifier, rotation and shearing operations. The basic properties of these transformations—such as cascadability, scaling, shift, phase modulation, coordinate multiplication and differentiation—are considered. We demonstrate that any linear canonical transformation is associated with affine transformations in phase space, defined by time-frequency or position-momentum coordinates. The affine transformation is described by a symplectic matrix, which defines the parameters of the transformation kernel. This alternative matrix description of linear canonical transformations is widely used along the chapter and allows simplifying the classification of such transformations, their eigenfunction identification, the interpretation of the related Wigner distribution and ambiguity function transformations, among many other tasks. Special attention is paid to the consideration of one- and two-dimensional linear canonical transformations, which are more often used in signal processing, optics and mechanics. Analytic expressions for the transforms of some selected functions are provided.

2.1 Introduction

In this chapter we introduce the class of linear canonical transformations and study some of the basic properties of these transformations. In one dimension, this class forms a three-parameter class of linear integral transformations, and includes such operations as the well-known Fourier transformation, the Fresnel transformation

---

M.J. Bastiaans (✉)
Department of Electrical Engineering, Eindhoven University of Technology,
P. O. Box 513, 5600 MB Eindhoven, Netherlands
e-mail: m.j.bastiaans@tue.nl

T. Alieva
Facultad de Ciencias Físicas, Universidad Complutense de Madrid,
Ciudad Universitaria s/n, Madrid 28040, Spain
e-mail: talieva@fis.ucm.es

J.J. Healy et al. (eds.), *Linear Canonical Transforms*, Springer Series
in Optical Sciences 198, DOI 10.1007/978-1-4939-3028-9_2

(used in optics, for instance, to describe the paraxial propagation of light in free space), and simple operations like scaling and multiplication by a quadratic-phase function. In the D-dimensional case, the class has $D(2D + 1)$ free parameters and includes additional operations, like rotation and shearing. Although we focus on the two-dimensional case, with an occasional restriction to one dimension, many of the results will hold for the general D-dimensional case. And although the results are normally presented using higher-dimensional vectors and matrices, the reduction to one dimension is throughout straightforward.

After a formal definition of the linear canonical transformation in Sect. 2.2, we turn to a description in a so-called phase space in Sect. 2.3, where a matrix is introduced with which the transformation is parameterized, and we derive some of the transformation's basic properties; this section contains also a number of linear canonical transformations for the one-dimensional case, which are more easily to visualize. Special cases of the transformation in two (and more) dimensions are considered in Sect. 2.4, while decompositions of a general linear canonical transformation into cascades of simpler transformations are studied in Sect. 2.5. In Sect. 2.6 we derive the linear canonical transforms of some selected functions, like a Gaussian signal, a harmonic signal, a periodic signal (with a short detour to Talbot imaging) and the Hermite–Gauss modes; the transformation of these Hermite–Gauss modes leads to a general class of Hermite–Gaussian type modes, with the Laguerre–Gauss modes as a special case, which are very important in optics. The eigenvalues of the transformation matrix are discussed in Sect. 2.7; the possible distributions of these eigenvalues lead to a classification of the linear canonical transformation based on simple nuclei, from which eigenfunctions may be derived. The final Sect. 2.8, deals with the effect of the linear canonical transformation on the second-order moments in phase space; we discuss, in particular, moment combinations that are invariant under a linear canonical transformation.

We conclude this Introduction with some remarks about notation. We will throughout denote column vectors by bold-face, lower-case symbols like $\mathbf{r}$ and $\mathbf{q}$, which in the two-dimensional case $D = 2$ read $\mathbf{r} = [x, y]^t$ and $\mathbf{q} = [u, v]^t$, while matrices and submatrices are denoted by bold-face, upper-case symbols, like $\mathbf{T} = [\mathbf{A}, \mathbf{B}; \mathbf{C}, \mathbf{D}]$; transposition of vectors and matrices is denoted by the superscript ^{t}. The identity matrix is denoted by $\mathbf{I}$. Submatrices may carry subscripts, like in $\mathbf{M} = [\mathbf{M_{rr}}, \mathbf{M_{rq}}; \mathbf{M_{qr}}, \mathbf{M_{qq}}]$. Scalars, including the entries of a 2×2 matrix, will appear in normal face, like in $\mathbf{T} = [a, b; c, d]$ and $\mathbf{M_{rq}} = [m_{xu}, m_{xv}; m_{yu}, m_{yv}]$. Complex conjugation is denoted by the superscript *, and the combined action of conjugation and transposition for vectors and matrices is denoted by the superscript †: $\mathbf{U}^\dagger = \mathbf{U}^{*t}$. Operators appear in calligraphic style, like $\mathscr{L}$ and $\mathscr{F}$, and the symbols $\mathbb{M}$ and $\mathbb{D}$ are used to denote the special operators related to coordinate multiplication and differentiation, respectively; brackets in connection with operators will be used to avoid ambiguities, if necessary. While a normal face font is used for signals and functions, a sans serif style is used for the Hermite–Gauss modes $\mathsf{H}_{m,n}(\mathbf{r})$ and Laguerre–Gauss modes $\mathsf{L}_{m,n}(\mathbf{r})$. Unless otherwise stated, all integrations and summations extend from $-\infty$ to $+\infty$, and the short-hand notation

$\mathbf{dr}$ in higher-dimensional integrals is used for $dx\,dy$. Subscripts $_i$ and $_o$, like in $f_i(\mathbf{r}_i)$, are throughout used to mark input and output signals and coordinates.

Finally: we will very often meet expressions of the form $(i\,\ell)^{1/2}$ or $(\det \mathrm{i}\,\mathbf{L})^{1/2}$. They shall be interpreted as $\exp[\mathrm{i}\,(\frac{1}{4}\pi)\,\mathrm{sgn}(\ell)]\,|\ell|^{1/2}$ or, with $\mathbf{L}$ a $D \times D$ matrix, as $\exp[\mathrm{i}\,(\frac{1}{4}\pi)D\ \mathrm{sgn}(\det \mathbf{L})]\,|\det \mathbf{L}|^{1/2}$.

2.2 Definition of the Linear Canonical Transformation

With the two-dimensional column vector $\mathbf{r}$ defined as $\mathbf{r} = [x, y]^t$, where ^{t} denotes transposition, and $\mathbf{dr} = dx\,dy$, the *linear canonical transformation* of the two-dimensional signal $f(\mathbf{r})$, $f_i(\mathbf{r}_i) \to f_o(\mathbf{r}_o) = \mathscr{L}f_i(\mathbf{r}_i)$, is defined as

$$f_o(\mathbf{r}_o) = (\det \mathrm{i}^{-1}\mathbf{L}_{io})^{1/2} \int \exp[\mathrm{i}\,\pi(\mathbf{r}_o^t\mathbf{L}_{oo}\mathbf{r}_o - 2\mathbf{r}_i^t\mathbf{L}_{io}\mathbf{r}_o + \mathbf{r}_i^t\mathbf{L}_{ii}\mathbf{r}_i)]f_i(\mathbf{r}_i)\,\mathbf{dr}_i\ , \quad (2.1)$$

where the 2×2 matrices $\mathbf{L}_{oo}$ and $\mathbf{L}_{ii}$ are symmetric: $\mathbf{L}_{oo} = \mathbf{L}_{oo}^t$ and $\mathbf{L}_{ii} = \mathbf{L}_{ii}^t$. We restrict ourselves to the case that the (not necessarily symmetric) 2×2 matrix $\mathbf{L}_{io}$ is non-singular. Although we will focus on the two-dimensional case, with an occasional restriction to one dimension, most of the results will hold for the general D-dimensional case. The reduction to one dimension is throughout straightforward; the one-dimensional version of Eq. (2.1), for instance, takes the form

$$f_o(x_o) = (\mathrm{i}^{-1}\ell_{io})^{1/2} \int \exp[\mathrm{i}\,\pi(\ell_{oo}x_o^2 - 2\ell_{io}x_ix_o + \ell_{ii}x_i^2)]f_i(x_i)\,dx_i \quad (\ell_{io} \neq 0)\ . \quad (2.2)$$

Note that in the D-dimensional case, a symmetric $D \times D$ matrix has $\frac{1}{2}D(D+1)$ degrees of freedom, and that the three matrices $\mathbf{L}_{oo}$, $\mathbf{L}_{ii}$ and $\mathbf{L}_{io}$ together thus have $D(2D+1)$ degrees of freedom: 3 for $D = 1$ and 10 for $D = 2$. If $\mathbf{L}_{oo} = \mathbf{L}_{ii}$ and $\mathbf{L}_{io} = \mathbf{L}_{io}^t$, the transformation is *symmetric* and the roles of $\mathbf{r}_o$ and $\mathbf{r}_i$ can be interchanged, as will be discussed in more detail in Sect. 2.3.1.2.

The linear canonical transformation is a *unitary* transformation in the sense that

$$\int f_o(\mathbf{r})\,h_o^*(\mathbf{r})\,\mathbf{dr} = \int f_i(\mathbf{r})\,h_i^*(\mathbf{r})\,\mathbf{dr}, \quad (2.3)$$

where * denotes complex conjugation, which relation is known as *Parseval's theorem* for lossless transformations; it yields the energy preservation law for $f(\mathbf{r}) = h(\mathbf{r})$.

We finally remark that an additional phase factor $\exp(i\varphi)$ may be added to the definition (2.1) without changing the main properties of the transformation. This is sometimes done to get a better connection with the actual physical phenomenon that the linear canonical transformation is describing.

Among the many works on the linear canonical transformation, to which we refer for further reading, we mention [7, 26, 32, 44, 46, 47, 51, 71–75].

2.3 Representation of the Linear Canonical Transformation in Phase Space

To represent the linear canonical transformation in phase space, we use the *Wigner distribution* $W(\mathbf{r},\mathbf{q})$ of the signal $f(\mathbf{r})$, defined as [11, 13, 18, 19, 27, 28, 43, 66–69]

$$W(\mathbf{r},\mathbf{q}) = \int f\left(\mathbf{r}+\tfrac{1}{2}\mathbf{r}'\right) f^*\left(\mathbf{r}-\tfrac{1}{2}\mathbf{r}'\right) \exp[-\mathrm{i}\,2\pi\,\mathbf{q}^t\mathbf{r}']\,\mathrm{d}\mathbf{r}' \,. \qquad (2.4)$$

The column vector $\mathbf{q}$ can be considered as the frequency variable associated with $\mathbf{r}$. Note that the Wigner distribution is real, $W(\mathbf{r},\mathbf{q}) = W^*(\mathbf{r},\mathbf{q})$, and that a constant phase factor in the signal is no longer visible in the Wigner distribution: the signals $f(\mathbf{r})$ and $f(\mathbf{r})\exp(\mathrm{i}\,\varphi)$ lead to the same Wigner distribution.

We derive the linear canonical transformation in terms of the *phase-space variables* $\mathbf{r}$ and $\mathbf{q}$ by substituting from (2.1) into (2.4) and get the relationship (see Appendix)

$$W_o(\mathbf{r}_o,\mathbf{q}_o) = W_o(\mathbf{A}\mathbf{r}_i+\mathbf{B}\mathbf{q}_i,\mathbf{C}\mathbf{r}_i+\mathbf{D}\mathbf{q}_i) = W_i(\mathbf{r}_i,\mathbf{q}_i)\,, \qquad (2.5)$$

where the input variables $(\mathbf{r}_i,\mathbf{q}_i)$ and the output variables $(\mathbf{r}_o,\mathbf{q}_o)$ are related by the simple matrix relation

$$\begin{bmatrix}\mathbf{r}_o\\ \mathbf{q}_o\end{bmatrix} = \begin{bmatrix}\mathbf{A} & \mathbf{B}\\ \mathbf{C} & \mathbf{D}\end{bmatrix}\begin{bmatrix}\mathbf{r}_i\\ \mathbf{q}_i\end{bmatrix} \equiv \mathbf{T}\begin{bmatrix}\mathbf{r}_i\\ \mathbf{q}_i\end{bmatrix}, \qquad (2.6)$$

in which the *transformation matrix* $\mathbf{T} = [\mathbf{A},\mathbf{B};\mathbf{C},\mathbf{D}]$ has been introduced, and where the matrices $\mathbf{A}$, $\mathbf{B}$, $\mathbf{C}$ and $\mathbf{D}$ are related to the matrices $\mathbf{L}_{oo}$, $\mathbf{L}_{io}$ and $\mathbf{L}_{ii}$ by

$$\begin{aligned}&\mathbf{A} = \mathbf{L}_{io}^{-1}\mathbf{L}_{ii}\,, \quad \mathbf{B} = \mathbf{L}_{io}^{-1}\,, \quad \mathbf{C} = \mathbf{L}_{oo}\mathbf{L}_{io}^{-1}\mathbf{L}_{ii} - \mathbf{L}_{io}^{t}\,, \quad \mathbf{D} = \mathbf{L}_{oo}\mathbf{L}_{io}^{-1}\,,\\ &\mathbf{L}_{oo} = \mathbf{D}\mathbf{B}^{-1}\,, \quad \mathbf{L}_{ii} = \mathbf{B}^{-1}\mathbf{A}\,, \quad \mathbf{L}_{io} = \mathbf{B}^{-1}\,.\end{aligned} \qquad (2.7)$$

Note that we can formulate the relationship between the frequency variables $\mathbf{q}$ and the original variables $\mathbf{r}$ also in the form

$$\begin{bmatrix}-\mathbf{q}_i\\ \mathbf{q}_o\end{bmatrix} = \begin{bmatrix}\mathbf{L}_{ii} & -\mathbf{L}_{io}\\ -\mathbf{L}_{io}^{t} & \mathbf{L}_{oo}\end{bmatrix}\begin{bmatrix}\mathbf{r}_i\\ \mathbf{r}_o\end{bmatrix}. \qquad (2.8)$$

In terms of the transformation matrix $\mathbf{T}$, the linear canonical transformation can be expressed in the form [26, 33, 40, 44, 74, 75]

$$f_o(\mathbf{r}_o) = \mathscr{L}(\mathbf{T})f_i(\mathbf{r}_i) = (\det \mathrm{i}\,\mathbf{B})^{-1/2} \times \int \exp[\mathrm{i}\,\pi(\mathbf{r}_o^t\mathbf{D}\mathbf{B}^{-1}\mathbf{r}_o - 2\mathbf{r}_i^t\mathbf{B}^{-1}\mathbf{r}_o + \mathbf{r}_i^t\mathbf{B}^{-1}\mathbf{A}\mathbf{r}_i)]f_i(\mathbf{r}_i)\,\mathrm{d}\mathbf{r}_i\,, \quad (2.9)$$

with $\det \mathbf{B} \neq 0$. In the limiting case $\mathbf{B} \Rightarrow \mathbf{0}$, for which $\mathbf{r}_o \Rightarrow \mathbf{A}\mathbf{r}_i$, see (2.6), we have

$$f_o(\mathbf{r}) = \mathscr{L}(\mathbf{T})f_i(\mathbf{r}) = |\det \mathbf{A}|^{-1/2} \exp[\mathrm{i}\,\pi\,\mathbf{r}^t\mathbf{C}\mathbf{A}^{-1}\mathbf{r}]f_i(\mathbf{A}^{-1}\mathbf{r})\,, \quad (2.10)$$

which follows readily when we write the exponent in (2.9) as

$$\exp[\mathrm{i}\,\pi\,\mathbf{r}_o^t\mathbf{C}\mathbf{A}^{-1}\mathbf{r}_o]\exp[\mathrm{i}\,\pi(\mathbf{r}_i - \mathbf{A}^{-1}\mathbf{r}_o)^t\mathbf{B}^{-1}\mathbf{A}(\mathbf{r}_i - \mathbf{A}^{-1}\mathbf{r}_o)]$$

and recall that $(\det \mathrm{i}\,\mathbf{B})^{-1/2}\exp[\mathrm{i}\,\pi\,\mathbf{r}^t\mathbf{B}^{-1}\mathbf{A}\mathbf{r}] \Rightarrow |\det \mathbf{A}|^{-1/2}\,\delta(\mathbf{r})$ when $\mathbf{B} \Rightarrow \mathbf{0}$. The singular case $\det \mathbf{B} = 0$, but $\mathbf{B} \neq \mathbf{0}$, will be dealt with in Sect. 2.4.4.1. For the one-dimensional versions of Eqs. (2.9) and (2.10), we refer to Sect. (2.3.2), where we will study one-dimensional transformations in more detail.

The symmetry of the matrices $\mathbf{L}_{oo}$ and $\mathbf{L}_{ii}$ reflects itself in the *symplecticity* of the transformation matrix $\mathbf{T}$:

$$\mathbf{T}^{-1} = \begin{bmatrix} \mathbf{A} & \mathbf{B} \\ \mathbf{C} & \mathbf{D} \end{bmatrix}^{-1} = \begin{bmatrix} \mathbf{D}^t & -\mathbf{B}^t \\ -\mathbf{C}^t & \mathbf{A}^t \end{bmatrix} = \mathbf{J}\mathbf{T}^t\mathbf{J} \quad \text{with} \quad \mathbf{J} \equiv \mathrm{i}\begin{bmatrix} \mathbf{0} & \mathbf{I} \\ -\mathbf{I} & \mathbf{0} \end{bmatrix}. \quad (2.11)$$

Note that $\mathbf{J} = \mathbf{J}^{-1} = -\mathbf{J}^t = \mathbf{J}^{*t} = \mathbf{J}^\dagger$.

The symplectic $2D \times 2D$ transformation matrix $\mathbf{T}$ has $D(2D + 1)$ degrees of freedom (3 for the one-dimensional case and 10 for the two-dimensional case), the same number as in the three $D \times D$ matrices $\mathbf{L}_{oo}$, $\mathbf{L}_{ii}$ and $\mathbf{L}_{io}$ together. In the one-dimensional case, the symplecticity condition (2.11) reduces to the much simpler relation $\det \mathbf{T} = ad - bc = 1$.

The input–output relationship (2.5) implies that the Wigner distribution of the output signal is simply a linearly distorted form of the Wigner distribution of the input signal, with the value of the Wigner distribution at each point in phase space being mapped to another point in phase space, without the need to calculate an integral. Since the determinant of the transformation matrix $\mathbf{T}$ is equal to unity, this pointwise geometrical distortion or deformation is area preserving; it distorts but does not concentrate or deconcentrate the Wigner distribution.

An isolated distribution around a point (t, f) in a *time-frequency* phase space can be considered as the representation of a short *musical note* at a certain time t with a certain frequency f. The Wigner distribution can thus be considered as a *musical score*, which tells us how a time signal can be composed as a superposition of notes.

In optics, such an isolated distribution around a point $(\mathbf{r}, \mathbf{q})$ in a *position-direction* phase space can be considered as an *optical ray* at a certain position $\mathbf{r}$ with a certain direction (i.e., spatial frequency) $\mathbf{q}$, and the Wigner distribution then tells us how the optical signal can be composed as a superposition of rays. This simultaneous position-direction description closely resembles the ray concept in geometrical optics, where the position and direction of a ray are also given simultaneously. The Wigner distribution thus yields a *ray pattern* of the optical signal and, in a way, $W(\mathbf{r},\mathbf{q})$ is the amplitude of a ray, passing through the point $\mathbf{r}$ with a direction $\mathbf{q}$.

Simple optical systems like a thins lens, a section of free space in the Fresnel approximation, and cascades of such systems are described by an input–output relation of the form (2.1) and belong to the realm of what is called *first-order optics* [33, 46]. The propagation of an optical signal through first-order optical systems can most elegantly be described by the coordinate transformation (2.5) of the signal's Wigner distribution. It is thus obvious why phenomena in first-order optics are often treated in a phase space [12, 64–66, 75].

We finally note that the concept of the Wigner distribution can directly be applied to stochastic signals; we only have to replace in its definition (2.4) the product $f(\mathbf{r}_1)f^*(\mathbf{r}_2)$ by the two-point correlation function $\langle f(\mathbf{r}_1)f^*(\mathbf{r}_2)\rangle$, where $\langle\cdot\rangle$ denotes ensemble averaging. The phase-space description of the linear canonical transformation is therefore not restricted to deterministic signals, but applies immediately to stochastic signals, as well [13, 18, 19].

Instead of with the Wigner distribution $W(\mathbf{r},\mathbf{q})$, we could also have chosen to work with the other well-known phase-space description, the *ambiguity function* $A(\mathbf{r}',\mathbf{q}')$ defined as [76, Chap. 7]

$$A(\mathbf{r}',\mathbf{q}') = \int f\left(\mathbf{r}+\tfrac{1}{2}\mathbf{r}'\right) f^*\left(\mathbf{r}-\tfrac{1}{2}\mathbf{r}'\right) \exp[-\mathrm{i}\,2\pi\,\mathbf{r}^t\mathbf{q}']\,\mathrm{d}\mathbf{r}\,, \qquad (2.12)$$

for which we have the input–output relation $A_o(\mathbf{A}\mathbf{r}'+\mathbf{B}\mathbf{q}',\mathbf{C}\mathbf{r}'+\mathbf{D}\mathbf{q}') = A_i(\mathbf{r}',\mathbf{q}')$, which is similar to (2.5).

2.3.1 Basic Properties

For easy reference, the basic properties of the linear canonical transformation and transforms, treated in this section, have been collected in Tables 2.1 and 2.2.

2.3.1.1 Cascadability

If two linear canonical transformations $\mathscr{L}(\mathbf{T}_1)$ and $\mathscr{L}(\mathbf{T}_2)$ are performed in cascade, $\mathscr{L}(\mathbf{T}_2)\,\mathscr{L}(\mathbf{T}_1)$, the resulting operation is again a linear canonical transformation $\mathscr{L}(\mathbf{T})$ with a transformation matrix $\mathbf{T} = \mathbf{T}_2\mathbf{T}_1$ that is the product of $\mathbf{T}_2$ and $\mathbf{T}_1$. An immediate consequence is that the *inverse* of the operation $\mathscr{L}(\mathbf{T})$ is parameterized by $\mathbf{T}^{-1}$: $\mathscr{L}^{-1}(\mathbf{T}) = \mathscr{L}(\mathbf{T}^{-1})$.

Table 2.1 Some basic properties of the linear canonical transformation

Operator		Transformation matrix	Remark
$\mathscr{L}(\mathbf{T})$		$\begin{bmatrix} \mathbf{A} & \mathbf{B} \\ \mathbf{C} & \mathbf{D} \end{bmatrix}$	$\begin{bmatrix} \mathbf{A} & \mathbf{B} \\ \mathbf{C} & \mathbf{D} \end{bmatrix}^{-1} = \begin{bmatrix} \mathbf{D}^t & -\mathbf{B}^t \\ -\mathbf{C}^t & \mathbf{A}^t \end{bmatrix}$ symplecticity condition
$\mathscr{L}(\mathbf{T}_2)\mathscr{L}(\mathbf{T}_1)$	cascadability	$\mathbf{T}_2\mathbf{T}_1$	
$\mathscr{L}^{-1}(\mathbf{T}) = \mathscr{L}(\mathbf{T}^{-1})$	inverse	$\begin{bmatrix} \mathbf{D}^t & -\mathbf{B}^t \\ -\mathbf{C}^t & \mathbf{A}^t \end{bmatrix}$	
$\mathscr{L}(\hat{\mathbf{T}}) = \mathscr{L}^*(\mathbf{T}^{-1})$	reverse	$\begin{bmatrix} \mathbf{D}^t & \mathbf{B}^t \\ \mathbf{C}^t & \mathbf{A}^t \end{bmatrix}$	$\mathscr{L}(\mathbf{T})f^*(\mathbf{r}) = [\mathscr{L}(\mathbf{T}^{-1})f(\mathbf{r})]^*$
$\mathscr{F}$	Fourier transformation	$\begin{bmatrix} \mathbf{0} & \mathbf{I} \\ -\mathbf{I} & \mathbf{0} \end{bmatrix}$	
$\mathscr{F}\mathscr{L}(\mathbf{T})\mathscr{F}^{-1} = \mathscr{L}(\mathbf{T}^{t-1})$ $\mathscr{F}^{-1}\mathscr{L}(\mathbf{T})\mathscr{F} = \mathscr{L}(\mathbf{T}^{t-1})$	duality	$\begin{bmatrix} \mathbf{D} & -\mathbf{C} \\ -\mathbf{B} & \mathbf{A} \end{bmatrix}$	$\mathbf{T} \Leftrightarrow \mathbf{T}^{t-1}$ implies $\mathbf{A} \Leftrightarrow \mathbf{D}$, $\mathbf{B} \Leftrightarrow -\mathbf{C}$

The property of cascadability is true when we consider linear canonical transformations as simple coordinate transformations in phase space. If we treat the concatenation of operators in their integral representations (2.1) or (2.9), we will encounter the possibility of an additional minus sign. This problem is known as the *metaplectic sign problem* and is carefully studied in [74, Sect. 9.1.4, Composition of transforms] and [75, Sect. C2. Linear canonical transforms] by considering the integral (with complex-valued variables r and s)

$$\int_{\Re} \exp[\mathrm{i}\,\pi(r^2q^2 + 2sq)]\,\mathrm{d}q = \sigma(r)\,\exp(\mathrm{i}\,\tfrac{1}{4}\pi)r^{-1}\exp(-\mathrm{i}\,\pi\,s^2r^{-2})\,, \tag{2.13a}$$

$$\text{with}\quad \sigma(r) = \begin{cases} +1 \text{ when } & 0 \le \arg r \le \tfrac{1}{2}\pi\,, \\ -1 \text{ when } & -\pi \le \arg r \le -\tfrac{1}{2}\pi\,. \end{cases} \tag{2.13b}$$

In this chapter we will ignore the metaplectic sign problem.

2.3.1.2 The Reverse System

If a system performs a linear canonical transformation $\mathscr{L}(\mathbf{T})$, it is sometimes advantageous to consider the behaviour of the so-called *reverse* system $\mathscr{L}(\hat{\mathbf{T}})$, with transformation matrix $\hat{\mathbf{T}}$; note that this is not the same as the inverse system $\mathscr{L}(\mathbf{T}^{-1})$. In the reverse system, the signal phases are reversed and the signals

Table 2.2 Some basic properties of linear canonical transforms

Input signal	Output signal	Remark
$f_i(\mathbf{r})$	$f_o(\mathbf{r}) = \mathscr{L}(\mathbf{T})f_i(\mathbf{r}) \qquad \mathbf{T} = \begin{bmatrix} \mathbf{A} & \mathbf{B} \\ \mathbf{C} & \mathbf{D} \end{bmatrix}$	(2.1)(2.9)(2.10)
$\sum_n a_n f_n(\mathbf{r})$	$\sum_n a_n \, \mathscr{L} f_n(\mathbf{r})$	Linearity
$f_i(\mathbf{r}), h_i(\mathbf{r})$	$\int f_i(\mathbf{r})\, h_i^*(\mathbf{r})\, \mathrm{d}\mathbf{r} = \int f_o(\mathbf{r})\, h_o^*(\mathbf{r})\, \mathrm{d}\mathbf{r}$	Parseval's theorem (2.3)
$f_i^*(\mathbf{r})$	$[\mathscr{L}(\mathbf{T}^{-1})f_i(\mathbf{r})]^* \qquad \mathbf{T}^{-1} = \begin{bmatrix} \mathbf{D}^t & -\mathbf{B}^t \\ -\mathbf{C}^t & \mathbf{A}^t \end{bmatrix}$	Complex conjugation (2.15)
$\mathbb{M}^n f_i(\mathbf{r})$	$(\mathbf{D}^t\mathbb{M} - \mathbf{B}^t\mathbb{D})^n f_o(\mathbf{r}) \qquad \mathbb{M} = \mathbf{r}$	Multiplication (2.23a)
$\mathbb{D}^n f_i(\mathbf{r})$	$(-\mathbf{C}^t\mathbb{M} + \mathbf{A}^t\mathbb{D})^n f_o(\mathbf{r}) \qquad \mathbb{D} = (\mathrm{i}\,2\pi)^{-1}\nabla^t$	Derivation (2.23b)
$f_i(\mathbf{r}) \exp(\mathrm{i}\,2\pi\, \mathbf{k}^t\mathbf{r})$	$f_o(\mathbf{r} - \mathbf{B}\mathbf{k}) \exp(\mathrm{i}\,2\pi\, \mathbf{k}^t\mathbf{D}^t\mathbf{r}) \exp(-\mathrm{i}\,\pi\, \mathbf{k}^t\mathbf{B}^t\mathbf{D}\mathbf{k})$	Modulation (2.25a)
$f_i(\mathbf{r} - \mathbf{k})$	$f_o(\mathbf{r}) \exp(\mathrm{i}\,2\pi\, \mathbf{k}^t\mathbf{C}^t\mathbf{r}) \exp(-\mathrm{i}\,\pi\, \mathbf{k}^t\mathbf{C}^t\mathbf{A}\mathbf{k})$	Shift (2.25b)
$\lvert \det \mathbf{W}\rvert^{-1/2} f_i(\mathbf{W}^{-1}\mathbf{r})$	$\mathscr{L}(\tilde{\mathbf{T}})f_i(\mathbf{r}) \qquad \tilde{\mathbf{T}} = \mathbf{T}\begin{bmatrix} \mathbf{W} & \mathbf{0} \\ \mathbf{0} & \mathbf{W}^{t-1} \end{bmatrix}$	Scaling (2.43)
$f_i(-\mathbf{r})$	$\mathscr{L}(-\mathbf{T})f_i(\mathbf{r}) = f_o(-\mathbf{r})$	Scaling by -1 (2.43) $\mathbf{W} = -\mathbf{I}$
$\exp(-\pi\, \mathbf{r}^t\mathbf{L}_i\mathbf{r})$	$[\det(\mathbf{A} + \mathrm{i}\,\mathbf{L}_i)]^{-1/2} \exp(-\pi\, \mathbf{r}^t\mathbf{L}_o\mathbf{r})$ with $\mathrm{i}\,\mathbf{L}_o = (\mathbf{C} + \mathrm{i}\,\mathbf{D}\mathbf{L}_i)(\mathbf{A} + \mathrm{i}\,\mathbf{B}\mathbf{L}_i)^{-1}$	Sect. 2.6.1 $\mathbf{L} = \mathbf{L}^t$
1	$(\det \mathbf{A})^{-1/2} \exp(\mathrm{i}\,\pi\, \mathbf{r}^t\mathbf{C}\mathbf{A}^{-1}\mathbf{r})$	Sect. 2.6.1
$\exp(\mathrm{i}\,2\pi\, k^t\mathbf{r})$	$(\det \mathbf{A})^{-1/2} \exp(\mathrm{i}\,\pi\, \mathbf{r}^t\mathbf{C}\mathbf{A}^{-1}\mathbf{r}) \times \exp(-\mathrm{i}\,\pi\, \mathbf{k}^t\mathbf{A}^{-1}\mathbf{B}\mathbf{k}) \exp(\mathrm{i}\,2\pi\, \mathbf{k}^t\mathbf{A}^{-1}\mathbf{r})$	Sect. 2.6.1

propagate in the opposite direction, which means that the frequency variable $\mathbf{q}$ has to be reversed, i.e., replaced by $-\mathbf{q}$, and that $f_o^*(\mathbf{r}_o)$ now acts as the input signal while $f_i^*(\mathbf{r}_i) = \mathscr{L}(\hat{\mathbf{T}})f_o^*(\mathbf{r}_o)$ is the output signal. The transformation matrix $\hat{\mathbf{T}}$ of the reverse system thus takes the form

$$\hat{\mathbf{T}} = \begin{bmatrix} \mathbf{I} & \mathbf{0} \\ \mathbf{0} & -\mathbf{I} \end{bmatrix} \mathbf{T}^{-1} \begin{bmatrix} \mathbf{I} & \mathbf{0} \\ \mathbf{0} & -\mathbf{I} \end{bmatrix} = \begin{bmatrix} \mathbf{A} & -\mathbf{B} \\ -\mathbf{C} & \mathbf{D} \end{bmatrix}^{-1} = \begin{bmatrix} \mathbf{D}^t & \mathbf{B}^t \\ \mathbf{C}^t & \mathbf{A}^t \end{bmatrix}. \qquad (2.14)$$

Substituting $f_o(\mathbf{r}_o) = \mathscr{L}(\mathbf{T})f_i(\mathbf{r}_i)$ into $f_i^*(\mathbf{r}_i) = \mathscr{L}(\hat{\mathbf{T}})f_o^*(\mathbf{r}_o)$, we immediately get $f_i^*(\mathbf{r}_i) = \mathscr{L}(\hat{\mathbf{T}})\,[\mathscr{L}(\mathbf{T})f_i(\mathbf{r}_i)]^* = \mathscr{L}(\hat{\mathbf{T}})\,\mathscr{L}^*(\mathbf{T})f_i^*(\mathbf{r}_i)$ and we conclude that $\mathscr{L}(\mathbf{T}) = \mathscr{L}^*(\hat{\mathbf{T}}^{-1})$ and also

$$\mathscr{L}(\mathbf{T})f^*(\mathbf{r}) = [\mathscr{L}(\mathbf{T}^{-1})f(\mathbf{r})]^* \,. \tag{2.15}$$

The system is symmetric if $\hat{\mathbf{T}} = \mathbf{T}$, i.e., if $\mathbf{A} = \mathbf{D}^t$, $\mathbf{B} = \mathbf{B}^t$ and $\mathbf{C} = \mathbf{C}^t$. This corresponds to $\mathbf{L}_{oo} = \mathbf{L}_{ii}$ and $\mathbf{L}_{io} = \mathbf{L}_{io}^t$, as we already mentioned before.

2.3.1.3 Formulation in Terms of Fourier Transforms

For $\mathbf{A} = \mathbf{D} = \mathbf{0}$ and $\mathbf{B} = -\mathbf{C} = \mathbf{I}$, the linear canonical transformation reduces to the common *Fourier transformation* [apart from the phase factor $(\det \mathrm{i}\,\mathbf{I})^{-1/2}$]

$$f_o(\mathbf{r}_o) = (\det \mathrm{i}\,\mathbf{I})^{-1/2} \int \exp[-\mathrm{i}\,2\pi\,\mathbf{r}_i^t\mathbf{r}_o] f_i(\mathbf{r}_i)\,\mathrm{d}\mathbf{r}_i \equiv (\det \mathrm{i}\,\mathbf{I})^{-1/2}\bar{f}_i(\mathbf{r}_o) \equiv \mathscr{F}f_i(\mathbf{r}_i)\,, \tag{2.16}$$

whereas the inverse Fourier transformation arises for $\mathbf{B} = -\mathbf{C} = -\mathbf{I}$; note that $(\det \mathrm{i}\,\mathbf{I})^{-1/2} = \mathrm{i}^{-D/2}$ for a D-dimensional signal. We thus conclude that the transformation matrices

$$\begin{bmatrix} \mathbf{0} & \mathbf{I} \\ -\mathbf{I} & \mathbf{0} \end{bmatrix} \quad \text{and} \quad \begin{bmatrix} \mathbf{0} & -\mathbf{I} \\ \mathbf{I} & \mathbf{0} \end{bmatrix} \tag{2.17}$$

correspond to a Fourier transformation and its inverse, respectively.

Using the Fourier transformation, we can easily express the linear canonical transformation in terms of the Fourier transforms $\bar{f}_{i,o}(\mathbf{q})$ of the signals $f_{i,o}(\mathbf{r})$:

$$\bar{f}_o(\mathbf{q}_o) = \mathscr{F}f_o(\mathbf{r}_o) = \mathscr{F}\mathscr{L}(\mathbf{T})f(\mathbf{r}_i) = \mathscr{F}\mathscr{L}(\mathbf{T})\mathscr{F}^{-1}\bar{f}_i(\mathbf{q}_i) = \mathscr{L}(\mathbf{T}^{t^{-1}})\bar{f}_i(\mathbf{q}_i)\,. \tag{2.18}$$

The transformation matrix $\mathbf{T}^{t^{-1}}$ in the Fourier domain takes the *dual* form

$$\mathbf{T}^{t^{-1}} = \begin{bmatrix} \mathbf{0} & \mathbf{I} \\ -\mathbf{I} & \mathbf{0} \end{bmatrix} \begin{bmatrix} \mathbf{A} & \mathbf{B} \\ \mathbf{C} & \mathbf{D} \end{bmatrix} \begin{bmatrix} \mathbf{0} & -\mathbf{I} \\ \mathbf{I} & \mathbf{0} \end{bmatrix} = \begin{bmatrix} \mathbf{D} & -\mathbf{C} \\ -\mathbf{B} & \mathbf{A} \end{bmatrix} \tag{2.19}$$

and is related to $\mathbf{T}$ by interchanging $\mathbf{A} \Leftrightarrow \mathbf{D}$ and $\mathbf{B} \Leftrightarrow -\mathbf{C}$. Note that interchanging $\mathscr{F}$ and $\mathscr{F}^{-1}$ leads to the same result.

2.3.1.4 Coordinate Multiplication and Differentiation

One of the main properties of the Fourier transformation is that a multiplication of $f(\mathbf{r})$ by its argument $\mathbf{r}$ corresponds to a differentiation of its Fourier transform $\bar{f}(\mathbf{q})$, and vice versa with a minus sign. With the operator $\mathbb{M}$ denoting multiplication by the argument and the operator $\mathbb{D}$ denoting differentiation with respect to the argument and dividing by $(\mathrm{i}\,2\pi)$, both resulting in a column vector, we can write

$$\mathbb{M}f(\mathbf{r}) = \mathbf{r}f(\mathbf{r}) \quad \Leftrightarrow \quad -\mathbb{D}\bar{f}(\mathbf{q}) = -(\mathrm{i}\,2\pi)^{-1}\nabla^t\bar{f}(\mathbf{q})\,, \tag{2.20a}$$

$$\mathbb{M}\bar{f}(\mathbf{q}) = \mathbf{q}\bar{f}(\mathbf{q}) \quad \Leftrightarrow \quad \mathbb{D}f(\mathbf{r}) = (\mathrm{i}\,2\pi)^{-1}\nabla^t f(\mathbf{r})\,, \tag{2.20b}$$

where the nabla operator ∇, which after transposition takes the form of a column vector, has been used to denote partial derivatives. We also define the operators $\mathbb{D}^t$ and $\mathbb{M}^t$ where the vectors are transposed, and easily verify that $\mathbb{D}^t\mathbb{M}f(\mathbf{r}) = (1 + \mathbb{M}^t\mathbb{D})f(\mathbf{r})$. The operator $\mathbb{D}^t\mathbb{M}-\mathbb{M}^t\mathbb{D}$ thus corresponds to the identity operator.

We now turn our attention to the linear canonical transformation $\mathscr{L}$ and determine the two transforms $\mathscr{L}[\mathbb{M}f(\mathbf{r})]$ and $\mathscr{L}[\mathbb{D}f(\mathbf{r})]$ in terms of $\mathscr{L}f(\mathbf{r})$. After some lengthy but straightforward calculation we get

$$\mathscr{L}[\mathbb{M}f(\mathbf{r})] = (\mathbf{D}^t\mathbb{M} - \mathbf{B}^t\mathbb{D})\mathscr{L}f(\mathbf{r})\,, \tag{2.21a}$$

$$\mathscr{L}[\mathbb{D}f(\mathbf{r})] = (-\mathbf{C}^t\mathbb{M} + \mathbf{A}^t\mathbb{D})\mathscr{L}f(\mathbf{r})\,, \tag{2.21b}$$

and we easily verify that for the Fourier transformation, i.e., $\mathbf{A} = \mathbf{D} = \mathbf{0}$ and $\mathbf{B} = -\mathbf{C} = \mathbf{I}$, we get $\mathscr{F}[\mathbb{M}f(\mathbf{r})] = -\mathbb{D}[\mathscr{F}f(\mathbf{r})]$ and $\mathscr{F}[\mathbb{D}f(\mathbf{r})] = \mathbb{M}[\mathscr{F}f(\mathbf{r})]$, see Eqs. (2.20). The operators $\mathbb{M}$ and $\mathbb{D}$ in the input domain thus lead to the operators $\mathbb{M}_\mathbf{T}$ and $\mathbb{D}_\mathbf{T}$ in the output domain, and the four operators are related by

$$\begin{bmatrix} \mathbb{M}_\mathbf{T} \\ \mathbb{D}_\mathbf{T} \end{bmatrix} = \begin{bmatrix} \mathbf{A} & \mathbf{B} \\ \mathbf{C} & \mathbf{D} \end{bmatrix} \begin{bmatrix} \mathbb{M} \\ \mathbb{D} \end{bmatrix}, \tag{2.22}$$

in accordance with (2.6).

We also verify that $\mathbb{D}^t_\mathbf{T}\mathbb{M}_\mathbf{T} - \mathbb{M}^t_\mathbf{T}\mathbb{D}_\mathbf{T}$ is the identity operator. Indeed: $\mathbb{D}^t_\mathbf{T}\mathbb{M}_\mathbf{T} - \mathbb{M}^t_\mathbf{T}\mathbb{D}_\mathbf{T} = (-\mathbb{M}^t\mathbf{C}^t + \mathbb{D}^t\mathbf{A}^t)(\mathbf{D}\mathbb{M} - \mathbf{B}\mathbb{D}) - (\mathbb{M}^t\mathbf{D}^t - \mathbb{D}^t\mathbf{B}^t)(-\mathbf{C}\mathbb{M} + \mathbf{A}\mathbb{D}) = \mathbb{M}^t(-\mathbf{C}^t\mathbf{D} + \mathbf{D}^t\mathbf{C})\mathbb{M} + \mathbb{M}^t(\mathbf{C}^t\mathbf{B} - \mathbf{D}^t\mathbf{A})\mathbb{D} + \mathbb{D}^t(\mathbf{A}^t\mathbf{D} - \mathbf{B}^t\mathbf{C})\mathbb{M} + \mathbb{D}^t(-\mathbf{A}^t\mathbf{B} + \mathbf{B}^t\mathbf{A})\mathbb{D} = \mathbb{D}^t\mathbb{M} - \mathbb{M}^t\mathbb{D}$, where we have used the symplecticity conditions $\mathbf{C}^t\mathbf{D} = \mathbf{D}^t\mathbf{C}$, $\mathbf{A}^t\mathbf{D} - \mathbf{B}^t\mathbf{C} = \mathbf{I}$ and $\mathbf{A}^t\mathbf{B} = \mathbf{B}^t\mathbf{A}$. We note that an alternative development of linear canonical transformations may be found in [74, Sect. 9.1.1, Posing the operator problem], where Eq. (2.22) is taken as the defining characteristic of these transformations and the integral form is subsequently defined.

The operators $\mathbb{M}$ and $\mathbb{D}$ can be applied an arbitrary number of times in cascade, which leads to the following relations, cf. (2.21):

$$\mathscr{L}[\mathbb{M}^n f(\mathbf{r})] = (\mathbf{D}^t\mathbb{M} - \mathbf{B}^t\mathbb{D})^n\mathscr{L}f(\mathbf{r})\,, \tag{2.23a}$$

$$\mathscr{L}[\mathbb{D}^n f(\mathbf{r})] = (-\mathbf{C}^t\mathbb{M} + \mathbf{A}^t\mathbb{D})^n\mathscr{L}f(\mathbf{r})\,. \tag{2.23b}$$

2.3.1.5 Coordinate Shift and Modulation

One of the other main properties of the Fourier transformation is that a modulation of $f(\mathbf{r})$ by $\exp(\mathrm{i}\,2\pi\,\mathbf{k}^t\mathbf{r})$ corresponds to a shift of the argument of its Fourier transform $\bar{f}(\mathbf{q})$, and vice versa with a minus sign:

$$f(\mathbf{r})\,\exp(\mathrm{i}\,2\pi\,\mathbf{k}^t\mathbf{r}) \quad \Leftrightarrow \quad \bar{f}(\mathbf{q}-\mathbf{k})\,, \tag{2.24a}$$

$$f(\mathbf{r}-\mathbf{k}) \quad \Leftrightarrow \quad \bar{f}(\mathbf{q})\,\exp(-\mathrm{i}\,2\pi\,\mathbf{k}^t\mathbf{q})\,. \tag{2.24b}$$

Similar relations hold for the linear canonical transformations. With $f_o(\mathbf{r}) = \mathscr{L}f_i(\mathbf{r})$ we write

$$\mathscr{L}[f_i(\mathbf{r})\,\exp(\mathrm{i}\,2\pi\,\mathbf{k}^t\mathbf{r})] = f_o(\mathbf{r}-\mathbf{B}\mathbf{k})\,\exp(\mathrm{i}\,2\pi\,\mathbf{k}^t\mathbf{D}^t\mathbf{r})\,\exp(-\mathrm{i}\,\pi\,\mathbf{k}^t\mathbf{B}^t\mathbf{D}\mathbf{k})\,, \tag{2.25a}$$

$$\mathscr{L}[f_i(\mathbf{r}-\mathbf{k})] = f_o(\mathbf{r})\,\exp(\mathrm{i}\,2\pi\,\mathbf{k}^t\mathbf{C}^t\mathbf{r})\,\exp(-\mathrm{i}\,\pi\,\mathbf{k}^t\mathbf{C}^t\mathbf{A}\mathbf{k})\,, \tag{2.25b}$$

and we verify that for the Fourier transformation we get indeed Eqs. (2.24).

The important *convolution property* of the Fourier transformation,

$$\int f(\mathbf{r}-\mathbf{k})\,h(\mathbf{k})\,\mathrm{d}\mathbf{k} = (f * h)(\mathbf{r}) \quad \Leftrightarrow \quad \bar{f}(\mathbf{q})\,\bar{h}(\mathbf{q})\,, \tag{2.26}$$

does not have a nice counterpart for a general linear canonical transformation; see [7], for instance. Nevertheless, for a transformation with the additional condition $\mathbf{A} = \mathbf{0}$, which is actually a Fourier transformation followed by a multiplication with a quadratic-phase function, we can write $f_o(\mathbf{r}_o) = \mathscr{L}f_i(\mathbf{r}_i) = (\det \mathrm{i}\,\mathbf{B})^{-1/2}\exp(\mathrm{i}\,\pi\,\mathbf{r}_o^t\mathbf{D}\mathbf{B}^{-1}\mathbf{r}_o)\bar{f}_i(\mathbf{B}^{-1}\mathbf{r}_o)$ and thus

$$\begin{aligned}\mathscr{L}[(f_i * h_i)(\mathbf{r})] &= (\det \mathrm{i}\,\mathbf{B})^{-1/2}\exp(\mathrm{i}\,\pi\,\mathbf{r}^t\mathbf{D}\mathbf{B}^{-1}\mathbf{r})\bar{f}_i(\mathbf{B}^{-1}\mathbf{r})\,\bar{h}_i(\mathbf{B}^{-1}\mathbf{r})\\ &= (\det \mathrm{i}\,\mathbf{B})^{1/2}\exp(-\mathrm{i}\,\pi\,\mathbf{r}^t\mathbf{D}\mathbf{B}^{-1}\mathbf{r})f_o(\mathbf{r})\,h_o(\mathbf{r})\,.\end{aligned} \tag{2.27}$$

By analogy with the alternative representation of the convolution in the Fourier domein, $f_1 * f_2 = \mathscr{F}^{-1}\{[\mathscr{F}f_1(\mathbf{r})][\mathscr{F}f_2(\mathbf{r})]\}$, we can introduce the *generalized canonical convolution* as $\mathscr{L}(\mathbf{T}_3)\{[\mathscr{L}(\mathbf{T}_1)f_1(\mathbf{r})][\mathscr{L}(\mathbf{T}_2)f_2(\mathbf{r})]\}$, which resembles the common convolution and fractional convolution operations [4, 9, 39]. In particular, if the transformation matrices used in the latter expression correspond to the ones of the fractional Fourier transformation, we obtain the generalized fractional convolution, whose applications for shift-variant filtering, encryption, etc., have been proposed [4, 9, 31, 39, 77]. Since two-dimensional canonical transformations include such operations as rotation, scaling, shearing, and fractional Fourier transformation, the generalized canonical convolution can be helpful for resolving the problem of scale-, rotation- and shear-invariant (or partially invariant) filtering and pattern recognition.

2.3.2 Simple Transformations for the One-dimensional Case

In this section we consider some simple transformations in the one-dimensional case. We explicitly state the two governing equations for one dimension, cf. (2.9) and (2.10):

$$f_o(x_o) = (\mathrm{i}\,b)^{-1/2} \int \exp[\mathrm{i}\,\pi\, b^{-1}(dx_o^2 - 2x_o x_i + a x_i^2)]\, f_i(x_i)\, \mathrm{d}x_i \quad (b \neq 0)\;, \tag{2.28a}$$

$$f_o(x) = |a|^{-1/2} \exp[\mathrm{i}\,\pi\, c a^{-1} x^2]\, f_i(a^{-1}x) \quad (b = 0)\;, \tag{2.28b}$$

and we study the result of a linear canonical transformation in phase space for several simple cases. As an illustration, we consider the effect on a rectangularly shaped Wigner distribution, see the top figure in Table 2.3.

2.3.2.1 Fourier Transformer

We already met the special case of a Fourier transformation,

$$f_o(x_o) = \mathrm{i}^{-1/2} \int \exp(-\mathrm{i}\,2\pi\, x_o x_i)\, f_i(x_i)\, \mathrm{d}x_i = \mathrm{i}^{-1/2} \bar{f}_i(x_o) = \mathscr{F} f_i(x_i)\;, \tag{2.29}$$

which is connected to the transformation matrix $[0, 1; -1, 0]$. Its effect on the Wigner distribution is a simple clockwise rotation in the xu plane through $\pi/2$. In other words: whatever happened in the original Wigner distribution on the u axis, now happens on the x axis, and whatever happened on the x axis, now happens on the $-u$ axis. We recall the well-known property of the Fourier transformation that $\mathscr{F}^4$ results in the identity operation.

2.3.2.2 Magnifier

Let us now turn our attention to some basic one-parameter transformations. We start with two examples of the $b = 0$ class. The matrix $[a, 0; 0, a^{-1}]$, associated with the transformation

$$f_o(x) = |a|^{-1/2} f_i(a^{-1}x) \equiv \mathscr{M}(a) f_i(x)\;, \tag{2.30}$$

leads to a scaling by a in the x direction and an inverse scaling by a^{-1} in the u direction. We denote this transformation by the operator $\mathscr{M}(a)$. In the example we have chosen $a = 2$. We recognize the well-known property of the Fourier transformation, that a scaling for the original variable x corresponds to an inverse scaling for its Fourier conjugate u.

2.3.2.3 Quadratic-Phase Modulator

Our second example for the $b = 0$ class is the transformation matrix $[1, 0; c, 1]$, with the input–output relation

$$f_o(x) = \exp(\mathrm{i}\,\pi\, c x_o^2) f_i(x) \equiv \mathscr{Q}(-c) f_i(x) \; . \tag{2.31}$$

The effect of this modulation by a quadratic-phase function leads to a shear in the u direction. We denote this transformation by the operator $\mathscr{Q}(-c)$. In the example we have chosen $-c = 1$.

2.3.2.4 Fresnel Transformer

The dual of the previous example arises when $a \Leftrightarrow d$ and $b \Leftrightarrow -c$; the resulting transformation matrix then reads $[1, b; 0, 1]$ and its effect is a shear in the x direction. We denote this transformation, which is also known as the *Fresnel transformation*,

$$f_o(x_o) = (\mathrm{i}\, b)^{-1/2} \int \exp[\mathrm{i}\,\pi\, b^{-1}(x_o - x_i)^2] f_i(x_i)\, \mathrm{d}x_i \equiv \mathscr{S}(b) f_i(x_i) \; , \tag{2.32}$$

by the operator $\mathscr{S}(b)$, expressible in terms of previously defined operators as $\mathscr{S}(b) = \mathscr{F}\mathscr{Q}(b)\mathscr{F}^{-1}$. In the example we have chosen $b = 1$. We note that the Fresnel transformation takes the form of a convolution with the quadratic-phase function $(\mathrm{i}\, b)^{-1/2} \exp(\mathrm{i}\,\pi\, b^{-1} x^2)$; in the Fourier domain such a convolution becomes a multiplication with the quadratic-phase function, $\bar{f}_o(u) = \exp(-\mathrm{i}\,\pi\, b u^2) \bar{f}_i(u)$, which shows the duality between the quadratic-phase modulation and the Fresnel transformation.

Now that we have introduced the Fresnel transformation, we can try and express a magnifier $\mathscr{M}(s)$ in terms of quadratic-phase modulators $\mathscr{Q}(-c)$ and Fresnel transformers $\mathscr{S}(b)$. We easily verify that

$$\mathscr{Q}(b_2^{-1} + b_1 b_2^{-2})\, \mathscr{S}(b_2)\, \mathscr{Q}(b_2^{-1} + b_1^{-1})\, \mathscr{S}(b_1) = \mathscr{M}(-b_2 b_1^{-1}) \; . \tag{2.33}$$

A decomposition of linear canonical transformations in terms of quadratic-phase modulators and Fresnel transformers is especially interesting in optics, where these two basic transformations correspond to a lens $\mathscr{Q}(f^{-1})$ with focal distance f and a section of free space $\mathscr{S}(z)$ with distance z, respectively. In (2.33), we then recognize the imaging condition $f^{-1} = b_1^{-1} + b_2^{-1}$, the magnification factor $s = -b_2 b_1^{-1}$ and a phase-compensating lens with focal distance $-sf$. We easily verify that the magnification factor is negative if b_1, b_2 and f are positive, and that we will see an inverted image of the object, as expected.

2.3.2.5 Fractional Fourier Transformer

When dealing with the Fourier transformation and its inverse, we saw that these transformations resulted in a clockwise rotation by $\frac{1}{2}\pi$ and $-\frac{1}{2}\pi$, respectively. If the rotation takes a different value γ, say, (with γ typically between 0 and π), the transformation matrix reads $[\cos\gamma, \sin\gamma; -\sin\gamma, \cos\gamma]$. We are now in the realm of the *fractional Fourier transformation* [42, 45, 47, 57], denoted by the operator $\mathscr{F}(\gamma)$:

$$\begin{aligned} f_o(x_o) &= \frac{\exp(\mathrm{i}\,\frac{1}{2}\gamma)}{\sqrt{\mathrm{i}\,\sin\gamma}} \int \exp\left(\mathrm{i}\,\pi\,\frac{x_o^2\cos\gamma - 2x_i x_o + x_i^2\cos\gamma}{\sin\gamma}\right) f_i(x_i)\,\mathrm{d}x_i \\ &\equiv \mathscr{F}(\gamma) f_i(x_i) \equiv F_\gamma(x_o) \qquad (\gamma \neq n\pi)\;; \end{aligned} \tag{2.34}$$

we recall that $(\mathrm{i}\,\sin\gamma)^{1/2}$ is defined as $\exp[\mathrm{i}\,(\frac{1}{4}\pi)\,\mathrm{sgn}(\sin\gamma)]|\sin\gamma|^{1/2}$. Note the additional phase factor $\exp(\mathrm{i}\,\frac{1}{2}\gamma)$, which has been added to get a better correspondence to the physical process of Fourier transformation; in particular, the phase additivity is preserved, $\mathscr{F}(\gamma_1)\,\mathscr{F}(\gamma_2) = \mathscr{F}(\gamma_1+\gamma_2)$, and $\mathscr{F}(\frac{1}{2}\pi) f(x_i) = \bar{f}(x_o)$. With the *fractional angle* γ going from 0 via $\frac{1}{2}\pi$ to π, $F_\gamma(x)$ goes from $F_0(x) = f(x)$ via $F_{\pi/2}(x) = \bar{f}(x)$ to $F_\pi(x) = f(-x)$. In the example we have chosen $\gamma = \frac{1}{4}\pi$.

A fractional Fourier transformer can easily be realized as a cascade of quadratic-phase modulators and Fresnel transformers. We have, for instance, the relations [19, 38]

$$\mathscr{F}(\gamma) = \mathscr{S}(\tan\tfrac{1}{2}\gamma)\,\mathscr{Q}(\sin\gamma)\,\mathscr{S}(\tan\tfrac{1}{2}\gamma)\,, \tag{2.35a}$$

$$\mathscr{F}(\gamma) = \mathscr{Q}(\tan\tfrac{1}{2}\gamma)\,\mathscr{S}(\sin\gamma)\,\mathscr{Q}(\tan\tfrac{1}{2}\gamma)\,. \tag{2.35b}$$

In coherent optics (with optical wavelength $\lambda_\circ$, say) this corresponds to cascades of thin convex lenses (with focal length $f > 0$) and sections of free space (with distance $z > 0$). Since we now work with real-world coordinates that are no longer dimensionless, we need additional magnifiers $\mathscr{M}(w)$ and $\mathscr{M}(w^{-1})$, where w has the dimension [m]. The real-world fractional Fourier transformer can then be described as the cascade $\mathscr{M}(w)\,\mathscr{F}(\gamma)\,\mathscr{M}(w^{-1})$, and similar relations hold for the thin lens and the section of free space. In real-world coordinates, the two Eqs. (2.35) read

$$\mathscr{M}(w)\,\mathscr{F}(\gamma)\,\mathscr{M}(w^{-1}) = \mathscr{S}(w^2\tan\tfrac{1}{2}\gamma)\,\mathscr{Q}(w^{-2}\sin\gamma)\,\mathscr{S}(w^2\tan\tfrac{1}{2}\gamma)\,, \tag{2.36a}$$

$$\mathscr{M}(w)\,\mathscr{F}(\gamma)\,\mathscr{M}(w^{-1}) = \mathscr{Q}(w^{-2}\tan\tfrac{1}{2}\gamma)\,\mathscr{S}(w^2\sin\gamma)\,\mathscr{Q}(w^{-2}\tan\tfrac{1}{2}\gamma)\,, \tag{2.36b}$$

and the corresponding cascades are depicted in Fig. 2.1(a) and (b), respectively. The fractional angle γ is related to z and f through the relation $\sin^2(\frac{1}{2}\gamma) = z/2f$ in both cascades, and the scaling parameter w is related to γ and $\lambda_\circ z$ through the

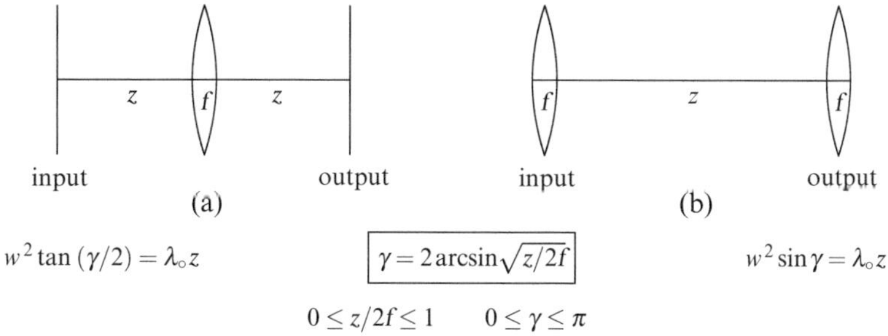

Fig. 2.1 Two optical realizations of the fractional Fourier transformer, consisting of lenses with focal distance f and sections of free space with a distance z

relations $w^2 \tan(\frac{1}{2}\gamma) = \lambda_\circ z$ in cascade (a) and $w^2 \sin\gamma = \lambda_\circ z$ in cascade (b). A common Fourier transformation $f_o(x) \propto w^{-1}\bar{f}_i(xw^{-2})$ occurs for $z = f$, while a simple coordinate reversion $f_o(x) \propto f_i(-x)$ occurs for $z = 2f$. The general formula (in real-world coordinates) reads

$$f_o(x_o) = \frac{\exp(\mathrm{i}\,\frac{1}{2}\gamma)}{w\sqrt{\mathrm{i}\,\sin\gamma}} \int \exp\left(\mathrm{i}\,\pi\,\frac{x_o^2\cos\gamma - 2x_i x_o + x_i^2\cos\gamma}{w^2\sin\gamma}\right) f_i(x_i)\,\mathrm{d}x_i\,. \qquad (2.37)$$

2.3.2.6 Hyperbolic Expander

Our final example is the *hyperbolic expander* [75, p. 183, Example: Hyperbolic expanders] with transformation matrix $[\cosh\gamma, \sinh\gamma; \sinh\gamma, \cosh\gamma]$. We denote the corresponding operation by $\mathscr{H}(\gamma)$ and remark that it can be considered as a magnifier $\mathscr{M}(\exp\gamma)$ embedded in between two fractional Fourier transformers $\mathscr{F}(\;\frac{1}{4}\pi)$ and $\mathscr{F}(\frac{1}{4}\pi)$. With this decomposition in mind, we readily recognize the effect of the hyperbolic expander in Table 2.3: the fractional Fourier transformer $\mathscr{F}(\frac{1}{4}\pi)$ rotates the Wigner distribution clockwise through $\frac{1}{4}\pi$, the magnifier $\mathscr{M}(\exp\gamma)$ scales by $\exp\gamma$ in the x direction and by $(\exp\gamma)^{-1}$ in the u direction, and the final fractional Fourier transformer $\mathscr{F}(-\frac{1}{4}\pi)$ rotates everything back to its original orientation. In the example we have chosen $\gamma = 0.5$, and thus $\exp\gamma = 1.64872$.

A hyperbolic expander can easily be realized again as a cascade of quadratic-phase modulators and Fresnel transformers. The result shows a nice resemblance to the fractional Fourier transformer; we only have to replace the quadratic-phase modulations by their inverses:

$$\mathscr{H}(\gamma) = \mathscr{S}(\tan\tfrac{1}{2}\gamma)\,\mathscr{Q}(-\sin\gamma)\,\mathscr{S}(\tan\tfrac{1}{2}\gamma)\,, \qquad (2.38a)$$

$$\mathscr{H}(\gamma) = \mathscr{Q}(-\tan\tfrac{1}{2}\gamma)\,\mathscr{S}(\sin\gamma)\,\mathscr{Q}(-\tan\tfrac{1}{2}\gamma)\,. \qquad (2.38b)$$

In optics this corresponds to replacing in Fig. 2.1 the convex lenses (with a positive focal length) by concave lenses (with a negative focal length).

2.3.2.7 Modified Iwasawa Decomposition

We end this section about the one-dimensional case with the important *modified Iwasawa decomposition* [55] (see also [75, Sects. 9.5 and 10.2]) of a linear canonical transformation $\mathscr{L}(\mathbf{T})$ as a cascade of a fractional Fourier transformer, a magnifier and a quadratic-phase modulator: $\mathscr{L}(\mathbf{T}) = \mathscr{Q}(g)\,\mathscr{M}(s)\,\mathscr{F}(\gamma)$, with

$$s = \sqrt{a^2+b^2}\,, \quad g = -(ac+bd)/s^2 \quad \text{and} \quad \exp(\mathrm{i}\,\gamma) = (a+\mathrm{i}\,b)/s\,;$$

see also [47, Sect. 9.7.1, Quadratic-phase systems as fractional Fourier transforms]. Note that this decomposition holds for all values of the matrix entries a, b, c and d. We can formulate an Iwasawa-type decomposition in the reversed order of the operators, by finding the Iwasawa decomposition of the inverse transformation $\mathscr{L}(\mathbf{T}^{-1})$ in its regular order, reversing the order of the operators, and replacing each operator by its inverse. We then get $\mathscr{L}(\mathbf{T}) = \mathscr{F}(\hat{\gamma})\,\mathscr{M}(\hat{s})\,\mathscr{Q}(\hat{g})$, with

$$\hat{s} = 1/\sqrt{d^2+b^2}\,, \quad \hat{g} = -(cd+ab)\,\hat{s}^2 \quad \text{and} \quad \exp(\mathrm{i}\,\hat{\gamma}) = (d+\mathrm{i}\,b)\,\hat{s}\,.$$

Many other possible decompositions, in particular as cascades of quadratic-phase modulators $\mathscr{Q}$, Fresnel transformers $\mathscr{S}$, magnifiers $\mathscr{M}$ and Fourier transformers $\mathscr{F}$, are known. If necessary, we can even restrict us to quadratic-phase modulators and Fresnel transformers. We refer to [47, Sect. 3.4, Linear canonical transforms], where some more detailed information about the linear canonical transformation for one-dimensional signals can be found.

2.4 Special Cases of the Linear Canonical Transformation

In this section we consider some special cases of linear canonical transformers:

1. The case $\mathbf{B} = \mathbf{0}$ and consequently $\mathbf{A}^{-1} = \mathbf{D}^t$,
2. The case $\mathbf{B}^{-1} = -\mathbf{C}^t$ and consequently $\mathbf{A}\mathbf{D}^t = \mathbf{0}$,
3. The case $\mathbf{C} = \mathbf{0}$ and consequently $\mathbf{A}^{-1} = \mathbf{D}^t$, and
4. The case that $\mathbf{T}$ is not only symplectic but also orthogonal, $\mathbf{T}^{-1} = \mathbf{T}^t$.

The reduction from the higher-dimensional to the one-dimensional case is throughout straightforward by simply substituting the general matrices $\mathbf{A}$, $\mathbf{B}$, $\mathbf{C}$ and $\mathbf{D}$ by their scalar versions $a\mathbf{I}$, $b\mathbf{I}$, $c\mathbf{I}$ and $d\mathbf{I}$, and leads to some additional formulas that are not included in Table 2.3. Matrix transposition is no longer relevant for the

Table 2.3 Some simple linear canonical transformations in one dimension

Operator	Transformation matrix	Example
$\mathscr{L}(\mathbf{T})$ $\quad \det \mathbf{T} = ad - bc = 1$	$\begin{bmatrix} a & b \\ c & d \end{bmatrix}$	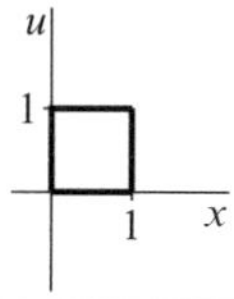
Fourier transformation $\mathscr{F}$ rotation in the xu plane through $\gamma = \frac{1}{2}\pi$	$\begin{bmatrix} 0 & 1 \\ -1 & 0 \end{bmatrix}$	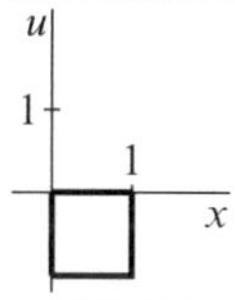
$\mathscr{F}\mathscr{L}(\mathbf{T})\mathscr{F}^{-1} = \mathscr{L}(\mathbf{T}^{t-1})$	$\begin{bmatrix} d & -c \\ -b & a \end{bmatrix}$	
Magnification $\mathscr{M}(a)$ scaling in the x direction by a scaling in the u direction by a^{-1}	$\begin{bmatrix} a & 0 \\ 0 & a^{-1} \end{bmatrix}$	$a = 2$ 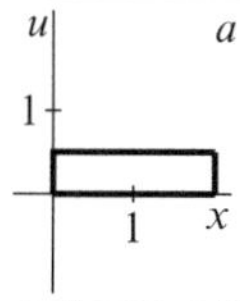
Quadratic-phase modulation $\mathscr{Q}(-c)$ shear in the u direction by c	$\begin{bmatrix} 1 & 0 \\ c & 1 \end{bmatrix}$	$c = -1$ 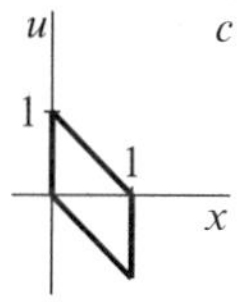
Fresnel transformation $\mathscr{S}(b) = \mathscr{F}\mathscr{Q}(b)\mathscr{F}^{-1}$ shear in the x direction by b	$\begin{bmatrix} 1 & b \\ 0 & 1 \end{bmatrix}$	$b = 1$ 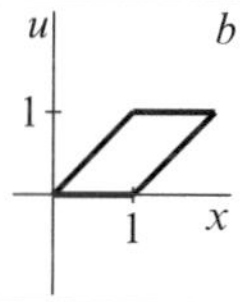
Fractional Fourier transformation $\mathscr{F}(\gamma) = \mathscr{S}(\tan\frac{1}{2}\gamma)\mathscr{Q}(\sin\gamma)\mathscr{S}(\tan\frac{1}{2}\gamma)$ $\mathscr{F}(\gamma) = \mathscr{Q}(\tan\frac{1}{2}\gamma)\mathscr{S}(\sin\gamma)\mathscr{Q}(\tan\frac{1}{2}\gamma)$	$\begin{bmatrix} \cos\gamma & \sin\gamma \\ -\sin\gamma & \cos\gamma \end{bmatrix}$	$\gamma = \frac{1}{4}\pi$ 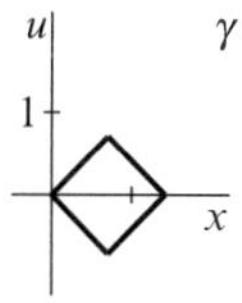
Hyperbolic expansion $\mathscr{H}(\gamma) = \mathscr{F}(\frac{1}{4}\pi)\mathscr{M}(\exp\gamma)\mathscr{F}(-\frac{1}{4}\pi)$ $\mathscr{H}(\gamma) = \mathscr{S}(\tan\frac{1}{2}\gamma)\mathscr{Q}(-\sin\gamma)\mathscr{S}(\tan\frac{1}{2}\gamma)$ $\mathscr{H}(\gamma) = \mathscr{Q}(-\tan\frac{1}{2}\gamma)\mathscr{S}(\sin\gamma)\mathscr{Q}(-\tan\frac{1}{2}\gamma)$	$\begin{bmatrix} \cosh\gamma & \sinh\gamma \\ \sinh\gamma & \cosh\gamma \end{bmatrix}$	$\gamma = 0.5$ 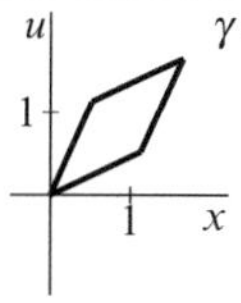
Iwasawa decomposition $\mathscr{L}(\mathbf{T}) = \mathscr{Q}(g)\mathscr{M}(s)\mathscr{F}(\gamma)$	$\begin{bmatrix} a & b \\ c & d \end{bmatrix}$	$s = \sqrt{a^2 + b^2}$ $g = -(ac + bd)/s^2$ $\exp(\mathrm{i}\gamma) = (a + \mathrm{i}b)/s$

one-dimensional case. A similar remark applies to the case of diagonal matrices **A**, **B**, **C** and **D**, in which case we are actually dealing with a transformation that is separable in multiple one-dimensional transformations.

2.4.1 *The Case* $\mathbf{B} = \mathbf{0}$ *and Consequently* $\mathbf{A}^{-1} = \mathbf{D}^t$

From the case $\mathbf{B} = \mathbf{0}$, see (2.10), we first consider the particular case of a D-dimensional *magnifier*

$$f_o(\mathbf{r}_o) = |\det \mathbf{A}|^{-1/2} f_i(\mathbf{A}^{-1}\mathbf{r}_o) \equiv \mathscr{M}(\mathbf{A}) f_i(\mathbf{r}_i) , \tag{2.39}$$

which results for $\mathbf{C} = \mathbf{0}$. If, in the two-dimensional case, the 2×2 matrix **A** takes the special form

$$\mathbf{A} = \begin{bmatrix} \cos\alpha & \sin\alpha \\ -\sin\alpha & \cos\alpha \end{bmatrix} \equiv \mathbf{U}_r(\alpha) = \mathbf{U}_r^{t-1}(\alpha) , \tag{2.40}$$

the magnifier $\mathscr{M}[\mathbf{U}_r(\alpha)]$ reduces to a *rotator*,

$$f_o(x_o, y_o) = f_i(x_o \cos\alpha - y_o \sin\alpha, x_o \sin\alpha + y_o \cos\alpha) \equiv \mathscr{R}(\alpha) f_i(x_i, y_i) . \tag{2.41}$$

Since an arbitrary 2×2 matrix **A** can always be expressed as a product of a positive-definite symmetric matrix $\mathbf{S} = (\mathbf{A}\mathbf{A}^t)^{1/2} = \mathbf{S}^t$ and a rotation matrix $\mathbf{U}_r(\alpha) = (\mathbf{A}\mathbf{A}^t)^{-1/2}\mathbf{A}$, we can easily consider the general magnifier as the cascade of a rotator and a *pure* magnifier: $\mathscr{M}(\mathbf{A}) = \mathscr{M}(\mathbf{S})\,\mathscr{R}(\alpha)$. The symmetric matrix **S** on its turn can be expressed in terms of its eigenvalues and eigenvectors as

$$\mathbf{S} = \begin{bmatrix} s_{xx} & s_{xy} \\ s_{xy} & s_{yy} \end{bmatrix} = \begin{bmatrix} \cos\varphi_s & \sin\varphi_s \\ -\sin\varphi_s & \cos\varphi_s \end{bmatrix} \begin{bmatrix} s_1 & 0 \\ 0 & s_2 \end{bmatrix} \begin{bmatrix} \cos\varphi_s & -\sin\varphi_s \\ \sin\varphi_s & \cos\varphi_s \end{bmatrix}$$
$$\equiv \mathbf{U}_r(\varphi_s)\, \mathbf{\Lambda}(s_1, s_2)\, \mathbf{U}_r(-\varphi_s) , \tag{2.42}$$

so that such a pure magnifier can be realized as a *separable* magnifier, i.e., a combination of two orthogonal one-dimensional magnifiers, oriented along the principal axes of the symmetry ellipse determined by **S** by embedding this combination in between two rotators: $\mathscr{M}(\mathbf{S}) = \mathscr{R}(\varphi_s)\,\mathscr{M}[\mathbf{\Lambda}(s_1, s_2)]\,\mathscr{R}(-\varphi_s)$. Each of the two one-dimensional magnifiers, with magnification factor s_1 and s_2, can then be decomposed, if necessary, in the form (2.33). The general magnifier $\mathscr{M}(\mathbf{A})$ can thus be decomposed as the cascade $\mathscr{M}(\mathbf{A}) = \mathscr{R}(\varphi_s)\,\mathscr{M}[\mathbf{\Lambda}(s_1, s_2)]\,\mathscr{R}(-\varphi_s + \alpha)$: a separable magnifier embedded in between two rotators (in general with different rotation angles). Of course, instead of $\mathscr{M}(\mathbf{A}) = \mathscr{M}(\mathbf{S})\,\mathscr{R}(\alpha)$, we may as well write $\mathscr{M}(\mathbf{A}) = \mathscr{R}(\alpha)\,\mathscr{M}(\mathbf{S}')$, where $\mathbf{S}' = (\mathbf{A}^t\mathbf{A})^{1/2}$.

Now that we have introduced the magnifier, we can easily derive the linear canonical transform of a scaled function and formulate the *scaling theorem*. We notice that scaling itself belongs to the class of linear canonical transformations, and we can thus use the cascadability property. We then observe that scaling of the input signal merely leads to a change of the parameterizing matrix. Indeed,

$$\mathscr{L}(\mathbf{T})\,|\det\mathbf{W}|^{-1/2}f(\mathbf{W}^{-1}\mathbf{r}) = \mathscr{L}(\mathbf{T})\,\mathscr{M}(\mathbf{W})f(\mathbf{r}) = \mathscr{L}(\tilde{\mathbf{T}})f(\mathbf{r})\;, \tag{2.43a}$$

where $\tilde{\mathbf{T}}$ takes the form

$$\tilde{\mathbf{T}} = \begin{bmatrix}\tilde{\mathbf{A}} & \tilde{\mathbf{B}}\\ \tilde{\mathbf{C}} & \tilde{\mathbf{D}}\end{bmatrix} = \begin{bmatrix}\mathbf{A} & \mathbf{B}\\ \mathbf{C} & \mathbf{D}\end{bmatrix}\begin{bmatrix}\mathbf{W} & \mathbf{0}\\ \mathbf{0} & \mathbf{W}^{t-1}\end{bmatrix} = \begin{bmatrix}\mathbf{AW} & \mathbf{BW}^{t-1}\\ \mathbf{CW} & \mathbf{DW}^{t-1}\end{bmatrix}. \tag{2.43b}$$

It may be interesting to see under what conditions scaling of the input signal by a matrix $\mathbf{W}_i$ produces only a scaling of the output signal by a (possibly different) matrix $\mathbf{W}_o$, i.e., $\mathscr{L}(\mathbf{T})\,\mathscr{M}(\mathbf{W}_i) = \mathscr{M}(\mathbf{W}_o)\,\mathscr{L}(\mathbf{T})$. This has been extensively studied in [7].

The second particular case is the D-dimensional *quadratic-phase modulator*

$$f_o(\mathbf{r}) = \exp[\mathrm{i}\,\pi\,\mathbf{r}^t\mathbf{C}\mathbf{r}]f_i(\mathbf{r}) \equiv \mathscr{Q}(-\mathbf{C})f_i(\mathbf{r})\;, \tag{2.44}$$

which results for $\mathbf{A} = \mathbf{D} = \mathbf{I}$ (and hence also $\mathbf{C} = \mathbf{C}^t$). Since $\mathbf{C}$ is symmetric, it can again be expressed in terms of its eigenvalues and eigenvectors and realized as a separable magnifier embedded in between two rotators: $\mathscr{Q}(-\mathbf{C}) = \mathscr{R}(\varphi_c)\,\mathscr{Q}[-\boldsymbol{\Lambda}(c_1,c_2)]\,\mathscr{R}(-\varphi_c)$.

The general case, represented by Eq. (2.10), which may be called a *generalized magnifier*, results as the cascade of a magnifier $\mathscr{M}(\mathbf{A})$ and a quadratic-phase modulator $\mathscr{Q}(-\mathbf{C}\mathbf{A}^{-1})$,

$$f_o(\mathbf{r}) = \mathscr{Q}(-\mathbf{C}\mathbf{A}^{-1})\,\mathscr{M}(\mathbf{A})f_i(\mathbf{r}) \tag{2.45}$$

and the transformation matrix has been decomposed as

$$\begin{aligned}\mathbf{T} = \begin{bmatrix}\mathbf{A} & \mathbf{0}\\ \mathbf{C} & \mathbf{A}^{t-1}\end{bmatrix} &= \begin{bmatrix}\mathbf{I} & \mathbf{0}\\ \mathbf{C}\mathbf{A}^{-1} & \mathbf{I}\end{bmatrix}\begin{bmatrix}\mathbf{A} & \mathbf{0}\\ \mathbf{0} & \mathbf{A}^{t-1}\end{bmatrix}\\ &= \begin{bmatrix}\mathbf{I} & \mathbf{0}\\ \mathbf{C}\mathbf{A}^{-1} & \mathbf{I}\end{bmatrix}\begin{bmatrix}(\mathbf{A}\mathbf{A}^t)^{1/2} & \mathbf{0}\\ \mathbf{0} & (\mathbf{A}\mathbf{A}^t)^{-1/2}\end{bmatrix}\begin{bmatrix}\mathbf{U}_r(\alpha) & \mathbf{0}\\ \mathbf{0} & \mathbf{U}_r(\alpha)\end{bmatrix}.\end{aligned} \tag{2.46}$$

We can of course change the order and write $\mathscr{M}(\mathbf{A})\,\mathscr{Q}(-\mathbf{A}^t\mathbf{C})$ instead. In one dimension, the positive-definite matrix $(\mathbf{A}\mathbf{A}^t)^{1/2}$ reduces to $|a|$, and the rotation matrix $\mathbf{U}_r(\alpha)$ reduces to $\operatorname{sgn} a$, which takes care of a possible negative sign in a. We thus have $\mathscr{Q}(-ca^{-1})\,\mathscr{M}(a) = \mathscr{Q}(-ca^{-1})\,\mathscr{M}(|a|)\,\mathscr{M}(\operatorname{sgn} a)$, but the decomposition of the magnifier $\mathscr{M}(a)$ into the cascade $\mathscr{M}(|a|)\,\mathscr{M}(\operatorname{sgn} a)$ is rather irrelevant.

For easy reference, some linear canonical transformations for the case $\mathbf{B} = \mathbf{0}$, treated in this section, have been collected in Table 2.4.

2.4.2 *The Case* $\mathbf{B}^{-1} = -\mathbf{C}^t$ *and Consequently* $\mathbf{A}\mathbf{D}^t = \mathbf{0}$

Let us first consider the special case that both **A** and **D** vanish, for which the integral (2.9) reduces to

$$f_o(\mathbf{r}_o) = (\det \mathrm{i}\,\mathbf{B})^{-1/2} \int \exp[-\mathrm{i}\,2\pi\, \mathbf{r}_i^t \mathbf{B}^{-1} \mathbf{r}_o] f_i(\mathbf{r}_i)\, \mathrm{d}\mathbf{r}_i\, . \tag{2.47}$$

We already studied the case $\mathbf{B} = \mathbf{I}$, which led to the common Fourier transformation. The general expression (2.47), with $\mathbf{B} \neq \mathbf{I}$, can be interpreted as a *scaled Fourier transformation*, i.e., a Fourier transformation with an additional scaling (and rotation if $\mathbf{B} \neq \mathbf{B}^t$):

$$f_o(\mathbf{r}_o) = (\det \mathrm{i}\,\mathbf{B})^{-1/2} \bar{f}_i(\mathbf{B}^{-1}\mathbf{r}_o) = \mathscr{M}(\mathbf{B})\, \mathscr{F} f_i(\mathbf{r}_i) = \mathscr{F} \mathscr{M}(\mathbf{B}^{t^{-1}}) f_i(\mathbf{r}_i)\, . \tag{2.48}$$

If, moreover, we allow $\mathbf{D} \neq \mathbf{0}$, we get an additional quadratic-phase modulation:

$$f_o(\mathbf{r}) = \mathscr{Q}(-\mathbf{D}\mathbf{B}^{-1})\, \mathscr{M}(\mathbf{B})\, \mathscr{F} f_i(\mathbf{r})\, . \tag{2.49}$$

The case $\mathbf{A} \neq \mathbf{0}$ can be considered as the dual of the case $\mathbf{D} \neq \mathbf{0}$, as described in Sect. 2.3.1.3; we only have to replace $\mathscr{F} f_i(\mathbf{r})$ by $f_i(\mathbf{r})$, $f_o(\mathbf{r})$ by $\mathscr{F}^{-1} f_o(\mathbf{r})$, **D** by **A** and **B** by $-\mathbf{C}$ in Eq. (2.49) and get

$$f_o(\mathbf{r}) = \mathscr{F} \mathscr{Q}(\mathbf{A}\mathbf{C}^{-1})\, \mathscr{M}(-\mathbf{C}) f_i(\mathbf{r})\, . \tag{2.50}$$

For easy reference, some linear canonical transformations for the case $\mathbf{B}^{-1} = -\mathbf{C}^t$, treated in this section, have been collected in Table 2.4.

2.4.3 *The Case* $\mathbf{C} = \mathbf{0}$ *and Consequently* $\mathbf{A}^{-1} = \mathbf{D}^t$

We consider the case $\mathbf{C} = \mathbf{0}$ as the dual of the case $\mathbf{B} = \mathbf{0}$, see Sect. 2.4.1. We thus start with the expression (2.45), replace the signals by their Fourier transforms, replace **A** by **D** and **C** by $-\mathbf{B}$, and get the general expression

$$\bar{f}_o(\mathbf{q}) = |\det \mathbf{D}|^{-1/2} \mathscr{Q}(\mathbf{B}\mathbf{D}^{-1}) \bar{f}_i(\mathbf{D}^{-1}\mathbf{q})\, , \tag{2.51}$$

which is the dual of (2.45). The case $\mathbf{A} = \mathbf{D} = \mathbf{I}$ (and hence also $\mathbf{B} = \mathbf{B}^t$) is special again and leads to a mere multiplication (but now in the frequency domain) by a quadratic phase-function: $\bar{f}_o(\mathbf{q}) = \exp[-\mathrm{i}\,\pi\, \mathbf{q}^t\mathbf{B}\mathbf{q}] \bar{f}_i(\mathbf{q})$. This is the Fourier domain version of a D-dimensional *Fresnel transformation*, which in terms of the variables $\mathbf{r}_o$ and $\mathbf{r}_i$ reads

$$f_o(\mathbf{r}_o) = (\det \mathrm{i}\,\mathbf{B})^{-1/2} \int \exp[\mathrm{i}\,\pi(\mathbf{r}_o - \mathbf{r}_i)^t\mathbf{B}^{-1}(\mathbf{r}_o - \mathbf{r}_i)] f_i(\mathbf{r}_i)\,\mathrm{d}\mathbf{r}_i$$
$$= \mathscr{F}^{-1}\mathscr{Q}(\mathbf{B})\,\mathscr{F}f_i(\mathbf{r}_i) \equiv \mathscr{S}(\mathbf{B})f_i(\mathbf{r}_i) \tag{2.52}$$

and takes the form of a *convolution* with a quadratic phase function. The general case with an additional scaling by $\mathbf{A}$,

$$f_o(\mathbf{r}_o) = (\det \mathrm{i}\,\mathbf{B})^{-1/2} \int \exp[\mathrm{i}\,\pi(\mathbf{r}_o - \mathbf{A}\mathbf{r}_i)^t(\mathbf{B}\mathbf{A}^t)^{-1}(\mathbf{r}_o - \mathbf{A}\mathbf{r}_i)] f_i(\mathbf{r}_i)\,\mathrm{d}\mathbf{r}_i$$
$$= \mathscr{S}(\mathbf{B}\mathbf{A}^t)\,\mathscr{M}(\mathbf{A})f_i(\mathbf{r}_i) = \mathscr{M}(\mathbf{A})\,\mathscr{S}(\mathbf{A}^{-1}\mathbf{B})f_i(\mathbf{r}_i)\,, \tag{2.53}$$

may be called a *scaled Fresnel transformation*.

For easy reference, some linear canonical transformations for the case $\mathbf{C} = \mathbf{0}$, treated in this section, have been collected in Table 2.4.

2.4.4 The Case $\mathbf{T}^{-1} = \mathbf{T}^t$: Phase-Space Rotators

We now concentrate on the important class of transformation matrices that are not only symplectic, $\mathbf{T}^{-1} = \mathbf{J}\mathbf{T}^t\mathbf{J}$, but also orthogonal, $\mathbf{T}^{-1} = \mathbf{T}^t$. We call such matrices *orthosymplectic*. We easily see that $\mathbf{A} = \mathbf{D}$ and $\mathbf{B} = -\mathbf{C}$, and that the combination $\mathbf{A} + \mathrm{i}\,\mathbf{B} = \mathbf{U}$ is a unitary matrix: $\mathbf{U}^{-1} = \mathbf{U}^\dagger$. We thus have

$$\mathbf{T} = \begin{bmatrix} \mathbf{A} & \mathbf{B} \\ -\mathbf{B} & \mathbf{A} \end{bmatrix} \quad \text{and} \quad (\mathbf{A} - \mathrm{i}\,\mathbf{B})^t = \mathbf{U}^\dagger = \mathbf{U}^{-1} = (\mathbf{A} + \mathrm{i}\,\mathbf{B})^{-1}\,. \tag{2.54}$$

In the one-dimensional case, with the scalar matrix entries $a = d = \cos\gamma$ and $b = -c = \sin\gamma$, the matrix $\mathbf{T}$ reduces to the rotation matrix $\mathbf{U}_r(\gamma)$. Note that this represents a *rotation in phase space*,

$$\begin{aligned} x_o &= x_i \cos\gamma \;\; + u_i \sin\gamma\,, \\ u_o &= -x_i \sin\gamma + u_i \cos\gamma\,, \end{aligned} \tag{2.55}$$

and that the corresponding operation is known as the *fractional Fourier transformation*, see Sect. 2.3.2. The extension from the one-dimensional case to a higher-dimensional *separable* fractional Fourier transformer (with *diagonal* matrices $\mathbf{A}$ and $\mathbf{B}$, and possibly different fractional angles γ for the different coordinates) is straightforward.

In the two-dimensional case, we observe three basic systems with an orthogonal transformation matrix: (1) the *separable fractional Fourier transformer* $\mathscr{F}(\gamma_x, \gamma_y)$, (2) the *rotator* $\mathscr{R}(\varphi)$, which we already met in Sect. 2.4.1, and (3) the *gyrator* $\mathscr{G}(\varphi)$. Their unitary representations $\mathbf{U}_f(\gamma_x, \gamma_y)$, $\mathbf{U}_r(\varphi)$ and $\mathbf{U}_g(\varphi)$ take the forms

Table 2.4 Some linear canonical transformations for the cases $\mathbf{B} = \mathbf{0}$, $\mathbf{B}^{-1} = -\mathbf{C}^t$ and $\mathbf{C} = \mathbf{0}$

Operator	Transformation matrix	Remark
Magnifier $\mathscr{M}(\mathbf{A})$	$\begin{bmatrix} \mathbf{A} & \mathbf{0} \\ \mathbf{0} & \mathbf{A}^{t^{-1}} \end{bmatrix}$	
Rotator $\mathscr{R}(\alpha)$	$\begin{bmatrix} \mathbf{U}_r(\alpha) & \mathbf{0} \\ \mathbf{0} & \mathbf{U}_r(\alpha) \end{bmatrix}$	$\mathbf{U}_r(\alpha) = \begin{bmatrix} \cos\alpha & \sin\alpha \\ -\sin\alpha & \cos\alpha \end{bmatrix}$
$\mathscr{M}(\mathbf{A}) = \mathscr{M}(\mathbf{S}_1)\mathscr{R}(\alpha) = \mathscr{R}(\alpha)\mathscr{M}(\mathbf{S}_2)$	$\begin{bmatrix} \mathbf{A} & \mathbf{0} \\ \mathbf{0} & \mathbf{A}^{t^{-1}} \end{bmatrix}$	$\mathbf{S}_1 = (\mathbf{A}\mathbf{A}^t)^{1/2} = \mathbf{S}_1^t$ $\mathbf{U}_r(\alpha) = \mathbf{S}_1^{-1}\mathbf{A} = \mathbf{A}\mathbf{S}_2^{-1}$ $\mathbf{S}_2 = (\mathbf{A}^t\mathbf{A})^{1/2} = \mathbf{S}_2^t$
Pure magnifier $\mathscr{M}(\mathbf{S}) = \mathscr{R}(\varphi_s)\mathscr{M}(\mathbf{\Lambda}_s)\mathscr{R}(-\varphi_s)$	$\begin{bmatrix} \mathbf{S} & \mathbf{0} \\ \mathbf{0} & \mathbf{S}^{-1} \end{bmatrix}$	$\mathbf{S} = \mathbf{S}^t = \mathbf{U}_r(\varphi_s)\mathbf{\Lambda}_s\mathbf{U}(-\varphi_s)$
Quadratic-phase modulator $\mathscr{Q}(-\mathbf{C}) = \mathscr{R}(\varphi_c)\mathscr{Q}(-\mathbf{\Lambda}_c)\mathscr{R}(-\varphi_c)$	$\begin{bmatrix} \mathbf{I} & \mathbf{0} \\ \mathbf{C} & \mathbf{I} \end{bmatrix}$	$\mathbf{C} = \mathbf{C}^t = \mathbf{U}_r(\varphi_c)\mathbf{\Lambda}_c\mathbf{U}(-\varphi_c)$
Generalized magnifier $\mathscr{Q}(-\mathbf{C}\mathbf{A}^{-1})\mathscr{M}(\mathbf{A}) = \mathscr{M}(\mathbf{A})\mathscr{Q}(-\mathbf{A}^t\mathbf{C})$	$\begin{bmatrix} \mathbf{A} & \mathbf{0} \\ \mathbf{C} & \mathbf{A}^{t^{-1}} \end{bmatrix}$	
Fourier transformation $\mathscr{F}$	$\begin{bmatrix} \mathbf{0} & \mathbf{I} \\ -\mathbf{I} & \mathbf{0} \end{bmatrix}$	
Scaled Fourier transformation $\mathscr{M}(\mathbf{B})\mathscr{F} = \mathscr{F}\mathscr{M}(\mathbf{B}^{t^{-1}})$	$\begin{bmatrix} \mathbf{0} & \mathbf{B} \\ -\mathbf{B}^{t^{-1}} & \mathbf{0} \end{bmatrix}$	
$\mathscr{Q}(-\mathbf{D}\mathbf{B}^{-1})\mathscr{M}(\mathbf{B})\mathscr{F}$	$\begin{bmatrix} \mathbf{0} & \mathbf{B} \\ -\mathbf{B}^{t^{-1}} & \mathbf{D} \end{bmatrix}$	
$\mathscr{F}\mathscr{Q}(\mathbf{A}\mathbf{C}^{-1})\mathscr{M}(-\mathbf{C})$	$\begin{bmatrix} \mathbf{A} & -\mathbf{C}^{t^{-1}} \\ \mathbf{C} & \mathbf{0} \end{bmatrix}$	
Fresnel transformation $\mathscr{S}(\mathbf{B}) = \mathscr{F}^{-1}\mathscr{Q}(\mathbf{B})\mathscr{F} = \mathscr{F}\mathscr{Q}(\mathbf{B})\mathscr{F}^{-1}$	$\begin{bmatrix} \mathbf{I} & \mathbf{B} \\ \mathbf{0} & \mathbf{I} \end{bmatrix}$	$\mathbf{B} = \mathbf{B}^t$
Scaled Fresnel transformation $\mathscr{S}(\mathbf{B}\mathbf{A}^t)\mathscr{M}(\mathbf{A}) = \mathscr{M}(\mathbf{A})\mathscr{S}(\mathbf{A}^{-1}\mathbf{B})$	$\begin{bmatrix} \mathbf{A} & \mathbf{B} \\ \mathbf{0} & \mathbf{A}^{t^{-1}} \end{bmatrix}$	

$$\mathbf{U}_f(\gamma_x,\gamma_y)=\begin{bmatrix}\exp(\mathrm{i}\,\gamma_x) & 0\\ 0 & \exp(\mathrm{i}\,\gamma_y)\end{bmatrix}, \quad \mathbf{U}_r(\varphi)=\begin{bmatrix}\cos\varphi & \sin\varphi\\ -\sin\varphi & \cos\varphi\end{bmatrix},$$

$$\text{and} \quad \mathbf{U}_g(\varphi)=\begin{bmatrix}\cos\varphi & \mathrm{i}\,\sin\varphi\\ \mathrm{i}\,\sin\varphi & \cos\varphi\end{bmatrix}. \tag{2.56}$$

Note that $\det[\mathbf{U}_f(\gamma_x,\gamma_y)]=\exp[\mathrm{i}\,(\gamma_x+\gamma_y)]$ and $\det[\mathbf{U}_r(\varphi)]=\det[\mathbf{U}_g(\varphi)]=1$. These three basic systems are additive in their parameters and correspond to *rotations in phase space*: the rotator $\mathscr{R}(\varphi)$ performs a rotation in the xy and the uv planes, the gyrator $\mathscr{G}(\varphi)$ in the xv and the yu planes, and the separable fractional Fourier transformer $\mathscr{F}(\gamma_x,\gamma_y)$ in the xu plane (through an angle γ_x) and the yv plane (through an angle γ_y), see also [75, Sect. 10.3, U(2) fractional Fourier transformers]. We remark that the *symmetric* fractional Fourier transformer $\mathscr{F}(\gamma,\gamma)$ is described by a *scalar* matrix $\mathbf{U}_f(\gamma,\gamma)=\exp(\mathrm{i}\,\gamma)\,\mathbf{I}$ and that it commutes with any other phase-space rotator $\mathscr{O}(\mathbf{U})$.

We easily verify, for instance, by expressing the unitary matrix $\mathbf{U}$ in the form

$$\mathbf{U}=\begin{bmatrix}\exp(\mathrm{i}\,\gamma_x)\cos\varphi & -\exp[\mathrm{i}\,(\gamma_y+\gamma)]\sin\varphi\\ \exp[\mathrm{i}\,(\gamma_x-\gamma)]\sin\varphi & \exp(\mathrm{i}\,\gamma_y)\cos\varphi\end{bmatrix}, \tag{2.57}$$

that the input–output relation for a phase-space rotator can be expressed in the form

$$\mathbf{r}_o-\mathrm{i}\,\mathbf{q}_o=\mathbf{U}(\mathbf{r}_i-\mathrm{i}\,\mathbf{q}_i)\,, \tag{2.58}$$

which is an easy alternative for (2.6).

Any two-dimensional phase-space rotator can be realized as a cascade of the three basic phase-space rotators, in which cascade only two different kinds are actually needed. The gyrator $\mathscr{G}(\varphi)$, for instance, can be realized as the cascade $\mathscr{R}(-\frac{1}{4}\pi)\,\mathscr{F}(\varphi,-\varphi)\,\mathscr{R}(\frac{1}{4}\pi)$, which represents in fact a separable fractional Fourier transformer oriented at an angle of $\frac{1}{4}\pi$. Some important decompositions of an arbitrary phase-space rotator are considered in the next section.

For easy reference, some linear canonical transformations for the case $\mathbf{T}^{-1}=\mathbf{T}^t$, treated in this section, have been collected in Table 2.5.

2.4.4.1 Decompositions of Phase-Space Rotators

If we start with the general expression (2.57) for the unitary matrix $\mathbf{U}$, we are immediately led to the decomposition

$$\mathscr{O}(\mathbf{U})=\mathscr{F}\left(\tfrac{1}{2}\gamma,-\tfrac{1}{2}\gamma\right)\mathscr{R}(-\varphi)\,\mathscr{F}\left(\gamma_x-\tfrac{1}{2}\gamma,\gamma_y+\tfrac{1}{2}\gamma\right), \tag{2.59}$$

Table 2.5 Some linear canonical transformations for the case $\mathbf{T}^{-1} = \mathbf{T}^t$

Operator	Transformation matrix	Remark
General phase-space rotator $\mathcal{O}(\mathbf{U}) = \mathcal{O}(\mathbf{X} + \mathrm{i}\,\mathbf{Y})$	$\begin{bmatrix} \mathbf{X} & \mathbf{Y} \\ -\mathbf{Y} & \mathbf{X} \end{bmatrix}$	$\mathbf{U} = \mathbf{X} + \mathrm{i}\,\mathbf{Y}\,; \mathbf{X} = \Re\mathbf{U}\,, \mathbf{Y} = \Im\mathbf{U}$ $\mathbf{U}^{-1} = \mathbf{U}^{\dagger}$
Rotator $\mathcal{R}(\varphi)$	$\begin{bmatrix} \mathbf{X}_r & \mathbf{0} \\ \mathbf{0} & \mathbf{X}_r \end{bmatrix}$	$\mathbf{U}_r(\varphi) = \begin{bmatrix} \cos\varphi & \sin\varphi \\ -\sin\varphi & \cos\varphi \end{bmatrix} = \mathbf{X}_r$
Separable fractional FT $\mathcal{F}(\gamma_x, \gamma_y)$	$\begin{bmatrix} \mathbf{X}_f & \mathbf{Y}_f \\ -\mathbf{Y}_f & \mathbf{X}_f \end{bmatrix}$	$\mathbf{U}_f(\gamma_x, \gamma_y) = \begin{bmatrix} \exp(\mathrm{i}\,\gamma_x) & 0 \\ 0 & \exp(\mathrm{i}\,\gamma_y) \end{bmatrix}$
Gyrator $\mathcal{G}(\varphi) = \mathcal{R}(-\frac{1}{4}\pi)\mathcal{F}(\varphi, -\varphi)\mathcal{R}(\frac{1}{4}\pi)$	$\begin{bmatrix} \mathbf{X}_g & \mathbf{Y}_g \\ -\mathbf{Y}_g & \mathbf{X}_g \end{bmatrix}$	$\mathbf{U}_g(\varphi) = \begin{bmatrix} \cos\varphi & \mathrm{i}\,\sin\varphi \\ \mathrm{i}\,\sin\varphi & \cos\varphi \end{bmatrix}$
$\mathcal{O}(\mathbf{U})\,\mathcal{F}(\gamma, \gamma) = \mathcal{F}(\gamma, \gamma)\,\mathcal{O}(\mathbf{U})$		$\mathbf{U}_f(\gamma, \gamma) = \exp(\mathrm{i}\,\gamma)\,\mathbf{I}$

Table 2.6 Some useful decompositions of phase-space rotators

Cascade of basic phase-space rotators	Equation	Remark
$\mathcal{O}(\mathbf{U}) = \mathcal{F}(\frac{1}{2}\gamma, -\frac{1}{2}\gamma)\,\mathcal{R}(-\varphi)\,\mathcal{F}(\gamma_x - \frac{1}{2}\gamma, \gamma_y + \frac{1}{2}\gamma)$	(2.59)	$u_{xx} = \exp(\mathrm{i}\,\gamma_x)\cos\varphi$ $u_{xy} = -\exp[\mathrm{i}\,(\gamma_y + \gamma)]\sin\varphi$ $u_{yx} = \exp[\mathrm{i}\,(\gamma_x - \gamma)]\sin\varphi$ $u_{yy} = \exp(\mathrm{i}\,\gamma_y)\cos\varphi$
$\mathcal{O}(\mathbf{U}) = \mathcal{R}(-\alpha)\,\mathcal{G}(-\beta)\,\mathcal{F}(-\psi, \psi)\,\mathcal{F}(\gamma_x, \gamma_y)$	(2.60)	$\sin 2\beta = \sin 2\varphi \sin\gamma$ $\cos 2\alpha = \cos 2\varphi / \cos 2\beta$ $\tan\psi = \tan\alpha \tan\beta$
$\mathcal{O}(\mathbf{U}) = \mathcal{R}(\varphi_2)\,\mathcal{F}(\gamma_1, \gamma_2)\,\mathcal{R}(\varphi_1)$	(2.61)	(2.61b) – (2.61g), see also [5]
$\mathcal{F}(\pm\frac{1}{4}\pi, \mp\frac{1}{4}\pi)\,\mathcal{G}(\pm\varphi) = \mathcal{R}(-\varphi)\,\mathcal{F}(\pm\frac{1}{4}\pi, \mp\frac{1}{4}\pi)$		
$\mathcal{F}(\pm\frac{1}{4}\pi, \mp\frac{1}{4}\pi)\,\mathcal{R}(\pm\varphi) = \mathcal{G}(\varphi)\,\mathcal{F}(\pm\frac{1}{4}\pi, \mp\frac{1}{4}\pi)$		
$\mathcal{R}(\pm\frac{1}{4}\pi)\,\mathcal{F}(\pm\varphi, \mp\varphi) = \mathcal{G}(-\varphi)\,\mathcal{R}(\pm\frac{1}{4}\pi)$		
$\mathcal{G}(\pm\frac{1}{4}\pi)\,\mathcal{F}(\pm\varphi, \mp\varphi) = \mathcal{R}(\varphi)\,\mathcal{G}(\pm\frac{1}{4}\pi)$		
$\mathcal{R}(\pm\frac{1}{4}\pi)\,\mathcal{G}(\pm\varphi) = \mathcal{F}(\varphi, -\varphi)\,\mathcal{R}(\pm\frac{1}{4}\pi)$		
$\mathcal{G}(\pm\frac{1}{4}\pi)\,\mathcal{R}(\pm\varphi) = \mathcal{F}(-\varphi, \varphi)\,\mathcal{G}(\pm\frac{1}{4}\pi)$		

where we recognize a rotator embedded in between two separable fractional Fourier transformers. This decomposition, along with the other two that are introduced in this section, have been collected in Table 2.6.

From the many other decompositions of a general phase-space rotator $\mathcal{O}(\mathbf{U})$ into the more basic ones, we mention in particular the cascade of a separable fractional Fourier transformer, a gyrator and a rotator [25],

$$\mathscr{O}(\mathbf{U}) = \mathscr{R}(-\alpha)\,\mathscr{G}(-\beta)\,\mathscr{F}(\gamma_x - \psi, \gamma_y + \psi)\ , \tag{2.60a}$$

which follows directly from the equality

$$\mathscr{F}\left(\tfrac{1}{2}\gamma, -\tfrac{1}{2}\gamma\right)\,\mathscr{R}(-\varphi)\,\mathscr{F}\left(-\tfrac{1}{2}\gamma, \tfrac{1}{2}\gamma\right) = \mathscr{R}(-\alpha)\,\mathscr{G}(-\beta)\,\mathscr{F}(-\psi, \psi);$$

the angles α, β and ψ in (2.60a) follow from γ and φ in (2.59) by

$$\sin 2\beta = \sin 2\varphi \sin\gamma\ , \tag{2.60b}$$

$$\cos 2\varphi = \cos 2\alpha \cos 2\beta\ , \tag{2.60c}$$

$$\tan\psi = \tan\alpha \tan\beta\ . \tag{2.60d}$$

We will use this decomposition later in Sect. 2.8.3.

Another useful decomposition of a general phase-space rotator $\mathscr{O}(\mathbf{U})$ takes the form of a separable fractional Fourier transformer embedded in between two rotators [5], see also [75, Sect. 10.3.3, SU(2)-Fourier transformer]:

$$\mathscr{O}(\mathbf{U}) = \mathscr{R}(\varphi_2)\,\mathscr{F}(\gamma_1, \gamma_2)\,\mathscr{R}(\varphi_1)\ . \tag{2.61a}$$

Without loss of generality, we may choose $0 \le \gamma_2 \le \gamma_1 < \pi$ and $0 \le \varphi_1 < \pi$, after which the four angles are unambiguous [5]. With $\mathbf{U} = \mathbf{X} + \mathrm{i}\,\mathbf{Y}$, the two fractional angles γ_1 and γ_2 follow easily from the relations

$$\exp[\mathrm{i}\,(\gamma_1 + \gamma_2)] = \det\mathbf{U}\ , \tag{2.61b}$$

$$\cos(\gamma_1 - \gamma_2) = \det\mathbf{X} + \det\mathbf{Y}\ ; \tag{2.61c}$$

and with $\mathbf{X} = [x_{11}, x_{12}; x_{21}, x_{22}]$ and $\mathbf{Y} = [y_{11}, y_{12}; y_{21}, y_{22}]$, the rotation angles φ_1 and φ_2 follow from the equations

$$x_{11} + x_{22} - y_{12} + y_{21} = 2\cos(\varphi_1 + \varphi_2 + \tfrac{1}{2}\gamma_1 + \tfrac{1}{2}\gamma_2)\,\cos\tfrac{1}{2}(\gamma_1 - \gamma_2), \tag{2.61d}$$

$$x_{12} - x_{21} + y_{11} + y_{22} = 2\,\sin(\varphi_1 + \varphi_2 + \tfrac{1}{2}\gamma_1 + \tfrac{1}{2}\gamma_2)\,\cos\tfrac{1}{2}(\gamma_1 - \gamma_2), \tag{2.61e}$$

$$-x_{11} + x_{22} + y_{12} + y_{21} = 2\,\sin(\varphi_1 - \varphi_2 + \tfrac{1}{2}\gamma_1 + \tfrac{1}{2}\gamma_2)\,\sin\tfrac{1}{2}(\gamma_1 - \gamma_2), \tag{2.61f}$$

$$x_{12} + x_{21} + y_{11} - y_{22} = 2\cos(\varphi_1 - \varphi_2 + \tfrac{1}{2}\gamma_1 + \tfrac{1}{2}\gamma_2)\,\sin\tfrac{1}{2}(\gamma_1 - \gamma_2). \tag{2.61g}$$

For the details we refer to [5]. We will use this decomposition in Sect. 2.5 to treat the case of a transformation matrix with $\det\mathbf{B} = 0$ but $\mathbf{B} \neq \mathbf{0}$. For completeness we mention [22] that the eigenvalues $\exp(\mathrm{i}\,\vartheta_1)$ and $\exp(\mathrm{i}\,\vartheta_2)$, say, of the unitary matrix $\mathbf{U} = \mathbf{U}_r(\varphi_2)\,\mathbf{U}_f(\gamma_1, \gamma_2)\,\mathbf{U}_r(\varphi_1)$, are related to the rotation angles φ_1 and φ_2 and the fractional angles γ_1 and γ_2 by the relationship

$$\vartheta_{1,2} = \tfrac{1}{2}(\gamma_1 + \gamma_2) \pm \arccos\left\{\cos(\varphi_1 + \varphi_2)\cos\left[\tfrac{1}{2}(\gamma_1 - \gamma_2)\right]\right\}. \tag{2.62}$$

For easy reference, we have collected in Table 2.6 also some useful relations between the rotator $\mathscr{R}(\varphi)$, the gyrator $\mathscr{G}(\varphi)$, and the antisymmetric fractional Fourier transformer $\mathscr{F}(\varphi, -\varphi)$, where the argument of one of them takes the value $\pm\frac{1}{4}\pi$.

2.5 Modified Iwasawa Decomposition

Any symplectic matrix can be decomposed in the *modified Iwasawa form* [55] (see also [75, Sects. 9.5 and 10.2]) as

$$\begin{bmatrix} \mathbf{A} & \mathbf{B} \\ \mathbf{C} & \mathbf{D} \end{bmatrix} = \begin{bmatrix} \mathbf{I} & \mathbf{0} \\ -\mathbf{G} & \mathbf{I} \end{bmatrix} \begin{bmatrix} \mathbf{S} & \mathbf{0} \\ \mathbf{0} & \mathbf{S}^{-1} \end{bmatrix} \begin{bmatrix} \mathbf{X} & \mathbf{Y} \\ -\mathbf{Y} & \mathbf{X} \end{bmatrix} \tag{2.63}$$

with

$$\mathbf{G} = -(\mathbf{C}\mathbf{A}^t + \mathbf{D}\mathbf{B}^t)(\mathbf{A}\mathbf{A}^t + \mathbf{B}\mathbf{B}^t)^{-1} = \mathbf{G}^t , \tag{2.64a}$$

$$\mathbf{S} = (\mathbf{A}\mathbf{A}^t + \mathbf{B}\mathbf{B}^t)^{1/2} = \mathbf{S}^t , \tag{2.64b}$$

$$\mathbf{X} + \mathrm{i}\,\mathbf{Y} = (\mathbf{A}\mathbf{A}^t + \mathbf{B}\mathbf{B}^t)^{-1/2}(\mathbf{A} + \mathrm{i}\,\mathbf{B}) = (\mathbf{X} + \mathrm{i}\,\mathbf{Y})^\dagger . \tag{2.64c}$$

The first matrix is the transformation matrix of a quadratic-phase modulator $\mathscr{Q}(\mathbf{G})$, the second one of a pure magnifier $\mathscr{M}(\mathbf{S})$, and the third one of a phase-space rotator $\mathscr{O}(\mathbf{U})$, with $\mathbf{U} \equiv \mathbf{X} + \mathrm{i}\,\mathbf{Y}$. The Iwasawa decomposition thus leads to the following cascade of any linear canonical transformer:

$$\mathscr{L}(\mathbf{T}) = \mathscr{Q}(\mathbf{G})\,\mathscr{M}(\mathbf{S})\,\mathscr{O}(\mathbf{U}) . \tag{2.65}$$

We can formulate an Iwasawa-type decomposition $\mathscr{L}(\mathbf{T}) = \mathscr{O}(\hat{\mathbf{U}})\,\mathscr{M}(\hat{\mathbf{S}})\,\mathscr{Q}(\hat{\mathbf{G}})$ in the reversed order of the operators, by finding the Iwasawa decomposition of the inverse transformation $\mathscr{L}(\mathbf{T}^{-1})$ in its regular order, reversing the order of the operators, and replacing each operator by its inverse; see Table 2.7.

If we substitute the cascade (2.60a) into (2.65), the Iwasawa decomposition can be expressed in the more detailed form

$$\mathscr{L}(\mathbf{T}) = \mathscr{Q}(\mathbf{G})\,\mathscr{M}(\mathbf{S})\,\mathscr{R}(-\alpha)\,\mathscr{G}(-\beta)\,\mathscr{F}(\gamma_x - \psi, \gamma_y + \psi) . \tag{2.66}$$

We will use this particular decomposition in Sect. 2.8.3.

If we substitute the cascade (2.61a) into (2.65), the Iwasawa decomposition reads

$$\mathscr{L}(\mathbf{T}) = \mathscr{Q}(\mathbf{G})\,\mathscr{M}(\mathbf{S})\,\mathscr{R}(\varphi_2)\,\mathscr{F}(\gamma_1, \gamma_2)\,\mathscr{R}(\varphi_1) , \tag{2.67}$$

Table 2.7 Some useful decompositions of the linear canonical transformation

Operator	Remark
$\mathscr{L}(\mathbf{T})$ $\quad \mathbf{T} = \begin{bmatrix} \mathbf{A}\,\mathbf{B} \\ \mathbf{C}\,\mathbf{D} \end{bmatrix}$	$\begin{bmatrix} \mathbf{A}\,\mathbf{B} \\ \mathbf{C}\,\mathbf{D} \end{bmatrix}^{-1} = \begin{bmatrix} \mathbf{D}^t & -\mathbf{B}^t \\ -\mathbf{C}^t & \mathbf{A}^t \end{bmatrix}$
Iwasawa decomposition $\mathscr{L}(\mathbf{T}) = \mathscr{Q}(\mathbf{G})\mathscr{M}(\mathbf{S})\mathscr{O}(\mathbf{U})$	$\mathbf{G} = -(\mathbf{CA}^t + \mathbf{DB}^t)(\mathbf{AA}^t + \mathbf{BB}^t)^{-1} = \mathbf{G}^t$ $\mathbf{S} = (\mathbf{AA}^t + \mathbf{BB}^t)^{1/2} = \mathbf{S}^t$ $\mathbf{U} = (\mathbf{AA}^t + \mathbf{BB}^t)^{-1/2}(\mathbf{A} + \mathrm{i}\,\mathbf{B}) = \mathbf{U}^\dagger$
Iwasawa decomposition in reversed order $\mathscr{L}(\mathbf{T}) = \mathscr{O}(\hat{\mathbf{U}})\mathscr{M}(\hat{\mathbf{S}})\mathscr{Q}(\hat{\mathbf{G}})$	$\hat{\mathbf{G}} = -(\mathbf{C}^t\mathbf{D} + \mathbf{A}^t\mathbf{B})(\mathbf{D}^t\mathbf{D} + \mathbf{B}^t\mathbf{B})^{-1} = \hat{\mathbf{G}}^t$ $\hat{\mathbf{S}} = (\mathbf{D}^t\mathbf{D} + \mathbf{B}^t\mathbf{B})^{-1/2} = \hat{\mathbf{S}}^t$ $\hat{\mathbf{U}} = (\mathbf{D} + \mathrm{i}\,\mathbf{B})(\mathbf{D}^t\mathbf{D} + \mathbf{B}^t\mathbf{B})^{-1/2} = \hat{\mathbf{U}}^\dagger$
$\mathscr{L}(\mathbf{T}) = \mathscr{Q}(-\mathbf{CA}^{-1})\mathscr{M}(\mathbf{A})\mathscr{S}(\mathbf{A}^{-1}\mathbf{B})$	$\det \mathbf{A} \neq 0$
$\mathscr{L}(\mathbf{T}) = \mathscr{S}(\mathbf{BD}^{-1})\mathscr{M}(\mathbf{D}^{t^{-1}})\mathscr{Q}(-\mathbf{D}^{-1}\mathbf{C})$	$\det \mathbf{D} \neq 0$
$\mathscr{L}(\mathbf{T}) = \mathscr{Q}(-\mathbf{DB}^{-1})\mathscr{M}(\mathbf{B})\mathscr{F}\mathscr{Q}(-\mathbf{B}^{-1}\mathbf{A})$	$\det \mathbf{B} \neq 0$ see below (i)
$\mathscr{L}(\mathbf{T}) = \mathscr{Q}[(\mathbf{I}-\mathbf{D})\mathbf{B}^{-1}]\mathscr{S}(\mathbf{B})\mathscr{Q}[\mathbf{B}^{-1}(\mathbf{I}-\mathbf{A})]$	$\det \mathbf{B} \neq 0$ $\mathbf{B} = \mathbf{B}^t$
$\mathscr{L}(\mathbf{T}) = \mathscr{S}(\mathbf{AC}^{-1})\mathscr{M}(-\mathbf{C}^{t^{-1}})\mathscr{F}\mathscr{S}(\mathbf{C}^{-1}\mathbf{D})$	$\det \mathbf{C} \neq 0$ see below (ii)
$\mathscr{L}(\mathbf{T}) = \mathscr{S}[(\mathbf{A}-\mathbf{I})\mathbf{C}^{-1}]\mathscr{Q}(-\mathbf{C})\mathscr{S}[\mathbf{C}^{-1}(\mathbf{D}-\mathbf{I})]$	$\det \mathbf{C} \neq 0$ $\mathbf{C} = \mathbf{C}^t$
$\mathscr{M}(\mathbf{B}^{-1})\mathscr{Q}(\mathbf{DB}^{-1})\mathscr{L}(\mathbf{T})\mathscr{Q}(\mathbf{B}^{-1}\mathbf{A}) = \mathscr{F}$	$\det \mathbf{B} \neq 0$ (i)
$\mathscr{M}(-\mathbf{C}^t)\mathscr{S}(-\mathbf{AC}^{-1})\mathscr{L}(\mathbf{T})\mathscr{S}(-\mathbf{C}^{-1}\mathbf{D}) = \mathscr{F}$	$\det \mathbf{C} \neq 0$ (ii)

which enables us to treat the case $\det \mathbf{B} = 0$. To do this, we write the explicit expression for the submatrix $\mathbf{B}$,

$$\mathbf{B} = \mathbf{S}\,\mathbf{U}_r(\varphi_2)\begin{bmatrix} \sin\gamma_1 & 0 \\ 0 & \sin\gamma_2 \end{bmatrix}\mathbf{U}_r(\varphi_1)\,, \tag{2.68}$$

and conclude that, since $\mathbf{S}$ and $\mathbf{U}_r$ are non-singular, the case $\det \mathbf{B} = 0$ arises only for $\sin\gamma_1 \sin\gamma_2 = 0$. The cascade (2.67) yields a clear physical interpretation of the linear canonical transformation. The cascade starts with a rotator $\mathscr{R}(\varphi_1)$ that rotates the coordinate system such that the new axes coincide with the axes of the separable fractional Fourier transformer $\mathscr{F}(\gamma_1, \gamma_2)$. This separable fractional Fourier transformer itself is responsible for a possible degeneration of the submatrix $\mathbf{B}$, but such a degeneration has a clear interpretation: it simply means that for one coordinate (or maybe even for both coordinates) the separable fractional Fourier transformer acts as an identity system. The cascade then continues with the rotator $\mathscr{R}(\varphi_2)$, followed by the pure magnifier $\mathscr{M}(\mathbf{S})$ and the quadratic-phase modulator $\mathscr{Q}(\mathbf{G})$. Equation (2.67) provides a useful representation of the linear canonical transformation, valid for any values of the transformation matrix. Note that this equation can also be used for numerical calculation of the canonical transform, using the algorithms developed for the fractional Fourier transformation [47, Sect. 6.7, Discrete computation of the fractional Fourier transform].

For completeness we recall that a different way to deal with a singular matrix $\mathbf{B}$ has been presented in [44]. It was shown that any symplectic matrix with a singular submatrix $\mathbf{B}$ can be decomposed as

$$\begin{bmatrix} \mathbf{A}\ \mathbf{B} \\ \mathbf{C}\ \mathbf{D} \end{bmatrix} = \begin{bmatrix} \mathbf{I}\ \mathbf{B}' \\ \mathbf{0}\ \mathbf{I} \end{bmatrix} \begin{bmatrix} \mathbf{A}-\mathbf{B}'\mathbf{C} & \mathbf{B}-\mathbf{B}'\mathbf{D} \\ \mathbf{C} & \mathbf{D} \end{bmatrix}, \tag{2.69}$$

in which $\mathbf{B}'$ is a non-singular *diagonal* matrix and $\det(\mathbf{B} - \mathbf{B}'\mathbf{D}) \neq 0$. The integral (2.9) can then be used for each of the two subsystems in this cascade separately, thus avoiding the singular case. The way to find the diagonal matrix $\mathbf{B}'$, however, is not easy.

2.5.1 *Other Decompositions*

The modified Iwasawa decomposition is valid for all values of the submatrices $\mathbf{A}$, $\mathbf{B}$, $\mathbf{C}$ and $\mathbf{D}$. Many other decompositions are possible, where sometimes an additional condition need be satisfied. A few of them are collected in Table 2.7, see also [47, Sect. 3.4.4, Decompositions]. Two of these decompositions have been repeated at the bottom of the table in a form that shows the possibility to bring an operator $\mathscr{L}(\mathbf{T})$ into a simple Fourier transformer $\mathscr{F}$, for instance—if $\det \mathbf{B} \neq 0$—by a pre-modulation by $\mathscr{Q}(\mathbf{B}^{-1}\mathbf{A})$, a post-modulation by $\mathscr{Q}(\mathbf{D}\mathbf{B}^{-1})$, and a scaling by the final magnifier $\mathscr{M}(\mathbf{B}^{-1})$.

2.6 Linear Canonical Transforms of Selected Functions

In this section we study the linear canonical transforms of

1. A Gaussian signal and a harmonic signal;
2. A periodic signal, with a short detour to Talbot imaging; and
3. Hermite–Gauss modes, in two dimensions leading to Hermite–Laguerre–Gauss modes—with the Laguerre–Gauss modes as a special case—and to the design of mode converters.

2.6.1 *Gaussian Signal and Harmonic Signal*

Let us consider the *Gaussian signal*

$$f_i(\mathbf{r}) = \exp(-\pi\, \mathbf{r}^t \mathbf{L}_i \mathbf{r}) \tag{2.70}$$

and determine its linear canonical transform

$$f_o(\mathbf{r}_o) = \mathscr{L}(\mathbf{T})f_i(\mathbf{r}_i) = (\det \mathrm{i}\,\mathbf{B})^{-1/2} \exp(\mathrm{i}\,\pi\,\mathbf{r}_o^t\mathbf{D}\mathbf{B}^{-1}\mathbf{r}_o)$$
$$\times \int \exp[-\pi\,\mathbf{r}_i^t(\mathbf{L}_i - \mathrm{i}\,\mathbf{B}^{-1}\mathbf{A})\mathbf{r}_i - \mathrm{i}\,2\pi\,\mathbf{r}_i^t\mathbf{B}^{-1}\mathbf{r}_o)]\,\mathrm{d}\mathbf{r}_i$$
$$= (\det \mathrm{i}\,\mathbf{B})^{-1/2}[\det(\mathbf{L}_i - \mathrm{i}\,\mathbf{B}^{-1}\mathbf{A})]^{-1/2} \exp(\mathrm{i}\,\pi\,\mathbf{r}_o^t\mathbf{D}\mathbf{B}^{-1}\mathbf{r}_o)$$
$$\times \exp[-\pi(\mathbf{B}^{-1}\mathbf{r}_o)^t(\mathbf{L}_i - \mathrm{i}\,\mathbf{B}^{-1}\mathbf{A})^{-1}(\mathbf{B}^{-1}\mathbf{r}_o)]\,, \tag{2.71}$$

where we have used the identity

$$\int \exp(-\pi\,\mathbf{s}^t\mathbf{P}\mathbf{s} - \mathrm{i}\,2\pi\,\mathbf{q}^t\mathbf{s})\,\mathrm{d}\mathbf{s} = (\det \mathbf{P})^{-1/2} \exp(-\pi\,\mathbf{q}^t\mathbf{P}^{-1}\mathbf{q})\,, \tag{2.72}$$

with $\mathbf{P}$ a symmetric matrix whose real part is positive definite. If we separate the part of the exponent that depends on $\mathbf{r}$ from the part that does not depend on $\mathbf{r}$, the transform can be written as [7]

$$\mathscr{L}(\mathbf{T}) \exp(-\pi\,\mathbf{r}^t\mathbf{L}_i\mathbf{r}) = [\det(\mathbf{A} + \mathrm{i}\,\mathbf{B}\mathbf{L}_i)]^{-1/2} \exp(-\pi\,\mathbf{r}^t\mathbf{L}_o\mathbf{r})\,, \tag{2.73}$$

where

$$\mathrm{i}\,\mathbf{L}_o = (\mathbf{C} + \mathrm{i}\,\mathbf{D}\mathbf{L}_i)(\mathbf{A} + \mathrm{i}\,\mathbf{B}\mathbf{L}_i)^{-1}\,. \tag{2.74}$$

We will meet the bilinear relationship (2.74) again in Sect. 2.8.4, cf. Eq. (2.128). For $\mathbf{A} = \mathbf{D} = \mathbf{0}$ and $\mathbf{B} = -\mathbf{C} = \mathbf{I}$, i.e., for a Fourier transformation, we get $\mathscr{F} \exp(-\pi\,\mathbf{r}^t\mathbf{L}_i\mathbf{r}) = (\det \mathrm{i}\,\mathbf{L}_i)^{-1/2} \exp(-\pi\,\mathbf{r}^t\mathbf{L}_i^{-1}\mathbf{r})$, as expected.

The linear canonical transform of the *constant signal* $f_i(\mathbf{r}) = 1$ arises for $\mathbf{L}_i = \mathbf{0}$, and reads $(\det \mathbf{A})^{-1/2} \exp(\mathrm{i}\,\pi\,\mathbf{r}^t\mathbf{C}\mathbf{A}^{-1}\mathbf{r})$; we have ignored convergence issues that may arise from the fact that we have lost the positive definiteness of the real part of $\mathbf{L}_i$. The transform of the *harmonic signal* $f_i(\mathbf{r}) = \exp(\mathrm{i}\,2\pi\,\mathbf{k}^t\mathbf{r})$ with frequency $\mathbf{k}$ then follows after applying the modulation property (2.25a) and takes the form

$$\mathscr{L}(\mathbf{T}) \exp(\mathrm{i}\,2\pi\,\mathbf{k}^t\mathbf{r}) = (\det \mathbf{A})^{-1/2} \exp(-\mathrm{i}\,\pi\,\mathbf{k}^t\mathbf{A}^{-1}\mathbf{B}\mathbf{k} + \mathrm{i}\,\pi\,\mathbf{r}^t\mathbf{C}\mathbf{A}^{-1}\mathbf{r}$$
$$+\mathrm{i}\,2\pi\,\mathbf{k}^t\mathbf{A}^{-1}\mathbf{r}). \tag{2.75}$$

In the limit $\mathbf{A} \Rightarrow \mathbf{0}$ and $\mathbf{B} = -\mathbf{C} = \mathbf{I}$, i.e., for a Fourier transformation, we get indeed the Dirac delta function $\delta(\mathbf{r}_o - \mathbf{k})$.

2.6.2 *Periodic Signal and Talbot Imaging*

Let us consider a *periodic signal* with periods p_x and p_y in the x and y directions. Such a signal can be represented as

$$f_i(\mathbf{r}) = \sum_{m,n=-\infty}^{\infty} a_{mn} \exp(\mathrm{i}\, 2\pi\, \mathbf{k}_{mn}^t \mathbf{r}) \;, \quad \mathbf{k}_{mn}^t = [m/p_x, n/p_y] \;. \tag{2.76}$$

The canonical transform of this periodic signal reads [7]

$$f_o(\mathbf{r}) = (\det \mathbf{A})^{-1/2} \sum_{m,n=-\infty}^{\infty} a_{mn} \exp(-\mathrm{i}\, \pi\, \mathbf{k}_{mn}^t \mathbf{A}^{-1}\mathbf{B}\mathbf{k}_{mn} + \mathrm{i}\, 2\pi\, \mathbf{k}_{mn}^t \mathbf{A}^{-1}\mathbf{r} + \mathrm{i}\, \pi\, \mathbf{r}^t\mathbf{C}\mathbf{A}\mathbf{r}) \;. \tag{2.77}$$

If $\mathbf{k}_{mn}^t \mathbf{A}^{-1}\mathbf{B}\mathbf{k}_{mn}$ is an even integer for all m and n, we get

$$f_o(\mathbf{r}) = (\det \mathbf{A})^{-1/2} \exp(\mathrm{i}\, \pi\, \mathbf{r}^t\mathbf{C}\mathbf{A}^{-1}\mathbf{r}) f_i(\mathbf{A}^{-1}\mathbf{r}) \;, \tag{2.78}$$

which corresponds to generalized *Talbot imaging*, cf. Eq. (2.10): an affine transformation of the input signal, possibly with an additional modulation by a quadratic-phase function.

Talbot imaging is well known in optics, where it appears for such a simple system as free space. Free space propagation is governed by the Fresnel transformation $\mathscr{S}(b\mathbf{I})$, and the imaging condition now requires that $\mathbf{k}_{mn}^t \mathbf{k}_{mn} b = (m^2/p_x^2 + n^2/p_y^2) b$ is an even integer. Since this equality has to hold for any integers m and n, we conclude that Talbot imaging appears for such a value of b that both b/p_x^2 and b/p_y^2 are even integers. In the case that $b/p_x^2 = \ell_x$ and $b/p_y^2 = \ell_y$ are integers, but not necessarily even, a kind of pseudo-imaging appears: the expansion coefficients a_{mn} are replaced by $-a_{mn}$ for those (m, n) combinations for which $m^2\ell_x + n^2\ell_y$ is an odd integer. Other simple examples of Talbot imaging are the separable fractional Fourier transformer $\mathscr{F}(\gamma_x, \gamma_y)$, for which the expression $\mathbf{k}_{mn}^t \mathbf{A}^{-1}\mathbf{B}\mathbf{k}_{mn}$ leads to the requirement that $\tan\gamma_x/p_x^2$ and $\tan\gamma_y/p_y^2$ should be even integers, and the gyrator $\mathscr{G}(\varphi)$, for which we get the condition that $\tan\varphi/p_x p_y$ is an integer [7].

2.6.3 *Hermite–Gaussian-Type Modes and Mode Conversion*

2.6.3.1 The One-dimensional Case

Based on the generating function $\exp(-s^2 + 2sz)$ for the Hermite polynomials $H_k(z)$, $k = 0, 1, \ldots$ [2, Sect. 22, Orthogonal polynomials; see Eq. (22.9.17)],

$$\sum_{k=0}^{\infty} H_k(z) \frac{s^k}{k!} = \exp(-s^2 + 2sz) \;, \tag{2.79}$$

the one-dimensional *Hermite–Gauss modes*

$$\mathsf{H}_k(x) = 2^{1/4} (2^k k!)^{-1/2} H_k(\sqrt{2\pi} x) \exp(-\pi\, x^2) \tag{2.80}$$

follow easily from their generating function

$$\sum_{k=0}^{\infty} \mathsf{H}_k(x) \left(\frac{2^k}{k!}\right)^{1/2} s^k = 2^{1/4} \exp(-s^2 + 2\sqrt{2\pi}\, sx - \pi\, x^2)\,. \tag{2.81}$$

We recall that the Hermite–Gauss modes $\mathsf{H}_k(x)$ form a complete orthonormal basis on the interval $-\infty \leq x \leq \infty$:

$$\int \mathsf{H}_m(x)\, \mathsf{H}_n(x)\, \mathrm{d}x = \delta_{m-n} = \begin{cases} 1 & \text{for } m = n \\ 0 & \text{for } m \neq n\,. \end{cases} \tag{2.82}$$

The linear canonical transforms $\mathscr{L}(\mathbf{T})\mathsf{H}_k(x) \equiv \mathsf{H}_k^{\mathbf{T}}(x)$ of these Hermite–Gauss modes follow from their generating function [6, 20, 22]

$$\sum_{k=0}^{\infty} \mathsf{H}_k^{\mathbf{T}}(x) \left(\frac{2^k}{k!}\right)^{1/2} s^k = \frac{2^{1/4}}{\sqrt{a+\mathrm{i}\,b}} \exp\left[-\pi\, \frac{d-\mathrm{i}\,c}{a+\mathrm{i}\,b}\, x^2\right]$$
$$\times \exp\left[-\left(s\sqrt{\frac{a-\mathrm{i}\,b}{a+\mathrm{i}\,b}}\right)^2 + 2\left(s\sqrt{\frac{a-\mathrm{i}\,b}{a+\mathrm{i}\,b}}\right)\frac{\sqrt{2\pi}\,x}{a^2+b^2}\right], \tag{2.83}$$

which function can be found by applying the integral (2.9) to the generating function (2.81) and evaluating the integral using the identity (2.72). We compare the right-hand side of (2.83) with the generating function (2.79) of the Hermite polynomials and write

$$\sum_{k=0}^{\infty} \mathsf{H}_k^{\mathbf{T}}(x) \left(\frac{2^k}{k!}\right)^{1/2} s^k$$
$$= \frac{2^{1/4}}{\sqrt{a+\mathrm{i}\,b}} \exp\left[-\pi\, \frac{d-\mathrm{i}\,c}{a+\mathrm{i}\,b}\, x^2\right] \sum_{k=0}^{\infty} H_k\left(\frac{\sqrt{2\pi}\,x}{\sqrt{a^2+b^2}}\right)\left(\sqrt{\frac{a-\mathrm{i}\,b}{a+\mathrm{i}\,b}}\right)^k \frac{s^k}{k!}, \tag{2.84}$$

from which we conclude that the transformed modes $\mathsf{H}_k^{\mathbf{T}}(x) = \mathscr{L}(\mathbf{T})\mathsf{H}_k(x)$ take the form [75, p. 284, Example: Canonical transforms of Hermite functions]

$$\mathsf{H}_k^{\mathbf{T}}(x)$$
$$= \frac{2^{1/4}(2^k k!)^{-1/2}}{\sqrt{a+\mathrm{i}\,b}} \left(\sqrt{\frac{a-\mathrm{i}\,b}{a+\mathrm{i}\,b}}\right)^k H_k\left(\frac{\sqrt{2\pi}\,x}{\sqrt{a^2+b^2}}\right) \exp\left[-\pi\, \frac{d-\mathrm{i}\,c}{a+\mathrm{i}\,b}\, x^2\right]. \tag{2.85}$$

In the special case of a fractional Fourier transformer, with $a = d = \cos\gamma$ and $b = -c = \sin\gamma$, the right-hand side of (2.85) can be written as $\mathsf{H}_k(x)\exp[-\mathrm{i}\,(k+\frac{1}{2})\gamma]$, from which we conclude that the Hermite–Gauss modes $\mathsf{H}_k(x)$ are eigenfunctions of the fractional Fourier transformer with eigenvalues $\exp(-\mathrm{i}\,k\gamma)$ [57]; note that we have taken into account the additional constant phase factor $\exp(i\frac{1}{2}\gamma)$. The fact that the Hermite–Gauss modes $\mathsf{H}_k(x)$ are eigenfunctions of the fractional Fourier transformation becomes also apparent when we inspect their Wigner distributions $W_{\mathsf{H}_k}(x,u)$ [16, 35], which depend on the combination x^2+u^2 only:

$$W_{\mathsf{H}_k}(x,u) = 2(-1)^k \exp[-2\pi(x^2+u^2)]\, L_k[4\pi(x^2+u^2)]\ , \tag{2.86}$$

where $L_k(\cdot)$, $k = 0, 1, \ldots$, are the *Laguerre polynomials* [2, Sect. 22, Orthogonal polynomials]. This Wigner distribution is indeed invariant under rotation in the xu plane.

2.6.3.2 The Two-dimensional Case

For the two-dimensional *separable Hermite–Gauss modes* $\mathsf{H}_{m,n}(\mathbf{r}) = \mathsf{H}_m(x)\mathsf{H}_n(y)$ we have, with $\mathbf{s} = [s_x, s_y]^t$, the generating function

$$\sum_{m=0}^{\infty}\sum_{n=0}^{\infty} \mathsf{H}_{m,n}(\mathbf{r}) \left(\frac{2^{m+n}}{m!n!}\right)^{1/2} s_x^m s_y^n = 2^{1/2}\exp(-\mathbf{s}^t\mathbf{s} + 2\sqrt{2\pi}\,\mathbf{s}^t\mathbf{r} - \pi\,\mathbf{r}^t\mathbf{r})\ , \tag{2.87}$$

which is simply the two-dimensional version of (2.81). The linear canonical transforms $\mathscr{L}(\mathbf{T})\mathsf{H}_{m,n}(\mathbf{r}) \equiv \mathsf{H}^{\mathbf{T}}_{m,n}(\mathbf{r})$ of these two-dimensional Hermite–Gauss modes follow from the two-dimensional version of the generating function (2.83) [6, 20–22]

$$\sum_{m=0}^{\infty}\sum_{n=0}^{\infty} \mathsf{H}^{\mathbf{T}}_{m,n}(\mathbf{r}) \left(\frac{2^{m+n}}{m!n!}\right)^{1/2} s_x^m s_y^n = \frac{2^{1/2}}{\sqrt{\det(\mathbf{A}+\mathrm{i}\,\mathbf{B})}} \exp[-\mathbf{s}^t(\mathbf{A}+\mathrm{i}\,\mathbf{B})^{-1}(\mathbf{A}-\mathrm{i}\,\mathbf{B})\mathbf{s}]$$
$$\times \exp[2\sqrt{2\pi}\,\mathbf{s}^t(\mathbf{A}+\mathrm{i}\,\mathbf{B})^{-1}\mathbf{r} - \pi\,\mathbf{r}^t(\mathbf{D}-\mathrm{i}\,\mathbf{C})(\mathbf{A}+\mathrm{i}\,\mathbf{B})^{-1}\mathbf{r}]\ . \tag{2.88}$$

Note that the complex symmetric matrix $(\mathbf{A}+\mathrm{i}\,\mathbf{B})^{-1}(\mathbf{A}-\mathrm{i}\,\mathbf{B})$ is unitary and that the real part $(\mathbf{A}\mathbf{A}^t+\mathbf{B}\mathbf{B}^t)^{-1}$ of the complex symmetric matrix $(\mathbf{D}-\mathrm{i}\,\mathbf{C})(\mathbf{A}+\mathrm{i}\,\mathbf{B})^{-1}$ is positive definite. From the generating function (2.88) we can derive derivative and recurrence relations for the transformed Hermite–Gauss modes [20] and also the direct expressions [6, 22]

$$\mathsf{H}^{\mathbf{T}}_{m,n}(\mathbf{r}) = \frac{2^{1/2}\,\mathscr{P}_x^m\,\mathscr{P}_y^n\,\exp[-\pi\,\mathbf{r}^t(\mathbf{D}-\mathrm{i}\,\mathbf{C})(\mathbf{A}+\mathrm{i}\,\mathbf{B})^{-1}\mathbf{r}]}{2^{m+n}\sqrt{\pi^{m+n}m!n!}\,\sqrt{\det(\mathbf{A}+\mathrm{i}\,\mathbf{B})}}\ , \tag{2.89}$$

$$\mathsf{H}^{\mathbf{T}}_{0,0}(\mathbf{r}) = \frac{2^{1/2}\exp[-\pi\,\mathbf{r}^t(\mathbf{D}-\mathrm{i}\,\mathbf{C})(\mathbf{A}+\mathrm{i}\,\mathbf{B})^{-1}\mathbf{r}]}{2^{m+n}\sqrt{\pi^{m+n}m!n!}\,\sqrt{\det(\mathbf{A}+\mathrm{i}\,\mathbf{B})}}\,, \tag{2.90}$$

where the operators $\mathscr{P}_x$ and $\mathscr{P}_y$ are determined by

$$\begin{bmatrix}\mathscr{P}_x\\ \mathscr{P}_y\end{bmatrix} = 2\pi(\mathbf{A}-\mathrm{i}\,\mathbf{B})^t[(\mathbf{D}-\mathrm{i}\,\mathbf{C})(\mathbf{A}+\mathrm{i}\,\mathbf{B})^{-1}]^*\begin{bmatrix}x\\ y\end{bmatrix} - (\mathbf{A}-\mathrm{i}\,\mathbf{B})^t\begin{bmatrix}\partial/\partial x\\ \partial/\partial y\end{bmatrix}, \tag{2.91}$$

with $\mathscr{P}_x^0$ and $\mathscr{P}_y^0$ the identity operators.

The generating functions (2.83) and (2.88) represent a general class of Hermite–Gaussian-type modes. If a member of this class, with defining matrix $\mathbf{T}_i = [\mathbf{A}_i, \mathbf{B}_i; \mathbf{C}_i, \mathbf{D}_i]$, say, undergoes a linear canonical transformation with transformation matrix $\mathbf{T}$, the input matrix $\mathbf{T}_i$ is transformed into the output matrix $\mathbf{T}_o = [\mathbf{A}_o, \mathbf{B}_o; \mathbf{C}_o, \mathbf{D}_o]$ by the simple matrix multiplication $\mathbf{T}_o = \mathbf{T}\mathbf{T}_i$.

2.6.3.3 Mode Conversion: Hermite–Laguerre–Gauss Modes

The transformation by phase-space rotators, for which $\mathbf{A}+\mathrm{i}\,\mathbf{B} = \mathbf{D}-\mathrm{i}\,\mathbf{C} = \mathbf{U}$, is important in optics for *mode conversion*. In that case, Eq. (2.91) reduces to $[\mathscr{P}_x, \mathscr{P}_y]^t = \mathbf{U}^{-1}(2\pi\,\mathbf{r} - \nabla^t)$, the transformed Hermite–Gauss modes $\mathsf{H}^{\mathbf{T}}_{m,n}(\mathbf{r})$ can be represented as [6]

$$\mathsf{H}^{\mathbf{T}}_{m,n}(\mathbf{r}) = \frac{2^{1/2}}{2^{m+n}\sqrt{\pi^{m+n}m!n!}\,\sqrt{\det\mathbf{U}}}\,\mathscr{P}_x^m(\mathbf{T})\,\mathscr{P}_y^n(\mathbf{T})\exp(-\pi\,\mathbf{r}^t\mathbf{r})\,, \tag{2.92}$$

and the generating function (2.88) reduces to the simpler form

$$\sum_{m=0}^{\infty}\sum_{n=0}^{\infty}\mathsf{H}^{\mathbf{T}}_{m,n}(\mathbf{r})\left(\frac{2^{m+n}}{m!n!}\right)^{1/2}s_x^m s_y^n = \frac{2^{1/2}}{\sqrt{\det\mathbf{U}}}\exp[-(\mathbf{U}^*\mathbf{s})^t(\mathbf{U}^*\mathbf{s})]$$
$$\times\exp[2\sqrt{2\pi}\,(\mathbf{U}^*\mathbf{s})^t\mathbf{r} - \pi\,\mathbf{r}^t\mathbf{r}]\,; \tag{2.93}$$

note that in this case the Gaussian part $\exp(-\pi\,\mathbf{r}^t\mathbf{r})$ of the Hermite–Gaussian-type modes does not change. Moreover, since $(\mathrm{d}/\mathrm{d}t - 2\pi\,t)^k\exp(-\pi\,t^2) = \exp(\pi\,t^2)(\mathrm{d}/\mathrm{d}t)^k\exp(-2\pi\,t^2)$, the transformed modes can as well be expressed in the more direct form

$$\mathsf{H}^{\mathbf{T}}_{m,n}(\mathbf{r}) = \frac{2^{1/2}(-1)^{m+n}\exp(\pi\,\mathbf{r}^t\mathbf{r})}{2^{m+n}\sqrt{\pi^{m+n}m!n!}\,\sqrt{\det\mathbf{U}}}\,\mathscr{U}_x^m\,\mathscr{U}_y^n\exp(-2\pi\,\mathbf{r}^t\mathbf{r})\,, \tag{2.94}$$

with the operators $\mathscr{U}_x$ and $\mathscr{U}_y$ defined as $[\mathscr{U}_x, \mathscr{U}_y]^t = \mathbf{U}^{-1}\nabla^t$. As an example we mention the phase-space rotator with

$$\mathbf{U} = \frac{1}{\sqrt{2}} \begin{bmatrix} \cos 2\alpha + \mathrm{i} & \sin 2\alpha \\ \sin 2\alpha & -\cos 2\alpha + \mathrm{i} \end{bmatrix} \equiv \mathbf{U}_{\mathrm{HLG}}(\alpha) \,, \tag{2.95}$$

which generates the recently introduced *Hermite–Laguerre–Gauss modes* [1] from the separable Hermite–Gauss modes.

For the special case $\alpha = \frac{1}{4}\pi$, the matrix $\mathbf{U}_{\mathrm{HLG}}(\alpha)$ takes the form

$$\mathbf{U}_{\mathrm{HLG}}\left(\tfrac{1}{4}\pi\right) = \frac{1}{\sqrt{2}} \begin{bmatrix} \mathrm{i} & 1 \\ 1 & \mathrm{i} \end{bmatrix} \equiv \mathbf{U}_{\mathrm{LG}} = \mathbf{U}_g\left(-\tfrac{1}{4}\pi\right) \mathbf{U}_f\left(\tfrac{1}{2}\pi, \tfrac{1}{2}\pi\right), \tag{2.96}$$

and the Hermite–Laguerre–Gauss modes reduce to the *Laguerre–Gauss modes*, whose generating function can be written in the form

$$2^{1/2}\mathrm{i}^{-1} \exp\{-\pi(x^2 + y^2) + 2\mathrm{i}\, s_x s_y - 2\sqrt{\pi}\,[\mathrm{i}\, s_x(x + \mathrm{i}\, y) - s_y(x - \mathrm{i}\, y)]\} \,. \tag{2.97}$$

We remark that the latter expression depends only on the combinations $x^2 + y^2$, $s_x(x + \mathrm{i}\, y)$ and $s_y(x - \mathrm{i}\, y)$, which shows the vortex behaviour of such modes. For completeness, we recall the explicit form of the Laguerre–Gauss modes [53, 60]

$$\begin{aligned} \mathsf{L}_{m,n}(r, \varphi) = 2^{1/2} \left[\frac{(\min\{m,n\})!}{(\max\{m,n\})!}\right]^{1/2} (\sqrt{2\pi}\, r)^{|m-n|} \exp[\mathrm{i}\,(m-n)\varphi] \\ \times L_{\min\{m,n\}}^{(|m-n|)}(2\pi\, r^2) \exp[-\pi\, r^2] \,, \end{aligned} \tag{2.98}$$

where $x = r\cos\varphi$ and $y = r\sin\varphi$, and where $L_n^{(\alpha)}(\cdot)$ are the *generalized* Laguerre polynomials [2, Sect. 22, Orthogonal polynomials]. Note that the vortex behaviour is clearly visible in the phase factor $\exp[\mathrm{i}\,(m-n)\varphi]$ and that the Laguerre–Gauss modes are eigenfunctions of the rotator: $\mathscr{R}(\alpha)\, \mathsf{L}_{m,n}(r, \varphi) = \mathsf{L}_{m,n}(r, \varphi - \alpha)$.

Other phase-space rotators exist that convert the separable Hermite–Gauss modes into Laguerre–Gauss modes [6, 20, 41, 53]. To find the operators that generate modes with a vortex behaviour, we require that in the generating function (2.88) the term $\mathbf{s}^t(\mathbf{A} + \mathrm{i}\,\mathbf{B})^{-1}\mathbf{r}$ depends only on the combinations $s_x(x + \mathrm{i}\, y)$ and $s_y(x - \mathrm{i}\, y)$, and the term $\mathbf{r}^t(\mathbf{D} - \mathrm{i}\,\mathbf{C})(\mathbf{A} + \mathrm{i}\,\mathbf{B})^{-1}\mathbf{r}$ only on the combination $x^2 + y^2$. These requirements lead to the class of linear canonical transformers that can be decomposed as $\mathscr{Q}(c\mathbf{I})\, \mathscr{R}(\alpha)\, \mathscr{G}(-\frac{1}{4}\pi)\, \mathscr{F}(\gamma_x, \gamma_y)$. The generating functions of the modes that arise at the output of this cascade have basically the same form as the generating function (2.97); Eq. (2.97) itself arises for the special choice $\gamma_x = \gamma_y = \frac{1}{2}\pi$ (and $\alpha = c = 0$) [1], while the case $\gamma_x = \gamma_y = 0$ has been reported in [53, Eq. (14)]. We remark that the important element in the cascade $\mathscr{Q}(c\mathbf{I})\, \mathscr{R}(\alpha)\, \mathscr{G}(-\frac{1}{4}\pi)\, \mathscr{F}(\gamma_x, \gamma_y)$ is the gyrator. Since the separable Hermite–Gauss modes are eigenfunctions of the separable fractional Fourier transformer, such a transformer that precedes the gyrator does not change the character of these modes. And once the gyrator has converted these modes to the Laguerre–Gauss mode, the succeeding rotator, for which the Laguerre–Gauss mode is an eigenfunction, and the final isotropic quadratic-phase modulator, which corresponds to a simple multiplication by $\exp(-\mathrm{i}\,\pi\, c\mathbf{r}^t\mathbf{r})$, do not destroy the vortex behaviour.

We remark that the Laguerre–Gauss modes $\mathsf{L}_{m,n}(\mathbf{r}) = \mathscr{G}(-\frac{1}{4}\pi)\,\mathsf{H}_{m,n}(\mathbf{r})$ are eigenfunctions of the symmetric fractional Fourier transformer $\mathscr{F}(\gamma,\gamma)$ with eigenvalue $\exp[-\mathrm{i}\,(m+n)\gamma]$,

$$\begin{aligned}\mathscr{F}(\gamma,\gamma)\,\mathsf{L}_{m,n}(\mathbf{r}) &= \mathscr{F}(\gamma,\gamma)\,\mathscr{G}\left(-\tfrac{1}{4}\pi\right)\,\mathsf{H}_{m,n}(\mathbf{r})\\ &= \mathscr{G}\left(-\tfrac{1}{4}\pi\right)\,\mathscr{F}(\gamma,\gamma)\,\mathsf{H}_{m,n}(\mathbf{r})\\ &= \mathscr{G}\left(-\tfrac{1}{4}\pi\right)\,\mathsf{H}_{m,n}(\mathbf{r})\,\exp[-\mathrm{i}\,(m+n)\gamma]\\ &= \mathsf{L}_{m,n}(\mathbf{r})\,\exp[-\mathrm{i}\,(m+n)\gamma]\,,\end{aligned} \tag{2.99}$$

where we have used the fact that the symmetric fractional Fourier transformer $\mathscr{F}(\gamma,\gamma)$ commutes with any other phase-space rotator $\mathscr{O}(\mathbf{U})$ and that the separable Hermite–Gauss modes $\mathsf{H}_{m,n}(\mathbf{r})$ are eigenfunctions of the separable fractional Fourier transformer $\mathscr{F}(\gamma_x,\gamma_y)$ with eigenvalue $\exp[-\mathrm{i}\,(m\gamma_x + n\gamma_y)]$.

2.7 Classification of the Linear Canonical Transformation Based on the Distribution of the Eigenvalues of Its Transformation Matrix

In this section we look for simple linear canonical transformations $\mathscr{L}(\mathbf{N})$, called nuclei, that are similar to a given transformation $\mathscr{L}(\mathbf{T})$ in the sense [37]

$$\mathscr{L}(\mathbf{T}) = \mathscr{L}(\mathbf{T}_\circ)\,\mathscr{L}(\mathbf{N})\,\mathscr{L}^{-1}(\mathbf{T}_\circ)\,. \tag{2.100}$$

An obvious guess would be to look for the eigenvalues and eigenvectors of the transformation matrix $\mathbf{T}$ and express it in its Jordan form [30], $\mathbf{T} = \mathbf{Q}\boldsymbol{\Lambda}_{\mathrm{J}}\mathbf{Q}^{-1}$, but it is not guaranteed that the matrices $\boldsymbol{\Lambda}_{\mathrm{J}}$ and $\mathbf{Q}$ are symplectic. Based on the eigenvalues of $\mathbf{T}$, we will be able to classify the linear canonical transformation $\mathscr{L}(\mathbf{T})$ and find a nucleus $\mathscr{L}(\mathbf{N})$ for each class [23]. A general proof for the existence of the decomposition (2.100) can be found in [37]; see, in particular, Theorem 41, which deals with the real symplectic Jordan form.

We remark that once we have found the representation (2.100), the eigenfunctions $\Phi(\mathbf{r})$ of the nucleus $\mathscr{L}(\mathbf{N})$, i.e., $\mathscr{L}(\mathbf{N})\,\Phi(\mathbf{r}) = \mu\Phi(\mathbf{r})$, immediately lead to the eigenfunctions $\Psi(\mathbf{r}) = \mathscr{L}(\mathbf{T}_\circ)\,\Phi(\mathbf{r})$ of the transformation $\mathscr{L}(\mathbf{T})$. Indeed,

$$\begin{aligned}\mathscr{L}(\mathbf{T})\Psi(\mathbf{r}) &= \mathscr{L}(\mathbf{T}_\circ)\,\mathscr{L}(\mathbf{N})\,\mathscr{L}^{-1}(\mathbf{T}_\circ)\,\Psi(\mathbf{r})\\ &= \mathscr{L}(\mathbf{T}_\circ)\,\mathscr{L}(\mathbf{N})\,\Phi(\mathbf{r}) = \mathscr{L}(\mathbf{T}_\circ)\,\mu\Phi(\mathbf{r}) = \mu\Psi(\mathbf{r}).\end{aligned}$$

The simpler the nucleus $\mathscr{L}(\mathbf{N})$, the simpler it is to find its eigenfunctions.

The different classes are defined by the possible distributions of the eigenvalues of the transformation matrix **T**. We note that if λ is an eigenvalue of a real symplectic matrix **T**, then λ^*, $1/\lambda$ and $1/\lambda^*$ are eigenvalues, too. Indeed, from the realness of **T**, we conclude that the characteristic equation $\det(\mathbf{T}-\lambda\mathbf{I}) = 0$ has real coefficients and that the eigenvalues are thus real or come in complex conjugated pairs: if λ is an eigenvalue, then λ^* is an eigenvalue, too. Moreover, from the symplecticity condition we get

$$\det(\mathbf{T}^{-1}-\lambda\mathbf{I}) = \det(\mathbf{J}\mathbf{T}^t\mathbf{J}-\lambda\mathbf{I}) = \det[\mathbf{J}(\mathbf{T}^t-\lambda\mathbf{I})\mathbf{J}) = \det(\mathbf{T}^t-\lambda\mathbf{I}) = \det(\mathbf{T}-\lambda\mathbf{I})$$

and we conclude that if λ is an eigenvalue, then $1/\lambda$ is an eigenvalue, too. So, for real symplectic matrices and $D \geq 2$, the eigenvalues come in complex quartets (if they are not unimodular and not real), or in complex conjugated pairs (if they are unimodular, but not real), or in real pairs (in particular: double if they are equal to $+1$ or -1). For $D = 1$, the two eigenvalues can of course only come as a single pair, either unimodular or real.

2.7.1 Nuclei for the One-dimensional Case

Let us first consider the one-dimensional case, in which the two eigenvalues follow from the characteristic equation $\lambda^2 - (a + d)\lambda + 1 = 0$ and three different distributions of the eigenvalues arise [23, 49]:

1. A pair of real eigenvalues σ and σ^{-1} ($\sigma \neq \pm 1$),
2. Two real eigenvalues $\lambda = 1$ or $\lambda = -1$, with only one eigenvector, and
3. A pair of unimodular, complex conjugated eigenvalues $\exp(i\gamma)$ and $\exp(-i\gamma)$.

The magnifier $\mathscr{M}(\sigma)$ is an obvious nucleus for class 1. Note that we can restrict ourselves to the case $\sigma > 0$; if the real eigenvalues are negative, we simply add an additional coordinate reverter $\mathscr{M}(-1)$ to the nucleus. The quadratic-phase modulator $\mathscr{Q}(-c)$ and the Fresnel transformer $\mathscr{S}(b)$ are obvious nuclei for class 2. Class 3 needs some more careful consideration, because the matrix of eigenvalues, $\mathbf{\Lambda}[\exp(\mathrm{i}\,\gamma), \exp(-\mathrm{i}\,\gamma)]$, is not symplectic and thus cannot act as the transformation matrix of a nucleus. A proper nucleus for this class might be the fractional Fourier transformer $\mathscr{F}(\gamma)$ [22].

For completeness we remark that for class 3—determined by the condition $|a + d| \leq 2$—the transformation $\mathscr{L}(\mathbf{T})$ can be decomposed as

$$\mathscr{L}(\mathbf{T}) = \mathscr{Q}(g)\,\mathscr{M}(s)\,\mathscr{F}(\gamma)\,\mathscr{M}(s^{-1})\,\mathscr{Q}(-g)\,, \tag{2.101}$$

where γ, s and g follow from $a+d = 2\cos\gamma$, $b = s\sin\gamma$ and $a-d = 2gb$. The opposite case, $|a+d| \geq 2$, corresponds in fact to class 1. To stress the symmetry of the two classes 1 and 3, we recall that for $|a+d| \geq 2$, the transformation $\mathscr{L}(\mathbf{T})$ can be decomposed as

$$\mathscr{L}(\mathbf{T}) = \mathscr{Q}(g)\,\mathscr{M}(s)\,\mathscr{H}(\gamma)\,\mathscr{M}(s^{-1})\,\mathscr{Q}(-g)\,, \tag{2.102}$$

where γ, s and g follow from $|a+d| = 2\cosh\gamma$, $b\,\mathrm{sgn}(a+d) = s\sinh\gamma$ and $a-d = 2gb$, and where $\mathscr{H}(\gamma) = \mathscr{F}(-\frac{1}{4}\pi)\,\mathscr{M}[\exp(\gamma)]\,\mathscr{F}(\frac{1}{4}\pi)$ is known as the *hyperbolic expander* [75, p. 183, Example: Hyperbolic expanders], see also Sect. 2.3.2. We conclude that not only the magnifier $\mathscr{M}(\sigma)$ but also the hyperbolic expander $\mathscr{H}(\ln\sigma)$ can be a proper nucleus for class 1 [22].

Class 2, for which $|a+d| = 2$, has been extensively studied in [49, Sect. IV]. If $a = d = \pm 1$ (and consequently $bc = 0$), the linear canonical transformer is obviously either an identity operator (for $b = c = 0$), or a quadratic-phase modulator $\mathscr{Q}(-c)$ (for $b = 0, c \neq 0$), or a Fresnel transformer $\mathscr{S}(b)$ (for $b \neq 0, c = 0$), possibly with an additional coordinate reversion (if $\lambda = -1$). For $a \neq d$ (and consequently $bc < 0$), the linear canonical transformation $\mathscr{L}(\mathbf{T})$ can always be decomposed with a quadratic-phase modulator or a Fresnel transformer as the nucleus $\mathscr{L}(\mathbf{N})$. For more details we refer to [23, Sect. 4] and [49, Sects. IV.D and IV.E].

2.7.2 Nuclei for the Two-dimensional Case

Concatenations of one-dimensional nuclei lead to *separable* two-dimensional nuclei. As a first example we mention the concatenation of a quadratic-phase modulator $\mathscr{Q}(-c_x)$ in the x direction and a fractional Fourier transformer $\mathscr{F}(\gamma_y)$ in the y direction. The corresponding two-dimensional nucleus has the transformation matrix

$$\begin{bmatrix} 1 & 0 \\ c_x & 1 \end{bmatrix} \oplus \begin{bmatrix} \cos\gamma_y & \sin\gamma_y \\ -\sin\gamma_y & \cos\gamma_y \end{bmatrix} = \begin{bmatrix} 1 & 0 & 0 & 0 \\ 0 & \cos\gamma_y & 0 & \sin\gamma_y \\ c_x & 0 & 1 & 0 \\ 0 & -\sin\gamma_y & 0 & \cos\gamma_y \end{bmatrix}.$$

As a second example we consider the class of phase-space rotators, described by their unitary matrix $\mathbf{U}$. It is well known that a unitary matrix has unimodular eigenvalues λ and can be diagonalized [36, Chap. 13]: $\mathbf{U} = \mathbf{P}\boldsymbol{\Lambda}\mathbf{P}^{-1}$. Moreover, it is not difficult to show that the eigenvalue $\exp(\mathrm{i}\,\gamma)$ of $\mathbf{U}$ corresponds to the complex conjugated pair of eigenvalues $\exp(\pm\mathrm{i}\,\gamma)$ of the symplectic matrix $\mathbf{T}$. The matrix $\mathbf{P}$ that diagonalizes the unitary matrix $\mathbf{U}$ can itself be made unitary and then corresponds to a symplectic matrix that diagonalizes the symplectic matrix $\mathbf{T}$. We thus conclude that the separable fractional Fourier transformer $\mathscr{F}(\gamma_x, \gamma_y)$, which is clearly a concatenation of two one-dimensional fractional Fourier transformers with transformation matrix

$$\begin{bmatrix} \cos\gamma_x & \sin\gamma_x \\ -\sin\gamma_x & \cos\gamma_x \end{bmatrix} \oplus \begin{bmatrix} \cos\gamma_y & \sin\gamma_y \\ -\sin\gamma_y & \cos\gamma_y \end{bmatrix} = \begin{bmatrix} \cos\gamma_x & 0 & \sin\gamma_x & 0 \\ 0 & \cos\gamma_y & 0 & \sin\gamma_y \\ -\sin\gamma_x & 0 & \cos\gamma_x & 0 \\ 0 & -\sin\gamma_y & 0 & \cos\gamma_y \end{bmatrix},$$

is an obvious nucleus of a phase-space rotator [22]. We recall that we already met decompositions based on $\mathbf{U} = \mathbf{P\Lambda P}^{-1}$ in Table 2.6, where we find such cascades as $\mathscr{G}(\gamma) = \mathscr{R}(-\frac{1}{4}\pi)\,\mathscr{F}(\gamma,-\gamma)\,\mathscr{R}(\frac{1}{4}\pi)$ and $\mathscr{R}(\gamma) = \mathscr{G}(\frac{1}{4}\pi)\,\mathscr{F}(\gamma,-\gamma)\,\mathscr{G}(-\frac{1}{4}\pi)$. From these cascades we conclude that the antisymmetric separable fractional Fourier transformer $\mathscr{F}(\gamma,-\gamma)$ can be decomposed as $\mathscr{R}(\frac{1}{4}\pi)\,\mathscr{G}(\gamma)\,\mathscr{R}(-\frac{1}{4}\pi)$ and as $\mathscr{G}(-\frac{1}{4}\pi)\,\mathscr{R}(\gamma)\,\mathscr{G}(\frac{1}{4}\pi)$, and that the gyrator and the rotator can thus also act as a nucleus in the special case that $\gamma_x = -\gamma_y$.

The four additional—inherently two-dimensional—classes correspond to the four possible eigenvalue distributions that can only occur in two dimensions [23, Sect. 5]:

4. A complex quartet of eigenvalues $\sigma\exp(\mathrm{i}\,\gamma)$, $\sigma\exp(-\mathrm{i}\,\gamma)$, $\sigma^{-1}\exp(-\mathrm{i}\,\gamma)$ and $\sigma^{-1}\exp(\mathrm{i}\,\gamma)$ $(\sigma \neq \pm 1)$,
5. Two identical pairs of unimodular, complex conjugated eigenvalues $\exp(\mathrm{i}\,\gamma)$ and $\exp(-\mathrm{i}\,\gamma)$, with only two linearly independent eigenvectors,
6. Two identical pairs of real eigenvalues σ and σ^{-1} $(\sigma \neq \pm 1)$, with only two linearly independent eigenvectors, and
7. Four real eigenvalues $\lambda = 1$ or $\lambda = -1$, with only one eigenvector.

We mention possible nuclei for these four classes.

A possible nucleus for class 4 is the (commuting) combination of a rotator and a magnifier $\mathscr{M}(\sigma\mathbf{I})\,\mathscr{R}(\gamma) = \mathscr{R}(\gamma)\,\mathscr{M}(\sigma\mathbf{I})$ with transformation matrix $[\sigma\,\mathbf{U}_r(\gamma), \mathbf{0}; \mathbf{0}, \sigma^{-1}\mathbf{U}_r(\gamma)]$. The input–output relation for such a nucleus reads

$$\sigma f_o(\sigma x, \sigma y) = f_i(x\cos\gamma - y\sin\gamma, x\sin\gamma + y\cos\gamma) \tag{2.103}$$

or in polar coordinates (with $x = r\cos\varphi$ and $y = r\sin\varphi$): $\sigma f_o(\sigma r, \varphi) = f_i(r, \varphi+\gamma)$. Like for class 1, an alternative nucleus for class 4 is the combination of a hyperbolic expander (instead of a magnifier) and a rotator.

A possible nucleus for class 5 is the (commuting) combination of a rotator and a quadratic-phase modulator $\mathscr{Q}(-c\mathbf{I})\,\mathscr{R}(\gamma) = \mathscr{R}(\gamma)\,\mathscr{Q}(-c\mathbf{I})$ with transformation matrix $[\mathbf{U}_r(\gamma), \mathbf{0}; c\,\mathbf{U}_r(\gamma), \mathbf{U}_r(\gamma)]$. The input–output relation for this nucleus reads

$$f_o(x, y) = f_i(x\cos\gamma - y\sin\gamma, x\sin\gamma + y\cos\gamma)\,\exp[\mathrm{i}\,\pi\,c(x^2 + y^2)] \tag{2.104}$$

or in polar coordinates again: $f_o(r,\varphi) = f_i(r,\varphi+\gamma)\,\exp(\mathrm{i}\,\pi\,cr^2)$. Like for class 2, an alternative nucleus for class 5 is the combination of a Fresnel transformer (instead of a quadratic-phase modulator) and a rotator.

For the remaining two classes we need a new element, the *shearer* $\mathscr{Z}$ (to be defined shortly). A possible nucleus for class 6 is then the (commuting) shearer-magnifier combination $\mathscr{M}(\sigma\mathbf{I})\mathscr{Z} = \mathscr{Z}\mathscr{M}(\sigma\mathbf{I})$ with transformation matrix

$$\begin{bmatrix} \sigma\,\mathbf{Z}_+ & \mathbf{0} \\ \mathbf{0} & \sigma^{-1}\mathbf{Z}_- \end{bmatrix}, \quad \text{where} \quad \mathbf{Z}_+ = \begin{bmatrix} 1 & 1 \\ 0 & 1 \end{bmatrix} \quad \text{and} \quad \mathbf{Z}_- = \begin{bmatrix} 1 & 0 \\ -1 & 1 \end{bmatrix}; \tag{2.105}$$

the shearer itself results for $\sigma = 1$. The input–output relation for this nucleus reads

$$\sigma\, f_o(\sigma x, \sigma y) = f_i(x - y, y)\,, \tag{2.106}$$

which represents—apart from a magnification with σ—a simple shearing of the x coordinate.

A possible nucleus for the final class 7 is the (non-commuting!) combination of a shearer and a quadratic-phase modulator $\mathscr{Q}(-c\mathbf{I})\mathscr{Z}$ with transformation matrix $[\mathbf{Z}_+, \mathbf{0}; c\,\mathbf{Z}_+, \mathbf{Z}_-]$. The input–output relation for this nucleus reads

$$f_o(x,y) = f_i(x-y,y)\,\exp[\mathrm{i}\,\pi\,c(x^2+y^2)]\,. \tag{2.107}$$

Again, an alternative nucleus is the combination of a Fresnel transformer (instead of a quadratic-phase modulator) and a shearer.

The results of this section have been combined in Table 2.8, where for each of the seven classes the corresponding nucleus can be extracted. Obvious concatenations of one-dimensional nuclei have not been stated explicitly. Two examples in reading the Table: (1) the nucleus for the (one-dimensional) class 3, i.e., a pair of unimodular, complex conjugated eigenvalues $\exp(\pm\mathrm{i}\,\gamma)$, is a fractional Fourier transformer $\mathscr{F}(\gamma)$. (2) possible nuclei for the (two-dimensional) class 7, i.e., four

Table 2.8 Seven classes of eigenvalue distributions λ for linear canonical transformers and their corresponding nuclei, composed of magnifiers $\mathscr{M}(\sigma)$ or hyperbolic expanders $\mathscr{H}(\ln\sigma)$, quadratic-phase modulators $\mathscr{Q}(\cdot)$ or Fresnel transformers $\mathscr{S}(\cdot)$, fractional Fourier transformers $\mathscr{F}(\gamma)$, rotators $\mathscr{R}(\gamma)$, and shearers $\mathscr{Z}$

λ	$\mathscr{M}(\sigma)$ or $\mathscr{H}(\ln\sigma)$	$\mathscr{Q}(\cdot)$ or $\mathscr{S}(\cdot)$	$\mathscr{F}(\gamma)$
	Class 1 σ, σ^{-1}	Class 2 $1, 1$	Class 3 $e^{i\gamma}, e^{-i\gamma}$
$\mathscr{R}(\gamma)$	Class 4 $\sigma e^{i\gamma}, \sigma e^{-i\gamma}, \sigma^{-1}e^{-i\gamma}, \sigma^{-1}e^{i\gamma}$	Class 5 $e^{i\gamma}, e^{i\gamma}, e^{-i\gamma}, e^{-i\gamma}$	
$\mathscr{Z}$	Class 6 $\sigma, \sigma, \sigma^{-1}, \sigma^{-1}$	Class 7 $1, 1, 1, 1$ $\mathscr{Z}$ and $\{\mathscr{Q}, \mathscr{S}\}$ do not commute !	

eigenvalues equal to 1 and with only one eigenvector, are a shearer $\mathscr{X}$ followed by a quadratic-phase modulator $\mathscr{Q}(\cdot)$ or followed by a Fresnel transformer $\mathscr{S}(\cdot)$, and the two subsystems do not commute.

2.7.3 The Search for Eigenfunctions

The search for eigenfunctions $\Psi(\mathbf{r})$ of linear canonical transformations now reduces to the search for eigenfunctions $\Phi(\mathbf{r})$ of the simple nuclei. The Dirac delta function $\delta(x-\xi)$ is an eigenfunction for any multiplication operator, with eigenvalue $\exp(\mathrm{i}\,\pi\, cx^2)$ in the particular case of a quadratic-phase modulator $\mathscr{Q}(-c)$; and the harmonic signal $\exp(\mathrm{i}\,2\pi\, ux)$ is an eigenfunction for any convolution operator, with eigenvalue $\exp(\mathrm{i}\,\pi\, bu^2)$ in the particular case of a Fresnel transformer $\mathscr{S}(b)$ (class 2). We also recall that the Hermite–Gauss modes (2.80) are eigenfunctions of the fractional Fourier transformer $\mathscr{F}(\gamma)$, with eigenvalues $\exp(-\mathrm{i}\,k\gamma)$ (class 3). And while powers x^k are evidently eigenfunctions of the nucleus $\mathscr{M}(\sigma)$ with eigenvalues $|\sigma|^{-1/2}\sigma^{-k}$ (class 1), signals of the form $r^k\exp(\mathrm{i}\,m\varphi)$ are eigenfunctions of the nucleus $\mathscr{M}(\sigma\mathbf{I})\,\mathscr{R}(\gamma)$ with eigenvalues $\sigma^{-1/2}\sigma^{-k}\exp(\mathrm{i}\,m\gamma)$ (class 4), and signals of the form $\delta(r-\rho)\exp(\mathrm{i}\,m\varphi)$ are eigenfunctions of the nucleus $\mathscr{Q}(-c\mathbf{I})\,\mathscr{R}(\gamma)$ with eigenvalues $\exp(\mathrm{i}\,m\gamma)\exp(\mathrm{i}\,\pi\, c\rho^2)$ (class 5). Proper eigenfunctions of the nuclei $\mathscr{M}(\sigma\mathbf{I})\,\mathscr{X}$ and $\mathscr{Q}(-c\mathbf{I})\,\mathscr{X}$ for the classes 6 and 7 are still to be found.

From the eigenfunctions $\Phi(\mathbf{r})$ for a nucleus $\mathscr{L}(\mathbf{N})$, we can generate eigenfunctions $\Psi(\mathbf{r})$ for the corresponding class of transformations $\mathscr{L}(\mathbf{T}_\circ)\,\mathscr{L}(\mathbf{N})\,\mathscr{L}^{-1}(\mathbf{T}_\circ)$ by letting the eigenfunctions $\Phi(\mathbf{r})$ propagate through $\mathscr{L}(\mathbf{T}_\circ)$: $\mathscr{L}(\mathbf{T}_\circ)\,\Phi(\mathbf{r}) = \Psi(\mathbf{r})$. We already met the Hermite–Gaussian-type modes $\mathsf{H}^{\mathbf{T}_\circ}_{m,n}(\mathbf{r})$ for class 3, see Sect. 2.6.3. As an example, we will find eigenfunctions for the one-dimensional hyperbolic expander $\mathscr{H}(\gamma)$ (class 1), see [24].

Let us start with the powers x^k, which are eigenfunctions of the magnifier $\mathscr{M}(\cdot)$, and recall that $H(\gamma) = \mathscr{F}(-\frac{1}{4}\pi)\,\mathscr{M}(\exp\gamma)\,\mathscr{F}(\frac{1}{4}\pi)$. We thus have to calculate the integral

$$f_o(x_o) = \mathscr{F}\left(-\tfrac{1}{4}\pi\right)x_i^k = \mathrm{i}^{\,1/2}2^{-1/4}\int x_i^k\exp[-\mathrm{i}\,\pi(x_o^2 - 2\sqrt{2}x_ox_i + x_i^2)]\,\mathrm{d}x_i\,, \tag{2.108}$$

for which we use the relationships

$$(-\mathrm{i}\,2\pi)^k\sqrt{p}\int x^k\exp(-\pi\, px^2 - \mathrm{i}\,2\pi\, ux)\,\mathrm{d}x = \frac{\mathrm{d}^k}{\mathrm{d}u^k}\exp(-\pi\, p^{-1}u^2) \tag{2.109}$$

and

$$\frac{\mathrm{d}^k}{\mathrm{d}u^k}\exp(-\pi\, p^{-1}u^2) = (-\sqrt{\pi\, p^{-1}})^k\exp(-\pi\, p^{-1}u^2)\,H_k(u\sqrt{\pi\, p^{-1}})\,, \tag{2.110}$$

in which we substitute $p = \mathrm{i}$, $x = x_i$ and $u = -x_o\sqrt{2}$. Note that Eq. (2.109) follows by differentiating the Fourier transform $\exp(-\pi p^{-1}u^2)$ of $\exp(-\pi p x^2)$, and holds for $\Re p > 0$ and for $[\Re p = 0, \Im p \neq 0]$, see, for instance, [50, (2.3.15.4)] and also [75, Sect. C.2, p. 279, Remark: The integral of complex Gaussians]; we may also refer to [50, (2.5.22.5)] and [34, (3.691.5) and (3.691.7)] for $k = 0$, and to [50, (2.5.22.3)] and [34, (3.851.1) and (3.851.3)] for $k = 1$. Equation (2.110) can be found, for instance, in [2, (7.1.19)]; see also Rodrigues' formula for Hermite polynomials [2, (22.11.7)]. We readily conclude that $f_o(x)$ is proportional to $H_k(\sqrt{2\pi}x\,\mathrm{i}^{-1/2})\exp(\mathrm{i}\pi x^2)$.

We note the remarkable resemblance between these eigenfunctions and the eigenfunctions of the fractional Fourier transformer—i.e., the Hermite–Gauss modes $\mathsf{H}_k(x)$, which are proportional to $H_k(\sqrt{2\pi}x)\exp(-\pi x^2)$—and we conclude that we can directly go from the fractional-Fourier-transformer case to the hyperbolic-expander case by simply replacing x by $(x\,\mathrm{i}^{-1/2})$.

2.8 The Effect of a Linear Canonical Transformation on the Second-order Moments in Phase Space

With $E = \iint W(\mathbf{r},\mathbf{q})\,\mathrm{d}\mathbf{r}\,\mathrm{d}\mathbf{q}$ denoting the total energy of a signal, the *normalized second-order moments* of its Wigner distribution are defined as

$$\frac{1}{E}\iint \begin{bmatrix}\mathbf{r}\\ \mathbf{q}\end{bmatrix}[\mathbf{r}^t,\mathbf{q}^t]\,W(\mathbf{r},\mathbf{q})\,\mathrm{d}\mathbf{r}\,\mathrm{d}\mathbf{q} \equiv \begin{bmatrix}\mathbf{M}_{\mathbf{rr}} & \mathbf{M}_{\mathbf{rq}}\\ \mathbf{M}_{\mathbf{rq}}^t & \mathbf{M}_{\mathbf{qq}}\end{bmatrix} \equiv \mathbf{M} \tag{2.111}$$

and constitute a real positive-definite symmetric moment matrix $\mathbf{M}$. It can easily be shown that when a signal undergoes a linear canonical transformation, $f_o(\mathbf{r}) = \mathscr{L}(\mathbf{T})f_i(\mathbf{r})$, the moment matrices $\mathbf{M}_i$ and $\mathbf{M}_o$ are related by the relationship [12, 59]

$$\mathbf{M}_o = \mathbf{T}\mathbf{M}_i\mathbf{T}^t\,. \tag{2.112}$$

We can easily prove the positive definiteness of the $2D \times 2D$ moment matrix $\mathbf{M}$ with the help of the input–output relationship (2.112). We therefore construct the transformation matrix $\mathbf{T}$ as follows:

- We choose $\mathbf{C} = \mathbf{0}$, with the immediate consequence $\mathbf{D} = \mathbf{A}^{t-1}$ and $\mathbf{A}\mathbf{B}^t = \mathbf{B}\mathbf{A}^t$ to satisfy the symplecticity condition.
- The matrix $\mathbf{A}$ is chosen as an upper triangular matrix with $\frac{1}{2}D(D+1)$ non-vanishing entries that can be chosen arbitrarily.
- The matrix $\mathbf{B}$ is chosen as an upper triangular matrix with D arbitrarily chosen entries in its top row.
- The remaining $\frac{1}{2}D(D-1)$ non-vanishing entries of $\mathbf{B}$ are determined from the $\frac{1}{2}D(D-1)$ equations that follow from the required symmetry of the matrix $\mathbf{A}\mathbf{B}^t$.

The top row $\mathbf{t}^t$ of the transformation matrix $\mathbf{T}$ can thus be constructed completely arbitrarily. We now consider the upper left entry $m_{xx,o}$ of the matrix $\mathbf{M}_o$ in the left-hand side of (2.112); this entry, which represents the square of an effective width, is positive:

$$(m_{xx})_o = \frac{1}{E}\int x^2 \left[\int W_o(\mathbf{r},\mathbf{q})\,\mathrm{d}\mathbf{q}\right]\mathrm{d}\mathbf{r} = \frac{1}{E}\int x^2\,|f_o(\mathbf{r})|^2\,\mathrm{d}\mathbf{r} > 0\,.$$

On the other hand, this entry equals $\mathbf{t}^t\mathbf{M}_i\mathbf{t}$, where the vector $\mathbf{t}$ can be chosen arbitrarily. We thus conclude that the quadratic form $\mathbf{t}^t\mathbf{M}_i\mathbf{t}$ is positive for any vector $\mathbf{t}$, with which we have proved that the moment matrix $\mathbf{M}_i$ is positive definite.

The moment matrix $2\pi\,\mathbf{M}$ can be represented in the form [18, Sect. 2.6, Second- and higher-order moments]

$$2\pi\,\mathbf{M} = 2\pi\begin{bmatrix}\mathbf{M}_{\mathbf{rr}} & \mathbf{M}_{\mathbf{rq}}\\ \mathbf{M}_{\mathbf{rq}}^t & \mathbf{M}_{\mathbf{qq}}\end{bmatrix} = \frac{1}{2}\begin{bmatrix}\mathbf{G}_1^{-1} & \mathbf{G}_1^{-1}\mathbf{H}\\ \mathbf{H}^t\mathbf{G}_1^{-1} & \mathbf{G}_2+\mathbf{H}^t\mathbf{G}_1^{-1}\mathbf{H}\end{bmatrix}, \tag{2.113}$$

where the matrices $\mathbf{G}_1$, $\mathbf{G}_2$ and $\mathbf{H}$ follow directly from the submatrices $\mathbf{M}_{\mathbf{rr}}$, $\mathbf{M}_{\mathbf{rq}}$ and $\mathbf{M}_{\mathbf{qq}}$:

$$\mathbf{G}_1 = (4\pi)^{-1}\mathbf{M}_{\mathbf{rr}}^{-1} = \mathbf{G}_1^t \tag{2.114a}$$

$$\mathbf{G}_2 = 4\pi\,(\mathbf{M}_{\mathbf{qq}} - \mathbf{M}_{\mathbf{rq}}^t\mathbf{M}_{\mathbf{rr}}^{-1}\mathbf{M}_{\mathbf{rq}}) = \mathbf{G}_2^t\,, \tag{2.114b}$$

$$\mathbf{H} = \mathbf{M}_{\mathbf{rr}}^{-1}\mathbf{M}_{\mathbf{rq}}\,. \tag{2.114c}$$

The matrices $\mathbf{G}_1$ and $\mathbf{G}_2$ are positive definite, which follows immediately from the positive definiteness of the quadratic form $(\mathbf{q}+\mathbf{H}\mathbf{r})^t\mathbf{G}_1^{-1}(\mathbf{q}+\mathbf{H}\mathbf{r}) + \mathbf{r}^t\mathbf{G}_2\mathbf{r} = 2[\mathbf{q}^t,\mathbf{r}^t]\,\mathbf{M}\,[\mathbf{q}^t,\mathbf{r}^t]^t$, and a possible asymmetry of the matrix $\mathbf{H}$ is responsible for the *twist* of the signal [10, 17, 29, 54, 56, 60, 61, 63]. We will study the twist later in Sect. 2.8.2.

In the one-dimensional case, the twist is irrelevant and the 2×2 moment matrix takes the form

$$\mathbf{M} = \frac{1}{4\pi}\begin{bmatrix}g_1^{-1} & hg_1^{-1}\\ hg_1^{-1} & g_2+h^2g_1^{-1}\end{bmatrix} = \frac{1}{4\pi\sigma}\begin{bmatrix}g^{-1} & hg^{-1}\\ hg^{-1} & g+h^2g^{-1}\end{bmatrix}, \tag{2.115}$$

with $g = \sqrt{g_1g_2} > 0$ and $\sigma = \sqrt{g_1/g_2} > 0$. Note that $\det\mathbf{M} = (4\pi\sigma)^{-2}$ and that σ is bounded by 1, $\sigma \le 1$, as a result of the uncertainty relation $m_{xx}m_{uu} \ge (4\pi)^{-2}$.

2.8.1 Moment Invariants for the Linear Canonical Transformation

Using the symplecticity condition $\mathbf{T}^{-1} = \mathbf{J}\mathbf{T}^t\mathbf{J}$, the moment relation (2.112) can be rewritten in the form of the similarity relation $\mathbf{M}_o\mathbf{J} = \mathbf{T}\mathbf{M}_i\mathbf{J}\mathbf{T}^{-1}$ [14], from which we conclude that the (real!) eigenvalues of the matrix $\mathbf{MJ}$ are invariant under a linear canonical transformation. The same holds, of course, for the coefficients of the characteristic equation $\det(\mathbf{MJ} - \lambda\mathbf{I}) = 0$, which appears to be an equation in λ^2. In the one-dimensional case this equation reads $\lambda^2 - \det\mathbf{M} = 0$, from which we conclude that $\det\mathbf{M}$ is the (only) invariant. In the two-dimensional case we have

$$\lambda^4 - [(m_{xx}m_{uu} - m_{xu}^2) + (m_{yy}m_{vv} - m_{yv}^2) + 2(m_{xy}m_{uv} - m_{xv}m_{yu})]\lambda^2 + \det\mathbf{M} = 0 \tag{2.116}$$

and we thus find the two (independent) invariants

$$I_1 = \sqrt{\det\mathbf{M}}\,, \tag{2.117a}$$

$$I_2 = (m_{xx}m_{uu} - m_{xu}^2) + (m_{yy}m_{vv} - m_{yv}^2) + 2(m_{xy}m_{uv} - m_{xv}m_{yu}). \tag{2.117b}$$

The latter moment combination is known in optics as the beam quality parameter [52]. Instead of $I_1^2 = \lambda_x^2\lambda_y^2$ and $I_2 = \lambda_x^2 + \lambda_y^2$, we might as well consider the eigenvalues $\pm\lambda_{x,y}$ of $\mathbf{MJ}$ themselves as invariants. We may arbitrarily choose $\lambda_x \geq \lambda_y > 0$, in which case $\lambda_x \pm \lambda_y = (I_2 \pm 2I_1)^{1/2}$. Note that $I_2 \geq 2I_1$ and that the equality sign arises for $\lambda_x = \lambda_y$.

In the special case of phase-space rotators, for which the symplectic transformation matrix is also orthogonal, $\mathbf{T}^t = \mathbf{T}^{-1}$, the relation $\mathbf{M}_o = \mathbf{T}\mathbf{M}_i\mathbf{T}^t = \mathbf{T}\mathbf{M}_i\mathbf{T}^{-1}$ between the moment matrices themselves takes the form of a similarity transformation, and we conclude that the eigenvalues of $\mathbf{M}$ (or the coefficients of its characteristic equation, like the determinant and the trace of $\mathbf{M}$) are invariant. Note, however, that some of these invariants are not new in the sense that they are identical to or depend on the ones that we already found.

Another way to find moment invariants for phase-space rotators is to consider the Hermitian matrix [8]

$$\mathbf{M}' = \frac{1}{E}\iint (\mathbf{r}-\mathrm{i}\,\mathbf{q})\,(\mathbf{r}-\mathrm{i}\,\mathbf{q})^\dagger\, W(\mathbf{r},\mathbf{q})\,\mathrm{d}\mathbf{r}\,\mathrm{d}\mathbf{q} = \mathbf{M}_{\mathbf{rr}} + \mathbf{M}_{\mathbf{qq}} + \mathrm{i}\,(\mathbf{M}_{\mathbf{rq}} - \mathbf{M}_{\mathbf{rq}}^t) \tag{2.118}$$

and to use (2.58) to get the relation $\mathbf{M}'_o = \mathbf{U}\mathbf{M}'_i\mathbf{U}^\dagger = \mathbf{U}\mathbf{M}'_i\mathbf{U}^{-1}$, which is again a similarity transformation. In the two-dimensional case, the matrix $\mathbf{M}'$ can be written as

$$\mathbf{M}' = \begin{bmatrix} Q_0 + Q_1 & Q_2 + \mathrm{i}\,Q_3 \\ Q_2 - \mathrm{i}\,Q_3 & Q_0 - Q_1 \end{bmatrix} = Q_0\mathbf{I} + Q\begin{bmatrix} \cos\vartheta & \exp(\mathrm{i}\,\gamma)\sin\vartheta \\ \exp(-\mathrm{i}\,\gamma)\sin\vartheta & -\cos\vartheta \end{bmatrix} \tag{2.119}$$

$$\text{with}\quad \begin{aligned} Q_0 &= \tfrac{1}{2}[(m_{xx}+m_{uu})+(m_{yy}+m_{vv})]\,,\\ Q_1 &= \tfrac{1}{2}[(m_{xx}+m_{uu})-(m_{yy}+m_{vv})] = Q\cos\vartheta\,,\\ Q_2 &= m_{xy}+m_{uv} = Q\sin\vartheta\,\cos\gamma\,,\\ Q_3 &= m_{xv}-m_{yu} = Q\sin\vartheta\,\sin\gamma\,, \end{aligned} \tag{2.120}$$

where the moment parameters Q_j ($j = 0, 1, 2, 3$) are the expectation values of the Hermitian operators [58, 60, 62] associated with the symmetric and antisymmetric fractional Fourier transformer, the gyrator, and the rotator, respectively; moreover, the matrix $\mathbf{M}'$ resembles the one introduced in [3], which is based on the operator approach. We then find two invariants from the coefficients of the characteristic equation $\det(\mathbf{M}' - \nu\mathbf{I}) = 0 = \nu^2 - 2Q_0\nu + Q_0^2 - Q^2 = (\nu - Q_0)^2 - Q^2$: the two parameters Q_0 and $Q^2 = Q_1^2 + Q_2^2 + Q_3^2$ or the two eigenvalues $\nu_{1,2} = Q_0 \pm Q$. Note that $2Q_0$, the trace of $\mathbf{M}'$, is also the trace of $\mathbf{M}$, and that $Q_0^2 - Q^2$ equals the determinant of $\mathbf{M}'$.

From the invariance of Q we conclude that the three-dimensional vector $(Q_1, Q_2, Q_3) = (Q\cos\vartheta, Q\sin\vartheta\cos\gamma, Q\sin\vartheta\sin\gamma)$ lives on a sphere with radius Q, known as the Poincaré sphere [3, 8, 48]. A phase-space rotator will only change the values of the angles ϑ and γ, but does not change the invariants Q_0 and Q. To transform a diagonal matrix $\mathbf{M}'$, with $\gamma = \vartheta = 0$, into the general form (2.119), we can use, for instance, the cascade $\mathscr{F}(\frac{1}{2}\gamma, -\frac{1}{2}\gamma)\,\mathscr{R}(-\frac{1}{2}\vartheta)\,\mathscr{F}(-\frac{1}{2}\gamma, \frac{1}{2}\gamma)$; we easily verify

$$\begin{aligned} &\mathbf{U}_f\left(\tfrac{1}{2}\gamma, -\tfrac{1}{2}\gamma\right)\,\mathbf{U}_r\left(-\tfrac{1}{2}\vartheta\right)\,\mathbf{U}_f\left(-\tfrac{1}{2}\gamma, \tfrac{1}{2}\gamma\right)\begin{bmatrix} Q_0+Q & 0\\ 0 & Q_0-Q\end{bmatrix}\mathbf{U}_f\left(\tfrac{1}{2}\gamma, -\tfrac{1}{2}\gamma\right)\\ &\quad\times\mathbf{U}_f\left(-\tfrac{1}{2}\gamma, \tfrac{1}{2}\gamma\right) = \begin{bmatrix} Q_0+Q\cos\vartheta & Q\exp(\mathrm{i}\,\gamma)\sin\vartheta\\ Q\exp(-\mathrm{i}\,\gamma)\sin\vartheta & Q_0-Q\cos\vartheta\end{bmatrix}. \end{aligned} \tag{2.121}$$

In the special case that the phase-space rotator is a symmetric fractional Fourier transformer, with a scalar matrix $\mathbf{U}$, the matrix $\mathbf{M}'$ itself is invariant, and so is the complete vector (Q_1, Q_2, Q_3). For the (antisymmetric) fractional Fourier transformer, the gyrator, and the rotator, one of the moment parameters Q_j ($j = 1, 2, 3$) is invariant, while the other two undergo a rotation-type transformation, see Table 2.9.

2.8.2 *The Twist As an Invariant for Transformations with* $\mathbf{B} = \mathbf{0}$

We will now consider an important parameter for two-dimensional signals that is invariant under a linear canonical transformation with $\mathbf{B} = \mathbf{0}$ (and $\mathbf{A}^{-1} = \mathbf{D}^t$), see (2.46). The invariant that we consider is known as the *twist* of the signal, which is

Table 2.9 Second-order moment invariants for linear canonical transformations

Transformation	Invariant	Remark
$\mathscr{L}(\mathbf{T})$	$I_1 = (\det \mathbf{M})^{1/2} = \lambda_x \lambda_y$	$\pm\, \lambda_{x,y}$ are the eigenvalues of $\mathbf{MJ}$
	$I_2 = (m_{xx}m_{uu} - m_{xu}^2) + (m_{yy}m_{vv} - m_{yv}^2) + 2(m_{xy}m_{uv} - m_{xv}m_{yu}) = \lambda_x^2 + \lambda_y^2$	
$\mathscr{L}(\mathbf{T})\ \mathbf{B} = \mathbf{0}$	$T = [(m_{xu} - m_{yv})m_{xy} + m_{xv}m_{yy} - m_{xx}m_{yu}]/(m_{xx}m_{yy} - m_{xy}^2)^{1/2}$	
$\mathscr{O}(\mathbf{U})$	$Q_0 = \frac{1}{2}[(m_{xx} + m_{uu}) + (m_{yy} + m_{vv})]$	
	$Q^2 = Q_1^2 + Q_2^2 + Q_3^2$	
$\mathscr{F}(\varphi, -\varphi)$	$Q_1 = \frac{1}{2}[(m_{xx} + m_{uu}) - (m_{yy} + m_{vv})]$	$(Q_2 + \mathrm{i}\, Q_3)_o = \exp(\mathrm{i}\, 2\varphi)\, (Q_2 + \mathrm{i}\, Q_3)_i$
$\mathscr{G}(\varphi)$	$Q_2 = m_{xy} + m_{uv}$	$(Q_3 + \mathrm{i}\, Q_1)_o = \exp(\mathrm{i}\, 2\varphi)\, (Q_3 + \mathrm{i}\, Q_1)_i$
$\mathscr{R}(-\varphi)$	$Q_3 = m_{xv} - m_{yu}$	$(Q_1 + \mathrm{i}\, Q_2)_o = \exp(\mathrm{i}\, 2\varphi)\, (Q_1 + \mathrm{i}\, Q_2)_i$
$\mathscr{F}(\gamma, \gamma)$	Q_1, Q_2, Q_3	

connected to the asymmetry of the matrix $\mathbf{H} = \mathbf{M}_{\mathbf{rr}}^{-1}\mathbf{M}_{\mathbf{rq}}$, see (2.114c). To measure the degree of twist, we use the asymmetry of the normalized matrix

$$\mathbf{M}_{\mathbf{rr}}^{1/2}(\mathbf{M}_{\mathbf{rr}}^{-1}\mathbf{M}_{\mathbf{rq}})\mathbf{M}_{\mathbf{rr}}^{1/2} = \mathbf{M}_{\mathbf{rr}}^{-1/2}\mathbf{M}_{\mathbf{rq}}\mathbf{M}_{\mathbf{rr}}^{1/2} = \mathbf{M}_{\mathbf{rr}}^{-1/2}(\mathbf{M}_{\mathbf{rq}}\mathbf{M}_{\mathbf{rr}})\mathbf{M}_{\mathbf{rr}}^{-1/2}$$

and we define the twist parameter T via the skew-symmetric matrix

$$\frac{\mathbf{M}_{\mathbf{rq}}\mathbf{M}_{\mathbf{rr}} - (\mathbf{M}_{\mathbf{rq}}\mathbf{M}_{\mathbf{rr}})^t}{(\det \mathbf{M}_{\mathbf{rr}})^{1/2}} \equiv \begin{bmatrix} 0 & T \\ -T & 0 \end{bmatrix};$$

hence

$$T = \frac{(m_{xu} - m_{yv})m_{xy} + m_{xv}m_{yy} - m_{xx}m_{yu}}{(m_{xx}m_{yy} - m_{xy}^2)^{1/2}} . \tag{2.122}$$

Note that the numerator in the above expression corresponds to the asymmetry of $\mathbf{M}_{\mathbf{rq}}\mathbf{M}_{\mathbf{rr}}$, i.e., to the upper off-diagonal element of $\mathbf{M}_{\mathbf{rq}}\mathbf{M}_{\mathbf{rr}} - (\mathbf{M}_{\mathbf{rq}}\mathbf{M}_{\mathbf{rr}})^t$. Using the input–output relation $\mathbf{M}_o = \mathbf{T}\mathbf{M}_i\mathbf{T}^t$, we easily derive

$$(\mathbf{M}_{\mathbf{rr}})_o = \mathbf{A}(\mathbf{M}_{\mathbf{rr}})_i\, \mathbf{A}^t , \tag{2.123a}$$

$$(\mathbf{M}_{\mathbf{rq}})_o = \mathbf{A}(\mathbf{M}_{\mathbf{rr}})_i\, \mathbf{C}^t + \mathbf{A}(\mathbf{M}_{\mathbf{rq}})_i\, \mathbf{A}^{-1} , \tag{2.123b}$$

$$(\mathbf{M}_{\mathbf{rq}}\mathbf{M}_{\mathbf{rr}})_o = \mathbf{A}(\mathbf{M}_{\mathbf{rr}})_i\, \mathbf{C}^t\mathbf{A}(\mathbf{M}_{\mathbf{rr}})_i\, \mathbf{A}^t + \mathbf{A}(\mathbf{M}_{\mathbf{rq}}\mathbf{M}_{\mathbf{rr}})_i\, \mathbf{A}^t . \tag{2.123c}$$

From (2.123a) we see that

$$(\det \mathbf{M}_{\mathbf{rr}})_o^{1/2} = (\det \mathbf{M}_{\mathbf{rr}})_i^{1/2} \det \mathbf{A} .$$

As the matrix $\mathbf{A}(\mathbf{M}_{\mathbf{rr}})_i\, \mathbf{C}^t\mathbf{A}(\mathbf{M}_{\mathbf{rr}})_i\, \mathbf{A}^t$ in (2.123c) is symmetric, the asymmetry of $(\mathbf{M}_{\mathbf{rq}}\mathbf{M}_{\mathbf{rr}})_o$ is equal to the asymmetry of $\mathbf{A}(\mathbf{M}_{\mathbf{rq}}\mathbf{M}_{\mathbf{rr}})_i\, \mathbf{A}^t$, which on its turn is equal to the asymmetry of $(\mathbf{M}_{\mathbf{rq}}\mathbf{M}_{\mathbf{rr}})_i\ \det \mathbf{A}$. Both the numerator and the denominator in the expression for T, see (2.122), scale with the same factor $\det \mathbf{A}$ and we thus conclude that the twist is invariant under a linear canonical transformation with $\mathbf{B} = \mathbf{0}$.

2.8.3 Williamson's Theorem, Canonical Form and the Twist

An interesting property follows from *Williamson's theorem* [25, 62, 70]: for any real positive-definite symmetric matrix $\mathbf{M}$, there exists a real symplectic matrix $\mathbf{T}_\circ$ such that $\mathbf{M} = \mathbf{T}_\circ \boldsymbol{\Delta}_\circ \mathbf{T}_\circ^t$, where $\boldsymbol{\Delta}_\circ = \mathbf{T}_\circ^{-1} \mathbf{M} \mathbf{T}_\circ^{t\,-1}$ takes the *canonical form*

$$\boldsymbol{\Delta}_\circ = \begin{bmatrix} \boldsymbol{\Lambda}_\circ & \mathbf{0} \\ \mathbf{0} & \boldsymbol{\Lambda}_\circ \end{bmatrix} \quad \text{with} \quad \boldsymbol{\Lambda}_\circ = \begin{bmatrix} \lambda_x & 0 \\ 0 & \lambda_y \end{bmatrix} \quad \text{and} \quad \lambda_x \geq \lambda_y > 0\,. \tag{2.124}$$

From the similarity transformation $\mathbf{MJ} = \mathbf{T}_\circ(\boldsymbol{\Delta}_\circ \mathbf{J})\mathbf{T}_\circ^{-1}$, we conclude that $\boldsymbol{\Delta}_\circ$ follows directly from the eigenvalues $\pm\lambda_x$ and $\pm\lambda_y$ of $\mathbf{MJ}$ and that $\mathbf{T}_\circ$ follows from the eigenvectors of $(\mathbf{MJ})^2$: $(\mathbf{MJ})^2 \mathbf{T}_\circ = \mathbf{T}_\circ \boldsymbol{\Delta}_\circ^2$. Any moment matrix $\mathbf{M}$ can thus be brought into the diagonal form $\boldsymbol{\Delta}_\circ$ by means of a realizable canonical transformation with ray transformation matrix $\mathbf{T}_\circ^{-1}$. We remark that the determination of the canonical eigenvalues λ_x and λ_y is easy; they follow immediately from the two moment invariants (2.117) $I_1 = \lambda_x \lambda_y$ and $I_2 = \lambda_x^2 + \lambda_y^2$.

The system $\mathscr{L}(\mathbf{T}_\circ)$ that connects the moment matrix $\mathbf{M}$ with its canonical form $\boldsymbol{\Delta}_\circ$ through $\mathbf{M} = \mathbf{T}_\circ \boldsymbol{\Delta}_\circ \mathbf{T}_\circ^t$, can be reduced—with the cascade (2.66) in mind—to the cascade of a gyrator and a generalized magnifier $\mathscr{Q}(\mathbf{G}_\circ)\,\mathscr{M}[\mathbf{S}_\circ \mathbf{U}_r(-\alpha_\circ)]\,\mathscr{G}(-\beta_\circ)$; note that the separable fractional Fourier transformer in (2.66) can be omitted because it does not affect the canonical form and that we have combined the rotator with the pure magnifier. The system $\mathscr{Q}(\mathbf{G}_\circ)\,\mathscr{M}[\mathbf{S}_\circ \mathbf{U}_r(-\alpha_\circ)]$ thus connects the moment matrix $\mathbf{M}$ with its *generalized canonical form* $\mathbf{M}^\circ$,

$$\mathbf{M}^\circ = \begin{bmatrix} \mathbf{M}^\circ_{\mathbf{rr}} & \mathbf{M}^\circ_{\mathbf{rq}} \\ -\mathbf{M}^\circ_{\mathbf{rq}} & \mathbf{M}^\circ_{\mathbf{rr}} \end{bmatrix}, \quad \mathbf{M}^\circ_{\mathbf{rr}} + \mathrm{i}\,\mathbf{M}^\circ_{\mathbf{rq}} = \mathbf{U}_g(-\beta_\circ)\,\boldsymbol{\Lambda}_\circ\,\mathbf{U}_g(\beta_\circ)\,,$$

$$\mathbf{M}^\circ_{\mathbf{rr}} + \mathrm{i}\,\mathbf{M}^\circ_{\mathbf{rq}} = \begin{bmatrix} \lambda_x \cos^2\beta_\circ + \lambda_y \sin^2\beta_\circ & \mathrm{i}\,\frac{1}{2}(\lambda_x - \lambda_y)\sin 2\beta_\circ \\ -\mathrm{i}\,\frac{1}{2}(\lambda_x - \lambda_y)\sin 2\beta_\circ & \lambda_x \sin^2\beta_\circ + \lambda_y \cos^2\beta_\circ \end{bmatrix}. \tag{2.125}$$

Applying the definition of the twist (2.122) to the generalized canonical form $\mathbf{M}^\circ$, we readily conclude that the gyrator angle $\beta_\circ$ is completely determined by the twist T and the two canonical eigenvalues λ_x and λ_y,

$$(\lambda_x - \lambda_y)\sin 2\beta_\circ = \frac{2T\sqrt{\lambda_x \lambda_y}}{\sqrt{(\lambda_x + \lambda_y)^2 - T^2}}\,, \tag{2.126}$$

and that the same holds for the generalized canonical form itself. Note that these three parameters are indeed invariant under a linear canonical transformation of the form $\mathscr{Q}(-\mathbf{CA}^{-1})\,\mathscr{M}(\mathbf{A})$, for which $\mathbf{B} = \mathbf{0}$.

Note that for the generalized canonical form $\mathbf{M}^\circ$, the moment vector (Q_1, Q_2, Q_3) reads $(Q\sin 2\beta_\circ, 0, Q\cos 2\beta_\circ)$ with $Q = \lambda_x - \lambda_y$, cf. (2.119), and that Q_3 corresponds to the left-hand side of (2.126). We easily verify that the maximum value of $|T|$ is reached for $\beta_\circ = \pm\frac{1}{4}\pi$ and that $|T|_{\max} = Q = \lambda_x - \lambda_y = \sqrt{I_2 - 2I_1}$. We also recall, see Sects. 2.6.3.3 and 2.8.1, that it is the gyrator $\mathscr{G}(\pm\frac{1}{4}\pi)$ that transforms the (vortex-free) Hermite–Gauss modes, with $(Q_1, Q_2, Q_3) = (Q, 0, 0)$, into the (maximum-vortex) Laguerre–Gauss modes, with $(Q_1, Q_2, Q_3) = (0, 0, \pm Q)$.

In the special case that $\lambda_x = \lambda_y = \lambda$, the canonical form $\mathbf{\Delta}_\circ = \lambda\mathbf{I}$ takes the form of a scalar matrix, and the system $\mathscr{L}(\mathbf{T}_\circ)$ that connects the moment matrix $\mathbf{M}$ to its canonical form reduces to the cascade $\mathscr{Q}(\mathbf{G}_\circ)\,\mathscr{M}(\mathbf{S}_\circ)$. Note that the Poincaré sphere for such a canonical form reduces to a single point: $Q = 0$. The case $\lambda_x = \lambda_y$ is known as the intrinsically isotropic case [25, 62] and the moment matrix $\mathbf{M}$ is now proportional to a symplectic matrix. Symplecticity of a moment matrix is preserved under a linear canonical transformation, and the moment relations $\mathbf{M}_o = \mathbf{T}\mathbf{M}_i\mathbf{T}^t$ and $(\mathbf{M}_o\mathbf{J}) = \mathbf{T}(\mathbf{M}_i\mathbf{J})\mathbf{T}^{-1}$, which deal with 4×4 matrices, can be replaced by an easier one that deals with 2×2 matrices; see Eq. (2.128) in the next section.

2.8.4 *The Special Case of a Symplectic Moment Matrix*

In this section we study the special case that in the moment representation (2.113) we have the additional conditions $\mathbf{H} = \mathbf{H}^t$, $\mathbf{G}_1 = \sigma\,\mathbf{G}$ and $\mathbf{G}_2 = \sigma^{-1}\mathbf{G}$, where $0 < \sigma \leq 1$; note that we are now dealing with a signal that has zero twist. The matrix moment then takes the special form [14, 15]

$$2\pi\,\mathbf{M} = \frac{1}{2\sigma}\begin{bmatrix} \mathbf{G}^{-1} & \mathbf{G}^{-1}\mathbf{H} \\ \mathbf{H}^t\mathbf{G}^{-1} & \mathbf{G} + \mathbf{H}^t\mathbf{G}^{-1}\mathbf{H} \end{bmatrix}, \tag{2.127}$$

and the input–output relations for the moments $\mathbf{M}_o = \mathbf{T}\mathbf{M}_i\mathbf{T}^t$ and $(\mathbf{M}_o\mathbf{J})\mathbf{T} = \mathbf{T}(\mathbf{M}_i\mathbf{J})$ can now be expressed as $[\mathbf{H}_o \pm \mathrm{i}\,\mathbf{G}_o][\mathbf{A} + \mathbf{C}(\mathbf{H}_i \pm \mathrm{i}\,\mathbf{G}_i)] = [\mathbf{C} + \mathbf{D}(\mathbf{H}_i \pm \mathrm{i}\,\mathbf{G}_i)]$, which leads to the *bilinear relationship*

$$\mathbf{H}_o \pm \mathrm{i}\,\mathbf{G}_o = [\mathbf{C} + \mathbf{D}(\mathbf{H}_i \pm \mathrm{i}\,\mathbf{G}_i)][\mathbf{A} + \mathbf{C}(\mathbf{H}_i \pm \mathrm{i}\,\mathbf{G}_i)]^{-1}\,. \tag{2.128}$$

This bilinear relationship, together with the invariance of $\det\mathbf{M} = (4\pi\,\sigma)^{-2}$, completely describes the propagation of a symplectic moment matrix $\mathbf{M}$ when the signal undergoes a linear canonical transformation. Note that the bilinear relationship is identical to the so-called **ABCD**-law for chirp-like signals of the form $\exp(\mathrm{i}\,\pi\,\mathbf{r}^t\mathbf{H}\mathbf{r})$; we have only replaced the (real) chirp matrix $\mathbf{H}$ by the (generally complex) matrix $\mathbf{H} \pm \mathrm{i}\,\mathbf{G}$, cf. (2.74). The bilinear relationship is also the basis for the treatment of complex Gaussian functions, $f(\mathbf{r}) = \exp[-\mathrm{i}\,\pi\,\mathbf{r}^t(\mathbf{G} - \mathrm{i}\,\mathbf{H})\mathbf{r}]$, under linear canonical transformations, see, for instance, [47, Sect. 3.4.6, Linear fractional transformations].

2.9 Conclusions

The mathematical formalism introduced in this chapter is used nowadays for numerous applications, including the description of paraxial light propagation through first-order optical systems, design and characterization of optical beams and systems, development of filtering and encryption techniques in signal processing.

For example, the linear canonical transformation's phase-space representation allows associating the transformation parameters with the ray transformation matrix known from geometrical optics and therefore establishing a relation between the ray and the wave description of light. The matrix description of linear canonical transformations drastically simplifies the design and analysis of the composed optical systems, as well as the calculation of the beam propagation through them. The use of the modified Iwasawa decomposition of the transformation matrix together with the detailed analysis of the phase-space rotator matrix provides a clear interpretation of the signal modification produced by the transformation. Thus, the central role of the fractional Fourier transformation among other linear canonical transformations is revealed. The affine transformation of the Wigner distribution and the ambiguity function produced by such transformations is the key for the establishing of phase-space tomography methods used for the characterization of classical and quantum light. The corresponding transformation of the second-order moments of the Wigner distribution, described in this chapter, is useful for global beam analysis.

The diversity of the linear canonical transformation parameters (ten in the two-dimensional case) is exploited, as it is discussed in the next chapters, in different phase retrieval, filtering and encryption techniques.

Appendix

Derivation of the Phase-Space Relation (2.5)

We start with (2.4) and substitute from (2.1):

$$\begin{aligned} W_o(\mathbf{r}_o, \mathbf{q}_o) &= \int f\left(\mathbf{r}_o + \tfrac{1}{2}\mathbf{r}'_o\right) f^*\left(\mathbf{r}_o - \tfrac{1}{2}\mathbf{r}'_o\right) \exp[-\mathrm{i}\,2\pi\,\mathbf{q}_o^t\mathbf{r}'_o]\,\mathrm{d}\mathbf{r}'_o \\ &= |\det \mathbf{L}_{io}| \iiint \exp[\mathrm{i}\,\pi(E_1 - E_2 - 2\mathbf{q}_o^t\mathbf{r}'_o)] f_i(\mathbf{r}_1) f_i^*(\mathbf{r}_2)\,\mathrm{d}\mathbf{r}_1\,\mathrm{d}\mathbf{r}_2\,\mathrm{d}\mathbf{r}'_o \end{aligned}$$

$$\text{with} \quad \begin{cases} E_1 = \left(\mathbf{r}_o + \tfrac{1}{2}\mathbf{r}'_o\right)^t \mathbf{L}_{oo}\left(\mathbf{r}_o + \tfrac{1}{2}\mathbf{r}'_o\right) - 2\mathbf{r}_1^t\mathbf{L}_{io}\left(\mathbf{r}_o + \tfrac{1}{2}\mathbf{r}'_o\right) + \mathbf{r}_1^t\mathbf{L}_{ii}\mathbf{r}_1 \,, \\ E_2 = \left(\mathbf{r}_o - \tfrac{1}{2}\mathbf{r}'_o\right)^t \mathbf{L}_{oo}\left(\mathbf{r}_o - \tfrac{1}{2}\mathbf{r}'_o\right) - 2\mathbf{r}_2^t\mathbf{L}_{io}\left(\mathbf{r}_o - \tfrac{1}{2}\mathbf{r}'_o\right) + \mathbf{r}_2^t\mathbf{L}_{ii}\mathbf{r}_2 \,. \end{cases}$$

We reorder the exponent $E_1 - E_2$ to get

$$E_1 - E_2 = (\mathbf{r}_o'^{\,t}\mathbf{L}_{oo}\mathbf{r}_o + \mathbf{r}_o^t\mathbf{L}_{oo}\mathbf{r}_o') - 2(\mathbf{r}_1 - \mathbf{r}_2)^t\mathbf{L}_{io}\mathbf{r}_o - (\mathbf{r}_1 + \mathbf{r}_2)^t\mathbf{L}_{io}\mathbf{r}_o' \\ +(\mathbf{r}_1^t\mathbf{L}_{ii}\mathbf{r}_1 - \mathbf{r}_2^t\mathbf{L}_{ii}\mathbf{r}_2) \; .$$

We substitute $\mathbf{r}_1 = \mathbf{r}_i + \frac{1}{2}\mathbf{r}_i'$ and $\mathbf{r}_2 = \mathbf{r}_i - \frac{1}{2}\mathbf{r}_i'$ and get

$$W_o(\mathbf{r}_o, \mathbf{q}_o) = |\det \mathbf{L}_{io}| \iiint \exp[\mathrm{i}\,\pi(E_1 - E_2 - 2\mathbf{q}_o^t\mathbf{r}_o')] \\ \times f_i\left(\mathbf{r}_i + \tfrac{1}{2}\mathbf{r}_i'\right) f_i^*\left(\mathbf{r}_i - \tfrac{1}{2}\mathbf{r}_i'\right) \mathrm{d}\mathbf{r}_i\, \mathrm{d}\mathbf{r}_i'\, \mathrm{d}\mathbf{r}_o'$$

with $E_1 - E_2 = (\mathbf{r}_o'^{\,t}\mathbf{L}_{oo}\mathbf{r}_o + \mathbf{r}_o^t\mathbf{L}_{oo}\mathbf{r}_o') - \mathbf{r}_i'^{\,t}\mathbf{L}_{io}\mathbf{r}_o - 2\mathbf{r}_i^t\mathbf{L}_{io}\mathbf{r}_o' + (\mathbf{r}_i'^{\,t}\mathbf{L}_{ii}\mathbf{r}_i + \mathbf{r}_i^t\mathbf{L}_{ii}\mathbf{r}_i')$.

We substitute $f_i\left(\mathbf{r}_i + \frac{1}{2}\mathbf{r}_i'\right) f_i^*\left(\mathbf{r}_i - \frac{1}{2}\mathbf{r}_i'\right) = \int W_i(\mathbf{r}_i, \mathbf{q}_i) \exp[\mathrm{i}\,2\pi\, \mathbf{r}_i'^{\,t}\mathbf{q}_i]\, \mathrm{d}\mathbf{q}_i$ and get

$$\begin{aligned} &W_o(\mathbf{r}_o, \mathbf{q}_o) \\ &= |\det \mathbf{L}_{io}| \iiiint W_i(\mathbf{r}_i, \mathbf{q}_i) \exp[\mathrm{i}\,\pi(E_1 - E_2 - 2\mathbf{q}_o^t\mathbf{r}_o'] \\ &\quad \times \exp[\mathrm{i}\,2\pi\, \mathbf{r}_i'^{\,t}\mathbf{q}_i]\, \mathrm{d}\mathbf{r}_i'\, \mathrm{d}\mathbf{r}_o'\, \mathrm{d}\mathbf{r}_i\, \mathrm{d}\mathbf{q}_i \\ &= |\det \mathbf{L}_{io}| \iint W_i(\mathbf{r}_i, \mathbf{q}_i) \left(\int \exp[\mathrm{i}\,2\pi(\mathbf{L}_{oo}\mathbf{r}_o - \mathbf{L}_{io}^t\mathbf{r}_i - \mathbf{q}_o)^t\mathbf{r}_o']\, \mathrm{d}\mathbf{r}_o' \right) \\ &\quad \times \left(\int \exp[\mathrm{i}\,2\pi(\mathbf{L}_{ii}\mathbf{r}_i - \mathbf{L}_{io}\mathbf{r}_o + \mathbf{q}_i)^t\mathbf{r}_i']\, \mathrm{d}\mathbf{r}_i' \right) \mathrm{d}\mathbf{r}_i\, \mathrm{d}\mathbf{q}_i \\ &= |\det \mathbf{L}_{io}| \iint W_i(\mathbf{r}_i, \mathbf{q}_i)\, \delta(\mathbf{L}_{oo}\mathbf{r}_o - \mathbf{L}_{io}^t\mathbf{r}_i - \mathbf{q}_o)\, \delta(\mathbf{L}_{ii}\mathbf{r}_i - \mathbf{L}_{io}\mathbf{r}_o + \mathbf{q}_i)\, \mathrm{d}\mathbf{r}_i\, \mathrm{d}\mathbf{q}_i \\ &= |\det \mathbf{L}_{io}| \int W_i(\mathbf{r}_i, \mathbf{L}_{io}\mathbf{r}_o - \mathbf{L}_{ii}\mathbf{r}_i)\, \delta(\mathbf{L}_{oo}\mathbf{r}_o - \mathbf{L}_{io}^t\mathbf{r}_i - \mathbf{q}_o)\, \mathrm{d}\mathbf{r}_i \\ &= W_i\left(\mathbf{L}_{io}^{t\;-1}\mathbf{L}_{oo}\mathbf{r}_o - \mathbf{L}_{io}^{t\;-1}\mathbf{q}_o, \mathbf{L}_{io}\mathbf{r}_o - \mathbf{L}_{ii}\mathbf{L}_{io}^{t\;-1}\mathbf{L}_{oo}\mathbf{r}_o + \mathbf{L}_{ii}\mathbf{L}_{io}^{t\;-1}\mathbf{q}_o\right) . \end{aligned}$$

After substituting from (2.7), we finally get

$$W_o(\mathbf{r}_o, \mathbf{q}_o) = W_i(\mathbf{D}^t\mathbf{r}_o - \mathbf{B}^t\mathbf{q}_o, -\mathbf{C}^t\mathbf{r}_o + \mathbf{A}^t\mathbf{q}_o)$$

and hence $W_o(\mathbf{A}\mathbf{r} + \mathbf{B}\mathbf{q}, \mathbf{C}\mathbf{r} + \mathbf{D}\mathbf{q}) = W_i(\mathbf{r}, \mathbf{q})$, which is identical to (2.5).

References

1. E.G. Abramochkin, V.G. Volostnikov, Generalized Gaussian beams. J. Opt. A Pure Appl. Opt. **6**, S157–S161 (2004)
2. M. Abramowitz, I.A. Stegun (eds.), *Pocketbook of Mathematical Functions* (Deutsch, Frankfurt am Main, 1984)
3. G.S. Agarwal, SU(2) structure of the Poincaré sphere for light beams with orbital angular momentum. J. Opt. Soc. Am. A **16**, 2914–2916 (1999)
4. O. Akay, G.F. Boudreaux-Bartels, Fractional convolution and correlation via operator methods and an application to detection of linear FM signals. IEEE Trans. Signal Process. **49**, 979–993 (2001)
5. T. Alieva, M.J. Bastiaans, Alternative representation of the linear canonical integral transform. Opt. Lett. **30**, 3302–3304 (2005)
6. T. Alieva, M.J. Bastiaans, Mode mapping in paraxial lossless optics. Opt. Lett. **30**, 1461–1463 (2005)
7. T. Alieva, M.J. Bastiaans, Properties of the linear canonical integral transformation. J. Opt. Soc. Am. A **24**, 3658–3665 (2007)
8. T. Alieva, M.J. Bastiaans, Phase-space rotations and orbital Stokes parameters. Opt. Lett. **34**, 410–412 (2009)
9. T. Alieva, M.J. Bastiaans, M.L. Calvo, Fractional transforms in optical information processing. EURASIP J. Appl. Signal Process. **2005**, 1498–1519 (2005)
10. D. Ambrosini, V. Bagini, F. Gori, M. Santarsiero. Twisted Gaussian Schell-model beams: a superposition model. J. Mod. Opt. **41**, 1391–1399 (1994)
11. M.J. Bastiaans, The Wigner distribution function applied to optical signals and systems. Opt. Commun. **25**, 26–30 (1978)
12. M.J. Bastiaans, Wigner distribution function and its application to first-order optics. J. Opt. Soc. Am. **69**, 1710–1716 (1979)
13. M.J. Bastiaans, Application of the Wigner distribution function to partially coherent light. J. Opt. Soc. Am. A **3**, 1227–1238 (1986)
14. M.J. Bastiaans, Second-order moments of the Wigner distribution function in first-order optical systems. Optik **88**, 163–168 (1991)
15. M.J. Bastiaans, ABCD law for partially coherent Gaussian light, propagating through first-order optical systems. Opt. Quant. Electron. **24**, 1011–1019 (1992)
16. M.J. Bastiaans, Application of the Wigner distribution function in optics, in *The Wigner Distribution: Theory and Applications in Signal Processing*, ed. by W. Mecklenbräuker, F. Hlawatsch (Elsevier Science, Amsterdam, 1997), pp. 375–426
17. M.J. Bastiaans, Wigner distribution function applied to twisted Gaussian light propagating in first-order optical systems. J. Opt. Soc. Am. A **17**, 2475–2480 (2000)
18. M.J. Bastiaans, Applications of the Wigner distribution to partially coherent light beams, in *Advances in Information Optics and Photonics*, ed. by A.T. Friberg, R. Dändliker (SPIE – The International Society for Optical Engineering, Bellingham, WA, 2008), pp. 27–56
19. M.J. Bastiaans, Wigner distribution in optics, in *Phase-Space Optics: Fundamentals and Applications*, ed. by M.E. Testorf, B.M. Hennelly, J. Ojeda-Castañeda (McGraw-Hill, New York, 2009), pp. 1–44
20. M.J. Bastiaans, T. Alieva, Generating function for Hermite-Gaussian modes propagating through first-order optical systems. J. Phys. A Math. Gen. **38**, L73–L78 (2005)
21. M.J. Bastiaans, T. Alieva, Propagation law for the generating function of Hermite-Gaussian-type modes in first-order optical systems. Opt. Express **13**, 1107–1112 (2005)
22. M.J. Bastiaans, T. Alieva, First-order optical systems with unimodular eigenvalues. J. Opt. Soc. Am. A **23**, 1875–1883 (2006)
23. M.J. Bastiaans, T. Alieva, Classification of lossless first-order optical systems and the linear canonical transformation. J. Opt. Soc. Am. A **24**, 1053–1062 (2007)

24. M.J. Bastiaans, T. Alieva, First-order optical systems with real eigenvalues. Opt. Commun. **272**, 52–55 (2007)
25. M.J. Bastiaans, T. Alieva, Signal representation on the angular Poincaré sphere, based on second-order moments. J. Opt. Soc. Am. A **27**, 918–927 (2010)
26. S.A. Collins Jr., Lens-system diffraction integral written in terms of matrix optics. J. Opt. Soc. Am. **60**, 1168–1177 (1970)
27. L.S. Dolin, Beam description of weakly-inhomogeneous wave fields. Izv. Vyssh. Uchebn. Zaved. Radiofiz. **7**, 559–563 (1964)
28. D. Dragoman, The Wigner distribution function in optics and optoelectronics. Prog. Opt. **37**, 1–56 (1997)
29. A.T. Friberg, B. Tervonen, J. Turunen, Interpretation and experimental demonstration of twisted Gaussian Schell-model beams. J. Opt. Soc. Am. A **11**, 1818–1826 (1994)
30. F.R. Gantmacher, *The Theory of Matrices* (Chelsea, New York, 1974–1977)
31. J. García, D. Mendlovic, Z. Zalevsky, A.W. Lohmann, Space-variant simultaneous detection of several objects by the use of multiple anamorphic fractional-Fourier transform filters. Appl. Opt. **35**, 3945–3952 (1996)
32. G. García-Calderón, M. Moshinsky, Wigner distribution functions and the representations of canonical transformations in quantum mechanics. J. Phys. A **13**, L185–L188 (1980)
33. J.W. Goodman, *Introduction to Fourier Optics*, 2nd edn. (McGraw-Hill, New York, 1996)
34. I.S. Gradshteyn, I.M. Ryzhik, A. Jeffrey (eds.), *Table of Integrals, Series, and Products*, 5th edn. (Academic, San Diego, CA, 1994)
35. A.J.E.M. Janssen, Positivity of weighted Wigner distributions. J. Math. Anal. **12**, 752–758 (1981)
36. G.A. Korn, T.M. Korn, *Mathematical Handbook for Scientists and Engineers: Definitions, Theorems, and Formulas for Reference and Review* (McGraw-Hill, London, 1968)
37. W.W. Lin, V. Mehrmann, H. Xu, Canonical forms for Hamiltonian and symplectic matrices and pencils. Linear Algebra Appl. **302–303**, 469–533 (1999)
38. A.W. Lohmann, Image rotation, Wigner rotation, and the fractional Fourier transform. J. Opt. Soc. Am. A **10**, 2181–2186 (1993)
39. A.W. Lohmann, Z. Zalevsky, D. Mendlovic, Synthesis of pattern recognition filters for fractional Fourier processing. Opt. Commun. **128**, 199–204 (1996)
40. R.K. Luneburg, *Mathematical Theory of Optics* (University of California Press, Berkeley, CA, 1966)
41. A.A. Malyutin, Use of the fractional Fourier transform in $\pi/2$ converters of laser modes. Quantum Electron. **34**, 165–171 (2004)
42. A.C. McBride, F.H. Kerr, On Namias fractional Fourier transforms. IMA J. Appl. Math. **39**, 159–175 (1987)
43. W. Mecklenbräuker, F. Hlawatsch (eds.), *The Wigner Distribution: Theory and Applications in Signal Processing* (Elsevier Science, Amsterdam, 1997)
44. M. Moshinsky, C. Quesne, Linear canonical transformations and their unitary representations. J. Math. Phys. **12**, 1772–1780 (1971)
45. V. Namias, The fractional order Fourier transform and its applications to quantum mechanics. J. Inst. Math. Appl. **25**, 241–265 (1980)
46. M. Nazarathy, J. Shamir, First-order optics—a canonical operator representation: lossless systems. J. Opt. Soc. Am. **72**, 356–364 (1982)
47. H.M. Ozaktas, Z. Zalevsky, M.A. Kutay, *The Fractional Fourier Transform with Applications in Optics and Signal Processing* (Wiley, New York, 2001)
48. M.J. Padgett, J. Courtial, Poincaré-sphere equivalent for light beams containing orbital angular momentum. Opt. Lett. **24**, 430–432 (1999)
49. S.C. Pei, J.J. Ding, Eigenfunctions of linear canonical transform. IEEE Trans. Signal Process. **50**, 11–26 (2002)
50. A.P. Prudnikov, Y.A. Brychkov, O.I. Marichev (eds.), *Elementary Functions*. Integrals and Series, vol. 1 (Gordon and Breach, New York, 1986)

51. C. Quesne, M. Moshinsky, Canonical transformations and matrix elements. J. Math. Phys **12**, 1780–1783 (1971)
52. J. Serna, R. Martínez-Herrero, P.M. Mejías, Parametric characterization of general partially coherent beams propagating through ABCD optical systems. J. Opt. Soc. Am. A **8**, 1094–1098 (1991)
53. R. Simon, G.S. Agarwal, Wigner representation of Laguerre-Gaussian beams. Opt. Lett. **25**, 1313–1315 (2000)
54. R. Simon, N. Mukunda, Twisted Gaussian Schell-model beams. J. Opt. Soc. Am. A **10**, 95–109 (1993)
55. R. Simon, N. Mukunda, Iwasawa decomposition in first-order optics: universal treatment of shape-invariant propagation for coherent and partially coherent beams. J. Opt. Soc. Am. A **15**, 2146–2155 (1998)
56. R. Simon, N. Mukunda, Twist phase in Gaussian-beam optics. J. Opt. Soc. Am. A **15**, 2373–2382 (1998)
57. R. Simon, K.B. Wolf, Fractional Fourier transforms in two dimensions. J. Opt. Soc. Am. A **17**, 2368–2381 (2000)
58. R. Simon, K.B. Wolf, Structure of the set of paraxial optical systems. J. Opt. Soc. Am. A **17**, 342–355 (2000)
59. R. Simon, N. Mukunda, E.C.G. Sudarshan, Partially coherent beams and a generalized ABCD-law. Opt. Commun. **65**, 322–328 (1988)
60. R. Simon, K. Sundar, N. Mukunda, Twisted Gaussian Schell-model beams. I. Symmetry structure and normal-mode spectrum. J. Opt. Soc. Am. A **10**, 2008–2016 (1993)
61. R. Simon, A.T. Friberg, E. Wolf, Transfer of radiance by twisted Gaussian Schell-model beams in paraxial systems. J. Eur. Opt. Soc. A Pure Appl. Opt **5**, 331–343 (1996)
62. K. Sundar, N. Mukunda, R. Simon, Coherent-mode decomposition of general anisotropic Gaussian Schell-model beams. J. Opt. Soc. Am. A **12**, 560–569 (1995)
63. K. Sundar, R. Simon, N. Mukunda, Twisted Gaussian Schell-model beams. II. Spectrum analysis and propagation characteristics. J. Opt. Soc. Am. A **10**, 2017–2023 (1993)
64. M.E. Testorf, J. Ojeda-Castañeda, A.W. Lohmann, (eds.), *Selected Papers on Phase-Space Optics, SPIE Milestone Series*, vol. MS 181 (SPIE, Bellingham, WA, 2006)
65. M.E. Testorf, B.M. Hennelly, J. Ojeda-Castañeda (eds.), *Phase-Space Optics: Fundamentals and Applications* (McGraw-Hill, New York, 2009)
66. A. Torre, *Linear Ray and Wave Optics in Phase Space* (Elsevier, Amsterdam, 2005)
67. A. Walther, Radiometry and coherence. J. Opt. Soc. Am. **58**, 1256–1259 (1968)
68. A. Walther, Propagation of the generalized radiance through lenses. J. Opt. Soc. Am. **68**, 1606–1610 (1978)
69. E. Wigner, On the quantum correction for thermodynamic equilibrium. Phys. Rev. **40**, 749–759 (1932)
70. J. Williamson, On the algebraic problem concerning the normal forms of linear dynamical systems. Am. J. Math. **58**, 141–163 (1936)
71. K.B. Wolf, Canonical transforms. I. Complex linear transforms. J. Math. Phys. **15**, 1295–1301 (1974)
72. K.B. Wolf, Canonical transforms. II. Complex radial transforms. J. Math. Phys. **15**, 2102–2111 (1974)
73. K.B. Wolf, Canonical transforms, separation of variables, and similarity solutions for a class of parabolic differential equations. J. Math. Phys. **17**, 601–613 (1976)
74. K.B. Wolf, *Integral Transforms in Science and Engineering*, Chap. 9 (Plenum Press, New York, 1979)
75. K.B. Wolf, *Geometric Optics on Phase Space* (Springer, Berlin, 2004)
76. P.M. Woodward, *Probability and Information Theory with Applications to Radar* (Pergamon, London, 1953)
77. B. Zhu, S. Liu, Optical image encryption based on the generalized fractional convolution operation. Opt. Commun. **195**, 371–381 (2001)

Chapter 3
Eigenfunctions of the Linear Canonical Transform

Soo-Chang Pei and Jian-Jiun Ding

Abstract In this chapter, the eigenfunctions and the eigenvalues of the linear canonical transform are discussed. The style of the eigenfunctions of the LCT is closely related to the parameters $\{a, b, c, d\}$ of the LCT. When $|a+d|<2$, the LCT eigenfunctions are the scaling and chirp multiplication of Hermite–Gaussian functions. When $|a+d|=2$ and $b=0$, the eigenfunctions are the impulse trains. When $|a+d|=2$ and $b\neq 0$, the eigenfunctions are the chirp multiplications of periodic functions. When $|a+d|>2$, the eigenfunctions are the chirp convolution and chirp multiplication of scaling-invariant functions, i.e., fractals. Moreover, the linear combinations of the LCT eigenfunctions with the same eigenvalue are also the eigenfunctions of the LCT. Furthermore, the two-dimensional case is also discussed. The eigenfunctions of the LCT are helpful for analyzing the resonance phenomena in the radar system and the self-imaging phenomena in optics.

3.1 Introduction

The linear canonical transform (LCT) [1–6] is a generalization of the Fourier transform and the fractional Fourier transform. Its formulas are

$$O_{\mathrm{LCT}}^{(a,b,c,d)}\left[x(t)\right]=\sqrt{\frac{1}{j2\pi b}}\exp\left(\frac{jdu^2}{2b}\right)\int_{-\infty}^{\infty}\exp\left(\frac{-jut}{b}\right)\exp\left(\frac{jat^2}{2b}\right)x(t)dt \quad \text{for } b\neq 0,$$

$$O_{\mathrm{LCT}}^{(a,b,c,d)}\left[x(t)\right]=\sqrt{d}\,\exp\left(jcdu^2/2\right)x(dt) \quad \text{for } b=0. \tag{3.1}$$

The eigenfunctions of the LCT are important for analyzing the characteristics of the LCT. They also play important roles in self-imaging phenomena analysis and resonance phenomena analysis, since many optical systems and electromagnet wave propagation system can be modeled by the LCT.

S.-C. Pei • J.-J. Ding (✉)
Department of Electrical Engineering, National Taiwan University, Taipei 10617, Taiwan
e-mail: pei@cc.ee.ntu.edu.tw; djj@cc.ee.ntu.edu.tw

J.J. Healy et al. (eds.), *Linear Canonical Transforms*, Springer Series in Optical Sciences 198, DOI 10.1007/978-1-4939-3028-9_3

It is well known that the eigenfunctions of the original Fourier transform (FT) is the Hermite–Gaussian function:

$$\phi_m(t) = \exp\left(-t^2/2\right) H_m(t), \tag{3.2}$$

where $H_m(t)$ is the mth order Hermite function. Moreover, the eigenvalues corresponding to $\phi_m(t)$ is $(-j)^m$:

$$FT\left[\phi_m(t)\right] = (-j)^m \phi_m(t).$$

In fact, the Hermite–Gaussian function is also the eigenfunction of the fractional Fourier transform (FRFT) [7, 8]:

$$FRFT_\alpha\left[\phi_m(t)\right] = \exp\left(-j\alpha m\right) \phi_m(t), \tag{3.3}$$

where [8]

$$FRFT_\alpha\left[x(t)\right] = \sqrt{\frac{1-j\cot\alpha}{2\pi}} \exp\left(\frac{ju^2\cot\alpha}{2}\right)$$
$$\int_{-\infty}^{\infty} \exp\left(-jut\operatorname{cosec}\alpha\right) \exp\left(\frac{jt^2\cot\alpha}{2}\right) x(t)dt. \tag{3.4}$$

Moreover, from the Talbot effect [9–12], the periodic function is the eigenfunction of the Fresnel transform [13]. The fractal [14] (i.e., the scaling-invariant function) is the eigenfunction of the scaling operation.

Note that the FT, the fractional Fourier transform, FRFT, the Fresnel transform, and the scaling operation are all the special cases of the LCT. They correspond to the LCT with parameters as follows:

$$\begin{aligned}&FT:\ \{a=0, b=1, c=-1, d=1\},\ FRFT:\ \{a=\cos\alpha, b=\sin\alpha, c=-\sin\alpha, d=\cos\alpha\},\\&\text{Fresnel}:\ \{a=1, b=\lambda z/2\pi, c=0, d=1\},\ \text{Scaling}:\ \{a=\sigma, b=0, c=0, d=1/\sigma\}.\end{aligned} \tag{3.5}$$

In this chapter, we discuss the eigenfunctions of the LCT.

3.2 Eigenfunctions of the LCT for the Case Where $|a+d| < 2$

The eigenfunctions and eigenvalues of the LCT can be derived from the following two theorems [15]:

Theorem 1. Suppose that $ad - bc = a_1d_1 - b_1c_1 = a_2b_2 - b_2c_2 = 1$ and

$$\begin{bmatrix} a & b \\ c & d \end{bmatrix} = \begin{bmatrix} a_1 & b_1 \\ c_1 & d_1 \end{bmatrix} \begin{bmatrix} a_2 & b_2 \\ c_2 & d_2 \end{bmatrix} \begin{bmatrix} d_1 & -b_1 \\ -c_1 & a_1 \end{bmatrix}, \tag{3.6}$$

then the following equality is satisfied:

$$a + d = a_2 + d_2. \tag{3.7}$$

This property can be proved by the facts that the eigenvalues of two similar matrices are the same and that the diagonal sum of a matrix is the sum of its eigenvalues.

Theorem 2. Suppose that $\{a, b, c, d\}$, $\{a_1, b_1, c_1, d_1\}$, and $\{a_2, b_2, c_2, d_2\}$ satisfy the relation in (3.6). If

$$O_{LCT}^{(a_2,b_2,c_2,d_2)}\left[e(t)\right] = \lambda e(t), \tag{3.8}$$

then, from (3.6) and the additivity property of the LCT,

$$\begin{aligned} O_{LCT}^{(a,b,c,d)}\left\{O_{LCT}^{(a_1,b_1,c_1,d_1)}\left[e(t)\right]\right\} &= O_{LCT}^{(a_1,b_1,c_1,d_1)}\left[O_{LCT}^{(a_2,b_2,c_2,d_2)}\left(O_{LCT}^{(d_1,-b_1,-c_1,a_1)}\left\{O_{LCT}^{(a_1,b_1,c_1,d_1)}\left[e(t)\right]\right\}\right)\right] \\ &= O_{LCT}^{(a_1,b_1,c_1,d_1)}\left[O_{LCT}^{(a_2,b_2,c_2,d_2)}\left(e(t)\right)\right] = \lambda O_{LCT}^{(a_1,b_1,c_1,d_1)}\left[e(t)\right]. \end{aligned} \tag{3.9}$$

That is, if $e(t)$ is the eigenfunction of the LCT with parameters $\{a_2, b_2, c_2, d_2\}$, then $O_{LCT}^{(a_1,b_1,c_1,d_1)}\left[e(t)\right]$ is the eigenfunction of the LCT with parameters $\{a, b, c, d\}$ and their corresponding eigenvalues are the same. Therefore, if the eigenfunctions and eigenvalues of the LCT with parameters $\{a_2, b_2, c_2, d_2\}$ are known, one can derive the eigenfunctions and eigenvalues of the LCT with parameters $\{a, b, c, d\}$ if $\{a, b, c, d\}$ and $\{a_2, b_2, c_2, d_2\}$ are similar.

In the case where $|a+d| < 2$, in Theorem 2, one can choose $\{a_2, b_2, c_2, d_2\}$ as $\{\cos\alpha, \sin\alpha, -\sin\alpha, \cos\alpha\}$. Note that the LCT with these parameters is the FRFT multiplied by some constant:

$$O_{LCT}^{(\cos\alpha,\sin\alpha,-\sin\alpha,\cos\alpha)}\left[x(t)\right] = \sqrt{\exp\left(-j\alpha\right)}O_{FRFT}^{\alpha}\left[x(t)\right]. \tag{3.10}$$

Then

$$\begin{bmatrix} a & b \\ c & d \end{bmatrix} = \begin{bmatrix} a_1 & b_1 \\ c_1 & d_1 \end{bmatrix} \begin{bmatrix} \cos\alpha & \sin\alpha \\ -\sin\alpha & \cos\alpha \end{bmatrix} \begin{bmatrix} d_1 & -b_1 \\ -c_1 & a_1 \end{bmatrix}. \tag{3.11}$$

Moreover, from Theorem 1, $a + d = 2\cos\alpha$ should be satisfied. Therefore,

$$\alpha = \cos^{-1}\left(\left(a + d\right)/2\right). \tag{3.12}$$

There are several choices for $\{a_1, b_1, c_1, d_1\}$. We suggest that it is proper to choose these parameters as

$$\begin{bmatrix} a_1 & b_1 \\ c_1 & d_1 \end{bmatrix} = \begin{bmatrix} \sigma & 0 \\ -\tau\sigma^{-1} & \sigma^{-1} \end{bmatrix}. \tag{3.13}$$

Then, the LCT with parameters $\{a_1, b_1, c_1, d_1\}$ is the combination of the scaling operation and the chirp multiplication operation:

$$O_{LCT}^{(a_1,b_1,c_1,d_1)}[x(t)] = \sigma^{-1/2}\exp\left(-j\tau t^2/2\sigma^2\right)x(t/\sigma). \tag{3.14}$$

Then, from (3.11)

$$\begin{bmatrix} a & b \\ c & d \end{bmatrix} = \begin{bmatrix} \cos\alpha + \tau\sin\alpha & \sigma^2\sin\alpha \\ -\left(\tau^2+1\right)\sin\alpha/\sigma^2 & \cos\alpha - \tau\sin\alpha \end{bmatrix}. \tag{3.15}$$

Therefore, the values of σ, τ, and α are

$$\begin{aligned} \alpha &= \cos^{-1}\left(\frac{a+d}{2}\right) = \sin^{-1}\left(\frac{sgn(b)}{2}\sqrt{4-(a+d)^2}\right), \\ \sigma^2 &= \frac{2\,|b|}{\sqrt{4-(a+d)^2}}, \quad \tau = \frac{sgn(b)\,(a-d)}{\sqrt{4-(a+d)^2}}. \end{aligned} \tag{3.16}$$

Since the LCT with parameters $\{a_2, b_2, c_2, d_2\}$ is the FRFT multiplied by some constant, as in (3.10), and the eigenfunctions of the FRFT are known to be Hermite–Gaussian functions, as in (3.3), from Theorem 2, we obtain [15, 16]:

Theorem 3. When $|a+d| < 2$, the eigenfunctions of the LCT are:

$$\phi_m^{(\sigma,\tau)}(t) = \sigma^{-1/2}\exp\left(\frac{-(1+i\tau)}{2\sigma^2}t^2\right)H_m\left(\frac{t}{\sigma}\right), \tag{3.17}$$

where σ and τ are defined in (3.16), $H_m(t)$ is the Hermite polynomial, and m is any nonnegative integer. In other words, when $|a+d| < 2$, the eigenfunction of the LCT is the scaling and chirp multiplication of Hermite–Gaussian functions. Moreover, the corresponding eigenvalues are $[\exp(-j\alpha)]^{1/2}\exp(-j\alpha m)$:

$$O_{LCT}^{(a,b,c,d)}\left[\phi_m^{(\sigma,\tau)}(t)\right] = \sqrt{\exp(-j\alpha)}\exp(-j\alpha m)\,\phi_m^{(\sigma,\tau)}(t), \tag{3.18}$$

where α is also defined in (3.16).

As the case of the FRFT, the eigenfunctions of the LCT in (3.17) also form a complete and orthogonal function set, i.e.,

$$\int_{-\infty}^{\infty}\phi_m^{(\sigma,\tau)}(t)\overline{\phi_n^{(\sigma,\tau)}(t)}\,dt = 0 \quad \text{if } m \neq n, \tag{3.19}$$

and any function can be expressed as a linear combination of $\phi_m^{(\sigma,\tau)}(t)$.

3.3 Eigenfunctions of the LCT for the Case Where $|a+d|=2$

Note that the eigenfunctions given in Theorem 3 can only be applied in the case where $|a+d|<2$. When $|a+d|>2$, in (3.12), α cannot be solved. When $|a+d|=2$, the values of σ and τ in (3.16) will be infinite. Therefore, the eigenfunctions of the LCT should have the form other than (3.17) in the case where $|a+d|\geq 2$.

To derive the eigenfunctions in the case where $|a+d|=2$, the method in Theorem 2 can also be applied. However, from Theorem 1

$$a_2 + d_2 = \pm 2 \tag{3.20}$$

must be satisfied. Therefore, we choose $\{a_2, b_2, c_2, d_2\} = \{1, 0, \eta, 1\}$ (i.e., the Fresnel transform) when $a_2 + d_2 = 2$ and $\{a_2, b_2, c_2, d_2\} = \{-1, 0, \eta, -1\}$ (i.e., the Fresnel transform + the space reverse operation) when $a_2 + d_2 = -2$.

From the theory of the Talbot effect [9–12], if the original light distribution is periodic, i.e., $f(x, 0) = f(x + q, 0)$, then the light distribution at the distance z is the same as the original one, i.e.,

$$f(x, z) = f(x, 0) \quad \text{where } z = 2q^2/\lambda, \tag{3.21}$$

and λ is the wavelength. Then, together with the relation between the Fresnel transform and the LCT, one can conclude that

$$O_{LCT}^{(1,q^2/\pi,0,1)}[e(t)] = e(t) \quad \text{if } e(t) = e(t+q). \tag{3.22}$$

More generally, from the fractional Talbot effect [11, 12], if $g(t) = g(t + q)$, then

$$O_{LCT}^{(1,q^2N/\pi M,0,1)}[g(t)] = \frac{1}{M}\sum_{p=0}^{M-1} g\left(u - \frac{pq}{M}\right)\sum_{n=0}^{M-1}\exp\left[j\frac{2\pi}{M}\left(pn - Nn^2\right)\right], \tag{3.23}$$

where N and M are some integers and N is prime to M. Therefore, if $[1, A_1, A_2, \ldots, A_{M-1}]^{\mathrm{T}}$ is the eigenvector of the following matrix:

$$\mathbf{J} = \begin{bmatrix} c_0 & c_{M-1} & c_{M-2} & \cdots & c_1 \\ c_1 & c_0 & c_{M-1} & \cdots & c_2 \\ c_2 & c_1 & c_0 & \cdots & c_3 \\ \vdots & \vdots & \vdots & \ddots & \vdots \\ c_{M-1} & c_{M-2} & c_{M-3} & \cdots & c_0 \end{bmatrix} \quad \text{where } c_p = \frac{1}{M}\sum_{n=0}^{M-1}\exp\left[j\frac{2\pi}{M}\left(pn - Nn^2\right)\right], \tag{3.24}$$

and its corresponding eigenvalue is ζ, then if $g(t) = g(t+q)$ and

$$\begin{aligned} &g(t) : g\left(t + q/M\right) : g\left(t + 2q/M\right) : \ldots\cdots : g\left(t + (M-1)\,q/M\right) \\ &= 1 : A_1 : A_2 : \ldots\cdots : A_{M-1} \quad \text{for } x \in \left(0, q/M\right), \end{aligned} \tag{3.25}$$

$g(t)$ will be the eigenfunction of the LCT with parameters $\{1, Nq^2/\pi M, 0, 1\}$ and the corresponding eigenvalue is ζ. Furthermore, it can be verified that the eigenvalues of **J** have the form of

$$\lambda_k = \exp\left(-j2\pi N k^2/M\right) \quad \text{where } k = 0,\ 1,\ 2, \ldots, M-1. \tag{3.26}$$

After the eigenvalues are known, the eigenvectors of **J** can be easily determined.

From the above results, one can conclude that the eigenfunctions of the LCT with parameters $\{1, b, 0, 1\}$ are periodic functions that satisfy

$$g(t) = g\left(t + q\right) \quad \text{where } q = \sqrt{|b|\,\pi M/N}, \tag{3.27}$$

and M and N are any positive integers that are prime to each other. Moreover, $g(t)$ should satisfy the symmetric relation in (3.25) and the corresponding eigenvalue is as in (3.26). Since there are infinite possible values for the period q, the LCT with parameters $\{1, b, 0, 1\}$ has infinite sets of eigenfunctions.

Then, we discuss the eigenfunctions of the LCT with parameters $\{-1, b, 0, -1\}$. Since

$$\begin{bmatrix} -1 & b \\ 0 & -1 \end{bmatrix} = \begin{bmatrix} 1 & -b \\ 0 & 1 \end{bmatrix} \begin{bmatrix} -1 & 0 \\ 0 & -1 \end{bmatrix}, \tag{3.28}$$

the eigenfunctions $g(t)$ of the LCT with parameters $\{-1, b, 0, -1\}$ should satisfy the following two constraints:

$$(1) g(t) = \pm g\left(-t\right). \tag{3.29}$$

(Remember that $O_{LCT}^{(-1,0,0,-1)}\left[g(t)\right] = \sqrt{-1}g\left(-t\right)$).

(2) $g(t)$ is the eigenfunction of the LCT with parameters $\{1, -b, 0, 1\}$.

Therefore, $g(t)$ should satisfy (3.27) and (3.25) and must be a symmetric or an asymmetric function that satisfy (3.29). Moreover, the corresponding eigenvalues are

$$\begin{aligned} \lambda_k &= \sqrt{-1}\ \exp\left(-j2\pi N k^2/M\right) \quad &\text{if } g(t) = g\left(-t\right), \\ \lambda_k &= -\sqrt{-1}\ \exp\left(-j2\pi N k^2/M\right) \quad &\text{if } g(t) = -g\left(-t\right). \end{aligned} \tag{3.30}$$

After the eigenfunctions of the LCT with parameters $\{1, b, 0, 1\}$ and $\{1, -b, 0, 1\}$ are determined, one can use Theorem 2 to derive the eigenfunctions of the LCT where $|a+d|=2$. When $a+d=2$ and $b\neq 0$, the *abcd* matrix of the LCT can be decomposed into

$$\begin{bmatrix} a & b \\ c & d \end{bmatrix} = \begin{bmatrix} 1 & 0 \\ \tau & 1 \end{bmatrix}\begin{bmatrix} 1 & \eta \\ 0 & 1 \end{bmatrix}\begin{bmatrix} 1 & 0 \\ -\tau & 1 \end{bmatrix} = \begin{bmatrix} 1-\tau\eta & \eta \\ -\tau^2\eta & 1+\tau\eta \end{bmatrix}, \tag{3.31}$$

$$\text{where } \eta = b \quad \text{and} \quad \tau = (d-a)/2b. \tag{3.32}$$

When $a+d=-2$ and $b\neq 0$, the *abcd* matrix of the LCT can be decomposed into

$$\begin{bmatrix} a & b \\ c & d \end{bmatrix} = \begin{bmatrix} 1 & 0 \\ \tau & 1 \end{bmatrix}\begin{bmatrix} -1 & \eta \\ 0 & -1 \end{bmatrix}\begin{bmatrix} 1 & 0 \\ -\tau & 1 \end{bmatrix} = \begin{bmatrix} -1-\tau\eta & \eta \\ -\tau^2\eta & -1+\tau\eta \end{bmatrix}, \tag{3.33}$$

$$\text{where } \eta = b \text{ and } \tau = (d-a)/2b. \tag{3.34}$$

Then, from Theorem 2 and the fact that the LCT with parameters $\{1, 0, \tau, 1\}$ is the multiplication of $\exp(j\tau t^2/2)$, we obtain the following two theorems:

Theorem 4. When $a+d=2$ and $b\neq 0$, the eigenfunctions of the LCT are

$$\phi_{(b,\eta)}(t) = \exp\left(j\frac{d-a}{4b}t^2\right)g(t), \tag{3.35}$$

where $g(t)$ is the periodic function that satisfies (3.27) and the symmetric relation in (3.25). The corresponding eigenvalues are also λ_k defined in (3.26):

$$O_{LCT}^{(a,b,c,d)}\left[\phi_{(b,\eta)}(t)\right] = \exp\left(-j2\pi Nk^2/M\right)\phi_{(b,\eta)}(t). \tag{3.36}$$

Theorem 5. When $a+d=-2$ and $b\neq 0$, the eigenfunctions of the LCT are

$$\phi_{(b,\eta)}(t) = \exp\left(j\frac{d-a}{4b}t^2\right)g(t), \tag{3.37}$$

where $g(t)$ is the periodic function that satisfies (3.27) and (3.25). Moreover, it should also satisfy the constraint that $g(t)=\pm g(-t)$. The corresponding eigenvalues are the same as those in (3.30):

$$\begin{aligned} O_{LCT}^{(a,b,c,d)}\left[\phi_{(b,\eta)}(t)\right] &= \sqrt{-1}\,\exp\left(-j2\pi Nk^2/M\right)\phi_{(b,\eta)}(t) \quad \text{if } g(t)=g(-t), \\ O_{LCT}^{(a,b,c,d)}\left[\phi_{(b,\eta)}(t)\right] &= -\sqrt{-1}\,\exp\left(-j2\pi Nk^2/M\right)\phi_{(b,\eta)}(t) \quad \text{if } g(t)=-g(-t). \end{aligned} \tag{3.38}$$

In both the cases, since there are infinite number of possible periods q for $g(t)$, as in (3.27), one can construct a complete and orthogonal eigenfunction set of the LCT from the subset of $\phi_{(b,\,\eta)}(t)$.

In the case where $|a+d|=2$ but $b=0$, one can prove that, since $ad-bc=1$ should be satisfied, the values of a, b, c, d must be one of the following form:

$$\begin{bmatrix} a & b \\ c & d \end{bmatrix} = \begin{bmatrix} 1 & 0 \\ c & 1 \end{bmatrix} \quad \text{or} \quad \begin{bmatrix} a & b \\ c & d \end{bmatrix} = \begin{bmatrix} -1 & 0 \\ c & -1 \end{bmatrix}. \tag{3.39}$$

Note that the LCT with parameters {1, 0, c, 1} is a chirp multiplication operation and the LCT with parameters {−1, 0, c, −1} is the combination of time reverse and chirp multiplication:

$$O_{LCT}^{(1,0,c,1)}\left[x(t)\right] = \exp\left(jcu^2/2\right)x(u), \tag{3.40}$$

$$O_{LCT}^{(-1,0,c,-1)}\left[x(t)\right] = \sqrt{-1}\,\exp\left(-jcu^2/2\right)x\left(-u\right). \tag{3.41}$$

The eigenfunctions of a multiplication are impulse trains and the eigenfunctions of a time reverse operation are symmetric or asymmetric functions. Therefore, we obtain:

Theorem 6. In the case where $|a+d|=2$ but $b=0$, the parameters of the LCT must have the form as in (3.39). When $\{a, b, c, d\}=\{1, 0, c, 1\}$, the eigenfunctions are

$$\phi_c(t) = \sum_{n=-\infty}^{\infty} A_n\delta\left(t-\sqrt{4n\pi\,|c|^{-1}+k}\right) + \sum_{m=-\infty}^{\infty} B_m\delta\left(t-\sqrt{4m\pi\,|c|^{-1}+k}\right), \tag{3.42}$$

where $0\leq k<4\pi/|c|$, A_n, B_m are any constants
and the corresponding eigenvalues are $\exp(jck/2)$:

$$O_{LCT}^{(1,0,c,1)}\left[\phi_c(t)\right] = \exp\left(jck/2\right)\phi_c(t). \tag{3.43}$$

When $\{a, b, \text{c}, \text{d}\}=\{-1, 0, \text{c}, -1\}$, the eigenfunctions are

$$\phi_{c,e}(t) = \sum_{n=-\infty}^{\infty} A_n\left\{\delta\left(t-\sqrt{4n\pi\,|c|^{-1}+k}\right)+\delta\left(t-\sqrt{4m\pi\,|c|^{-1}+k}\right)\right\}, \tag{3.44}$$

$$\text{or}\quad \phi_{c,o}(t) = \sum_{n=-\infty}^{\infty} A_n\left\{\delta\left(t-\sqrt{4n\pi\,|c|^{-1}+k}\right)-\delta\left(t-\sqrt{4m\pi\,|c|^{-1}+k}\right)\right\}, \tag{3.45}$$

where $0 \le k < 4\pi/|c|$ and A_n are any constants. The corresponding eigenvalues are $\pm(-1)^{1/2}\exp(jck/2)$:

$$O_{LCT}^{(-1,0,c,-1)}\left[\phi_{c,e}(t)\right] = \sqrt{-1}\,\exp\left(jck/2\right)\phi_{c,e}(t), \tag{3.46}$$

$$O_{LCT}^{(-1,0,c,-1)}\left[\phi_{c,o}(t)\right] = -\sqrt{-1}\,\exp\left(jck/2\right)\phi_{c,o}(t). \tag{3.47}$$

Moreover, since two delta functions are orthogonal if the locations of the impulses are different, one can always find a subset of (3.42) (or (3.44) and (3.45)) that forms a complete and orthogonal eigenfunction set for the LCT.

3.4 Eigenfunctions of the LCT for the Case Where |a + d| > 2

In the case where $|a+d|>2$, Theorem 2 can also be applied to derive the eigenfunctions of the LCT. However, since $a+d=a_2+d_2$, the parameters $\{a_2, b_2, c_2, d_2\}$ should be chosen properly such that $|a_2+d_2|>2$. We suggest that these parameters can be chosen as

$$\begin{aligned} &\{a_2,b_2,c_2,d_2\} = \left\{\sigma^{-1},0,0,\sigma\right\} \quad \text{when } a+d>2, \\ &\{a_2,b_2,c_2,d_2\} = \left\{-\sigma^{-1},0,0,-\sigma\right\} \quad \text{when } a+d<-2. \end{aligned} \tag{3.48}$$

The LCT with parameters $\{\sigma^{-1}, 0, 0, \sigma\}$ is a scaling operation. The LCT with parameters $\{-\sigma^{-1}, 0, 0, -\sigma\}$ is the combination of time reverse and the scaling operation:

$$O_{LCT}^{(\sigma^{-1},0,0,\sigma)}\left[x(t)\right] = \sqrt{\sigma}x\left(\sigma t\right), \quad O_{LCT}^{(-\sigma^{-1},0,0,-\sigma)}\left[x(t)\right] = \sqrt{-\sigma}x\left(-\sigma t\right). \tag{3.49}$$

It is known that the eigenfunctions of the scaling operation are scaling-invariant function (i.e., the fractal) [14]. Some simple examples of scaling-invariant functions are:

(a) constant:

$$O_{LCT}^{(\sigma^{-1},0,0,\sigma)}\left[C\right] = \sqrt{\sigma}C, \tag{3.50}$$

(b) delta function:

$$O_{LCT}^{(\sigma^{-1},0,0,\sigma)}\left[\delta(t)\right] = \sqrt{\sigma^{-1}}\delta(u), \tag{3.51}$$

(c) t^n:

$$O_{LCT}^{(\sigma^{-1},0,0,\sigma)}\left[t^n\right] = \sigma^{n+1/2}u^n. \tag{3.52}$$

Furthermore, the eigenfunctions of the LCT with parameters $\{-\sigma^{-1}, 0, 0, -\sigma\}$ is the symmetric (or asymmetric) scaling-invariant function. From Theorem 2, the LCT with parameters $\{a_1, b_1, c_1, d_1\}$ for these scaling-invariant functions will be the eigenfunctions of the LCT where $|a + d| > 2$.

In the case where $a + d > 2$, we suggest that $\{a_1, b_1, c_1, d_1\}$ can be chosen as:

$$\begin{bmatrix} a_1 & b_1 \\ c_1 & d_1 \end{bmatrix} = \begin{bmatrix} 1 & 0 \\ \tau & 1 \end{bmatrix} \begin{bmatrix} 1 & \eta \\ 0 & 1 \end{bmatrix} = \begin{bmatrix} 1 & \eta \\ \tau & \tau\eta + 1 \end{bmatrix}.$$

That is, the LCT with parameters $\{a_1, b_1, c_1, d_1\}$ is the combination of chirp convolution and chirp multiplication operation:

$$O_{LCT}^{(a_1,b_1,c_1,d_1)}[x(t)] = \frac{1}{\sqrt{j2\pi\eta}} \exp\left(\frac{j\tau u^2}{2}\right) \int_{-\infty}^{\infty} \exp\left(\frac{j(u-t)^2}{2\eta}\right) x(t)dt. \quad (3.53)$$

Then

$$\begin{bmatrix} a & b \\ c & d \end{bmatrix} = \begin{bmatrix} 1 & \eta \\ \tau & \tau\eta + 1 \end{bmatrix} \begin{bmatrix} \sigma^{-1} & 0 \\ 0 & \sigma \end{bmatrix} \begin{bmatrix} \tau\eta + 1 & -\eta \\ -\tau & 1 \end{bmatrix}$$
$$= \begin{bmatrix} \frac{\tau\eta + 1}{\sigma} - \tau\eta\sigma & \eta\left(\sigma - \frac{1}{\sigma}\right) \\ -\left(\tau^2\eta + \tau\right)\left(\sigma - \frac{1}{\sigma}\right) & (\tau\eta + 1)\,\sigma - \frac{\tau\eta}{\sigma} \end{bmatrix}, \quad (3.54)$$

where

$$\sigma = \left(a + d \pm \sqrt{(a+d)^2 - 4}\right)/2, \quad \eta = \frac{sb}{\sqrt{(a+d)^2 - 4}},$$
$$\tau = \frac{-2sc}{s\,(d-a) + \sqrt{(a+d)^2 - 4}}, \quad s = \operatorname{sgn}\left(\sigma - \sigma^{-1}\right). \quad (3.55)$$

Therefore, we obtain the following theorem:

Theorem 7. In the case where $a + d > 2$, the eigenfunctions of the LCT are

$$\phi_{\sigma,\tau,\eta}(t) = \exp\left(\frac{j\tau u^2}{2}\right) \int_{-\infty}^{\infty} \exp\left(\frac{j(u-t)^2}{2\eta}\right) g(t)dt, \quad (3.56)$$

where g(t) is a scaling-invariant function:

$$\sqrt{\sigma}g\,(\sigma t) = \lambda g(t), \quad (3.57)$$

and σ, τ, η are defined in (3.55). Moreover, the corresponding eigenvalue is also λ:

$$O_{LCT}^{(a,b,c,d)}\left[\phi_{\sigma,\tau,\eta}(t)\right] = \lambda\phi_{\sigma,\tau,\eta}(t). \quad (3.58)$$

Similarly, for the case where $a + d < -2$, the abcd matrix can be decomposed into

$$\begin{bmatrix} a & b \\ c & d \end{bmatrix} = \begin{bmatrix} 1 & \eta \\ \tau & \tau\eta + 1 \end{bmatrix} \begin{bmatrix} -\sigma^{-1} & 0 \\ 0 & -\sigma \end{bmatrix} \begin{bmatrix} \tau\eta + 1 & -\eta \\ -\tau & 1 \end{bmatrix}. \tag{3.59}$$

The values of τ, η are the same as those in (3.55), but σ and s are changed into

$$\sigma = \left(-a - d \pm \sqrt{(a+d)^2 - 4}\right)/2, \quad s = \operatorname{sgn}\left(\sigma^{-1} - \sigma\right). \tag{3.60}$$

Thus, we have the following theory:

Theorem 8. When $a + d < -2$, the eigenfunctions of the LCT are

$$\phi_{\sigma,\tau,\eta}(t) = \exp\left(\frac{j\tau u^2}{2}\right) \int_{-\infty}^{\infty} \exp\left(\frac{j(u-t)^2}{2\eta}\right) g(t) dt, \tag{3.61}$$

$$\text{where } \sqrt{\sigma} g(\sigma t) = \lambda g(t), \quad g(t) = \pm g(-t) \tag{3.62}$$

τ and η are defined the same as in (3.55) but σ and s are defined in (3.60). Moreover, the corresponding eigenvalues are $\pm(-1)^{1/2}\lambda$:

$$O_{LCT}^{(a,b,c,d)} \left[\phi_{\sigma,\tau,\eta}(t)\right] = \sqrt{-1}\lambda \phi_{\sigma,\tau,\eta}(t) \quad \text{if } g(t) = g(-t), \tag{3.63}$$

$$O_{LCT}^{(a,b,c,d)} \left[\phi_{\sigma,\tau,\eta}(t)\right] = -\sqrt{-1}\lambda \phi_{\sigma,\tau,\eta}(t) \quad \text{if } g(t) = -g(-t). \tag{3.64}$$

As other cases, when $|a + d| > 2$, one can also find a subset of eigenfunctions that forms a complete and orthogonal eigenfunctions set of the LCT.

In Table 3.1, we make a summation for the eigenfunctions of the LCT in all the cases.

3.5 Self-Linear Canonical Functions

As the theories about the self-functions of the FT and the FRFT [17–22], the summation of the eigenfunctions of the LCT with the same eigenvalues is also an eigenfunction of the LCT.

For example, for the case where $|a + d| < 2$, the functions in (3.17) are the eigenfunctions of the LCT and the corresponding eigenvalues are $[\exp(-j\alpha)]^{1/2}\exp(-j\alpha m)$. It can be verified that when

$$\alpha = \arg\left[\frac{a+d}{2} + j\left(\frac{\operatorname{sgn}(b)}{2}\sqrt{4-(a+d)^2}\right)\right] = 2\pi\frac{P}{Q}, \tag{3.65}$$

Table 3.1 The eigenfunctions of the LCT in all cases

	Style of eigenfunctions	Explicit formulas for eigenfunctions and eigenvalues
Case A $\lvert a+d \rvert < 2$	Scaling and chirp multiplication of Hermite Gaussian functions	(3.16), (3.17)
Case B $a+d=2,\ b \neq 0$	Chirp multiplication of periodic functions	(3.35), (3.36)
Case C $a+d=-2,\ b \neq 0$	Chirp multiplication of (a)symmetric periodic functions	(3.37), (3.38)
Case D $a+d=2,\ b=0$	Impulse trains	(3.42), (3.43)
Case E $a+d=-2,\ b=0$	(A)symmetric impulse trains	(3.44), (3.45), (3.46), (3.47)
Case F $a+d>2$	Chirp convolution and chirp multiplication of scaling-invariant functions	(3.56), (3.58)
Case G $a+d<-2$	Chirp convolution and chirp multiplication of (a)symmetric scaling-invariant functions	(3.61), (3.63), (3.64)

where P and Q are some integers, then $\phi_m^{(\sigma,\tau)}(t)$, $\phi_{m+Q}^{(\sigma,\tau)}(t)$, $\phi_{m+2Q}^{(\sigma,\tau)}(t)$, ... have the same eigenvalue. Therefore, if

$$\psi(t) = \sum_{k=0}^{\infty} \zeta_k \exp\left(\frac{-(1+i\tau)}{2\sigma^2}t^2\right) H_{m+kQ}\left(\frac{t}{\sigma}\right) \quad \text{where } m = 0, 1, \ldots, Q-1, \tag{3.66}$$

and ζ_k are any constants, then $\psi(t)$ is also the eigenfunction of the 2-D LCT:

$$O_{LCT}^{(a,b,c,d)}\left[\psi(t)\right] = \left[\exp\left(-j2\pi\frac{P}{Q}\right)\right]^{1/2} \exp\left(-j2\pi\frac{P}{Q}m\right)\psi(t). \tag{3.67}$$

Furthermore, we can use this fact to define the "self-linear canonical function."

Theorem 9. For any function f(t), if

$$\begin{aligned} f_n(t) = {} & f(t) + \exp\left(j\phi_1\right) O_{LCT}^{(a_1,b_1,c_1,d_1)}\left[f(t)\right] + \exp\left(j\phi_2\right) O_{LCT}^{(a_2,b_2,c_2,d_2)}\left[f(t)\right] + \cdots \\ & + \exp\left(j\phi_{Q-1}\right) O_{LCT}^{(a_{Q-1},b_{Q-1},c_{Q-1},d_{Q-1})}\left[f(t)\right], \end{aligned} \tag{3.68}$$

$$\text{where } \phi_S = \pi\frac{S}{Q} + 2\pi\frac{S}{Q}n, \quad n = 0, 1, \ldots, Q-1, \quad S = 0, 1, \ldots, Q-1, \tag{3.69}$$

$$\begin{bmatrix} a_S & b_S \\ c_S & d_S \end{bmatrix} = \begin{bmatrix} \cos(2\pi S/Q) + \tau \sin(2\pi S/Q) & \sigma^2 \sin(2\pi S/Q) \\ -\left(\tau^2+1\right)\sin(2\pi S/Q)/\sigma^2 & \cos(2\pi S/Q) - \tau \sin(2\pi S/Q) \end{bmatrix}, \tag{3.70}$$

Q is any integer, and τ and σ are any real constants, then $f_n(t)$ is the eigenfunction of the LCT with parameters {a, b, c, d}:

$$O_{LCT}^{(a,b,c,d)}[f_n(t)] = \left[\exp\left(-j2\pi\frac{P}{Q}\right)\right]^{-1/2} \exp\left(-j2\pi\frac{Pn}{Q}\right) f_n(t), \tag{3.71}$$

where P is any integer and

$$\begin{bmatrix} a & b \\ c & d \end{bmatrix} = \begin{bmatrix} \cos(2\pi P/Q) + \tau \sin(2\pi P/Q) & \sigma^2 \sin(2\pi P/Q) \\ -\left(\tau^2+1\right)\sin(2\pi P/Q)/\sigma^2 & \cos(2\pi P/Q) - \tau \sin(2\pi P/Q) \end{bmatrix}. \tag{3.72}$$

This theorem can be viewed as the extension of the self-Fourier functions and the self-fractional Fourier functions in [17–22]. $f_n(t)$ in (3.68) can be named as the self-linear canonical function.

Theorem 9 can be proven from the fact that f(t) can be expressed as a linear combination of $\phi_m^{(\sigma,\tau)}(t)$ (defined in (3.17))

$$f(t) = \sum_{m=0}^{\infty} \beta_m \phi_m^{(\sigma,\tau)}(t) \quad \text{where } \beta_m = \int_{-\infty}^{\infty} f(t)\overline{\phi_m^{(\sigma,\tau)}(t)}\, dt, \tag{3.73}$$

$$\exp(j\phi_S)\, O_{LCT}^{(a_S,b_S,c_S,d_S)}\left[\phi_m^{(\sigma,\tau)}(t)\right] = \exp\left(j2\pi\frac{S}{Q}(n-m)\right)\phi_m^{(\sigma,\tau)}(t), \tag{3.74}$$

$$f_n(t) = \sum_{m=0}^{\infty} \beta_m \sum_{S=0}^{Q-1} \exp\left(j2\pi\frac{S}{Q}(n-m)\right)\phi_m^{(\sigma,\tau)}(t) = \sum_{k=0}^{\infty} N\beta_{n+kN}\phi_{n+kN}^{(\upsilon,\iota)}(t). \tag{3.75}$$

3.6 Eigenfunctions in the 2D Case

Recently, the eigenfunctions of the two-dimensional nonseparable linear canonical transform (2-D NSLCT) were derived successfully [23, 24]. The 2-D NSLCT is the two-dimensional counterpart of the 1-D LCT and is even useful in optical system analysis [25, 26]. Its formula is:

$$O_{NSLCT}^{(\mathbf{A},\mathbf{B},\mathbf{C},\mathbf{D})}[f(\mathbf{x})] = (-\det(\mathbf{B}))^{-1/2} \int \exp\left[j\left(\mathbf{w}\mathbf{D}\mathbf{B}^{-1}\mathbf{w}^{\mathbf{T}} - 2\mathbf{x}\mathbf{B}^{-1}\mathbf{w}^{\mathbf{T}} + \mathbf{x}\mathbf{B}^{-1}\mathbf{A}\mathbf{x}^{\mathbf{T}}\right)/2\right] f(\mathbf{x})\, d\mathbf{x},$$

where

$$\mathbf{x} = [x\ y]^T,\ \ \mathbf{w} = [f\ h]^T,$$
$$\mathbf{A} = \begin{bmatrix} a_{11} & a_{12} \\ a_{21} & a_{22} \end{bmatrix},\ \mathbf{B} = \begin{bmatrix} b_{11} & b_{12} \\ b_{21} & b_{22} \end{bmatrix},\ \mathbf{C} = \begin{bmatrix} c_{11} & c_{12} \\ c_{21} & c_{22} \end{bmatrix},\ \mathbf{D} = \begin{bmatrix} d_{11} & d_{12} \\ d_{21} & d_{22} \end{bmatrix}. \tag{3.76}$$

The 2-D NSLCT has 16 parameters and is very complicated. However, using the method similar to that in Theorem 2, the eigenfunctions and eigenvalues of the 2-D NSLCT can still be derived successfully [23, 24].

To derive the eigenfunctions of the 2-D NSLCT, first, we determine the eigenvalues of the ABCD matrix, which is a 4×4 matrix with the following form:

$$\begin{bmatrix} \mathbf{A} & \mathbf{B} \\ \mathbf{C} & \mathbf{D} \end{bmatrix}. \tag{3.77}$$

Then, according to whether the eigenvalues of the ABCD matrix are real and whether these eigenvalues have unitary amplitudes, we divide the derivation into five classes [24]. In different cases, the methods for deriving the eigenfunctions are different.

For example, in the case where all the four eigenfunctions of the ABCD matrix are real (i.e., Case 1 in [24]), the ABCD matrix can be decomposed into

$$\begin{bmatrix} \mathbf{A} & \mathbf{B} \\ \mathbf{C} & \mathbf{D} \end{bmatrix} = \begin{bmatrix} \mathbf{A_1} & \mathbf{B_1} \\ \mathbf{C_1} & \mathbf{D_1} \end{bmatrix} \begin{bmatrix} \mathbf{A_2} & \mathbf{B_2} \\ \mathbf{C_2} & \mathbf{D_2} \end{bmatrix} \begin{bmatrix} \mathbf{A_1} & \mathbf{B_1} \\ \mathbf{C_1} & \mathbf{D_1} \end{bmatrix}^{-1}, \tag{3.78}$$

where

$$\begin{bmatrix} \mathbf{A_1} & \mathbf{B_1} \\ \mathbf{C_1} & \mathbf{D_1} \end{bmatrix} = \begin{bmatrix} \mathbf{I} & \mathbf{0} \\ \mathbf{E_a} & \mathbf{I} \end{bmatrix} \begin{bmatrix} \mathbf{I} & \mathbf{\Phi_1} \\ \mathbf{0} & \mathbf{I} \end{bmatrix} \begin{bmatrix} \mathbf{E_2} & \mathbf{0} \\ \mathbf{0} & (\mathbf{E_2^T})^{-1} \end{bmatrix},\ \ \mathbf{E_a} = \begin{bmatrix} \eta_1 & \eta_2 \\ \eta_2 & \eta_3 \end{bmatrix},\ \ \mathbf{\Phi_1} = \begin{bmatrix} \tau_1 & 0 \\ 0 & 0 \end{bmatrix}$$
$$\eta_1 = \frac{\eta_2^2 (b_{12} - b_{21}) + \eta_2 (a_{11} + d_{11} - a_{22} - d_{22}) + \eta_3 (d_{12} + a_{21}) + c_{12} - c_{21}}{\eta_3 (b_{12} - b_{21}) + a_{12} + d_{21}},$$
$$\tau_1 = \frac{b_{12} - b_{21}}{a_{21} + d_{12} + \eta_1 (b_{21} - b_{12})},\quad \eta_2 \text{ and } \eta_3 \text{ are free to choose,} \tag{3.79}$$

the columns of $\mathbf{E_2}$ are the eigenvectors of the matrix $\mathbf{A_3}$:

$$\begin{bmatrix} \mathbf{A_3} & \mathbf{B_3} \\ \mathbf{C_3} & \mathbf{D_3} \end{bmatrix} = \begin{bmatrix} \mathbf{I} & -\mathbf{\Phi_1} \\ \mathbf{0} & \mathbf{I} \end{bmatrix} \begin{bmatrix} \mathbf{I} & \mathbf{0} \\ -\mathbf{E_a} & \mathbf{I} \end{bmatrix} \begin{bmatrix} \mathbf{A} & \mathbf{B} \\ \mathbf{C} & \mathbf{D} \end{bmatrix} \begin{bmatrix} \mathbf{I} & \mathbf{0} \\ \mathbf{E_a} & \mathbf{I} \end{bmatrix} \begin{bmatrix} \mathbf{I} & \mathbf{\Phi_1} \\ \mathbf{0} & \mathbf{I} \end{bmatrix}, \tag{3.80}$$

and the diagonal entries of $\mathbf{A_2}$ are the eigenvalues of $\mathbf{A_3}$. In [24], we proved that $\mathbf{A_2}$, $\mathbf{B_2}$, $\mathbf{C_2}$, and $\mathbf{D_2}$ in (3.79) are all real diagonal matrices and the 2-D NSLCT with parameters $\{\mathbf{A_2}, \mathbf{B_2}, \mathbf{C_2}, \mathbf{D_2}\}$ are the 2-D separable LCT. Therefore, if

$$O_{LCT}^{(a_1,b_1,c_1,d_1)} [e_1(x)] = \lambda_1 e_1(x),\quad O_{LCT}^{(a_2,b_2,c_2,d_2)} [e_2(x)] = \lambda_2 e_2(x), \tag{3.81}$$

where a_1, a_2, b_1, b_2, c_1, c_2, d_1, d_2 are the diagonal entries of $\mathbf{A_2}$, $\mathbf{B_2}$, $\mathbf{C_2}$, and $\mathbf{D_2}$, then

$$E(x, y) = O_{NSLCT}^{(\mathbf{A_1},\mathbf{B_1},\mathbf{C_1},\mathbf{D_1})}\left[e_1(x)e_2(x)\right], \tag{3.82}$$

is the eigenfunction of the 2-D NSLCT and the corresponding eigenvalue is $\lambda_1\lambda_2$.

3.7 Conclusion

The eigenfunctions and eigenvalues of the LCT are discussed in this chapter. The discussion is divided into three cases: $|a+d|<2$, $|a+d|=2$, and $|a+d|>2$. In these cases, the eigenfunctions of the LCT are closely related to Hermite–Gaussian functions, periodic functions, and scaling-invariant functions, respectively. The eigenfunctions of the LCT in the 2-D case are also discussed briefly.

Moreover, the eigenfunctions of the gyrator transform, which is a special case of the 2-D NSLCT, was discussed in [27]. The eigenfunctions of the offset LCT, which is the time-shifting and modulation of the original LCT, was discussed in [28].

Since the optical system [4, 13, 29, 30], the radar system [31], and the gradient index fiber system [32] can be modeled by the LCT, the eigenfunctions of the LCT are helpful for analyzing the self-imaging phenomenon in the optical system and the resonance phenomena in the radar system and the gradient-index fiber system. These are discussed in [15, 16, 25, 26, 28].

References

1. K.B. Wolf, Construction and properties of linear canonical transforms, in Integral Transforms in Science and Engineering, Ch. 9 (Plenum, New York, 1979)
2. M. Moshinsky, C. Quesne, Linear canonical transformations and their unitary representations. J. Math. Phys. **12**(8), 1772–1783 (1971)
3. S. Abe, J.T. Sheridan, Generalization of the fractional Fourier transformation to an arbitrary linear lossless transformation: an operator approach. J. Phys. A **27**, 4179–4187 (1994)
4. L.M. Bernardo, ABCD matrix formalism of fractional Fourier optics. Opt. Eng. **35**(3), 732–740 (1996)
5. J.J. Healy, B.M. Hennelly, J.T. Sheridan, Additional sampling criterion for the linear canonical transform. Opt. Lett. **33**, 2599–2601 (2008)
6. J.J. Healy, J.T. Sheridan, Fast linear canonical transform. J. Opt. Soc. Am. A **27**, 21–30 (2010)
7. V. Namias, The fractional order Fourier transform and its application to quantum mechanics. J. Inst. Math. Appl. **25**, 241–265 (1980)
8. H.M. Ozaktas, Z. Zalevsky, M.A. Kutay, *The Fractional Fourier Transform with Applications in Optics and Signal Processing* (Wiley, New York, 2000)
9. K. Paiorski, The self-imaging phenomenon and its applications, in Progress in Optics, ed. by E. Wolf, vol. 27, part 1 (North-Holland, Amsterdam, 1989)
10. J.T. Winthrop, C.R. Worthington, Theory of Fresnel images. 1. Plane periodic objects in monochromatic light. J. Opt. Soc. Am. **55**, 373–381 (1965)

11. A.W. Lohmann, An array illuminator based on the Talbot effect. Optik (Stuttgart) **79**, 41–45 (1988)
12. J. Leger, G.J. Swanson, Efficient array illuminator using binary-optics phase plates as fractional Talbot planes. Opt. Lett. **15**, 288–290 (1990)
13. J.W. Goodman, *Introduction to Fourier Optics*, 3rd edn. (Roberts & Co., Englewood, 2005)
14. G.W. Wronell, *Signal Processing with Fractals* (Prentice-Hall, Upper Saddle River, 1996)
15. S.C. Pei, J.J. Ding, Eigenfunctions of linear canonical transform. IEEE Trans. Signal Process. **50**(1), 11–26 (2002)
16. D.F.V. James, G.S. Agarwal, The generalized Fresnel transform and its applications to optics. Opt. Commun. **126**, 207–212 (1996)
17. M.J. Caola, Self-Fourier functions. J. Phys. A: Math. Gen. **24**, 1143–1144 (1991)
18. G. Cincotti, F. Gori, M. Santarsiero, Generalized self-Fourier functions. J. Phys. A: Math. Gen. **25**, 1191–1194 (1992)
19. D. Mendlovic, H.M. Ozaktas, A.W. Lohmann, Self Fourier functions and fractional Fourier transform. Opt. Commun. **105**, 36–38 (1994)
20. T. Alieva, On the self-fractional Fourier functions. J. Phys. A: Math. Gen. **29**, 377–379 (1996)
21. T. Alieva, A.M. Barbe, Self-fractional Fourier functions and selection of modes. J. Phys. A: Math. Gen. **30**, 211–215 (1997)
22. T. Alieva, A.M. Barbe, Self-imaging in fractional Fourier transform systems. Opt. Commun. **152**, 11–15 (1998)
23. M.J. Bastiaans, T. Alieva, Classification of lossless first order optical systems and the linear canonical transformation. J. Opt. Soc. Am. A **24**, 1053–1062 (2007)
24. J.J. Ding, S.C. Pei, Eigenfunctions and self-imaging phenomena of the two dimensional nonseparable linear canonical transform. J. Opt. Soc. Am. A **28**(2), 82–95 (2011)
25. G.B. Folland, *Harmonic Analysis in Phase Space* (Princeton University Press, Princeton, 1989)
26. A. Koç, H.M. Ozaktas, L. Hesselink, Fast and accurate computation of two-dimensional non-separable quadratic-phase integrals. J. Opt. Soc. Am. A **27**, 1288–1302 (2010)
27. S.C. Pei, J.J. Ding, Properties, digital implementation, applications, and self image phenomena of the Gyrator transform, in European Signal Processing Conference (August 2009), pp. 441–445
28. S.C. Pei, J.J. Ding, Eigenfunctions of the offset Fourier, fractional Fourier, and linear canonical transforms. J. Opt. Soc. Am. A **20**(3), 522–532 (2003)
29. M.J. Bastiaans, Propagation laws for the second-order moments of the Wigner distribution function in first-order optical systems. Optik (Stuttgart) **82**, 173–181 (1989)
30. H.M. Ozaktas, D. Mendlovic, Fractional Fourier optics. J. Opt. Soc. Am. A **12**, 743–751 (1995)
31. P. Pellat-Finet, G. Bonnet, Fractional order Fourier transform and Fourier optics. Opt. Commun. **111**, 141–154 (1994)
32. D. Mendlovic, H.M. Ozaktas, A.W. Lohmann, Gradedindex fibers, Wigner distribution and the fractional Fourier transform. Appl. Opt. **33**, 6188–6193 (1994)

Chapter 4
Uncertainty Principles and the Linear Canonical Transform

Ran Tao and Juan Zhao

Abstract In this chapter some uncertainty principles for the linear canonical transform (LCT) have been introduced. For the Heisenberg's principles there exist different bounds for real and complex signals. Based on the LCT moments properties the lower bounds related to the covariance of time and frequency have been derived, which can reduce to different bounds for real and complex signals because real signals have zero covariance. Furthermore, some extensions of uncertainty principles including the logarithmic, entropic and Renyi entropic uncertainty principles are deduced based on the relationship between the LCT and the Fourier transform.

4.1 Introduction

The uncertainty principles are very important in signal processing. The classic Heisenberg's principle provides the lower bound on the spreads of the signal energy in the time and frequency domains [1, 2]. Recently researchers have investigated the Heisenberg's principles for the linear canonical transform (LCT) [3–8] and its special cases such as fractional Fourier transform (FrFT) [3, 9–13]. It is shown that there exist different bounds for real and complex signals in the LCT and FrFT domains.

The Heisenberg's uncertainty principles are closely related to the signal moments. The LCT's uncertainty relations can be derived by using properties of the LCT moments. The lower bounds are related to the covariance of time and frequency, which can explain the phenomenon of different bounds for real and complex signals existing in the Heisenberg's uncertainty relations proposed before because real signals have zero covariance.

On the other hand, there are some different uncertainty principles such as the logarithmic, entropic and Renyi entropic uncertainty principles, which are extensions of the Heisenberg's uncertainty principle. These uncertainty relations in

R. Tao (✉) • J. Zhao
Department of Electronic Engineering, Beijing Institute of Technology, Beijing 100081, China

J.J. Healy et al. (eds.), *Linear Canonical Transforms*, Springer Series in Optical Sciences 198, DOI 10.1007/978-1-4939-3028-9_4

the LCT and FrFT domains have been presented in [13–16]. It is shown that these lower bounds are only relevant to the transform parameters and could possibly tend to zeros.

4.2 Heisenberg's Uncertainty Principles

The classical Heisenberg's uncertainty principle in the Fourier transform (FT) domain is very important in signal processing, particularly time-frequency analysis, which states that a signal cannot be both time-limited and band-limited.

Assume that $x(t)$ is a complex signal such that $t|x(t)| \in L_2(R)$ and the energy of the signal is

$$E = \int_{-\infty}^{+\infty} |x(t)|^2 dt = \int_{-\infty}^{+\infty} |X(\omega)|^2 d\omega.$$

For the convenience of analysis, we express the signal $x(t)$ and its FT $X(\omega)$ in terms of their respective amplitudes and phases, i.e.,

$$x(t) = |x(t)|e^{j\phi(t)}, X(\omega) = |X(\omega)|e^{j\psi(\omega)}.$$

The first and second order moments in the time and frequency domains are defined as [2]

$$\langle t \rangle \triangleq \frac{1}{E} \int_{-\infty}^{+\infty} t|x(t)|^2 dt \tag{4.1}$$

$$\langle t^2 \rangle \triangleq \frac{1}{E} \int_{-\infty}^{+\infty} t^2|x(t)|^2 dt \tag{4.2}$$

$$\langle \omega \rangle \triangleq \frac{1}{E} \int_{-\infty}^{+\infty} \omega|X(\omega)|^2 d\omega = \frac{1}{E} \int_{-\infty}^{+\infty} \phi'(t)|x(t)|^2 dt \tag{4.3}$$

$$\langle \omega^2 \rangle \triangleq \frac{1}{E} \int_{-\infty}^{+\infty} \omega^2|X(\omega)|^2 d\omega = \frac{1}{E} \int_{-\infty}^{\infty} |x'(t)|^2 dt, \tag{4.4}$$

where $(\cdot)'$ denotes derivative. Then

$$\Delta t^2 \Delta\omega^2 \geq 1/4 \tag{4.5}$$

where

$$\Delta t^2 \triangleq \frac{1}{E}\int_{-\infty}^{+\infty} (t - \langle t\rangle)^2 |x(t)|^2 dt = \langle t^2\rangle - \langle t\rangle^2 \tag{4.6}$$

$$\Delta \omega^2 \triangleq \frac{1}{E}\int_{-\infty}^{+\infty} (\omega - \langle \omega\rangle)^2 |X(\omega)|^2 d\omega = \langle \omega^2\rangle - \langle \omega\rangle^2 \tag{4.7}$$

In [2] Cohen defined the first mixed moment of time and frequency

$$\langle t\phi'(t)\rangle \triangleq \frac{1}{E}\int_{-\infty}^{+\infty} t\phi'(t)|x(t)|^2 dt = -\frac{1}{E}\int_{-\infty}^{+\infty} \omega\psi'(\omega)|X(\omega)|^2 d\omega \tag{4.8}$$

and the covariance of time and frequency

$$\mathrm{Cov}_x \triangleq \langle t\phi'(t)\rangle - \langle t\rangle\langle \omega\rangle \tag{4.9}$$

It should be noted that $\langle \omega\rangle = 0$ and $\langle t\phi'(t)\rangle = 0$ for real signals according to (4.3) and (4.8), which implies that $\mathrm{Cov}_x = 0$ when $x(t)$ is real.

Furthermore, Cohen derived the following uncertainty principle, which is related to the covariance of time and frequency [2].

Theorem 4.1. *For an arbitrary signal* $x(t) = |x(t)|e^{j\phi(t)}$, *then*

$$\Delta t^2 \Delta \omega^2 \geq (\mathrm{Cov}_x)^2 + \frac{1}{4} \tag{4.10}$$

Since the covariances can be zero for real and complex signals, (4.10) will reduce to (4.5). That is to say, the lower bounds in the FT domain are the same for real and complex signals.

4.2.1 *The LCT Moments and Spreads*

From (4.5) to (4.7), it is obvious that the Heisenberg's uncertainty principles are closely related to the signal moments. In this section we will present the properties of the LCT moments, which will be useful for deriving Heisenberg's uncertainty relations for the LCT.

Assume the LCT of $x(t)$ with parameter $M = (a, b, c, d)$ is $X_M(u) = |X_M(u)|e^{j\varphi(u)}$. The energy of the signal is

$$E = \int_{-\infty}^{+\infty} |x(t)|^2 dt = \int_{-\infty}^{+\infty} |X(\omega)|^2 d\omega = \int_{-\infty}^{+\infty} |X_M(u)|^2 du$$

According to the definitions of signal moments in the time and frequency domains, the first and second order moments in the LCT domain can be defined similarly, which are, respectively, denoted by [7]

$$\langle u_M \rangle \triangleq \frac{1}{E}\int_{-\infty}^{+\infty} u|X_M(u)|^2 du \tag{4.11}$$

$$\langle u_M^2 \rangle \triangleq \frac{1}{E}\int_{-\infty}^{+\infty} u^2|X_M(u)|^2 du \tag{4.12}$$

According to the additivity property of the LCT, the FT of $X_M(u)$ is $X_{FM}(u)$, where $F = (0, 1, -1, 0)$ and $FM = (c, d, -a, -b)$. The first and second order moments of $X_{FM}(u)$ are, respectively, denoted by

$$\langle u_{FM} \rangle = \frac{1}{E}\int_{-\infty}^{+\infty} u|X_{FM}(u)|^2 du = \frac{1}{E}\int_{-\infty}^{+\infty} \varphi'(u)|X_M(u)|^2 du \tag{4.13}$$

$$\langle u_{FM}^2 \rangle = \frac{1}{E}\int_{-\infty}^{+\infty} u^2|X_{FM}(u)|^2 du = \frac{1}{E}\int_{-\infty}^{\infty} |X_M'(u)|^2 du \tag{4.14}$$

The first mixed moment of the LCT and its FT can also be defined as

$$\langle u_M \varphi'(u_M) \rangle \triangleq \frac{1}{E}\int_{-\infty}^{+\infty} u\varphi'(u)|X_M(u)|^2 du \tag{4.15}$$

and the covariance of the LCT and its FT is

$$\mathrm{Cov}_{X_M} \triangleq \langle u_M \varphi'(u_M) \rangle - \langle u_M \rangle \langle u_{FM} \rangle \tag{4.16}$$

Then we consider the spreads of the signal in the LCT domain, which can be expressed in terms of the first and second order moments in the LCT domains, i.e.,

$$\Delta u_M^2 \triangleq \frac{1}{E}\int_{-\infty}^{+\infty} (u - \langle u_M \rangle)^2 |X_M(u)|^2 du = \langle u_M^2 \rangle - \langle u_M \rangle^2 \tag{4.17}$$

The relations between the moments in the LCT domain and those in the time and frequency domains for complex signals are given in Table 4.1. For real signals the properties of the LCT moments can be simplified as shown in Table 4.2 because $\langle \omega \rangle = 0$ and $\langle t\phi'(t) \rangle = 0$. The results of moments in the FrFT domain [17] are special cases of Tables 4.1 and 4.2.

The derivations of the second order moment and first mixed moment are given in Appendix. In addition, these relationships can be written in the following matrix form, i.e.,

$$\begin{pmatrix} \langle u_M^2 \rangle & \langle u_M \varphi'(u_M) \rangle \\ \langle u_M \varphi'(u_M) \rangle & \langle u_{FM}^2 \rangle \end{pmatrix} = \begin{pmatrix} a & b \\ c & d \end{pmatrix} \begin{pmatrix} \langle t^2 \rangle & \langle t\phi'(t) \rangle \\ \langle t\phi'(t) \rangle & \langle \omega^2 \rangle \end{pmatrix} \begin{pmatrix} a & c \\ b & d \end{pmatrix} \tag{4.18}$$

Table 4.1 Some properties of the LCT moments and spreads for complex signals [7]

Classes	Property
First order moment	$\langle u_M \rangle = a\langle t \rangle + b\langle \omega \rangle$
	$\langle u_{FM} \rangle = c\langle t \rangle + d\langle \omega \rangle$
Second order moment	$\langle u_M^2 \rangle = a^2\langle t^2 \rangle + b^2\langle \omega^2 \rangle + 2ab\langle t\phi'(t) \rangle$
	$\langle u_{FM}^2 \rangle = c^2\langle t^2 \rangle + d^2\langle \omega^2 \rangle + 2cd\langle t\phi'(t) \rangle$
First mixed moment	$\langle u_M \varphi'(u_M) \rangle = ac\langle t^2 \rangle + bd\langle \omega^2 \rangle + (ad+bc)\langle t\phi'(t) \rangle$
	$\langle u_M \varphi'(u_M) \rangle + \langle t\phi'(t) \rangle = \frac{d}{b}\langle u_M^2 \rangle - \frac{a}{b}\langle t^2 \rangle$
The spread	$\Delta u_M^2 = a^2 \Delta t^2 + b^2 \Delta \omega^2 + 2ab\mathrm{Cov}_x$
The covariance	$\mathrm{Cov}_{x_M} = ac\Delta t^2 + bd\Delta\omega^2 + (ad+bc)\mathrm{Cov}_x$
	$\mathrm{Cov}_{x_M} + \mathrm{Cov}_x = \frac{d}{b}\Delta u_M^2 - \frac{a}{b}\Delta t^2$

Table 4.2 Some properties of the LCT moments and spreads for real signals [7]

Classes	Property
First order moment	$\langle u_M \rangle = a\langle t \rangle$
	$\langle u_{FM} \rangle = c\langle t \rangle$
Second order moment	$\langle u_M^2 \rangle = a^2\langle t^2 \rangle + b^2\langle \omega^2 \rangle$
	$\langle u_{FM}^2 \rangle = c^2\langle t^2 \rangle + d^2\langle \omega^2 \rangle$
First mixed moment	$\langle u_M \varphi'(u_M) \rangle = ac\langle t^2 \rangle + bd\langle \omega^2 \rangle$
	$\langle u_M \varphi'(u_M) \rangle = \frac{d}{b}\langle u_M^2 \rangle - \frac{a}{b}\langle t^2 \rangle$
The spread	$\Delta u_M^2 = a^2 \Delta t^2 + b^2 \Delta \omega^2$
The covariance	$\mathrm{Cov}_{x_M} = ac\Delta t^2 + bd\Delta\omega^2$
	$\mathrm{Cov}_{x_M} = \frac{d}{b}\Delta u_M^2 - \frac{a}{b}\Delta t^2$

4.2.2 Uncertainty Principles for Complex and Real Signals

The Heisenberg' uncertainty relations for the LCT and FrFT have been investigated in [3–13], which are different from the uncertainty relations in the FT domain shown by (4.5) that there exist different bounds for complex and real signals. In the following the uncertainty relations for the LCT are given.

- For complex signal $x(t)$, then

$$\Delta t^2 \Delta u_M^2 \geq \frac{b^2}{4} \tag{4.19}$$

  and the equality is achieved iff

$$x(t) = \left(\frac{1}{\pi\sigma^2}\right)^{\frac{1}{4}} e^{-\frac{(t-t_0)^2}{2\sigma^2}} e^{j(-\frac{a}{2b}t^2 + \Omega t + \varphi)} \tag{4.20}$$

  where $t_0, \sigma, \Omega, \varphi$ are arbitrary real constants.

Furthermore, for $M_1 = (a_1, b_1, c_1, d_1)\,, M_2 = (a_2, b_2, c_2, d_2)$, then

$$\Delta u_{M_1}^2 \Delta u_{M_2}^2 \geq \frac{(a_1 b_2 - a_2 b_1)^2}{4} \tag{4.21}$$

and the equality is achieved iff

$$x(t) = \left(\frac{1}{\pi \Delta t^2}\right)^{\frac{1}{4}} e^{-\frac{(t-t_0)^2}{2\Delta t^2}} e^{j(-\frac{K}{2}t^2 + \Omega t + \varphi)}, \tag{4.22}$$

where $\Delta t^2 = \frac{b_1^2}{2\sigma^2} + \frac{b_2^2}{(a_1 b_2 - a_2 b_1)^2} 2\sigma^2$, $K = [\frac{a_1 b_1}{2\sigma^2} + \frac{a_2 b_2}{(a_1 b_2 - a_2 b_1)^2} 2\sigma^2]/\Delta t^2$ and $t_0, \sigma, \Omega, \varphi$ are arbitrary real constants.

- For real signal $x(t)$, then

$$\Delta t^2 \Delta u_M^2 \geq (a \Delta t^2)^2 + \frac{b^2}{4} \tag{4.23}$$

and

$$\Delta u_{M_1}^2 \Delta u_{M_2}^2 \geq \left(a_1 a_2 \Delta t^2 + \frac{b_1 b_2}{4 \Delta t^2}\right)^2 + \frac{(a_1 b_2 - a_2 b_1)^2}{4} \tag{4.24}$$

and the above two equalities are achieved iff

$$x(t) = \left(\frac{1}{\pi \sigma^2}\right)^{\frac{1}{4}} e^{-\frac{(t-t_0)^2}{2\sigma^2}} \tag{4.25}$$

where t_0, σ are arbitrary real constants.

The uncertainty principles in the FrFT domain [3, 9–13] are special cases of the above results. These results show that the uncertainty principle for complex signals can be achieved by a complex chirp signal with Gaussian envelope and the uncertainty principle for real signals can be achieved by a Gaussian signal.

To explain the phenomenon of different lower bounds for real and complex signals existing in the Heisenberg's uncertainty relations proposed before. In the following we present two Heisenberg's uncertainty relations in the LCT domains, which are related to the covariance of time and frequency.

Theorem 4.2 gives a tighter lower bound on the uncertainty product of the time and LCT spreads [7].

Theorem 4.2. *For an arbitrary signal* $x(t) = |x(t)|e^{j\phi(t)}, M = (a, b, c, d), X_M(u) = |X_M(u)|e^{j\varphi(u)}$ *, then*

$$\Delta t^2 \Delta u_M^2 \geq (a \Delta t^2 + b \mathrm{Cov}_x)^2 + \frac{b^2}{4} \tag{4.26}$$

Proof. Using the property of the LCT spread given in Table 4.1 and Theorem 4.1, we can obtain

$$\begin{aligned}\Delta t^2 \Delta u_M^2 &= \Delta t^2 (a^2 \Delta t^2 + b^2 \Delta \omega^2 + 2ab\mathrm{Cov}_x) \\ &= (a\Delta t^2 + b\mathrm{Cov}_x)^2 + b^2[\Delta t^2 \Delta \omega^2 - (\mathrm{Cov}_x)^2] \\ &\geq (a\Delta t^2 + b\mathrm{Cov}_x)^2 + \tfrac{b^2}{4}\end{aligned}$$

The theorem is proved. □

From Theorem 4.2, it is clear that the covariance of complex signals can make the first term of the right part of (4.26) be zero. It will lead to different bounds for real and complex signals in the LCT domain [3, 4]. That is to say, $(a\Delta t^2)^2 + \frac{b^2}{4}$ for real signals and $\frac{b^2}{4}$ for complex signals because real signals have zero covariance.

When $M = (\cos\alpha, \sin\alpha, -\sin\alpha, \cos\alpha)$, we can obtain the following uncertainty relation in the FrFT domain, i.e.,

$$\Delta t^2 \Delta u_\alpha^2 \geq (\cos\alpha \Delta t^2 + \sin\alpha \mathrm{Cov}_x)^2 + \frac{\sin^2\alpha}{4} \tag{4.27}$$

Furthermore, we consider the uncertainty relation in two LCT domains, which is given by the following theorem [7].

Theorem 4.3. *For an arbitrary signal* $x(t) = |x(t)|e^{j\phi(t)}, M = (a, b, c, d), X_M(u) = |X_M(u)|e^{j\varphi(u)}$ *, then*

$$\Delta u_{M_1}^2 \Delta u_{M_2}^2 \geq [a_1 a_2 \Delta t^2 + b_1 b_2 \Delta\omega^2 + (a_1 b_2 + b_1 a_2)\mathrm{Cov}_x]^2 + \frac{(a_1 b_2 - b_1 a_2)^2}{4} \tag{4.28}$$

Proof. Using the additivity and inversity properties of the LCT, we have

$$\begin{aligned}X_{M_2}(u) - L_{(a_2,b_2,c_2,d_2)}(x(t)) &= L_{(a_2,b_2,c_2,d_2)} L_{(d_1,-b_1,-c_1,a_1)}(X_{M_1}(u)) \\ &= L_{(a_3,b_3,c_3,d_3)}(X_{M_1}(u))\end{aligned}$$

where

$$\begin{pmatrix} a_3 & b_3 \\ c_3 & d_3 \end{pmatrix} = \begin{pmatrix} a_2 & b_2 \\ c_2 & d_2 \end{pmatrix} \begin{pmatrix} d_1 & -b_1 \\ -c_1 & a_1 \end{pmatrix}$$

From Theorem 4.2, we can obtain

$$\Delta u_{M_1}^2 \Delta u_{M_2}^2 \geq [a_3 \Delta u_{M_1}^2 + b_3 \mathrm{Cov}_{X_{M_1}}]^2 + \frac{b_3^2}{4}$$

Substituting the property of the LCT spread and covariance given in Table 4.1 into the right part of the above inequality, (4.28) can be derived. □

The tighter lower bound given in Theorem 4.3 is also related to the covariance of time and frequency of the signal, which can similarly explain the phenomenon of different lower bounds for real and complex signals given in [5, 6] because real signals have zero covariance.

Similarly, when $M = (\cos\alpha, \sin\alpha, -\sin\alpha, \cos\alpha)$, the uncertainty relation in the FrFT domain is given below.

$$\Delta u_\alpha^2 \Delta u_\beta^2 \geq [\cos\alpha\cos\beta\Delta t^2 + \sin\alpha\sin\beta\Delta\omega^2 + \sin(\alpha+\beta)\mathrm{Cov}_x]^2 + \frac{\sin^2(\alpha-\beta)}{4} \tag{4.29}$$

Theorem 4.4. *The tighter lower bound given in Theorems 4.2 and 4.3 can be achieved when*

$$x(t) = \left(\frac{1}{2\pi\Delta t^2}\right)^{\frac{1}{4}} e^{-\frac{(t-t_0)^2}{4\Delta t^2}} e^{j(-\frac{K}{2}t^2+\Omega t+\varphi)} \tag{4.30}$$

where t_0, Ω, φ are arbitrary real constants.

Proof. In the following, by computation we will show that the tighter lower bound given in Theorems 4.2 and 4.3 can be achieved by the signal given in (4.30). For the signal given in (4.30), using the definitions of the moments and spreads, we can obtain

$$\begin{aligned}
\langle\omega\rangle &= -Kt_0 + \Omega \\
\langle\omega^2\rangle &= \frac{1}{4\Delta t^2} + K^2\langle t^2\rangle + \Omega^2 + 2K\Omega t_0 \\
\Delta\omega^2 &= \frac{1}{4\Delta t^2} + K^2\Delta t^2 \\
\langle t\phi'(t)\rangle &= -K\langle t^2\rangle + \Omega t_0 \\
\mathrm{Cov}_x &= -K\Delta t^2 \\
\Delta u_M^2 &= \Delta t^2(a-bK)^2 + \frac{b^2}{4\Delta t^2}
\end{aligned}$$

For the time and LCT domains, we have

$$\Delta t^2\Delta u_M^2 = (a\Delta t^2 - bK\Delta t^2)^2 + \frac{b^2}{4} = (a\Delta t^2 + b\mathrm{Cov}_x)^2 + \frac{b^2}{4}$$

i.e., the inequality (4.26) becomes equality. If $K = a/b$ (or $K = 0, \Omega = 0, \varphi = 0$) with $\Delta t^2 = \sigma^2/2$, we can obtain (4.19) (or (4.23)) and (4.20) (or (4.25)).

Furthermore, for two LCT domains, we can obtain

$$\begin{aligned}
\Delta u_{M_1}^2\Delta u_{M_2}^2 &= \left[\Delta t^2(a_1-b_1K)^2 + \frac{b_1^2}{4\Delta t^2}\right]\left[\Delta t^2(a_2-b_2K)^2 + \frac{b_2^2}{4\Delta t^2}\right] \\
&= [a_1a_2\Delta t^2 + b_1b_2K^2\Delta t^2 - (a_1b_2
\end{aligned}$$

$$+ b_1a_2)K\Delta t^2]^2 + \frac{b_1^2b_2^2}{(4\Delta t^2)^2} + \frac{(a_1b_2 - b_1b_2K)^2 + (a_2b_1 - b_1b_2K)^2}{4}$$

$$= \left[a_1a_2\Delta t^2 + b_1b_2\left(\frac{1}{4\Delta t^2} + K^2\Delta t^2\right) - (a_1b_2 + b_1a_2)K\Delta t^2\right]^2$$

$$+ \frac{(a_1b_2 - b_1a_2)^2}{4}$$

$$= [a_1a_2\Delta t^2 + b_1b_2\Delta\omega^2 + (a_1b_2 + b_1a_2)\mathrm{Cov}_x]^2 + \frac{(a_1b_2 - b_1a_2)^2}{4}$$

i.e., the inequality (4.28) becomes equality. It can be verified that Theorem 4.3 reduces to the result given by (4.21) (or (4.24)) and (4.22) (or (4.25)) when $\Delta t^2 = \frac{b_1^2}{4\sigma^2} + \frac{b_2^2}{(a_1b_2-a_2b_1)^2}\sigma^2$ (or $\Delta t^2 = \sigma^2/2$) and $K = [\frac{a_1b_1}{4\sigma^2} + \frac{a_2b_2}{(a_1b_2-a_2b_1)^2}\sigma^2]/\Delta t^2$ (or $K = 0, \Omega = 0, \varphi = 0$), where σ is arbitrary real constant. □

4.3 Extensions of Uncertainty Principles

There are some extensions of the Heisenberg's uncertainty principle, such as the logarithmic, entropic and Renyi entropic uncertainty principles [1, 18–20]. These uncertainty relations in the FrFT and LCT domains have been presented in [13–16]. In this section we will derive these uncertainty principles in the LCT domain based on the relationship between the LCT and the FT, and the signals achieving these lower bounds are also given.

Firstly, we give the following relationship between the LCT and the FT, which will be useful to derive the extensions of uncertainty principle in the LCT domain. Let $g(t) = x(t)e^{j\frac{a}{2b}t^2}$, $M = (a, b, c, d)$, then

$$X_M(u) = e^{-j\frac{\pi}{4}}e^{j\frac{d}{2b}u^2}\sqrt{1/b}G(u/b) \tag{4.31}$$

where $G(u)$ is the FT of $g(t)$. Equation (4.31) can be easily obtained according to the definition of the LCT.

4.3.1 Logarithmic Uncertainty Principles

The logarithmic uncertainty principles in the LCT domain are shown by the following theorem.

Theorem 4.5. *For arbitrary unit energy signal* $x(t)$, $M_1 = (a_1, b_1, c_1, d_1), M_2 = (a_2, b_2, c_2, d_2)$, *then*

$$\int_{-\infty}^{+\infty} |X_{M_1}(u)|^2 \ln |u| du + \int_{-\infty}^{+\infty} |X_{M_2}(u)|^2 \ln |u| du \geq \frac{\Gamma'(1/4)}{\Gamma(1/4)} + \ln |a_1 b_2 - a_2 b_1| \tag{4.32}$$

where $\Gamma(\cdot)$ *is the Gamma function and* ln *stands for the natural logarithm. And the equality is achieved if* $x(t)$ *satisfies*

$$|a_1 b_2 - a_2 b_1|^{\alpha} \int_{-\infty}^{+\infty} |u|^{-\alpha} |X_{M_2}(u)|^2 du = C_\alpha \int_{-\infty}^{+\infty} |u|^{\alpha} |X_{M_1}(u)|^2 du \tag{4.33}$$

where $C_\alpha = \Gamma((1-\alpha)/4)/\Gamma((1+\alpha)/4), 0 \leq \alpha < 1$.

Proof. Let $g(t) = x(t) e^{j\frac{a}{2b}t^2}$, using (4.31), we have

$$\begin{aligned} \int_{-\infty}^{+\infty} |G(u)|^2 \ln |u| du &= \int_{-\infty}^{+\infty} |\frac{1}{b}||G(\frac{u'}{b})|^2 \ln |\frac{u'}{b}| du' \\ &= \int_{-\infty}^{+\infty} |X_M(u')|^2 \ln |\frac{u'}{b}| du' \\ &= \int_{-\infty}^{+\infty} |X_M(u)|^2 \ln |u| du - \ln |b| \end{aligned}$$

According to the following logarithmic uncertainty relation in the FT domain [1, 18],

$$\int_{-\infty}^{+\infty} |g(t)|^2 \ln |t| dt + \int_{-\infty}^{+\infty} |G(u)|^2 \ln |u| du \geq \frac{\Gamma'(1/4)}{\Gamma(1/4)} \tag{4.34}$$

and noting $|x(t)|^2 = |g(t)|^2$, we have

$$\int_{-\infty}^{+\infty} |x(t)|^2 \ln |t| dt + \int_{-\infty}^{+\infty} |X_M(u)|^2 \ln |u| du \geq \frac{\Gamma'(1/4)}{\Gamma(1/4)} + \ln |b| \tag{4.35}$$

Using the similar approach in Theorem 4.3, inequality (4.32) can be derived.

From [18] the inequality (4.34) becomes equality when $g(t)$ satisfies

$$\int_{-\infty}^{+\infty} |u|^{-\alpha} |G(u)|^2 du = C_\alpha \int_{-\infty}^{+\infty} |t|^{\alpha} |g(t)|^2 dt$$

Similarly, the inequality (4.35) becomes equality if $x(t)$ satisfies

$$|b|^{\alpha} \int_{-\infty}^{+\infty} |u|^{-\alpha} |X_M(u)|^2 du = C_{\alpha} \int_{-\infty}^{+\infty} |t|^{\alpha} |x(t)|^2 dt \tag{4.36}$$

For two LCT domains, it is clear that the inequality (4.32) becomes equality when $x(t)$ satisfies (4.33). □

4.3.2 Entropic Uncertainty Principles

Entropic uncertainty principle is extension of the Heisenberg's uncertainty principle in information theory.

Definition 4.1. The Shannon entropy is defined as

$$E(\rho) = -\int_{-\infty}^{+\infty} \rho(t) \ln \rho(t) dt$$

As derived in Theorem 4.5, we can obtain the following entropic uncertainty principles in the LCT domain.

Theorem 4.6. *For arbitrary unit energy signal $x(t)$, $M_1 = (a_1, b_1, c_1, d_1)$, $M_2 = (a_2, b_2, c_2, d_2)$, then*

$$E(|X_{M_1}(u)|^2) + E(|X_{M_2}(u)|^2) \geq \ln(\pi e) + \ln |a_1 b_2 - a_2 b_1| \tag{4.37}$$

And the equality is achieved if

$$x(t) = \left(\frac{1}{\pi \Delta t^2}\right)^{\frac{1}{4}} e^{-\frac{(t-t_0)^2}{2\Delta t^2}} e^{j(-\frac{K}{2}t^2 + \Omega t + \varphi)} \tag{4.38}$$

where $\Delta t^2 = \frac{b_1^2}{2\sigma^2} + \frac{b_2^2}{(a_1 b_2 - a_2 b_1)^2} 2\sigma^2$, $K = [\frac{a_1 b_1}{2\sigma^2} + \frac{a_2 b_2}{(a_1 b_2 - a_2 b_1)^2} 2\sigma^2]/\Delta t^2$,*and* $t_0, \sigma, \Omega, \varphi$ *are arbitrary real constants.*

Proof. Let $g(t) = x(t) e^{j\frac{a}{2b}t^2}$, using (4.31), we have

$$\begin{aligned}
\int_{-\infty}^{+\infty} |G(u)|^2 \ln |G(u)|^2 du &= \int_{-\infty}^{+\infty} |G(\frac{u'}{b})|^2 |\frac{1}{b}| \ln |G\left(\frac{u'}{b}\right)|^2 du' \\
&= \int_{-\infty}^{+\infty} |X_M(u')|^2 \ln(|X_M(u')|^2 |b|) du' \\
&= \int_{-\infty}^{+\infty} |X_M(u')|^2 \ln(|X_M(u')|^2) du' + \ln |b|
\end{aligned}$$

According to the following entropic uncertainty relation in the FT domain [1, 19, 20]

$$E(|g(t)|^2) + E(|G(u)|^2) \geq \ln(\pi e) \tag{4.39}$$

and noting $|x(t)|^2 = |g(t)|^2$, we have

$$E(|x(t)|^2) + E(|X_M(u)|^2) \geq \ln(\pi e) + \ln|b| \tag{4.40}$$

Similarly to Theorem 4.5, inequality (4.37) can be derived.

From [20] inequalities (4.39) can be achieved if $g(t)$ is a Gaussian signal. It is clear that the inequalities (4.40) become equalities when

$$x(t) = \left(\frac{1}{2\pi\sigma^2}\right)^{\frac{1}{4}} e^{-\frac{(t-t_0)^2}{4\sigma^2}} e^{j(-\frac{a}{2b}t^2+\Omega t+\varphi)}$$

where t_0, Ω, φ are arbitrary real constants. Using the result of Theorem 1 in [6], we can similarly obtain that the inequalities (4.37) can be achieved if the signal $x(t)$ is given by (4.38). □

4.3.3 Renyi Entropic Uncertainty Principles

Renyi entropy is widely used in statistical physics and in signal or image processing, which is a generalization of Shannon entropy.

Definition 4.2. The Renyi λ-entropy is defined as [20]

$$R_\lambda(f) = \frac{1}{1-\lambda}\ln\left(\int_{-\infty}^{+\infty} f(t)^\lambda dt\right)$$

for $\lambda \neq 1$.

When $\lambda \to 1$, the Renyi entropy converges to the Shannon entropy. Theorem 4.7 gives the uncertainty relations of Renyi entropy in the LCT domains.

Theorem 4.7. *For arbitrary unit energy signal $x(t)$, $M_1 = (a_1, b_1, c_1, d_1)$, $M_2 = (a_2, b_2, c_2, d_2)$, then*

$$R_\theta(|X_{M_1}(u)|^2)+R_\lambda(|X_{M_2}(u)|^2) \geq \frac{\ln(\theta/\pi)}{2(\theta-1)}+\frac{\ln(\lambda/\pi)}{2(\lambda-1)}+\ln|a_1b_2-a_2b_1| \tag{4.41}$$

where $\frac{1}{2} < \theta \leq 1$, $\frac{1}{\theta}+\frac{1}{\lambda} = 2$. And the equality is achieved if the signal is given by (4.38).

Proof. Let $g(t) = x(t)e^{j\frac{a}{2b}t^2}$, using (4.31), we can obtain

$$\ln\left(\int_{-\infty}^{+\infty}(|G(u)|^2)^{\lambda}du\right) = \ln\left(\int_{-\infty}^{+\infty}|\frac{1}{b}|(|G(\frac{u'}{b})|^2)^{\lambda}du'\right)$$

$$= \ln\left(\int_{-\infty}^{+\infty}(|X_M(u')|^2)^{\lambda}|b|^{\lambda-1}du'\right)$$

$$= \ln\left(\int_{-\infty}^{+\infty}(|X_M(u')|^2)^{\lambda}du'\right) - (1-\lambda)\ln|b|$$

Using the following Renyi entropic uncertainty relation in the FT domain [20]

$$R_{\theta}(|g(t)|^2) + R_{\lambda}(|G(u)|^2) \geq \frac{\ln(\theta/\pi)}{2(\theta-1)} + \frac{\ln(\lambda/\pi)}{2(\lambda-1)} \tag{4.42}$$

and noting $|x(t)|^2 = |g(t)|^2$, we can obtain

$$R_{\theta}(|x(t)|^2) + R_{\lambda}(|X_M(u)|^2) \geq \frac{\ln(\theta/\pi)}{2(\theta-1)} + \frac{\ln(\lambda/\pi)}{2(\lambda-1)} + \ln|b| \tag{4.43}$$

Since inequality (4.42) can be achieved if $g(t)$ is a Gaussian signal [20], using the similar method in Theorem 4.6, inequality (4.41) can be derived and the equality is achieved if the signal is given by (4.38). □

4.4 Conclusions

The uncertainty principles are very important and some uncertainty principles for the LCT have been summarized in this chapter. The Heisenberg' uncertainty principles for complex signals and real signals are reviewed and the lower bounds related to the covariance of time and frequency have been derived based on the LCT moments properties, which can explain the phenomenon of different lower bounds for real and complex signals existing in the Heisenberg's uncertainty relations proposed before because real signals have zero covariance. Furthermore some extensions of uncertainty principles such as the logarithmic, entropic and Renyi entropic uncertainty principles are deduced based on the relationship between the LCT and the Fourier transform.The uncertainty principles for all the special cases of the LCT can be obtained from these uncertainty principles in the LCT domain.

Appendix

Here we give the derivations of the second order moment and first mixed moment shown in Table 4.1.

Proof. According to the following result (see Lemma 3 in [6]), we have

$$\langle u_M^2 \rangle = a^2 \langle t^2 \rangle + b^2 \langle \omega^2 \rangle + jab(I^* - I)/E \tag{4.44}$$

where $I = \int_{-\infty}^{+\infty} tx'(t)x^*(t)dt$ and $(\cdot)^*$ denotes conjugate. Noting that

$$\mathrm{Im}(I) = \mathrm{Im}\left(\int_{-\infty}^{+\infty} tx'(t)x^*(t)dt\right) = \int_{-\infty}^{+\infty} t\phi'(t)|x(t)|^2 dt$$

(4.44) becomes

$$\langle u_M^2 \rangle = a^2 \langle t^2 \rangle + b^2 \langle \omega^2 \rangle + 2ab\langle t\phi'(t) \rangle \tag{4.45}$$

and

$$\langle u_{FM}^2 \rangle = \frac{1}{E}\int_{-\infty}^{\infty} |X_M'(u)|^2 du = c^2 \langle t^2 \rangle + d^2 \langle \omega^2 \rangle + 2cd\langle t\phi'(t) \rangle \tag{4.46}$$

Then consider the first mixed moment, because $x(t) = L_{(d,-b,-c,a)}(X_M(u))$, from (4.45), we can obtain

$$\langle u_M \varphi'(u_M) \rangle = \frac{1}{2bd}(d^2 \langle u_M^2 \rangle + b^2 \langle u_{FM}^2 \rangle - \langle t^2 \rangle)$$

Substituting (4.45) and (4.46) into the above equation derives

$$\langle u_M \varphi'(u_M) \rangle = ac\langle t^2 \rangle + bd\langle \omega^2 \rangle + (ad + bc)\langle t\phi'(t) \rangle \tag{4.47}$$

In addition, (4.47) reduces to (4.8) when $(a, b, c, d) = (0, 1, -1, 0)$. Substituting (4.45) into (4.47), (4.47) can be expressed in a different form, i.e.,

$$\langle u_M \varphi'(u_M) \rangle + \langle t\phi'(t) \rangle = \frac{d}{b}\langle u_M^2 \rangle - \frac{a}{b}\langle t^2 \rangle \tag{4.48}$$

□

References

1. G.B. Folland, A. Sitaram, The uncertainty principle: a mathematical survey. J. Fourier Anal. Appl. **3**(3), 207–238 (1997)
2. L. Cohen, The uncertainty principle in signal analysis, in *Proceedings of the IEEE-SP International Symposium on Time-Frequency and Time-Scale Analysis*, pp. 182–185, October 1994
3. H.M. Ozaktas, M.A. Kutay, Z. Zalevsky, *The Fractional Fourier Transform with Applications in Optics and Signal Processing* (Wiley, New York, 2000)
4. A. Stern, Uncertainty principles in linear canonical transform domains and some of their implications in optics. J. Opt. Soc. Am. A **25**(3), 647–652 (2008)
5. K.K. Sharma, S.D. Joshi, Uncertainty principles for real signals in linear canonical transform domains. IEEE Trans. Signal Process. **56**(7), 2677–2683 (2008)
6. J. Zhao, R. Tao, Y.L. Li, Y. Wang, Uncertainty principles for linear canonical transform. IEEE Trans. Signal Process. **57**(7), 2856–2858 (2009)
7. J. Zhao, R. Tao, Y. Wang, On signal moments and uncertainty relations associated with linear canonical transform. Signal Process. **90**(9), 2686–2689 (2010)
8. G. L. Xu, X.T. Wang, X.G. Xu, On uncertainty principle for the linear canonical transform of complex signals. IEEE Trans. Signal Process. **58**(9), 4916–4918 (2010)
9. O. Aytur, H.M. Ozaktas, Non-orthogonal domains in phase space of quantum optics and their relation to fractional Fourier transforms. Opt. Commun. **120**(3), 166–170 (1995)
10. H.M. Ozaktas, O. Aytur, Fractional Fourier domains. Signal Process. **46**(1), 119–124 (1995)
11. D. Mustard, Uncertainty principle invariant under fractional Fourier transform. J. Aust. Math. Soc. B **33**, 180–191 (1991)
12. S. Shinde, V.M. Gadre, An uncertainty principle for real signals in the fractional Fourier transform domain. IEEE Trans. Signal Process. **49**(11), 2545–2548 (2001)
13. G.L. Xu, X.T. Wang, X.G. Xu, The logarithmic, Heisenberg's and short-time uncertainty principles associated with fractional Fourier transform. Signal Process. **89**(3), 339–343 563265, 1–7(2009)
14. G.L. Xu, X.T. Wang, X.G. Xu, Uncertainty inequalities for linear canonical transform. IET Signal Process. **3**(5), 392–402 (2009)
15. G.L. Xu, X.T. Wang, X.G. Xu, New inequalities and uncertainty relations on linear canonical transform revisit. EURASIP J. Adv. Signal Process. **563265**, 1–7 (2009)
16. G.L. Xu, X.T. Wang, X.G. Xu, Generalized entropic uncertainty principle on fractional Fourier transform. Signal Process. **89**(12), 2692–2697 (2009)
17. T. Alieva, M.J. Bastiaans, On fractional Fourier transform moments. IEEE Signal Process. Lett. **7**(11), 320–323 (2000)
18. W. Beckner, Pitt's inequality and the uncertainty principle. Proc. Am. Math. Soc. **123**(6), 1897–1905 (1995)
19. A. Dembo, T.M. Cover, J.A. Thomas, Information theoretic inequalities. IEEE Trans. Inf. Theory **37**(6), 1501–1518 (1991)
20. S. Zozor, C. Vignat, Non-Gaussian asymptotic minimizers in entropic uncertainty principles and the dimensional effect, in *IEEE International Symposium on Information Theory*, pp. 2085–2089, July 2006

Chapter 5
The Linear Canonical Transformations in Classical Optics

Tatiana Alieva, José A. Rodrigo, Alejandro Cámara, and Martin J. Bastiaans

Abstract In this chapter we consider the application of the linear canonical transformations (LCTs) for the description of light propagation through optical systems. It is shown that the paraxial approximation of ray and wave optics leads to matrix and integral forms of the two-dimensional LCTs. The LCT description of the first-order optical systems consisting of basic optical elements: lenses, mirrors, homogeneous and quadratic refractive index medium intervals and their compositions is discussed. The applications of these systems for the characterization of the completely and partially coherent monochromatic light are considered. For this purpose the phase space beam representation in the form of the Wigner distribution (WD), which reveals local beam coherence properties, is used. The phase space tomography method of the WD reconstruction is discussed. The physical meaning and application of the second-order WD moments for global beam analysis, classification, and comparison are reviewed. At the similar way optical systems used for manipulation and characterization of optical pulses are described by the one-dimensional LCTs.

5.1 Introduction

In this chapter we consider the application of the linear canonical transformations (LCTs) in classical optics. The phase space representation of the LCTs based on matrix formalism, which has been introduced in [26], will be widely used here. It significantly simplifies the solution of different problems making a bridge between ray and wave optics.

We start considering the propagation of monochromatic electromagnetic waves in isotropic, but not necessarily homogeneous, charge-free media. In this case the

T. Alieva (✉) • J.A. Rodrigo • A. Cámara
Facultad de Ciencias Físicas, Universidad Complutense de Madrid, Ciudad Universitaria s/n, Madrid 28040, Spain
e-mail: talieva@fis.ucm.es

M.J. Bastiaans
Department of Electrical Engineering, Eindhoven University of Technology, P.O. Box 513, 5600 MB Eindhoven, The Netherlands

J.J. Healy et al. (eds.), *Linear Canonical Transforms*, Springer Series in Optical Sciences 198, DOI 10.1007/978-1-4939-3028-9_5

wave equations for the electric **E** and magnetic **H** field vectors can be written for every Cartesian component E_x, E_y, and E_z (or correspondingly H_x, H_y, and H_z) in the form of the Helmholtz equation. The paraxial approximation of the Helmholtz equation allows using the LCTs for the description of beam propagation in homogeneous and certain type of weakly inhomogeneous media.

Next section is devoted to the analysis of ray and wave propagation through, or reflected from widely used optical elements such as lenses and mirrors. Again, the corresponding transformations of the complex field amplitude lead to the LCTs in the paraxial approximation. The combinations of these elements with intervals of homogeneous medium (or inhomogeneous medium with quadratic refractive index) form the so-called first-order optical systems considered. The ray matrix formalism drastically simplifies their analysis and design.

In Sect. 5.4 we turn to the consideration of partially coherent beam and their propagation through first-order optical systems. We benefit from describing the beam in the phase space using the Wigner distribution (WD) (see [26]) since it allows identical treatment of completely and partially coherent light. Notice that the WD is real but not necessarily a positive function. This fact makes impossible its direct measurement. Instead of the WD itself, the WD projections, corresponding to the beam intensity distributions at the output plane of the systems described by the LCTs can be easily measured and used for beam characterization. This approach is discussed in Sect. 5.5. A global beam analysis according to the ISO recommendations is based on the central second-order moments of the WD. The physical meaning of these moments and their use for comparison and classification of optical beams are considered in Sect. 5.6.

In the last section we show that the LCTs are also applied for the description of the propagation of quasi-monochromatic plane waves in a homogeneous medium with frequency dispersion. The approaches which are similar to ones used for the description of the spatial beam structure are applied for pulse analysis. The chapter ends with concluding remarks.

5.2 Helmholtz Equation in Paraxial Approximation: Ray and Wave Pictures

5.2.1 Helmholtz Equation

Let us consider the wave equation written in the Cartesian coordinates $\boldsymbol{r} = [x\,, y\,, z]^{\mathrm{t}}$ for the electric field vector $\mathbf{E}(x, y, z, t)$ in a charge-free weakly inhomogeneous medium characterized by the permittivity ε and the permeability μ, which can be easily derived from the Maxwell equations (see, for example, [36, 55]),

$$\nabla^2\mathbf{E} = \varepsilon\mu\frac{\partial^2}{\partial t^2}\mathbf{E}, \tag{5.1}$$

where $\nabla^2 = \frac{\partial^2}{\partial x^2} + \frac{\partial^2}{\partial y^2} + \frac{\partial^2}{\partial z^2}$ is the Laplacian operator. The dependence on the time variable t can be excluded if we consider the propagation of the monochromatic wave $\mathbf{E}(\boldsymbol{r}, t) = \mathbf{E}(\boldsymbol{r}) \exp(-\mathrm{i}\omega t)$ with angular frequency ω

$$\nabla^2 \mathbf{E} + k^2 \mathbf{E} = 0. \tag{5.2}$$

Here $k = \omega\sqrt{\varepsilon\mu} = \omega/\upsilon$ is the wave number in the medium, which is related to the wave number in vacuum, $k_0 = \omega/c$, by the relation $k = k_0 n$ being υ and c the phase velocity in the medium and vacuum, correspondingly, and $n = c/\upsilon$ the refractive index of the medium.

The Eq. (5.2) can be treated independently for every vector component E_x, E_y, and E_z if $n(\boldsymbol{r})$ does not change significantly in distances of the order of the wavelength $\lambda = 2\pi/k$. Thus, using the notation $\Psi(\boldsymbol{r})$ for one of the components of the electric field vector we obtain the Helmholtz equation

$$\nabla^2 \Psi + k_0^2 n^2 \Psi = 0, \tag{5.3}$$

which is a basic equation of the scalar diffraction theory. The same equation is valid for the components of the magnetic field vector. In isotropic homogeneous medium n is a constant.

There are two approximations which simplify solving the Helmholtz equation: the eikonal and paraxial wave approximations. The eikonal approximation is the base of geometrical optics, where the light propagation is described in form of rays. The paraxial wave approximation leads to the LCTs, as we will see further, and it is widely applied in wave optics for the description of the interference and diffraction phenomena. Applying the paraxial approximation to the eikonal equation simplifies the use of the LCTs by introducing the ray transformation matrix formalism, which allows avoiding, in many cases, unnecessary integral calculations.

5.2.2 *Eikonal Approximation*

It is easy to prove that a plane wave, $\Psi(\boldsymbol{r}) = \Psi_0 \exp(\mathrm{i}n\mathbf{k}_0^{\mathrm{t}}\boldsymbol{r} + \mathrm{i}\varphi)$, is a solution of the Helmholtz equation for a homogeneous medium ($n = \text{const}$), where $k_0 = \omega/c = 2\pi/\lambda_0$ and λ_0 is the wavelength in the vacuum. The direction of the wave vector, $\mathbf{k}_0$, is defined by the boundary conditions.

In order to find the solution in the general case of inhomogeneous medium, $n \neq$ const, let us use the probe function

$$\Psi(\boldsymbol{r}) = a(\boldsymbol{r}) \exp(\mathrm{i}k_0 \widehat{L}(\boldsymbol{r})), \tag{5.4}$$

where the amplitude $a(\boldsymbol{r})$ and the phase $k_0\widehat{L}(\boldsymbol{r})$ are real functions that depend on the position vector $\boldsymbol{r}$. Introducing this expression in the Eq. (5.3) and applying the Laplacian operator, we conclude that the real and the imaginary parts of the following equation

$$\nabla^2 a(\boldsymbol{r}) + k_0^2 a(\boldsymbol{r}) \left[n^2(\boldsymbol{r}) - (\nabla\widehat{L}(\boldsymbol{r}))^2\right] + \mathrm{i}k_0 \left[2\nabla a(\boldsymbol{r}) \cdot \nabla\widehat{L}(\boldsymbol{r}) + a(\boldsymbol{r})\nabla^2\widehat{L}(\boldsymbol{r})\right] = 0 \tag{5.5}$$

have to be zero. In particular, we obtain that

$$\nabla^2 a(\boldsymbol{r}) + k_0^2 a(\boldsymbol{r}) \left[n^2(\boldsymbol{r}) - (\nabla\widehat{L}(\boldsymbol{r}))^2\right] = 0. \tag{5.6}$$

In the approximation of the geometrical optics we suppose that the amplitude $a(\boldsymbol{r})$ changes significantly for distances d much larger than the wavelength λ. This means that $\left|\nabla^2 a(\boldsymbol{r})/k_0^2 a(\boldsymbol{r})\right| \sim (\lambda/d)^2 \approx 0$. Then, the Eq. (5.6) is reduced to eikonal equation

$$\left[\nabla\widehat{L}(\boldsymbol{r})\right]^2 = n^2(\boldsymbol{r}) \tag{5.7}$$

which plays a fundamental role in geometrical optics.

The phase function $\widehat{L}(\boldsymbol{r})$ is called eikonal, which comes from the Greek word *eikonal,* meaning image. Let us consider its decomposition in Taylor series around a point $\boldsymbol{r}_0$, i.e. $\widehat{L}(\boldsymbol{r}) \approx \widehat{L}(\boldsymbol{r}_0) + (\boldsymbol{r} - \boldsymbol{r}_0)\nabla\widehat{L}(\boldsymbol{r}_0) + \cdots$. If $|\boldsymbol{r} - \boldsymbol{r}_0| < \lambda$, then $a(\boldsymbol{r}) \approx a(\boldsymbol{r}_0)$ and we obtain the following approximation for $\Psi(\boldsymbol{r})$

$$\Psi(\boldsymbol{r})\,|_{\boldsymbol{r}_0} \approx a(\boldsymbol{r}_0) \exp\left\{\mathrm{i}k_0 \left[\widehat{L}(\boldsymbol{r}_0) - \boldsymbol{r}_0\nabla\widehat{L}(\boldsymbol{r}_0)\right]\right\} \exp(\mathrm{i}k_0\boldsymbol{r}\nabla\widehat{L}(\boldsymbol{r}_0)). \tag{5.8}$$

While first exponential term corresponds to a constant phase, the last one has the form of a plane wave propagating in the direction of the unitary vector **s** defined by $\mathbf{s_0} = \nabla\widehat{L}(\boldsymbol{r}_0)/n(\boldsymbol{r}_0)$, where we have taken into account the eikonal equation (5.7). Since this approximation is only valid in the neighborhood of $\boldsymbol{r}_0$, which might be considered as a sphere of radius λ with the center in that point, the wave is plane only locally. In another point, $\boldsymbol{r}_1$, the propagation direction and its amplitude are changed.

By analogy with a plane wave, the vector $\nabla\widehat{L}(\boldsymbol{r}_0)$, similar to the wave vector **k**, is normal to the geometrical wavefront defined by the surface $\widehat{L}(\boldsymbol{r}) = \text{const}$. Moreover, considering the relation provided by the imaginary part of the Eq. (5.5) we can write

$$a(\boldsymbol{r}) \left[2\nabla a(\boldsymbol{r}) \cdot \nabla\widehat{L}(\boldsymbol{r}) + a(\boldsymbol{r})\nabla^2\widehat{L}(\boldsymbol{r})\right] = \nabla \cdot (a^2(\boldsymbol{r})\nabla\widehat{L}(\boldsymbol{r})) = 0, \tag{5.9}$$

which is the energy transport equation for geometrical optics, since the squared amplitude is associated with the energy of the wave. We observe that the product $a^2(\boldsymbol{r})\nabla\widehat{L}(\boldsymbol{r})$ is analogue of the Poynting vector for the plane waves.

The above considerations allow describing light in eikonal approximation by rays which define, for every point $\boldsymbol{r}$, the direction of the energy propagation,

$\mathbf{s} = \nabla\widehat{L}(\boldsymbol{r})/n(\boldsymbol{r})$. The ray follows a certain curve in space that is characterized by the position vector $\boldsymbol{r}(s)$, where s is the arc longitude of the curve. Its tangential in every point coincides with the ray vector $\mathbf{s}$ and can also be expressed as $\mathbf{s} = \mathrm{d}\boldsymbol{r}/\mathrm{d}s$. From the tangent definition and the eikonal equation (see the Eq. (5.7)) we obtain the ray equation

$$n(\boldsymbol{r})\frac{\mathrm{d}\boldsymbol{r}}{\mathrm{d}s} = \nabla\widehat{L}(\boldsymbol{r}). \tag{5.10}$$

For practical use it is more convenient to write the whole ray equation in terms of the refraction index of the medium, $n(\boldsymbol{r})$. This can be achieved differentiating the ray equation with respect to the curve parameter, s, taking into account that $\frac{\mathrm{d}}{\mathrm{d}s} = \frac{\mathrm{d}\boldsymbol{r}}{\mathrm{d}s}\cdot\nabla = \mathbf{s}\cdot\nabla$, and the following relation

$$\frac{\mathrm{d}}{\mathrm{d}s}\left[\nabla\widehat{L}\right] = \mathbf{s}\cdot\nabla\left[\nabla\widehat{L}\right] = \frac{1}{2n}\nabla n^2 = \nabla n. \tag{5.11}$$

Thus, we obtain another form for the ray equation which we will use further

$$\frac{\mathrm{d}}{\mathrm{d}s}\left[n\left(\boldsymbol{r}\right)\frac{\mathrm{d}\boldsymbol{r}}{\mathrm{d}s}\right] = \nabla n\left(\boldsymbol{r}\right). \tag{5.12}$$

It is easy to see that in a homogeneous medium, $n = \text{const} \Rightarrow \nabla n = 0$ and $\mathbf{s} = \text{const}$, the ray follows a linear trajectory. Below we consider the solutions of the ray equation in the stratified media which yield in the paraxial approximation to the matrix description of the ray propagation.

5.2.3 *Ray Propagation in Linearly Stratified Medium*

Let us consider the ray propagation in a stratified medium, that is, a medium in which the refraction index depends on only one Cartesian coordinate, $n\left(\boldsymbol{r}\right) = n(x)$ for example. In this case the ray equation is reduced to

$$\frac{\mathrm{d}}{\mathrm{d}s}\left[n(x)\frac{\mathrm{d}\boldsymbol{r}}{\mathrm{d}s}\right] = \mathbf{i}_x\frac{\mathrm{d}n(x)}{\mathrm{d}x}, \tag{5.13}$$

being $\mathbf{i}_x$, $\mathbf{i}_y$, and $\mathbf{i}_z$ the unitary vectors in the directions of x, y, and z correspondingly. To determine the ray trajectories we introduce the vector

$$\mathbf{a} = \mathbf{i}_x \times n(x)\frac{\mathrm{d}\boldsymbol{r}}{\mathrm{d}s} = n(x)\,\mathbf{i}_x\times\mathbf{s}. \tag{5.14}$$

Differentiating **a** with respect to the arc longitude s and using the relations $\mathrm{d}\mathbf{i}_x/\mathrm{d}s = 0$ and Eq. (5.13), we obtain $\frac{\mathrm{d}\mathbf{a}}{\mathrm{d}s} = 0$. It means that the vector **a** is constant along the ray trajectory. Then taking into account the Eq. (5.14) we conclude that the vectors **s** and $\mathbf{i}_x$ are always perpendicular to a constant vector **a**. Therefore, the ray curve is contained at the plane spanned by **i** and a vector that is perpendicular to both **a** and **i**, i.e. $\mathbf{i} \times \mathbf{a}$. This plane and the modulus of **a** are defined by an initial condition, vector $\mathbf{s}_i$.

Suppose that the ray belongs to the plane $\boldsymbol{r}^{\mathrm{t}} \cdot \mathbf{i}_y = 0$ and $\mathbf{s}_i$ is known at the point $\boldsymbol{r}_i = [x_i,\ 0,\ z_i]^{\mathrm{t}}$. Considering the angle, θ, formed by the vector **s** and the X axis we obtain the Snell law for a stratified medium

$$a = |\mathbf{a}| = n(x)\sin\theta = n(x_i)\sin\theta_i. \tag{5.15}$$

Introducing $\mathrm{d}z = \sin\theta\mathrm{d}s$ and $\mathrm{d}x = \cos\theta\mathrm{d}s$ in the previous relation one obtains the equation that relates the positions of the x and z coordinates of the ray trajectory:

$$\frac{\mathrm{d}z}{\mathrm{d}x} = \frac{\sin\theta\mathrm{d}s}{\cos\theta\mathrm{d}s} = \tan\theta = \pm\frac{a}{\sqrt{n^2(x) - a^2}}, \tag{5.16}$$

where the sign $\pm$ is defined by the sign of the projection of $\mathbf{s}_0$ on the X axis at the point $\boldsymbol{r}_i$, $s_{i,x} = \cos\theta_i$. Notice that the sign of $\mathrm{d}z/\mathrm{d}x$ can change during ray propagation (see Fig. 5.1). Thus at the point with coordinate x_r such that $n^2(x_r) = a^2$ the derivative goes to infinity $(\mathrm{d}z/\mathrm{d}x)_{x_r} = \infty$ and the ray reflects from the given layer (see [66] for details). Therefore, the signs of s_x and $\mathrm{d}z/\mathrm{d}x$ change.

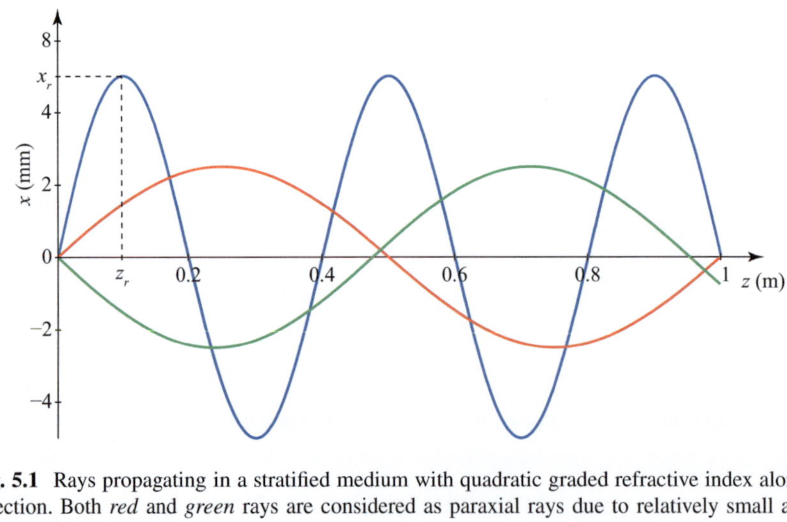

Fig. 5.1 Rays propagating in a stratified medium with quadratic graded refractive index along X direction. Both *red* and *green* rays are considered as paraxial rays due to relatively small angle $|\pi/2 - \theta_i| \ll 1$, while the *blue* ray is out of this approximation. The point (x_r, z_r) represents a reflection of the ray from the layer where $n^2(x_r) = a^2$

Note that in the particular case $a = 0$ the ray follows a lineal trajectory parallel to the X axis, since $\theta_i = 0$ or $\theta_i = \pi$, and $\mathrm{d}z/\mathrm{d}x = 0$.

To find the ray trajectory from the Eq. (5.16) we have to know the function $n(x)$ and the ray direction $\mathbf{s}_0$ at a given point $\boldsymbol{r}_0$. Let us consider the beam propagation through the medium with a refractive index $n(x)$ satisfying the equation $n^2(x) = n_0^2 - g^2x^2$, which takes a maximum value n_0 at $x = 0$. Note that such quadratic graded refractive index (GRIN) media as well as axially stratified quadratic GRIN ones, considered in Sect. 5.2.4 are applied in fiber optics telecommunication and imaging. The ray equation, Eq. (5.16), is written in this case as

$$\frac{\mathrm{d}x}{\sqrt{n_0^2 - a^2 - g^2x^2}} = \frac{\sigma \mathrm{d}z}{a}, \tag{5.17}$$

where $\sigma = \mathrm{sign}(\cos\theta_i)$. Notice that for $\theta_i = \pi/2$ and $x_i = 0$ the ray follows the lineal trajectory $x = 0$ which coincides with the Z axis. Indeed, this initial condition yields $a = n_0$ that holds only for $x = 0$.

For other cases $n_{\max} > a$ and it is easy to prove that the equation has a solution

$$x(z) = \frac{\sqrt{n_0^2 - a^2}}{g} \sin\left(\frac{\sigma g}{a}(z - c)\right). \tag{5.18}$$

The constant c is defined from the initial conditions as

$$c = z_i - \frac{\sigma a}{g} \arcsin\left(\frac{gx}{\sqrt{n_0^2 - a^2}}\right), \tag{5.19}$$

that yields to the final expression for ray trajectory

$$x(z) = x_i \cos\left[\frac{g}{a}(z - z_i)\right] + \cos\theta_i \frac{n(x_i)}{g} \sin\left[\frac{g}{a}(z - z_i)\right]. \tag{5.20}$$

We observe that the ray trajectories in such media are periodic in z. The amplitude and the period of the ray oscillation depend on the initial conditions. In general the rays emergent from the same point but with different angles θ_i have different periods $T = 2\pi a/g$, defined by a. In Fig. 5.1 we observe the propagation of three rays emergent from the point $(0, 0, 0)$ with different angles θ_i. The red and the green rays have similar value of a but different initial direction of propagation (due to the sign of $\cos\theta_i$). The trajectories of these rays oscillate slower the one which belongs to blue ray with smaller value of a.

5.2.3.1 Paraxial Approximation

Let us now consider the ray propagation in the stratified media, $n(x)$, in paraxial approximation, when the ray deviation from X or Z axes is small that corresponds to the cases $|\theta| \ll 1$ and $2\,|\pi/2 - \theta| \ll 1$, respectively. These cases occur for adequate initial conditions and sufficiently small changes of the refractive index.

In the first case we can use only the first term of the Taylor expansion for the sine function $\sin\theta \approx \theta$, which allows rewriting the Snell law as

$$a = n(x)\theta = n(x_i)\theta_i. \tag{5.21}$$

Since $|a| \ll 1$ and therefore $n^2(x) \gg a^2$, the ray equation (5.16) is reduced to $\mathrm{d}z = n(x_i)\theta_i \frac{\mathrm{d}x}{n(x)}$ that yields the solution

$$z_o = z_i + n(x_i)\theta_i \int_{x_i}^{x_o} \frac{\mathrm{d}x}{n(x)} = z_i + n(x_i)\theta_i B. \tag{5.22}$$

The coordinate z in this case is close to z_i defined by the initial position of the ray. The latter equation together with the Snell law in paraxial approximation, Eq. (5.21) allow writing the relation between the position, z, and direction of propagation, θ, of the ray at the input $x = x_i$ and output $x = x_o$ planes in the matrix form

$$\begin{bmatrix} z_o \\ q_o \end{bmatrix} = \begin{bmatrix} 1 & B \\ 0 & 1 \end{bmatrix} \begin{bmatrix} z_i \\ q_i \end{bmatrix},$$

where $q_o = n(x_o)\theta_o = n(x_i)\theta_i = q_i$. In particular, for homogeneous media $n(x) = n = \text{const}$ the matrix parameter is given by $B = x/n$, that as expected corresponds to the line ray trajectory: $z = z_i + x\theta_i$ and $\theta = \theta_i$. Correspondingly the ray propagation through a medium containing J parallel layers of thickness d_j and the refractive index n_j, is described by the ray transformation matrix

$$\mathbf{t} = \begin{bmatrix} 1 & \sum_{j=1}^{J} d_j/n_j \\ 0 & 1 \end{bmatrix}. \tag{5.23}$$

Further it will be used for the parametrization of the kernel of integral transform which describes (also in paraxial approximation) the wave propagation through this medium.

Let us now consider the ray propagation through the medium with refractive index n, given by $n^2(x) = n_0^2 - g^2x^2$ for the initial condition $\theta_i \approx \pi/2$. In this case we can ignore the differences in the period of ray oscillations discussed in Sect. 5.2.4

and approximate it by $T \approx 2\pi n_0/g$. By introducing an angle $\alpha_i = \pi/2 - \theta_i$ we simplify the solution (5.20):

$$x_o = x_i \cos\left[\frac{g}{n_0}(z_o - z_i)\right] + \alpha_i \frac{n_0}{g} \sin\left[\frac{g}{n_0}(z_o - z_i)\right]. \tag{5.24}$$

Moreover, calculating the derivative $\mathrm{d}x/\mathrm{d}z = (\tan\theta)^{-1} = \tan\alpha \approx \alpha$, where α is the angle that the ray forms with axes z, we obtain the equation for the direction of ray propagation:

$$\alpha_o = -\frac{x_i g}{n_0} \sin\left[\frac{g}{n_0}(z_o - z_i)\right] + \alpha_i \cos\left[\frac{g}{n_0}(z_o - z_i)\right]. \tag{5.25}$$

The last two equations again can be written in the matrix form as

$$\begin{bmatrix} x_o \\ q_o \end{bmatrix} = \begin{bmatrix} \cos\gamma & g^{-1}\sin\gamma \\ -g\sin\gamma & \cos\gamma \end{bmatrix} \begin{bmatrix} x_i \\ q_i \end{bmatrix}, \tag{5.26}$$

where $q = n_0\alpha$ and $\gamma = g(z_o - z_i)/n_0$. By this way we connect the position and the direction of the ray at two parallel planes defined by $z = z_o$ and $z = z_i$. The condition of paraxiality is satisfied for the red and green rays in Fig. 5.1, which propagate almost with the same period.

5.2.4 Ray Propagation in Axially Stratified Medium

When the refractive index depends on two variables $n(x, y)$ the ray equation is written as:

$$\frac{\mathrm{d}}{\mathrm{d}s}(n\mathbf{s}) = \frac{\mathrm{d}}{\mathrm{d}s}\left(n\frac{\mathrm{d}\mathbf{r}}{\mathrm{d}s}\right) - \mathbf{i}_r\frac{\mathrm{d}n}{\mathrm{d}x} + \mathbf{i}_y\frac{\mathrm{d}n}{\mathrm{d}y}. \tag{5.27}$$

Using cylindrical coordinates for the unitary vector $\mathbf{s} = [\cos\theta\sin\varphi, \sin\theta\sin\varphi, \cos\varphi]^{\mathrm{t}}$ and taking into account that the right part of this equation does not depend on z, we find that

$$\frac{\mathrm{d}}{\mathrm{d}s}\left(n\frac{\mathrm{d}z}{\mathrm{d}s}\right) = 0. \tag{5.28}$$

It corresponds to the Snell law for axially stratified medium

$$n\frac{\mathrm{d}z}{\mathrm{d}s} = ns_z = n\cos\varphi = n(x_0, y_0)\cos\varphi_0 = a = \text{const.} \tag{5.29}$$

Here we have defined the constant a and introduced the angle φ between the vector $\mathbf{s}$ and the axes z. Since the normal to the surface describing the changes of n is perpendicular to z, we may express the Snell law in a common form $n(x,y)\sin\alpha = n(x_i, y_i)\sin\alpha_i$ using the angle α between such a normal and the ray vector: $\alpha = \pi/2 - \varphi$. Note that if $a = 0$ (and correspondingly $\varphi_i = \pm\pi/2$, and $\mathrm{d}z/\mathrm{d}s = 0$) any curve describing the ray belongs to the plane XY. For the cases $a \neq 0$, since $\mathrm{d}s = (n\mathrm{d}z)/a$, the ray equations for the other two components of the vector **s** have a form similar to one obtained for linear stratified media:

$$\frac{\mathrm{d}}{\mathrm{d}s}\left(n\frac{\mathrm{d}x}{\mathrm{d}s}\right) = \frac{\mathrm{d}n}{\mathrm{d}x},$$
$$\frac{\mathrm{d}}{\mathrm{d}s}\left(n\frac{\mathrm{d}y}{\mathrm{d}s}\right) = \frac{\mathrm{d}n}{\mathrm{d}y}. \tag{5.30}$$

Nevertheless, in this case the ray curve is not necessarily plane since x and y depend on z.

Considering the medium with quadratic GRIN $n(x,y)$ given by:

$$n^2(x,y) = n_0^2 - g_x^2 x^2 - g_y^2 y^2, \tag{5.31}$$

following the same steps as in Sect. 5.2.3 we find the solution for ray trajectory

$$x(z) = x_i \cos\left[\frac{g_x}{a}(z - z_i)\right] + \cos\theta_i \sin\varphi_i \frac{n(x_i, y_i)}{g_x} \sin\left[\frac{g_x}{a}(z - z_i)\right],$$
$$y(z) = y_i \cos\left[\frac{g_y}{a}(z - z_i)\right] + \sin\theta_i \sin\varphi_i \frac{n(x_i, y_i)}{g_y} \sin\left[\frac{g_y}{a}(z - z_i)\right]. \tag{5.32}$$

For initial conditions $y_i = 0$ and $\sin\theta_i = 0$ the ray curve belongs to the plane XZ (see red ray in Fig. 5.2). If $x_i = 0$ and $\cos\theta_i = 0$, it belongs to the plane YZ. While in the case $y_i = 0$ and $\cos\theta_i = 0$, $x_i = \sin\theta_i \sin\varphi_i\, n(x_i, y_i)/g$ and $g_x = g_y = g$, the ray follows the helical trajectory (see green ray in Fig. 5.2): $x^2(z) + y^2(z) = x_i^2$.

In general the rays which do not belong to the meridional plane containing the optical axes are known as skew rays. They are characterized by the nonzero parameter $xq_y - yq_x$ (see [101] for details).

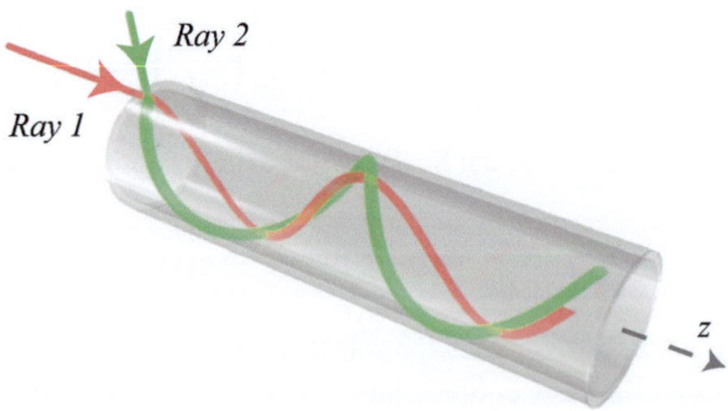

Fig. 5.2 Ray propagation (*green* and *red color*) through a medium with axially graded quadratic refractive index

In the paraxial approximation, which corresponds to small angles φ_i and small variations of n, the rays have almost the same period because $a \approx n_0$ (see Eq. (5.29)). Introducing the variables $q_{x,y} = n_0 s_{x,y}$, where $\mathrm{d}x/\mathrm{d}s \approx \mathrm{d}x/\mathrm{d}z = s_x = \cos\theta \sin\varphi$ and $\mathrm{d}y/\mathrm{d}s \approx \mathrm{d}y/\mathrm{d}z = s_y = \sin\theta \sin\varphi$, the relations between the positions and directions of the ray at planes defined by $z = z_o$ and $z = z_i$ can be written in the matrix form

$$\begin{bmatrix} x_o \\ y_o \\ q_{x(o)} \\ q_{y(o)} \end{bmatrix} = \begin{bmatrix} \cos\gamma_x & 0 & g_x^{-1}\sin\gamma_x & 0 \\ 0 & \cos\gamma_y & 0 & g_y^{-1}\sin\gamma_y \\ -g_x\sin\gamma_x & 0 & \cos\gamma_x & 0 \\ 0 & -g_y\sin\gamma_y & 0 & \cos\gamma_y \end{bmatrix} \begin{bmatrix} x_i \\ y_i \\ q_{x(i)} \\ q_{y(i)} \end{bmatrix}, \tag{5.33}$$

where $\gamma_{x,y} = g_{x,y}(z_o - z_i)/n_0$. Further we will suppose that $z_i = 0$ and $z_o = z$.

This ray transformation matrix resembles the one used in [26] for parametrization of the separable fractional Fourier transform kernel. It is not a coincidence, since there is a deep connection between the ray and wave descriptions of light propagation in paraxial approximation, as we will see further.

5.2.5 *Paraxial Wave Approximation*

Now we consider a coherent wave that propagates along a certain direction which without loss of generality is denoted by the coordinate z. The paraxial approximation assumes that the wave function, $\Psi(x, y, z)$, changes more rapidly in the direction of the wave propagation, z coordinate, than in the traversal, XY, plane. In this sense the paraxial waves are similar to a plane wave defined as $\Psi_0 \exp(\mathrm{i}kz)$ and sometimes are refereed as quasi-plane waves. In addition, if the electromagnetic wave is band-limited in the XY plane, the wave is said to form a beam.

In order to simplify the Helmholtz equation by taking advantage of this approximation, it is convenient to separate the rapid variation part, $\exp(\mathrm{i}k_0 n_0 z)$, from the slow variation part, the complex field amplitude $f(\mathbf{r}, z)$, in the field function Ψ as

$$\Psi(\mathbf{r}, z) = f(\mathbf{r}, z)\exp(\mathrm{i}k_0 n_0 z). \tag{5.34}$$

Here $\mathbf{r} = [x, y]^{\mathrm{t}}$ is the position vector in the traversal plane and n_0 is the average value of the refractive index of the medium.

Introducing this expression in Eq. (5.3), one obtains the following equation for the complex amplitude after some easy calculus

$$\nabla^2 f(\mathbf{r}, z) + k_0^2 f(\mathbf{r}, z)\left[n^2(\mathbf{r}) - n_0^2\right] + \mathrm{i}2k_0 n_0 \frac{\partial}{\partial z} f(\mathbf{r}, z) = 0. \tag{5.35}$$

Since the rapid variations of $\Psi(\mathbf{r}, z)$ with respect to z are described by the exponential term of Eq. (5.34), the changes of f in this direction occur at distances $d \gg \lambda$, and therefore

$$k_0 n_0 \frac{\partial}{\partial z} f(\mathbf{r}, z) \gg \frac{\partial^2}{\partial z^2} f(\mathbf{r}, z). \tag{5.36}$$

Ignoring the smallest term we derive the paraxial equation for the complex field amplitude

$$\Delta_\perp f(\mathbf{r}, z) + k_0^2 f(\mathbf{r}, z)\left[n^2(\mathbf{r}) - n_0^2\right] + \mathrm{i}2k_0 n_0 \frac{\partial}{\partial z} f(\mathbf{r}, z) = 0, \tag{5.37}$$

where we have introduced the traversal Laplacian operator $\Delta_\perp = \frac{\partial^2}{\partial x^2} + \frac{\partial^2}{\partial y^2}$. This equation is known in quantum mechanics as Schrödinger equation, where z plays the role of time and $n_0^2 - n^2(\mathbf{r})$ corresponds to the potential energy. Hence, it is not surprising that many problems in quantum mechanics and paraxial optics have the same solutions. We will cover two special cases of the potential energy $n_0^2 = n^2(\mathbf{r})$ and $n_0^2 - n^2(\mathbf{r}) = g_x^2 x^2 + g_y^2 y^2$ that will lead to different LCT descriptions of beam propagation. The consideration of more general case can be found in [53].

5.2.5.1 Propagation in Homogeneous Medium: Fresnel Transform

The first special case, $n(\mathbf{r}) = n_0$, corresponds to a homogeneous medium. The paraxial equation is reduced to

$$\Delta_\perp f(\mathbf{r}, z) + \mathrm{i}2k_0 n_0 \frac{\partial}{\partial z} f(\mathbf{r}, z) = 0. \tag{5.38}$$

To solve this differential equation, the boundary conditions have to be known. Assuming that the complex field amplitude at the plane $z = 0$ is given by $f_i(\mathbf{r}_i, 0)$, then the solution of this equation can be written as the well-known Fresnel integral

$$f_o(\mathbf{r}_o, z) = \frac{1}{\mathrm{i}\lambda z} \int f(\mathbf{r}_\mathrm{i}, 0) \exp\left[\frac{\mathrm{i}\pi}{\lambda z} (\mathbf{r}_o - \mathbf{r}_i)^2\right] \mathrm{d}\mathbf{r}_i, \tag{5.39}$$

as it can be easily proved by its substitution in the Eq. (5.38), where we have used $k = k_0 n_0 = 2\pi/\lambda$. Here and further the integration $\int$ extends from $-\infty$ to $+\infty$.

5.2.5.2 Angular Spectrum

It is worth mentioning that the same integral solution can be obtained using the angular spectrum decomposition method. This approach is based on the search of the solution of the Helmholtz equation (5.3) in the Fourier domain. For this purpose $\Psi(\mathbf{r}, z)$ is expressed through the 2D Fourier transform (FT) with respect to the vector $\mathbf{r}$:

$$\Psi(\mathbf{r}, z) = \int \overline{\Psi}(\mathbf{p}, z) \exp(\mathrm{i}2\pi \mathbf{r}^{\mathrm{t}}\mathbf{p})\mathrm{d}\mathbf{p}, \tag{5.40}$$

where $\overline{\Psi}(\mathbf{p}, z)$ is the beam angular spectrum at the plane $z = \text{const}$ defined as

$$\overline{\Psi}(\mathbf{p}, z) = \int \Psi(\mathbf{r}, z) \exp(-\mathrm{i}2\pi \mathbf{r}^{\mathrm{t}}\mathbf{p})\mathrm{d}\mathbf{r}. \tag{5.41}$$

The vector $\mathbf{p} = [u, v]^{\mathrm{t}}$ is formed by the spatial frequency components along x and y directions, respectively. Substitution of $\Psi(\mathbf{r}, z)$ in the form (5.40) into Eq. (5.3) yields the Helmholtz equation written for the angular spectrum:

$$\frac{\partial^2}{\partial z^2}\overline{\Psi}(\mathbf{p}, z) + \left[k^2 - 4\pi^2(u^2 + v^2)\right]\overline{\Psi}(\mathbf{p}, z) = 0. \tag{5.42}$$

This differential equation has a simple solution given by

$$\overline{\Psi_o}(\mathbf{p}, z) = \overline{\Psi_i}(\mathbf{p}, 0) \exp\left[\mathrm{i}z\sqrt{k^2 - 4\pi^2(u^2 + v^2)}\right], \tag{5.43}$$

where $\overline{\Psi_i}(\mathbf{p}, 0)$ is the angular spectrum of the beam at the plane $z = 0$, which corresponds to the boundary condition of the problem.

We underline that the propagation in free space is equivalent to a linear filtering. Notice that, neglecting the evanescent waves, which correspond to the condition $k^2 - 4\pi^2(u^2 + v^2) \geq 0$, the propagation only affects the phase of the angular spectrum of the traveling waves.

Expression (5.40) represents the field function, Ψ, as a superposition of plane waves traveling in directions given by the direction cosines $q_x = \lambda u$, $q_y = \lambda v$, and $q_z = \sqrt{1 - q_x^2 - q_y^2}$. For a quasi-plane wave propagating in the z direction, the direction cosine in the transverse plane, q_x and q_y, are much smaller than q_z, i.e. $q_z \gg q_x, q_y$ and thus $\lambda^2\left(u^2 + v^2\right) \ll 1$. Then the square root in the Eq. (5.43) can be approximated by its first term in its Taylor series,

$$\sqrt{k^2 - 4\pi^2(u^2 + v^2)} \approx k\left[1 - 2\pi^2(u^2 + v^2)/k^2\right] = k - \pi\lambda(u^2 + v^2), \tag{5.44}$$

deriving the angular spectrum evolution equation in the paraxial regime

$$\overline{\Psi_o}(\mathbf{p}, z) = \overline{\Psi_i}(\mathbf{p}, 0) \exp(\mathrm{i}kz) \exp\left[-\mathrm{i}\pi\lambda z(u^2 + v^2)\right]. \tag{5.45}$$

Finally, substituting this expression into Eq. (5.40) and applying the well-known formula [80]

$$\int \exp\left(\mathrm{i}bx + \mathrm{i}ax^2\right)\mathrm{d}x = \sqrt{\frac{\pi}{|a|}}\exp\left[-\mathrm{i}\frac{b^2}{4a}x + \mathrm{i}\frac{\pi}{4}\mathrm{sign}\,(a)\right], \tag{5.46}$$

where a and b are real numbers and $a \neq 0$, we obtain that $\Psi(\mathbf{r}, z) = f(\mathbf{r}, z)\exp(\mathrm{i}kz)$ where the complex field amplitude $f(\mathbf{r}, z)$ is given by the Eq. (5.39).

5.2.5.3 Beam Propagation in a Quadratic GRIN Medium

The second special case of the Eq. (5.37) studied in this chapter is associated with a quadratic GRIN medium $n_0^2 - n^2(\mathbf{r}) = g_x^2x^2 + g_y^2y^2$. As we have seen in Sect. 5.2.4 the ray propagation through such media is described by the rotation matrix (5.33).

The equation for the complex field amplitude

$$\Delta_\perp f(\mathbf{r}, z) - k_0^2 f(\mathbf{r}, z)\left[g_x^2x^2 + g_y^2y^2\right] + \mathrm{i}2k_0n_0\frac{\partial}{\partial z}f(\mathbf{r}, z) = 0, \tag{5.47}$$

coincides with the well-known quantum mechanics equation for the harmonic oscillator [70, 78]. The Hermite–Gaussian (HG) modes, beams with amplitudes given by

$$f(\mathbf{r}, z) = \mathrm{HG}_m\left(\frac{x}{w_x}\right)\mathrm{HG}_n\left(\frac{y}{w_y}\right) = 2^{1/2}\left(2^m m! 2^n n!\right)^{-1/2}$$
$$\times H_m\left(\sqrt{2\pi}\frac{x}{w_x}\right)H_n\left(\sqrt{2\pi}\frac{y}{w_y}\right)\exp\left[-\pi\left(\frac{x^2}{w_x^2} + \frac{y^2}{w_y^2}\right)\right], \tag{5.48}$$

which have been discussed in [26], are eigenfunctions of this equation with eigenvalues $\exp\left[-\mathrm{i}\left(m + \frac{1}{2}\right)g_xz/n_0 - \mathrm{i}g_yz\left(n + \frac{1}{2}\right)g_yz/n_0\right]$. Since the HG modes form a complete orthonormal set, the general solution of the equation can be found as their linear composition

$$f(\mathbf{r}, z) = \sum_{m,n=0}^{\infty} f_{mn}\mathrm{HG}_m\left(\frac{x}{w_x}\right)\mathrm{HG}_n\left(\frac{y}{w_y}\right)\exp\left[-\mathrm{i}\frac{g_xz}{n_0}\left(m + \frac{1}{2}\right) - \mathrm{i}\frac{g_yz}{n_0}\left(n + \frac{1}{2}\right)\right], \tag{5.49}$$

where $w_{x,y} = \sqrt{\lambda_0 g_{x,y}^{-1}}$ and the coefficients are found from the boundary conditions at the plane $z = 0$ as

$$f_{mn} = \frac{1}{w_xw_y}\int f(\mathbf{r}_i, 0)\mathrm{HG}_m\left(\frac{x_i}{w_x}\right)\mathrm{HG}_n\left(\frac{y_i}{w_y}\right)\mathrm{d}\mathbf{r}_i. \tag{5.50}$$

Substitution of f_{mn} from Eq. (5.50) into Eq. (5.49) yields

$$f_o(\mathbf{r}_o, z) = \int f(\mathbf{r}_i, 0)\mathscr{K}(\mathbf{r}_i, \mathbf{r}_o; , z)\mathrm{d}\mathbf{r}_i, \tag{5.51}$$

where the kernel $\mathscr{K}(\mathbf{r}_i, \mathbf{r}_o; , z)$ is defined by $\mathscr{K}(\mathbf{r}_i, \mathbf{r}_o; , z) = \mathscr{K}_x(x_i, x_o; z)\, \mathscr{K}_y(y_i, y_o; z)$ with

$$\mathscr{K}_x(x_i, x_o; z) = \sum_{m=0}^{\infty} \mathrm{HG}_m\left(\frac{x_i}{w_x}\right)\mathrm{HG}_m\left(\frac{x_o}{w_x}\right)\exp\left[-\mathrm{i}\left(m+\frac{1}{2}\right)g_x z/n_0\right]. \tag{5.52}$$

Using the Meller formula

$$\sum_{m=0}^{\infty} \frac{1}{m!}\left(\frac{\xi}{2}\right)^2 H_m(x)H_m(y) = (1-\xi^2)^{-1/2}\exp\left[\frac{2xyq - \xi^2(x^2+y^2)}{1-\xi^2}\right], \tag{5.53}$$

for $\xi = \exp(\mathrm{i}g_x z/n_0)$ the kernel is transformed into the fractional Fourier transform kernel $\mathscr{K}(\mathbf{r}_i, \mathbf{r}_o; , z) = K_{\gamma_x}(x_i, x_o; w_x)K_{\gamma_y}(y_i, y_o; w_x) = K_{\gamma_x,\gamma_y}(\mathbf{r}_i, \mathbf{r}_o; w_x, w_y)$, where

$$K_{\gamma_x}(x_i, x_o; w_x) = \frac{1}{w_x\sqrt{\mathrm{i}\sin\gamma_x}}\exp\frac{\mathrm{i}\pi\left(x_o^2\cos\gamma_x - 2x_o x_i + x_i^2\cos\gamma_x\right)}{w_x^2\sin\gamma_x}, \tag{5.54}$$

$w_{x,y} = \sqrt{\lambda_0/g_{x,y}}$and the angles are defined as $\gamma_{x,y} = g_{x,y}z/n_0$. Thus, the paraxial beam propagation through the quadratic GRIN media is described by the separable fractional FT (see [26]). Fibers made from such media are widely used in practice. The special case $g_x = g_y$ corresponds to the so-called *selfoc* (self-focusing) fibers and the propagation is described by the symmetric fractional FT with $\gamma_x = \gamma_y = \gamma$ and $w_x = w_y = w$. They received such name because at a propagation distance corresponding to $\gamma = 2\pi n$ the complex field amplitude equals the input one apart from a scaling factor and constant phase shift.

It has been noted in [26] (see also [1, 4, 8, 9, 16, 79, 109]) that there are other complete orthogonal sets of modes which are eigenfunctions of the symmetric fractional FT. For instance, check the Laguerre–Gauss (LG), Hermite–Laguerre–Gauss (HLG) [2], and Ince–Gauss (IG) [16] modes. Thus, the $K_\gamma(\mathbf{r}_i, \mathbf{r}_o; w)$ kernel can be also written as a superposition of modes of these sets. Probably, the most important modes from this list are the LG modes defined as

$$\mathrm{LG}_{m,n}\left(\mathbf{r}; w_x, w_y\right) = \left(w_x w_y\right)^{-1/2}\sqrt{\frac{\min(m,n)!}{\max(m,n)!}}\left[\sqrt{2\pi}\left(\frac{x}{w_x} + \mathrm{i}\,\mathrm{sgn}\,(m-n)\frac{y}{w_y}\right)\right]^l$$
$$\times L_p^l\left[2\pi\left(\frac{x^2}{w_x^2}+\frac{y^2}{w_y^2}\right)\right]\exp\left[-\pi\left(\frac{x^2}{w_x^2}+\frac{y^2}{w_y^2}\right)\right], \tag{5.55}$$

where min(·, ·) and max(·, ·) are the minimum and maximum functions, $p = \min(m, n)$, $l = |m - n|$, and $L_p^l(\cdot)$ is the Laguerre polynomial of azimuthal index l and radial index p. These beams are used for optical microparticle manipulation, optical communications, laser ablation, metrology, etc. [108]. The main characteristic of these beams is the singularity in the phase distribution which has a vortex (helical) structure. This property is characterized by a topological charge defined by $\pm l$ [11, 28]. The symmetric fractional FT kernel is expressed as a combination of the LG modes as $K_\gamma(\mathbf{r}_i, \mathbf{r}_o; w) = \sum_{m,n=0}^{\infty} \mathrm{LG}_{m,n}(\mathbf{r}_i; w, w)\, \mathrm{LG}_{m,n}(\mathbf{r}_o; w, w) \exp[-\mathrm{i}(m + n + 1)\gamma]$.

5.3 First-Order Optical Systems

In the previous section we have studied two types of LCTs, the Fresnel transform associated with a homogeneous medium and the fractional FT associated with a quadratic GRIN medium. Both describe in the paraxial approximation the evolution of the complex field amplitude during the beam propagation. We have also derived that in the paraxial approximation of geometric optics, the ray transformation in such media can be expressed in a matrix form. These matrices parametrize the kernels of the corresponding LCTs. In this section we consider the basic optical elements for image formation and signal processing systems—lenses and mirrors—whose action in paraxial approximation is also described in the form of LCTs. We show that the parameters of the kernel describing the integral transformation are expressed by the parameters of the corresponding 4 × 4 ray transformation matrix **T**, which provides the relation between the position and direction of the ray before ($\mathbf{r}_i$ and $\mathbf{q}_i$) and after ($\mathbf{r}_o$ and $\mathbf{q}_o$) the optical elements:

$$\begin{bmatrix} \mathbf{r}_o \\ \mathbf{q}_o \end{bmatrix} = \mathbf{T} \begin{bmatrix} \mathbf{r}_i \\ \mathbf{q}_i \end{bmatrix}. \tag{5.56}$$

5.3.1 *Lens Description*

5.3.1.1 Spherical Lens

We start from the consideration of the glass spherical lens which can be defined as a portion of optically transparent material with refractive index n_l, limited by two spherical surfaces with radius R_1 and R_2, whose centers belong to the optical axis. The radius has positive (negative) sign if the center of the spherical surface is on the right (left) side with respect to the point of the surface intersection with the optical axes. The distance between the intersection points of the two surfaces with the axes is d.

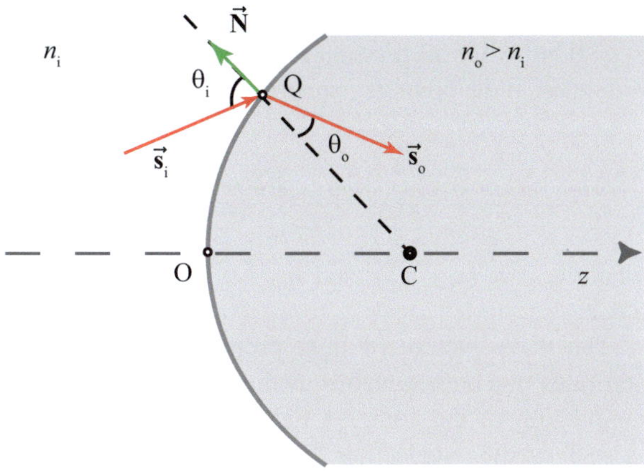

Fig. 5.3 Ray refraction in paraxial approximation on spherical surface separating media with different refractive index

In order to describe the ray transformation produced by a lens let us consider the refraction of a ray on the first surface. The incident and the refracted rays at a point $Q = (x, y, z)$ are characterized by the unitary vectors $\mathbf{s}_i$ and $\mathbf{s}_o$ given by $\mathbf{s}_{i,o} = \left[q_{x(i,o)}, q_{y(i,o)}, q_{z(i,o)}\right] / n_{i.o}$, respectively, where $n_{i,o}$ is the refraction index of the corresponding medium, see Fig. 5.3. The coordinate origin coincides with intersection point $O = (0, 0, 0)$. In paraxial approximation a spherical surface is substituted by paraboloidal (symmetrical in revolution) one given by

$$z - \frac{x^2 + y^2}{2R} = 0, \tag{5.57}$$

where we have taken into account that $x^2 + y^2 \ll R^2$ and $z - R < 0$ (convex surface).

Applying the Snell law at the point Q of the surface $f = z - \left(x^2 + y^2\right) / (2R)$

$$n_i\mathbf{s} - n_o\mathbf{s}_o = h\mathbf{N}, \tag{5.58}$$

where $\mathbf{N} = (\partial f/\partial x, \partial f/\partial y, \partial f/\partial z)$ is a normal to the surface at this point and h is a constant, we obtain

$$h\mathbf{N} = h\left(-\frac{x}{R}, -\frac{y}{R}, 1\right) = \left(q_{x(i)} - q_{x(o)}, q_{y(i)} - q_{y(o)}, q - q_{z(o)}\right). \tag{5.59}$$

Using again the paraxial approximation, $h = q_{z(i)} - q_{z(o)} \approx n_i - n_o$, we can express the relation between the ray vector components as

$$\begin{aligned} q_{x(o)} &= q_{x(i)} + x\frac{n_i - n_o}{R}, \\ q_{y(o)} &= q_{y(i)} + y\frac{n_i - n_o}{R}. \end{aligned} \tag{5.60}$$

Since the position of the ray does not change ($\mathbf{r}_o = \mathbf{r}_i$) and assuming that $z_o = z_i = 0$ holds for all paraxial rays, we find the ray transformation matrix, $\mathbf{T}$ for the refraction at the boundary spherical surface

$$\mathbf{T} = \begin{bmatrix} \mathbf{I} & \mathbf{0} \\ \frac{n_o - n_i}{R}\mathbf{I} & \mathbf{I} \end{bmatrix}, \tag{5.61}$$

where $\mathbf{r}_{i,o} = (x_{i,o}, y_{i,o})^t$ and $\mathbf{q}_{i,o} = \left(q_{x(i,o)}, q_{y(i,o)}\right)^t$ and $\mathbf{I}$ stands for unity matrix 2×2.

The wave picture of light propagation through the spherical surface, which separates two homogeneous media with different refractive index, yields the phase modulation of the complex field amplitude due to the different phase velocities in such media. Neglecting the light reflection on the frontier and using again the paraxial approximation of the sphere by a paraboloid, the relation between the complex field amplitudes on both sides of the surface is expressed by

$$f_o(\mathbf{r}_o, 0) = f_i(\mathbf{r}_o, 0) \exp\left[\mathbf{i}\frac{k_0(n_i - n_o)}{2R} r_o^2\right], \tag{5.62}$$

where k_0 is a wave number in vacuum and $r_o^2 = (x_o^2 + y_o^2)$. We underline that this relation corresponds to the quadratic phase modulation LCT (see [26]).

Taking into account the Eqs. (5.61) and (5.62) as well as the propagation of paraxial rays and waves in homogeneous media (see Eqs. (5.23) and (5.39)) we can now define the action of the lens in both cases.

The lens can be modeled as a system composed of three elements: a spherical surface, a piece of homogeneous medium, and another spherical surface. Then, the ray propagation through the lens is described by the product of the three matrices corresponding to these systems:

1. The first spherical boundary of radius R_1 separates the input medium with refractive index n_m from the lens medium with refractive index n_l. The associated matrix is

$$\mathbf{T}_1 = \begin{bmatrix} \mathbf{I} & \mathbf{0} \\ \frac{n_m - n_l}{R_1}\mathbf{I} & \mathbf{I} \end{bmatrix}. \tag{5.63}$$

2. The ray propagation for a distance d inside the homogeneous medium with refractive index n_l—the lens—is described by the ray transformation matrix

$$\mathbf{T}_2 = \begin{bmatrix} \mathbf{I} & d/n_l\mathbf{I} \\ \mathbf{0} & \mathbf{I} \end{bmatrix}. \tag{5.64}$$

3. The second spherical boundary of radius R_2 separates the lens medium with refractive index n_l from the output medium with refractive index n_m. The corresponding matrix is

$$\mathbf{T}_3 = \begin{bmatrix} \mathbf{I} & \mathbf{0} \\ -\frac{n_m - n_l}{R_2}\mathbf{I} & \mathbf{I} \end{bmatrix}. \tag{5.65}$$

Therefore, the relation between the position, $\mathbf{r}_i$, and direction, $\mathbf{q}_i$, of rays at the lens input, $z = 0$, and output, $z = d$, planes is described by the ray transformation matrix obtained after the matrix multiplication

$$\mathbf{T} = \mathbf{T}_3\mathbf{T}_2\mathbf{T}_1 = \begin{bmatrix} A\mathbf{I} & B\mathbf{I} \\ C\mathbf{I} & D\mathbf{I} \end{bmatrix}$$

$$= \begin{bmatrix} \left[1 + \frac{d(n_m - n_l)}{n_l R_1}\right]\mathbf{I} & \frac{d}{n_l}\mathbf{I} \\ \left[(n_m - n_l)\left(\frac{1}{R_1} - \frac{1}{R_2}\right) - \frac{d(n_m - n_l)^2}{n_l R_1 R_2}\right]\mathbf{I} & \left[1 - \frac{d(n_m - n_l)}{n_l R_2}\right]\mathbf{I} \end{bmatrix}. \tag{5.66}$$

The focal distance of the lens related to the parameter C is given by

$$\mathrm{f} = -n_m\left[(n_m - n_l)\left(\frac{1}{R_1} - \frac{1}{R_2}\right) - \frac{d(n_m - n_l)^2}{n_l R_1 R_2}\right]^{-1}. \tag{5.67}$$

The matrix description of the ray propagation through a spherical lens is summarized by Eq. (5.66). Now we will derive an analogous expression from the point of view of wave optics. The lens again is modeled as the concatenation of an input spherical boundary, a propagation in a homogeneous medium and an output spherical boundary. The relation between the complex field amplitudes at the input $(\mathbf{r}_i, 0)$ and output $(\mathbf{r}_o, d)$ planes of the lens can be found following these steps:

1. Phase modulation of the input amplitude, $f_i(\mathbf{r}_i, 0)$, at the input surface boundary

$$f_1(\mathbf{r}_i, 0) = f_i(\mathbf{r}_i, 0)\exp\left[\mathrm{i}\frac{k_0(n_m - n_l)}{2R_1}\mathbf{r}_i^2\right]. \tag{5.68}$$

2. Calculation of the Fresnel transformation of $f_1(\mathbf{r}_i, 0)$ which describes the propagation through the homogeneous medium of the refractive index n_l

$$f_2(\mathbf{r}_o, d) = \frac{n_l}{\mathrm{i}\lambda_0 d}\int f_1(\mathbf{r}_i, 0)\exp\left[\frac{\mathrm{i}\pi n_l}{\lambda_0 d}(\mathbf{r}_1 - \mathbf{r}_i)^2\right]\mathrm{d}\mathbf{r}_i. \tag{5.69}$$

3. Phase modulation of the $f_2(\mathbf{r}_o, d)$ field by the output surface boundary

$$f_o(\mathbf{r}_o, d) = f_2(\mathbf{r}_o, d)\exp\left[\mathrm{i}\frac{k_0(n_l - n_m)}{2R_2}\mathbf{r}_o^2\right]. \tag{5.70}$$

Following the latter scheme we conclude that $f_o(\mathbf{r}_o, d)$ is defined by the LCT of $f_i(\mathbf{r}_i, 0)$

$$f_o(\mathbf{r}_o, d) = \frac{n_m}{i\lambda_0 B} \int f_i(\mathbf{r}_i, 0) \exp \frac{\mathrm{i}\pi n_m \left(r_o^2 D - 2\mathbf{r}_o \mathbf{r}_i + r_i^2 A\right)}{\lambda_0 B} \mathrm{d}\mathbf{r}_i$$

whose kernel is parametrized by the ray transformation matrix (5.66). The expression for $\Psi_o(\mathbf{r}_o, d)$ is obtained by adding a constant phase factor describing the propagation inside the lens: $\Psi_o(\mathbf{r}_o, d) = \exp(\mathrm{i}k_0 n_l d) f_o(\mathbf{r}_o, d)$. Note that absorption inside the lens and reflections on the surfaces are neglected in this approximation.

5.3.1.2 Thin Lens Approximation

If the width of the lens d is very small comparing with the radii R_1 and R_2 we can further simplify the lens action by substituting $d = 0$ in the ray transformation matrix Eq. (5.66). This yields

$$\mathbf{T} = \begin{bmatrix} \mathbf{I} & \mathbf{0} \\ -\frac{n_m}{\mathrm{f}}\mathbf{I} & \mathbf{I} \end{bmatrix}, \tag{5.71}$$

where the focal length is defined by

$$\mathrm{f} = \left[(n_l/n_m - 1) \left(\frac{1}{R_1} - \frac{1}{R_2} \right) \right]^{-1}. \tag{5.72}$$

The positive (negative) sign of f corresponds to convergent (divergent) lens. This approximation is known as a *thin lens* approximation while the lens described by the Eq. (5.66) is denoted as *thick lens.*

Since the form of the ray transformation matrix is similar to Eq. (5.61), the action of the thin lens corresponds to a quadratic phase modulation analogous to Eq. (5.62)

$$f_o(\mathbf{r}, 0) = f_i(\mathbf{r}, 0) \exp\left[-\mathrm{i}\frac{k_0}{2\mathrm{f}} r^2 \right]. \tag{5.73}$$

Here the propagation through the homogeneous medium that forms the lens is neglected since we assume that the lens is thin. In this case the input and output planes share the same transversal coordinates $\mathbf{r}$.

Since the thin lens approximation is widely used in optics the LCT associated with quadratic phase modulation is known as a lens transform (see [26]). In the rest of this chapter we will consider thin lenses, except when stated otherwise. Moreover, we will assume that the refractive index of the medium surrounding the lens is air and, hence, $n_m = 1$.

5.3.1.3 Cylindrical and Generalized Lenses

In the previous sections we have studied the spherical lenses which are symmetric with respect to rotation around the optical axes. Now we will study other quadratic surfaces through which the beams propagate following an LCT description. Indeed, instead considering the paraboloid of revolution (paraxial approximation of a spherical surface) we may analyze the elliptic or hyperbolic paraboloids. For example, we can assume that one of the lens surfaces is plane and the other one is represented by an elliptical or hyperbolic paraboloid described by

$$z - \frac{x^2}{2R_x} - \frac{y^2}{2R_y} = 0. \tag{5.74}$$

The elliptical or hyperbolic cases are obtained when R_x/R_y is positive or negative, respectively. Following similar steps as in the previous subsection it can be shown that the ray transformation matrix of this type of lenses, in the thin lens approximation, is given by

$$\mathbf{T} = \begin{bmatrix} \mathbf{I} & \mathbf{0} \\ \mathbf{C} & \mathbf{I} \end{bmatrix}, \tag{5.75}$$

where submatrix $\mathbf{C}$ is not scalar but diagonal:

$$\mathbf{C} = -\begin{bmatrix} \mathrm{f}_x^{-1} & 0 \\ 0 & \mathrm{f}_y^{-1} \end{bmatrix}. \tag{5.76}$$

Here the focal lengths are $\mathrm{f}_{x,y} = R_{x,y}/(n_l - 1)$ where remember that n_l is the refraction index of the lens.

Correspondingly, the phase modulation of the complex field amplitude of the beam propagating through the lens is described by elliptical ($\mathrm{f}_x/\mathrm{f}_y > 0$) or hyperbolic ($\mathrm{f}_x/\mathrm{f}_y < 0$) functions

$$f_o(\mathbf{r}, 0) = f_i(\mathbf{r}, 0) \exp\left[-\mathrm{i}\frac{k_0}{2}\left(\frac{x^2}{\mathrm{f}_x} + \frac{y^2}{\mathrm{f}_y}\right)\right]. \tag{5.77}$$

One special case of significant relevance is when $R_y, \mathrm{f}_y \rightarrow \infty$, which describes a cylindrical lens that modulates the phase only in the x direction. Notice that a general elliptical or hyperbolic lens can be represented as a superposition of two cylindrical ones.

Ideally, a cascade of thin lenses can be considered as an optical element which can also be treated as a thin lens. Let us assume that all the cascade elements are centered with respect to the optical axis, but every element can have a different orientation of its transversal principal axis in the XY plane. If a cylindrical lens is rotated at angle ϕ with respect to the X axis, the corresponding submatrix $\mathbf{C}$ is expressed as (see [26, 77, 89]):

$$\mathbf{C} = \begin{bmatrix} \cos\phi & \sin\phi \\ -\sin\phi & \cos\phi \end{bmatrix} \begin{bmatrix} 0 & 0 \\ 0 & -\mathrm{f}_{y^{-1}} \end{bmatrix} \begin{bmatrix} \cos\phi & -\sin\phi \\ \sin\phi & \cos\phi \end{bmatrix} = -\mathrm{f}_y^{-1} \begin{bmatrix} \sin^2\phi & \sin\phi\cos\phi \\ \sin\phi\cos\phi & \cos^2\phi \end{bmatrix}. \tag{5.78}$$

Useful optical elements are obtained when several cylindrical lenses are assembled in such a way that their transversal axes are rotated one with respect to another. In the general case the lens can be composed by the series of m cylindrical lenses of focal distance f_i oriented according to the angles ϕ_i, where $i = 1, \ldots, m$. The ray transformation matrix associated with this *generalized lens* is described by the Eq. (5.75) whose symmetric submatrix $\mathbf{C}$ is defined by

$$\mathbf{C} = \begin{bmatrix} c_{xx} & c_{xy} \\ c_{xy} & c_{yy} \end{bmatrix} = -\frac{1}{2} \begin{bmatrix} \sum_{i=1}^{m} \mathrm{f}_i^{-1}(1 + \cos 2\phi_i) & \sum_{i=1}^{m} \mathrm{f}_i^{-1} \sin 2\phi_i \\ \sum_{i=1}^{m} \mathrm{f}_i^{-1} \sin 2\phi_i & \sum_{i=1}^{m} \mathrm{f}_i^{-1}(1 - \cos 2\phi_i) \end{bmatrix}. \tag{5.79}$$

The previous result has been obtained applying the ray transformation matrix multiplication law, from where it is easy to obtain that for the cylindrical lens cascade system $\mathbf{C} = \mathbf{C}_m + \cdots + \mathbf{C}_2 + \mathbf{C}_1$, being $\mathbf{C}_i$ the submatrix associated with the i-th rotated cylindrical lens. One example is a spherical lens, which can be represented as a superposition of two cylindrical lenses with equal focal distances, $\mathrm{f}_1 = \mathrm{f}_2$, whose principal axes are orthogonal, $\phi_1 = \phi_2 + \pi/2$.

5.3.1.4 Thin Lens from Wave Point of View

The phase modulation associated with the action of a generalized lens is expressed by:

$$f_o(\mathbf{r}, z) = f_i(\mathbf{r}, z) \exp\left(\mathrm{i}\frac{k_0}{2} \left[c_{xx}x^2 + 2c_{xy}xy + c_{yy}y^2 \right] \right), \tag{5.80}$$

where the c_{ij} coefficients are the submatrix $\mathbf{C}$ elements (see Eq. (5.79)). A generalized lens composed by a pair of two identical convergent cylindrical lenses is displayed in Fig. 5.4(a). The phase distribution associated with it is displayed for different orientation of the lens principal axes: Fig. 5.4(b) 90° and (c) 45°. As an example of hyperbolical lens composed by convergent and divergent cylindrical lenses of the same focal length modulus, $|\mathrm{f}_1| = |\mathrm{f}_2|$ is displayed in Fig. 5.4(d). The phase distribution associated with such hyperbolical lens is shown for the lens-rotation configurations: Fig. 5.4(e) 90° and (f) 45°.

5.3.2 *Mirrors*

The reflection from quadratic surfaces can also be described by an LCT. In order to find the corresponding ray transformation matrix we will use the same notations and approximations which were applied when the ray refraction at the spherical surface

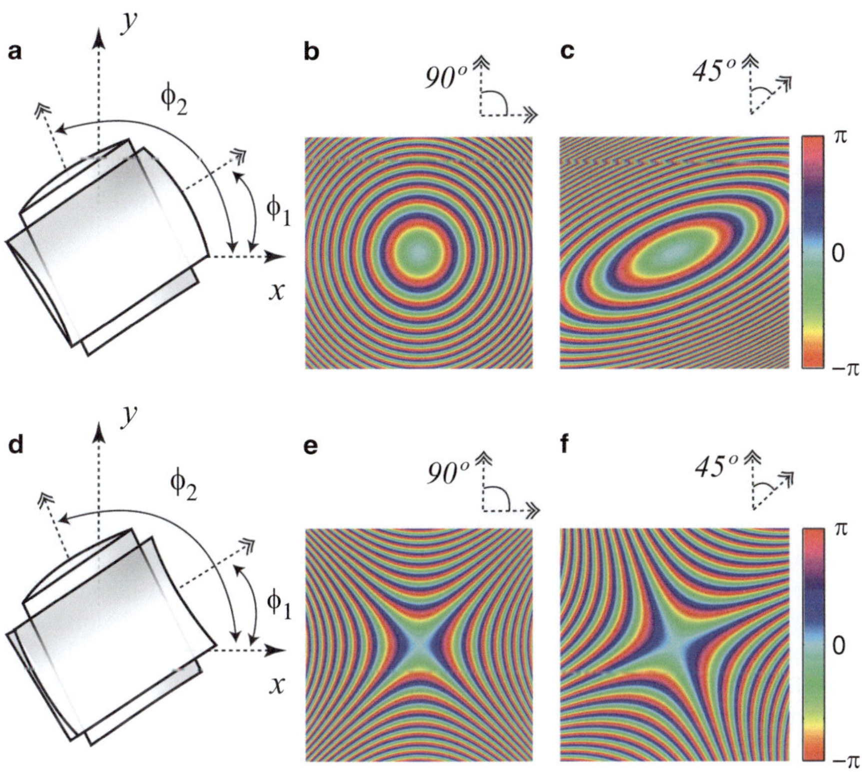

Fig. 5.4 (**a**) Generalized lens assembled as a set of two identical convergent cylindrical lenses rotated respect the X axis the angles ϕ_1 and ϕ_2. Phase distribution associated with the assemble for the case (**b**) $\phi_1 = \pi/2, \phi_2 = 0$ and (**c**) $\phi_1 = \pi/4$, $\phi_2 = 0$. (**d**) Generalized lens assembled as a set of two cylindrical lenses, one divergent and one convergent, in the same situation as in (**a**). Phase distribution associated with the assemble for the case (**e**) $\phi_1 = \pi/2$, $\phi_2 = 0$ and (**f**) $\phi_1 = \pi/4$, $\phi_2 = 0$. In these examples the focal lengths of the cylindrical lenses are the same, $f = |f_1| = |f_2|$

has been studied (see Sect. 5.3.1). Taking into account that the reflection inverts the sign of the z component of the ray vector $q_{z(o)} = -q_{z(i)}$, and that $n_o = n_i$ the Eq. (5.58) is transformed into the law of reflection

$$\mathbf{s}_i - \mathbf{s}_o = 2\mathbf{N}, \tag{5.81}$$

where $\mathbf{s}_i$ and $\mathbf{s}_o$ are the unitary vectors of the incident and reflected ray. The constant h in this case is given by $h = q_{z(i)} - q_{z(o)} = 2q_{z(i)} \approx 2n_i$. It is easy to prove that the ray reflection matrix can be obtained from the ray refraction matrix, Eq. (5.61) by substitution $(n_o - n_i)$ into $2n_i$

$$\mathbf{T} = \begin{bmatrix} \mathbf{I} & \mathbf{0} \\ \frac{2n_i}{R}\mathbf{I} & \mathbf{I} \end{bmatrix}. \tag{5.82}$$

Notice that the case $R \to \infty$ corresponds to the plane mirror.

Similarly to the refracting surfaces, the reflecting surfaces can also be made in the form of elliptical and hyperbolic paraboloids or cylinders. Therefore, a general quadratic mirror reflection matrix has an identical form to the generalized lens matrix, Eq. (5.75), associated with the corresponding phase modulation function, Eq. (5.80). Nevertheless, we have to take into account that the reflection always inverts the direction of the beam propagation along axes z. The quadratic surface mirrors play an important role in optics, specially in the technology related with resonators [63].

5.3.3 Programmable Optical Elements Working as Lenses and Mirrors

Relevant applications in optics demand setups whose parameters can be changed almost in real time. Programmable spatial light modulator devices (SLMs) are often used in such setups. In particular they can work as dynamic phase modulating optical element, as a thin lens or mirror.

There exist several technologies in which the SLMs are based. A liquid crystal display SLM is one of the most used. Here we consider, as an example, pixelated SLM based on the LCOS (liquid crystal on silicon chip) technology because it offers important advantages such as high spatial resolution (e.g., 1920×1080 pixels, with a pixel size about 8 μm), compact design, versatility, and an affordable cost. This type of SLMs can electronically address a digital or analogical signal via a conventional PC, which makes easier its implementation in complex optical setups.

The signal to be implemented in the SLM can be sent as an image that represents the desired phase modulation. For instance, if we want to implement a generalized thin lens in the SLM, the signal to be sent to the SLM is given by the argument of the lens phase modulation function: $\arg\left[\Psi(x,y)\right] = \frac{k_0}{2}\left[c_{xx}x^2 + 2c_{xy}xy + c_{yy}y^2\right]$. In general an LCOS-SLM device modulates the phase of an incident (linear polarized and monochromatic) beam up to one wavelength of phase shift (thus up to 2π), which is quantized into N levels (often $N = 256$). This digital image of the signal $\mathrm{mod}(\arg\left[\Psi(x,y)\right], 2\pi)$ is real-time addressed into the pixelated SLM device, which displays it as the phase-only modulation function $\Psi(x,y) = \exp\left[\mathrm{i}\frac{k_0}{2}\left(c_{xx}x^2 + 2c_{xy}xy + c_{yy}y^2\right)\right]$. This kind of programmable SLM device can modify the displayed phase signal at video rate value, about 30 frames per second (fps). For instance, it is possible to implement a temporal varifocal lens whose focal length varies in time at 30 fps.

The pixelated structure of the SLM display produces some drawbacks, which have to be mentioned. In particular, the pixelated structure itself works as a diffraction grating that modulates and splits the input beam into several spatially multiplexed signals. The true modulated signal can be separated from the other ones, considered as an additive noise, by spatial filtering. For this task a well-known 4-f setup [55] that consists in two glass-type spherical lenses working as a telescope is usually applied. The beam focused in the back focal plane of the first lens (Fourier plane) is filtered using an aperture or iris diaphragm. From this filtered Fourier spectrum the signal is properly reconstructed by using the second lens of the 4-f setup. For successful filtering of a linear carrier phase $\varphi_c = 2\pi(u_0 x + v_0 y)$ is added into the digital signal, $\mathrm{mod}(\Psi(x, y) + \varphi_c, 2\pi)$ in order to isolate the component of the Fourier spectrum associated with the signal. The values of the spatial frequency u_0 and v_0 are limited by the pixel size of the SLM. Another important drawback is the energy lost due to the diffraction from the pixelated structure itself and the corresponding filtering process.

The pixel size, δ, of an SLM display limits the sampling of the signal to be addressed. Indeed, according to the Nyquist–Shannon sampling theorem, only the signals with a maximum spatial frequency value given by $1/2\delta$ can be correctly sampled by the SLM display. This is an important fact because it limits the minimum value of the focal length and the effective pupil diameter of the lens that the SLM can address. Notice that a digital lens behaves as a discrete Fresnel lens according to $\mathrm{mod}(\arg[\Psi(x, y)], 2\pi)$ when implemented into the pixelated SLM. Only the Fresnel lens zones with a size higher than 2δ are correctly displayed into the SLM. Since the Fresnel zones far away from the lens center get thinner, the effective lens pupil diameter is smaller as the focal length decreases. This drawback can be minimized by using SLMs which provide a wider range of phase-shift modulation, above the conventional 2π one.

Other programmable SLM devices such as deformable mirrors can implement the elliptical or hyperbolical mirrors but they neither can address short focal lengths because of the insufficient deformation (curvature) reachable by their membrane. In addition, they are less versatile and much more expensive than the LCOS-SLM devices. Nevertheless, a deformable mirror can change its phase modulation functions faster than the LCOS-SLM devices (higher frames per second) without the pixelated structure, and, thus, with higher light efficiency. Besides, the deformable mirror is polarization independent and it does not suffer from chromatic aberration because it works in reflection geometry. This fact is very important in relevant optical applications such as imaging and beam characterization.

Regardless of their differences, both SLM devices have a similar active area with a diameter about 10–20 mm and both can work in the visible and near infrared ranges. The deformable mirror technology is often used as an adaptive optics for advanced wavefront control within vision science and astronomical imaging, in which compensation of spatial aberrations of the light beam is involved, while the LCOS-SLM technology has found wider application in fields such as imaging, laser beam shaping, and image processing.

5.3.4 *Composite Systems*

In the previous subsections we have found that beam propagation through the homogeneous and weakly inhomogeneous quadratic GRIN media, and the action of various optical elements such as the spherical, parabolical, and hyperbolical lenses and mirrors are described in paraxial approximation by certain LCT: Fresnel transformation, fractional FT and quadratic phase modulation. Here we show following the [74, 75] that the combination of these basic systems allows constructing one corresponding to the LCT in more general form. The ray transformation matrix formalism is used throughout since it significantly simplifies the system design for the realization of a certain LCT [82]. We underline that, since the LCT description of the basic systems is only valid in paraxial approximation, these assumptions are kept for the composite system, which, for that reason, are called first-order optical systems. Other common names are Gaussian or paraxial systems.

In optics, apart from the imaging systems described by the imager transform the phase-space rotators (see [26]) are widely used for information processing tasks. We remind that the Fourier transformer, which allows performing the filtering operation, belongs to this class. The growing interest to the application of the fractional FT [70, 78, 79] and gyrator transform [83, 84, 93] requires design a system with tunable transformation parameters. Here we discuss only general approach for the construction of such systems. The optical implementation of the LCTs is considered in detail in [65].

It has been shown in [26] that the LCT in the most general form is characterized by a 4×4 symplectic matrix with ten free parameters. As we have seen, the matrices associated with the LCTs describing the basic optical elements correspond to the ray transformation matrices. It is easy to prove that all of them are symplectic with determinant equals one (see Table 5.1). From the cascadability property of the LCT (see [26, Sect. 3.1.1]) it follows that we can design an optical system for its realization using as bricks the basic elements: homogeneous medium sections, quadratic GRIN-medium sections, lenses, mirrors, etc. Then the ray transformation matrix $\mathbf{T}$ describing the composite system is described as a product of the corresponding matrices $\mathbf{T}_n$ of the N basic elements ($n = 1, 2, \ldots, N$), arranged in the reverse order of element appearance in the cascade

$$\mathbf{T} = \mathbf{T}_N \mathbf{T}_{N-1} \ldots \mathbf{T}_1 .$$

The relation between the complex field amplitude at the input and output planes of the composed system is given in the form of the LCT [42, 66, 71, 106, 107]

$$\begin{aligned} f_o(\mathbf{r}_o) &= \mathscr{L}(\mathbf{T}) f_i(\mathbf{r}_i) = \lambda_0^{-1} (\det \mathrm{i}\,\mathbf{B})^{-1/2} \\ &\times \int \exp\left[\mathrm{i}\,\frac{k_0}{2}(\mathbf{r}_o^t \mathbf{D}\mathbf{B}^{-1}\mathbf{r}_o - 2\mathbf{r}_i^t \mathbf{B}^{-1}\mathbf{r}_o + \mathbf{r}_i^t \mathbf{B}^{-1}\mathbf{A}\mathbf{r}_i)\right] f_i(\mathbf{r}_i)\, \mathrm{d}\mathbf{r}_i \,, \end{aligned} \tag{5.83}$$

with $\det \mathbf{B} \neq 0$ and

$$f_o(\mathbf{r}) = \mathscr{L}(\mathbf{T}) f_i(\mathbf{r}) = \lambda_0^{-1} |\det \mathbf{A}|^{-1/2} \exp\left[\mathrm{i}\,\frac{k_0}{2}\mathbf{r}^t \mathbf{C}\mathbf{A}^{-1}\mathbf{r}\right] f_i(\mathbf{A}^{-1}\mathbf{r}) \,, \tag{5.84}$$

Table 5.1 Summary of ray transformation matrices of basic first-order optical systems

System	Transformation matrix
Propagation in free-space or in a homogeneous medium (propagation distance z and constant refractive index n)	$\begin{bmatrix} \mathbf{I} & \frac{z}{n}\mathbf{I} \\ \mathbf{0} & \mathbf{I} \end{bmatrix}$
Reflection from a spherical surface (planar mirror if $R \to \infty$)	$\begin{bmatrix} \mathbf{I} & \mathbf{0} \\ \frac{2n}{R}\mathbf{I} & \mathbf{I} \end{bmatrix}$
Transmission through a thin spherical lens (focal length f given by Eq. (5.72))	$\begin{bmatrix} \mathbf{I} & \mathbf{0} \\ -\frac{1}{\mathrm{f}}\mathbf{I} & \mathbf{I} \end{bmatrix}$
Transmission through a thin generalized lens (submatrix **C** is symmetric, Eq. (5.79))	$\begin{bmatrix} \mathbf{I} & \mathbf{0} \\ \mathbf{C} & \mathbf{I} \end{bmatrix}$
Transmission through a graded index selfoc media (z is the length of the inhomogeneous medium of refractive index $n = n_0\sqrt{1 - w^2r^2}$, ray trajectories $r(z)$ inside it are periodical with a frequency w)	$\begin{bmatrix} \cos(wz)\,\mathbf{I} & \frac{1}{n_0 w}\sin(wz)\,\mathbf{I} \\ -n_0 w \sin(wz)\,\mathbf{I} & \cos(wz)\,\mathbf{I} \end{bmatrix}$

for the limiting case $\mathbf{B} \Rightarrow \mathbf{0}$, for which $\mathbf{r}_o \Rightarrow \mathbf{A}\mathbf{r}_i$. The widely used composite systems can be divided into two classes: imagers and phase-space rotators (PhSRs) described correspondingly by the ray transformation matrices

$$\mathbf{T}_{\text{Imager}} = \begin{bmatrix} \mathbf{A} & \mathbf{0} \\ \mathbf{C} & \mathbf{A}^{-1} \end{bmatrix}, \mathbf{T}_{\text{PhSR}} = \begin{bmatrix} \mathbf{A} & s\mathbf{X} \\ -s^{-1}\mathbf{X} & \mathbf{A} \end{bmatrix},$$

where s is a dimension normalization constant. According to the modified Iwasawa decomposition (see [26, 92, 94, 96]) any optical system can be constructed (but not in the most efficient way) as a cascade of phase-space rotator and imager. The imagers are used for image formation and describe (of course, in paraxial approximation) the functioning of microscope and telescope. The phase-space rotators are mostly applied for optical signal processing tasks, that include signal analysis, filtering and correlation operations.

As it has been shown in [26] the key phase-space rotator is the separable fractional FT. We have seen that the beam propagation in the quadratic GRIN medium yields in the paraxial approximation to the fractional FT relation between its complex field amplitudes at the input and output planes. The use of such systems may have benefits for the LCTs implementation in photonics. Nevertheless, their application for performing other LCTs by introducing additional phase modulating elements as well as the modification of the transformation parameter is rather difficult. Then more often the phase-space rotator system consists of homogeneous medium sections and phase modulators (thin lenses or mirrors). This is the case of typical schemes for optical Fourier transformer (see [26, Fig. 2.1]). Certain conditions, such as minimal number of elements, easier change of transformation parameter, etc., have to be satisfied in order to design a system optimal for a given

LCT realization. An example of the construction of a versatile phase-space rotator system, which can be applied for the beam characterization (see Sects. 5.5 and 5.6) is discussed in [65, 82, 85].

5.4 Completely and Partially Coherent Light

In the previous sections we have considered the propagation of the monochromatic spatially coherent light through the first-order optical systems, here we turn to the analysis of the partially coherent light.

5.4.1 Description of Partially Coherent Light

The models of spatially coherent and incoherent light, which are often used for the description of imaging systems [55], are two extreme cases of the partially coherent light. In spite of these models significantly simplify the solution of different problems, they cannot always be applied. Moreover, several imaging modalities are benefited from or based on the use of the partially coherent light [15, 36, 51, 54]. Statistical phenomena of light emission and detection, local fluctuations of the refractive index of the medium where light propagates or surface from which it reflects yield to the partially coherent light model. The theory of the optical coherence can be found in the seminal books [36, 54, 68] among many others.

5.4.1.1 Mutual Intensity

Here we consider the statistical description of the linearly polarized monochromatic light using again the paraxial wave approximation. Taking into account the different temporal fluctuations of the phase in each point of space [36, 50, 56], we now cannot describe the spatial structure of the optical field as a complex function of $\mathbf{r}$ as we did in the previous sections, $f(\mathbf{r}) = a(\mathbf{r}) \exp[\mathrm{i}\varphi(\mathbf{r})]$. In this case the wide-sense statistically stationary field is characterized by the two-point correlation function, known as mutual intensity (MI)

$$\Gamma_f(\mathbf{r}_1, \mathbf{r}_2) = \langle f(\mathbf{r}_1) f^*(\mathbf{r}_2) \rangle, \tag{5.85}$$

which is a function of four variables: the vectors $\mathbf{r}_1$ and $\mathbf{r}_2$. Here the brackets $\langle\cdot\rangle$ stand for ensemble averaging that coincides with time averaging for the ergodic processes, and the asterisk * stands for the complex conjugation operation.

The MI describes the correlation between the optical field oscillations at two points $\mathbf{r}_1$ and $\mathbf{r}_2$. It is a non-negative definite Hermitian function of $\mathbf{r}_1$ and $\mathbf{r}_2$ [18, 67], i.e.,

$$\Gamma_f(\mathbf{r}_1, \mathbf{r}_2) = \Gamma_f^*(\mathbf{r}_2, \mathbf{r}_1) \quad \text{and} \quad \int g(\mathbf{r}_1)\, \Gamma_f(\mathbf{r}_1, \mathbf{r}_2)\, g^*(\mathbf{r}_2)\, \mathrm{d}\mathbf{r}_1\, \mathrm{d}\mathbf{r}_2 \geq 0 \tag{5.86}$$

for any smooth function $g(\mathbf{r})$.

The MI can be also written in the form, known as Mercer or coherent mode expansion,

$$\Gamma_f(\mathbf{r}_1, \mathbf{r}_2) = \sum_{n=0}^{\infty} \Lambda_n f_n(\mathbf{r}_1) f_n^*(\mathbf{r}_2), \tag{5.87}$$

where the eigenvalues of the MI, Λ_n, are real and non-negative and the eigenfunctions $f_n(\mathbf{r})$ form the complete orthonormal set [68]. The search of the eigenfunctions of the MI which satisfy the relation

$$\int \Gamma_f(\mathbf{r}_1, \mathbf{r}_2) f_n(\mathbf{r}_2) d\mathbf{r}_2 = \Lambda_n f_n(\mathbf{r}_1),$$

is not easy, but their knowledge simplifies the calculation of the evolution of the MI during field propagation.

If $\mathbf{r}_1 = \mathbf{r}_2 = \mathbf{r}$ the MI reduces to the intensity distribution $I(\mathbf{r}) = \Gamma_f(\mathbf{r}, \mathbf{r}) = \left\langle |f(\mathbf{r})|^2 \right\rangle = \sum_{n=0}^{\infty} \Lambda_n |f(\mathbf{r})|^2$, that explains the name "mutual intensity." For quantitative estimation of field correlation between two points $\mathbf{r}_1$ and $\mathbf{r}_2$ the complex degree of spatial coherence is applied

$$\mu(\mathbf{r}_1, \mathbf{r}_2) = \frac{\Gamma_f(\mathbf{r}_1, \mathbf{r}_2)}{\sqrt{\Gamma_f(\mathbf{r}_1, \mathbf{r}_1)\Gamma_f(\mathbf{r}_2, \mathbf{r}_2)}},$$

whose modulus equals one for the completely coherent light and it is proportional to the Dirac delta function $\delta(\mathbf{r}_1 - \mathbf{r}_2)$ for the incoherent case.

Notice that the two-dimensional (2D) problem of the description of the coherent paraxial field is converted to the 4D one, when the partially coherent light model has to be applied. The determination of the entire MI requires the application of interferometric or tomographic techniques [36, 39, 49, 81]. The tomographic method based on the registration and processing of the diffraction pattern obtained after the beam propagation through the first-order optical systems is considered in Sect. 5.5.

For coherent fields, the brackets in the Eq. (5.85) can be omitted and MI is transformed into the complex-amplitude product $\Gamma_f(\mathbf{r}_1, \mathbf{r}_2) = f(\mathbf{r}_1) f^*(\mathbf{r}_2)$. The incoherent light is described by the MI given by

$$\Gamma_f(\mathbf{r}_1, \mathbf{r}_2) = p(\mathbf{r}_1)\, \delta(\mathbf{r}_1 - \mathbf{r}_2), \tag{5.88}$$

where $\delta(\mathbf{r})$ is the Dirac delta function and $p(\mathbf{r}_1)$ is proportional to the beam intensity distribution $I(\mathbf{r}_1)$.

The complexity of the MI reduces if it is separable in a certain coordinate system. Thus, the so-called quasi-stationary light is described by the MI given by

$$\Gamma_f(\mathbf{r}_1,\mathbf{r}_2) = I((\mathbf{r}_1 - \mathbf{r}_2)/2)\,\mu(\mathbf{r}_1 - \mathbf{r}_2). \tag{5.89}$$

Here $I(\mathbf{r})$ is a real non-negative function corresponding to the intensity distribution and $\mu(\mathbf{r})$ is an Hermitian dimensionless function, $\mu(\mathbf{r}) = \mu^*(-\mathbf{r})$, such that $|\mu(\mathbf{r})| \leq \mu(0) = 1$.

Another important case corresponds to the field separable in the Cartesian coordinates given by

$$\Gamma_f(\mathbf{r}_1,\mathbf{r}_2) = \Gamma_{f_x}(x_1,x_2)\Gamma_{f_y}(y_1,y_2), \tag{5.90}$$

where $\Gamma_{f_\xi}(\xi_1,\xi_2) = \langle f(\xi_1)f^*(\xi_2)\rangle$ and the problem of the 2D beam description is reduced to the characterization of two 1D signals. Finally, we mention the rotationally symmetric beams whose MI depends on only three variables written in the polar coordinates as

$$\tilde{\Gamma}_f(r_1,\theta_1,r_2,\theta_2) = \tilde{\Gamma}_f(r_1,\theta_1+\theta_0,r_2,\theta_2+\theta_0), \tag{5.91}$$

where $x_{1,2} = r_{1,2}\cos\theta_{1,2}$ and $y_{1,2} = r_{1,2}\sin\theta_{1,2}$. The use of these simplified but rather realistic models of partially coherent light makes possible the experimental measurements of the MI as well as the numerical simulation of its evolution during beam propagation.

In many practical applications the partially coherent beams are described by Gaussian functions. The MI of the most general Gaussian beam with ten degrees of freedom, which is known as *Twisted Gaussian Schell-Model* (TGSM), is expressed by three *real* 2×2 submatrices $\mathbf{G}_0$, $\mathbf{G}_1$, and $\mathbf{H}$, in the form

$$\begin{aligned}\Gamma(\mathbf{r}_1,\mathbf{r}_2) &= 2\sqrt{\det\mathbf{G}_1}\exp\left[-\frac{1}{2}\pi(\mathbf{r}_1-\mathbf{r}_2)^{\mathrm{t}}\mathbf{G}_0(\mathbf{r}_1-\mathbf{r}_2)\right]\\ &\times\exp\left\{-\pi\mathbf{r}_1^{\mathrm{t}}\left[\mathbf{G}_1-\mathrm{i}\frac{1}{2}(\mathbf{H}+\mathbf{H}^{\mathrm{t}})\right]\mathbf{r}_1\right\}\\ &\times\exp\left\{-\pi\mathbf{r}_2^{\mathrm{t}}\left[\mathbf{G}_1+\mathrm{i}\frac{1}{2}(\mathbf{H}+\mathbf{H}^{\mathrm{t}})\right]\mathbf{r}_2\right\}\\ &\times\exp\left\{-\mathrm{i}\pi\mathbf{r}_1^{\mathrm{t}}(\mathbf{H}-\mathbf{H}^{\mathrm{t}})\mathbf{r}_2\right\},\end{aligned} \tag{5.92}$$

where the matrices $\mathbf{G}_0$ and $\mathbf{G}_1$ are *positive definite* and *symmetric*. Note that the asymmetry of the matrix $\mathbf{H}$ is a measure for the *twist* or vorticity [22, 48, 91] of the Gaussian light. When $\mathbf{H}$ is symmetric, i.e., $\mathbf{H}-\mathbf{H}^{\mathrm{t}} = \mathbf{0}$, the general Gaussian light reduces to zero-twist *Gaussian Schell-model* light with nine degrees of freedom [57, 86]. In that case, the beam can be considered as spatially stationary with a Gaussian power spectrum $2\sqrt{\det\mathbf{G}_1}\,\exp\left[-\frac{1}{2}\pi(\mathbf{r}_1-\mathbf{r}_2)^{\mathrm{t}}\mathbf{G}_0(\mathbf{r}_1-\mathbf{r}_2)\right]$, modulated by a Gaussian function $\exp[-\pi\mathbf{r}^{\mathrm{t}}(\mathbf{G}_1 - \mathrm{i}\mathbf{H})\mathbf{r}]$. Other simplifications lead to the rotationally invariant case, i.e., $\mathbf{H} = h\mathbf{I}$, $\mathbf{G}_0 = g_0\mathbf{I}$, $\mathbf{G}_1 = g_1\mathbf{I}$, and the completely coherent case, i.e., $\mathbf{H} = \mathbf{H}^{\mathrm{t}}$, $\mathbf{G}_0 = \mathbf{0}$.

5.4.1.2 Wigner Distribution

Instead of the pure space description of a stochastic process by means of its MI, we can describe an optical signal in space and spatial-frequency coordinates simultaneously using the Wigner distribution function (WD), discussed in [12, 19, 26, 41, 103, 104].

The WD of an optical signal, $W_f(\mathbf{r}, \mathbf{p})$, can be defined in terms of its MI:

$$W_f(\mathbf{r}, \mathbf{p}) = \int \Gamma_f\left(\mathbf{r} + \frac{1}{2}\mathbf{r}', \mathbf{r} - \frac{1}{2}\mathbf{r}'\right) \exp(-\mathrm{i}2\pi\mathbf{p}^{\mathrm{t}}\mathbf{r}')\,\mathrm{d}\mathbf{r}'. \tag{5.93}$$

Here $\mathbf{r}$ is a position vector and $\mathbf{p} = \mathbf{q}/\lambda_0 = (u, v)^{\mathrm{t}}$ is a spatial frequency vector, where $\mathbf{q}$ is related to the direction of the ray propagation as it has been explained in Sect. 5.2. Since the WD is the FT of a Hermitian function, Eq. (5.86), it is a real function, though not always positive. Notice that the MI of the beam can be obtained directly from the WD using the inverse FT:

$$\Gamma_f(\mathbf{r}_1, \mathbf{r}_2) = \int W_f\left[\frac{1}{2}(\mathbf{r}_1 + \mathbf{r}_2), \mathbf{p}\right] \exp\left[\mathrm{i}2\pi\mathbf{p}^{\mathrm{t}}(\mathbf{r}_1 - \mathbf{r}_2)\right]\mathrm{d}\mathbf{p}. \tag{5.94}$$

For a coherent beam, its phase, apart from an additive constant, can be recovered from the MI. In particular, if $\mathbf{r}_\circ$ is a point in which the signal carries energy, i.e., $f(\mathbf{r}_\circ) \neq 0$, the phase can be obtained from

$$\varphi(\mathbf{r}) + \varphi_\circ = \arg\left[\Gamma_f(\mathbf{r}, \mathbf{r}_\circ)\right], \tag{5.95}$$

where $\varphi_\circ$ is an unknown constant.

The WD provides a unified description of completely and partially coherent optical beams in phase-space formed by position and spatial frequency coordinates. Comparing with the ray picture in geometrical optics the WD gives the next order approximation of beam representation in space-frequency space which takes into account the interference phenomena. The WD can be interpreted as the local power distribution of the beam as a function of position-spatial frequency, in spite of not being completely correct since the WD can take negative values. This description is particularly useful for the analysis of the spatially non-stationary optical fields.

The integral of the WD over the frequency variable,

$$\int W_f(\mathbf{r},\mathbf{p})\,\mathrm{d}\mathbf{p} = \Gamma_f(\mathbf{r},\mathbf{r}), \tag{5.96}$$

corresponds to the beam intensity distribution, whereas the integral over the space variable yields its directional power spectrum $\bar{\Gamma}_f(\mathbf{p},\mathbf{p})$:

$$\int W_f(\mathbf{r},\mathbf{p})\,\mathrm{d}\mathbf{r} = \bar{\Gamma}_f(\mathbf{p},\mathbf{p}), \tag{5.97}$$

where $\bar{\Gamma}_f(\mathbf{p}_1,\mathbf{p}_2)$ is an FT of the $\Gamma_f(\mathbf{r}_1,\mathbf{r}_2)$ with respect to the vectors $\mathbf{r}_1$ and $\mathbf{r}_2$. Both $\Gamma_f(\mathbf{r},\mathbf{r})$ and $\bar{\Gamma}_f(\mathbf{p},\mathbf{p})$ are non-negative and can be easily measured as we will see in the next section. The *beam power*, E, is given by integration of the WD over the entire space-frequency domain, $E = \int W(\mathbf{r},\mathbf{p})\,\mathrm{d}\mathbf{r}\,\mathrm{d}\mathbf{p} = \int \Gamma(\mathbf{r},\mathbf{r})\,\mathrm{d}\mathbf{r} = \int \bar{\Gamma}(\mathbf{p},\mathbf{p})\,\mathrm{d}\mathbf{p}$.

The mapping of the MI into the WD allows working with a real function but for the 2D optical field it is again a function of four variables that makes difficult its graphical representation, experimental recovery, and numerical manipulation. The hypothesis about the field spatial structure yielding the simplified form of the MI also reduces the complexity of the corresponding WD. Indeed, the knowledge of the MI coherent mode expansion, Eq. (5.87) allows expressing the WD as a sum of the WDs of the corresponding modes, $W_{f_n}(\mathbf{r},\mathbf{p})$,

$$W_f(\mathbf{r},\mathbf{p}) = \sum_{n=0}^{\infty} \Lambda_n W_{f_n}(\mathbf{r},\mathbf{p}). \tag{5.98}$$

We recall that the modes are orthogonal and therefore do not interfere between each other.

The WD for the quasi-homogeneous light, Eq. (5.89) has the form

$$W_f(\mathbf{r},\mathbf{p}) = I(\mathbf{r})\,\bar{\mu}(\mathbf{p}), \tag{5.99}$$

where $\bar{\mu}(\mathbf{p})$ is the FT of $\mu(\mathbf{r})$. Taking into account Eq. (5.97) this kind of light is completely characterized by its intensity distribution and directional power spectrum. In the case of the incoherent light the WD depends only on position vector $\mathbf{r}$, thus $W_f(\mathbf{r},\mathbf{p}) = \bar{\mu}_0 I(\mathbf{r})$, where $\bar{\mu}_0$ is a dimensional constant. In other words, the light is radiated equally in all directions with intensity distribution $I(\mathbf{r})$.

It is easy to prove that the WD for the beams separable in the Cartesian coordinates, Eq. (5.90), is also separable as

$$W_f(\mathbf{r},\mathbf{p}) = W_{f_x}(x,u)W_{f_y}(y,v). \tag{5.100}$$

Table 5.2 MI and WD of some simple 2D example signals $f(\mathbf{r})$

Signal type	$f(\mathbf{r})$	$\Gamma_f(\mathbf{r}_1, \mathbf{r}_2)$	$W_f(\mathbf{r}, \mathbf{p})$
Point source	$\delta(\mathbf{r} - \mathbf{r}_0)$	$\delta(\mathbf{r}_1 - \mathbf{r}_0)\,\delta(\mathbf{r}_2 - \mathbf{r}_0)$	$\delta(\mathbf{r} - \mathbf{r}_0)$
Plane wave	$\exp(\mathrm{i}2\pi\mathbf{p}_0^t\mathbf{r})$	$\exp[\mathrm{i}2\pi\mathbf{p}_0^t(\mathbf{r}_1 - \mathbf{r}_2)]$	$\delta(\mathbf{p} - \mathbf{p}_0)$
Spherical wave	$\exp(\mathrm{i}\pi a\mathbf{r}^t\mathbf{r})$	$\exp[\mathrm{i}\pi a(\mathbf{r}_1^t\mathbf{r}_1 - \mathbf{r}_2^t\mathbf{r}_2)]$	$\delta(\mathbf{p} - a\mathbf{r})$

While the rotationally symmetrical beams, defined by Eq. (5.91), has the WD that being expressed in the polar coordinates ($x = r\cos\theta_r$, $y = r\sin\theta_r$, $u = p\cos\theta_p$ $v = p\sin\theta_p$) satisfies the following relation:

$$\tilde{W}_f(r, \theta_r, p, \theta_p) = \tilde{W}_f(r, \theta_r + \theta_0, p, \theta_p + \theta_0). \tag{5.101}$$

Finally, regarding the partially coherent light, we mention the WD for the TGSM beam [21, 95], whose MI is given by Eq. (5.92)

$$W(\mathbf{r}, \mathbf{p}) = 4\sqrt{\frac{\det \mathbf{G}_1}{\det \mathbf{G}_2}} \exp\left(-2\pi \begin{bmatrix} \mathbf{r} \\ \mathbf{p} \end{bmatrix}^t \begin{bmatrix} \mathbf{G}_1 + \mathbf{H}\mathbf{G}_2^{-1}\mathbf{H}^t & -\mathbf{H}\mathbf{G}_2^{-1} \\ -\mathbf{G}_2^{-1}\mathbf{H}^t & \mathbf{G}_2^{-1} \end{bmatrix} \begin{bmatrix} \mathbf{r} \\ \mathbf{p} \end{bmatrix}\right), \tag{5.102}$$

where we have introduced the real and positive definite symmetric 2×2 matrix $\mathbf{G}_2 = \mathbf{G}_0 + \mathbf{G}_1$.

The WDs of the coherent fields, often used as a simple but rather academic models of light, are gathered in Table 5.2.

The two-dimensional HG beam is separable in Cartesian coordinates as

$$\mathrm{HG}_{m,n}(\mathbf{r}, w_x, w_y) \equiv \mathrm{HG}_m(x; w_x)\,\mathrm{HG}_n(y; w_y), \tag{5.103}$$

where

$$\mathrm{HG}_{m,n}(\mathbf{r}; w_x, w_y) = \sqrt{2}\,\frac{H_m\left(\sqrt{2\pi}\,\frac{x}{w_x}\right) H_n\left(\sqrt{2\pi}\,\frac{y}{w_y}\right)}{\sqrt{2^m m! w_x 2^n n! w_y}} \exp\left[-\pi\left(\frac{x^2}{w_x^2} + \frac{y^2}{w_y^2}\right)\right]. \tag{5.104}$$

Its WD, according to (see [26, 90]), can be written as

$$W(\mathbf{r}, \mathbf{p}) = 4(-1)^{m+n} \exp\left[-2\pi\left(\frac{x^2}{w_x^2} + \frac{y^2}{w_y^2} + w_x^2u^2 + w_y^2v^2\right)\right]$$
$$\times L_m\left[4\pi\left(\frac{x^2}{w_x^2} + w_x^2u^2\right)\right] L_n\left[4\pi\left(\frac{y^2}{w_y^2} + w_y^2v\right)\right]. \tag{5.105}$$

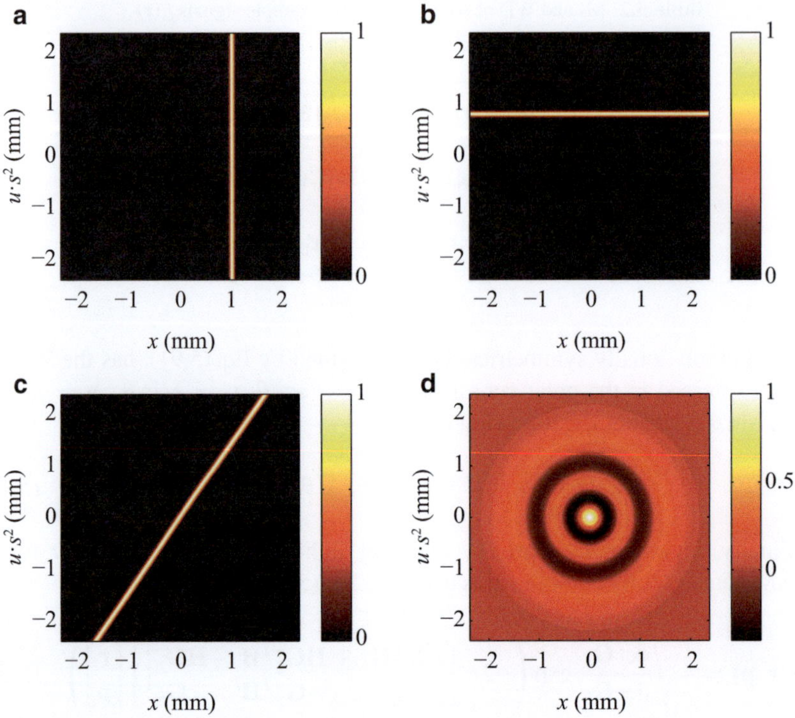

Fig. 5.5 The Wigner distribution of a point source (**a**) with $x_0 = 1\,\mathrm{mm}$, a plane wave (**b**) with $u_0 = 0.8\,\mathrm{mm}^{-1}$, a spherical wave (**c**) with $a = 1.4\,\mathrm{mm}^{-2}$, and the 1D Hermite Gaussian beam $\mathrm{HG}_4(x/w)$ (**d**) with $w = 0.73\,\mathrm{mm}$. Note that the Wigner distribution of the last beam takes positive and negative values

Taking advantage of that the considered examples correspond to signals separable in Cartesian coordinates, we represent the WD for their x-part in Fig. 5.5.

5.4.2 Partially Coherent Beam Propagation Through First-Order Optical Systems

Up to now, we have defined the beam at a given plane, let us say input plane. In order to describe it in other planes, the information about the system (or medium) in which it propagates is required. Note that we can also formulate the inverse problem: To recover the parameters of the medium or the system from the knowledge of the beam properties in two different planes, these two tasks are connected.

Let us consider a beam propagating through a deterministic system. Such a system is characterized by its coherent point-spread function (PSF) $h(\mathbf{r}_o, \mathbf{r}_i)$, which defines the response of the system to an impulse stimulus (a Dirac delta function).

This means that if the input of the system is a Dirac delta function, $\delta(\mathbf{r}_o - \mathbf{r}_i)$, the output will be the PSF of the system, $h(\mathbf{r}_i, \mathbf{r}_o)$. Using that $f_i(\mathbf{r}) = \int \delta(\mathbf{r}-\mathbf{r}_i) f_i(\mathbf{r}_i)\, d\mathbf{r}_i$ holds for the completely coherent case, we obtain a simple relation between the input signal $f_i(\mathbf{r}_i)$ and the output signal $f_o(\mathbf{r}_o)$ via the PSF of the system:

$$f_o(\mathbf{r}_o) = \int h(\mathbf{r}_i, \mathbf{r}_o) f_i(\mathbf{r}_i)\, d\mathbf{r}_i. \tag{5.106}$$

Similarly, the MI at the input plane, $\Gamma_i(\mathbf{r}_{1i}, \mathbf{r}_{2i})$, is transformed into the MI at the output plane, $\Gamma_o(\mathbf{r}_{1o}, \mathbf{r}_{2o})$, according to

$$\Gamma_o(\mathbf{r}_{1o}, \mathbf{r}_{2o}) = \int h(\mathbf{r}_{1i}, \mathbf{r}_{1o})\, \Gamma_i(\mathbf{r}_{1i}, \mathbf{r}_{2i})\, h^*(\mathbf{r}_{2i}, \mathbf{r}_{2o})\, d\mathbf{r}_{1i}\, d\mathbf{r}_{2i}. \tag{5.107}$$

The PSF of the first-order optical system is the kernel of the corresponding LCT, Eq. (5.83). Therefore, the following expression

$$\begin{aligned}\Gamma_o(\mathbf{r}_{1o}, \mathbf{r}_{2o}) = {} & \frac{\exp\left[ik_o\left(\mathbf{r}_{1o}^{t}\mathbf{D}\mathbf{B}^{-1}\mathbf{r}_{1o} - \mathbf{r}_{2o}^{t}\mathbf{D}\mathbf{B}^{-1}\mathbf{r}_{2o}\right)/2\right]}{\lambda_0^2 \det(i\mathbf{B})} \\ & \times \int \exp\left[ik_o\left(\mathbf{r}_{1i}^{t}\mathbf{B}^{-1}\mathbf{A}\mathbf{r}_{1i} - \mathbf{r}_{2i}^{t}\mathbf{B}^{-1}\mathbf{A}\mathbf{r}_{2i}\right.\right. \\ & \left.\left. - 2\mathbf{r}_{1i}^{t}\mathbf{B}^{-1}\mathbf{r}_{1o} + 2\mathbf{r}_{2i}^{t}\mathbf{B}^{-1}\mathbf{r}_{2o}\right)/2\right] \\ & \times \Gamma_i(\mathbf{r}_{1i}, \mathbf{r}_{2i})\, d\mathbf{r}_{1i}\, d\mathbf{r}_{2i}\end{aligned} \tag{5.108}$$

describes the MI evolution in this case.

In imaging systems the transformation of the MI can be reduced to a quadratic phase term apart from an affine transformation of its coordinates (e.g., scaling, rotation, etc.), as it follows:

$$\Gamma_o(\mathbf{r}_{1o}, \mathbf{r}_{2o}) = \frac{\exp\left[ik_0\left(\mathbf{r}_{1o}^{t}\mathbf{C}\mathbf{A}^{-1}\mathbf{r}_{1o} - \mathbf{r}_{2o}^{t}\mathbf{C}\mathbf{A}^{-1}\mathbf{r}_{2o}\right)/2\right]}{|\det(\mathbf{A})|} \Gamma_i(\mathbf{A}^{-1}\mathbf{r}_{1o}, \mathbf{A}^{-1}\mathbf{r}_{2o}). \tag{5.109}$$

We observe that the MI and the complex degree of the spatial coherence are changing during beam propagation through the first-order optical system. Nevertheless, in the case of coherent fields the modulus of the coherence degree $|\mu(\mathbf{r}_1, \mathbf{r}_2)|$ remains invariant.

As it has been shown in [26] the transformation of the WD for both cases is written in a simple form that corresponds to an affine transformation of the phase space coordinates. The relationship between the input WD, $W_{f_i}(\mathbf{r}, \mathbf{p})$, and the output WD, $W_{f_o}(\mathbf{r}, \mathbf{p})$, of a beam that propagates through a first-order optical system is given by

$$W_{f_o}(\mathbf{A}\mathbf{r} + \lambda_0\mathbf{B}\mathbf{p}, \lambda_0^{-1}\mathbf{C}\mathbf{r} + \mathbf{D}\mathbf{p}) = W_{f_i}(\mathbf{r}, \mathbf{p}). \tag{5.110}$$

This relation underlines again the close connection between the ray and wave phase-space field representations. It is easy to see that the beams separable in the Cartesian coordinates remains separable during its propagation through the system described by ray transformation matrix with diagonal **A**, **B**, **C**, and **D**. Similarly, the rotationally symmetric beams preserve their symmetry if the submatrices are scalar.

In particular, Eq. (5.110) is helpful for the determination of the WD of the LG modes defined by Eq. (5.55). It has been noted in [26] that the LG mode can be obtained from the HG one by applying the gyrator transform. Then from the expression for the WD of the HG mode, Eq. (5.105), we derive the WD of the LG mode [90] preforming the corresponding rotation of the phase-space coordinates:

$$\begin{aligned} W_{LG_{mn}}(\mathbf{r},\mathbf{p}) = 4\left(-1\right)^{m+n} L_m &\left[2\pi\left(\frac{x^2}{w_x^2}+\frac{y^2}{w_y^2}+u^2w_x^2\right.\right. \\ &\left.\left.+v^2w_y^2+2xv\frac{w_y}{w_x}-2yu\frac{w_x}{w_y}\right)\right] \\ &\times L_n\left[2\pi\left(\frac{x^2}{w_x^2}+\frac{y^2}{w_y^2}+u^2w_x^2+v^2w_y^2-2xv\frac{w_y}{w_x}+2yu\frac{w_x}{w_y}\right)\right] \\ &\times \exp\left[-2\pi\left(\frac{x^2}{w_x^2}+\frac{y^2}{w_y^2}+u^2w_x^2+v^2w_y^2\right)\right]. \end{aligned} \tag{5.111}$$

This approach of the WD calculation can be extended to another pair of the LCT related completely or partially coherent beams with known expression of the WD for one of them. We will further use it (see Sect. 5.6) for the determination of the moments of the beam expanded in the different families of Gaussian functions.

5.4.2.1 Van Cittert–Zernike Theorem

We have underlined that a coherent beam remains coherent during its propagation through any deterministic system and therefore through any first-order optical system. Here we consider the propagation of an incoherent beam described by the MI Eq. (5.88) through the first-order optical system. Substituting this particular expression into Eq. (5.108) we find the MI at the output plane of the first-order system

$$\begin{aligned} \Gamma_{f_o}(\mathbf{r}_{1o},\mathbf{r}_{2o}) = &\frac{\exp\left[ik_0\left(\mathbf{r}_{1o}^t\mathbf{D}\mathbf{B}^{-1}\mathbf{r}_{1o}-\mathbf{r}_{2o}^t\mathbf{D}\mathbf{B}^{-1}\mathbf{r}_{2o}\right)/2\right]}{\lambda_0^2\det\left(\mathrm{i}\mathbf{B}\right)} \\ &\times\int\exp\left[-ik_0\mathbf{r}_i^t\mathbf{B}^{-1}\left(\mathbf{r}_{1o}-\mathbf{r}_{2o}\right)\right]p_i(\mathbf{r}_i)\,\mathrm{d}\mathbf{r}_i. \end{aligned} \tag{5.112}$$

This relation is known as van Cittert–Zernike theorem in honor of the scientists (P.H. van Cittert and F. Zernike) who first formulated it for the case of propagation in free space at distance z, described by the scalar submatrices $\mathbf{D} = \mathbf{I}$, $\mathbf{B} = z\mathbf{I}$.

We remark that the MI and the degree of spatial coherence change during the propagation. Thus the propagation through a GRIN media yields to periodic variation of these characteristics while the propagation in homogeneous media provides the increasing of the coherence degree. For $\mathbf{B} \neq \mathbf{0}$ the MI at the output plane is proportional to the FT of the intensity distribution of the incident incoherent beam, apart from an affine coordinate transformation. Note that at the output plane the modulus of the MI is stationary, $\left|\Gamma_{f_o}(\mathbf{r}_{1o}, \mathbf{r}_{2o})\right| = \left|\Gamma_{f_o}(\mathbf{r}_{1o} - \mathbf{r}_{2o})\right|$, and the intensity distribution $\Gamma_{f_o}(\mathbf{r}_o, \mathbf{r}_o) = I_o$ is uniform. The last fact certainly does not correspond to physical reality and it is a consequence of the use of Dirac delta function for the description of incoherent light. From more rigorous consideration, it follows that light is coherent at least in the area λ^2 that brings to more realistic approximation of the incoherent light. In that case the degree of coherence can be represented by an extremely narrow Gaussian function. In spite of these details the van Cittert–Zernike theorem is widely used for many applications. Here we mention only some of them.

The thermal light sources (incandescent bulbs, gas discharge lamps, stars, etc.) are rather accurate described by the incoherent light model. Nevertheless, for a certain distance z from the plane source with a finite area a^2, the light acquires a certain degree of coherence which allows its description by a coherent model in the area roughly estimated by $A = (\lambda z/a)^2$. Based on this fact, the phase contrast techniques used for observation of weakly adsorbing specimens in optical and X-ray microscopy have been established. We also have to mention the stellar Michelson interferometer used in astronomy for the determination of the angular difference between the stars or their size. It is based on the measurement of the degree of the spatial coherence of light coming from the star which according to the Eq. (5.112) is proportional to the FT of the intensity distribution and therefore to the size of the source. As we will see in the next section, the beam coherence properties can be estimated without an interferometer, but by acquisition of the intensity distributions at the output planes of the appropriate first-order optical systems.

5.5 Phase Space Tomography

5.5.1 *Fundamentals*

As we have discussed, for the complete characterization of a monochromatic scalar paraxial beam its MI, or complex field amplitude in the completely coherent case, has to be identified. This task is important for the development of new imaging techniques oriented to the extraction of quantitative information about the object under study, which is encoded into the phase or statistical properties of the light propagated through or scattered from it. Here we consider the application of first-order optical systems for characterization of partially coherent beams.

This problem is rather difficult because the partially coherent light is described by functions of four variables (MI or WD). Moreover, these functions are not always positive and therefore cannot be directly measured by light intensity detectors. However, the measurable intensity distribution, see Eq. (5.96), corresponds to a projection of the WD. Of course, one 2D projection is not enough for the recovery of a 4D function (WD). A complete reconstruction of the WD can be achieved from a proper set of projections. Such set can be acquired using phase-space rotator setup. Indeed, the light propagation through first-order system described by a phase-space rotator yields the rotation of the WD of the input beam (see [26]) and therefore the output intensity distribution corresponds to a certain WD projection. The optical phase-space rotator setup with tunable parameters allows registering a set of different WD projections sufficient for the WD reconstruction. Then, applying the inverse Radon transform to this projection set, the WD as well as the MI can be recovered.

It has been mentioned [26] that any phase-space rotator can be expressed as a separable fractional FT embedded into two image rotators, $\mathscr{R}(\beta)\mathscr{F}(\gamma_x, \gamma_y)\mathscr{R}(\alpha)$. Since the action of the last image rotator, $\mathscr{R}(\beta)$, corresponds to the rotation of the coordinate system at the detection plane, the phase-space rotator which can be used for WD reconstruction is given in the form $\mathscr{F}(\gamma_x, \gamma_y)\,\mathscr{R}(\alpha)$. Taking into account that the WD is a 4D function and the projections are 2D functions, only two additional independent degrees of freedom are needed for the reconstruction. Thus, one of the rotation parameters can be fixed. Usually $\alpha = 0$ is chosen, in which case the phase-space rotator corresponds to the separable fractional FT.

Although the tomographic method for the WD reconstruction from its projections was proposed almost 20 years ago [81], it has not been widely used for the characterization of an arbitrary beam for two principal reasons: The lack of robustness and easily controlled optical setup, both required for acquisition of WD projections and the processing of a huge volume of data to obtain the WD. While the first problem can be solved using the programmable setup described in [85], which allows for acquisition of projections at video rate, the processing of the obtained data requires significant computational resources unavailable in a conventional laboratory.

To recover the MI of a beam without any information about its statistical properties and partially overcome the computational problems mentioned above, we may use a set of WD projections different from the fractional FT power spectra. It has been shown in [40] that using the phase-space rotator $\mathscr{F}(\gamma_x, \gamma_y)\,\mathscr{R}(\alpha)$, where γ_x is fixed (for example, $\gamma_x = 0$) and the other two angles γ_y and α change independently in a π-interval, we are able to perform both the projection acquisition and processing tasks in parallel. The recovering of the MI for pairs of points belonging to lines parallel to the y axis only requires the measurement of the WD projections corresponding to $\alpha = 0$. Measuring all projections for variable γ_y corresponding to a certain value of α, we can start processing the information while measuring the next projection subset (corresponding to another value of α).

Certain hypothesis about the beam symmetry and its coherence state can significantly simplify both the beam characterization process and the optical setup suitable for the acquisition of the WD projection set. For instance, in the case of coherent beams the problem is reduced to a phase retrieval problem which can be solved by applying iterative algorithms similar to the Gerchberg–Saxton algorithm [47, 52], or the propagation intensity equation [99]. In this case only several WD projections are needed. The number and diversity of the projections used for phase recovery are key factors to reach successful algorithm convergence and accurateness. Phase retrieval techniques are considered in detail in [60]. In the case of partially coherent fields, the simplest case corresponds to quasi-homogeneous fields, see Eq. (5.89). In this case only two WD projections corresponding to Fourier conjugated planes are needed. Indeed, by measuring the intensity distribution of the beam, the function I is easily determined while at the Fourier plane the intensity distribution, according to the WD rotation rule, yields the $\bar{\mu}(\mathbf{r})$. The reduced number of the WD projections is required for the characterization of beam which is rotationally symmetric, Eqs. (5.91) and (5.100), or separable in Cartesian coordinates Eqs. (5.90) and (5.101). The last case is considered in detail in the next subsection.

5.5.2 *WD Reconstruction for Beams Separable in Cartesian Coordinates*

Let us show that for complete characterization of beams, separable in Cartesian coordinates, it is sufficient to measure the WD projections corresponding to the symmetric ($\gamma_x = \gamma_y = \gamma$) or antisymmetric ($\gamma_x = -\gamma_y = \gamma$) fractional FT power spectra for γ running in a π-interval.

If the mutual intensity of the beam is separable with respect to the x and y Cartesian coordinates, i.e. $\Gamma_f(\boldsymbol{x}_1, \boldsymbol{x}_2) = \Gamma_{f_x}(x_1, x_2)\Gamma_{f_y}(y_1, y_2)$, then its fractional FT spectra, $S_f^{\gamma_x,\gamma_y}(\mathbf{r})$, are also separable:

$$
\begin{aligned}
S_f^{\gamma_x,\gamma_y}(\mathbf{r}) = \Gamma_{F^{\gamma_x,\gamma_y}}(\mathbf{r},\mathbf{r}) &= \iint \Gamma_{f_x}(x_1,x_2)K_{\gamma_x}(x_1,x;s)K_{\gamma_x}(x_2,x;s)\mathrm{d}x_1\mathrm{d}x_2 \\
&\times \iint \Gamma_{f_y}(y_1,y_2)K_{\gamma_y}(y_1,y;s)K_{\gamma_y}(y_2,y;s)\mathrm{d}y_1\mathrm{d}y_2 \\
&= S_{f_x}^{\gamma_x}(x)S_{f_y}^{\gamma_y}(y).
\end{aligned} \tag{5.113}
$$

Here we have used the Eqs. (5.108) and (5.109) and the separability of the kernel of the fractional FT, $K_{\gamma_x,\gamma_y}(\mathbf{r}_i, \mathbf{r}_o; w_x, w_y) = K_{\gamma_x}(x_i, x_o; w_x)\, K_{\gamma_y}(y_i, y_o; w_y)$, where $K_{\gamma_\xi}(\xi_i, \xi_o; w_\xi)$ is given by Eq. (5.54) and for simplicity $w_x = w_y = s$. The parameter s has a dimension of length whose value depends on the optical system. Since the WD function of a separable field is separable: $W_f(\mathbf{r}, \mathbf{p}) = W_{f_x}(x, u)W_{f_y}(y, v)$ we consider the reconstruction of x-part of the beam.

To reconstruct $W_{f_x}(x,u)$, and therefore the corresponding MI part, $\Gamma_{f_x}(x_1,x_2)$, the set of WD projections $S_{f_x}^{\gamma_x}(x)$ for angles γ_x covering the interval $[\gamma_0, \gamma_0+\pi]$ is required. These projections can be obtained by integrating the 2D fractional FT spectra, $S_f^{\gamma_x,\gamma_y}(\mathbf{r})$, over the y coordinate. Indeed, using the Parseval theorem for the one-dimensional fractional FT [79] we derive that

$$\int S_{f_\xi}^{\alpha}(\xi)\mathrm{d}\xi = \int S_{f_\xi}^{\beta}(\xi)\mathrm{d}\xi = A_\xi, \tag{5.114}$$

where A_ξ is a constant. Then the equations

$$\int S_f^{\gamma_x,\gamma_y}(\mathbf{r})\mathrm{d}y = A_y S_{f_x}^{\gamma_x}(x), \tag{5.115}$$

and

$$\iint W_{f_x}(x,u)W_{f_y}(y,v)\mathrm{d}y\mathrm{d}v = A_y W_{f_x}(x,u) \tag{5.116}$$

hold for any angle value γ_y. The $W_{f_y}(y,v)$ and its projections satisfy analogous relations. Therefore, the problem of the WD reconstruction is reduced to the one-dimensional case. The rotation of the WD under the fractional FT is schematically represented as:

$$\begin{array}{ccc} \Gamma_{f_x}(x_1,x_2) & \longrightarrow & W_f(x,u) \\ \downarrow & & \downarrow \\ \Gamma_{F^{\gamma_x}}(x_1,x_2) & \longrightarrow & W_{F^{\gamma_x}}(x,u) = W_f(x',u'), \end{array} \tag{5.117}$$

where the coordinates x' and u' are given by

$$\begin{aligned} x' &= x\cos\gamma_x - s^2 u\sin\gamma_x, \\ u' &= s^{-2}x\sin\gamma_x + u\cos\gamma_x. \end{aligned} \tag{5.118}$$

Then the squared modulus of the fractional FT, $S_{f_x}^{\gamma_x}(x)$:

$$S_{f_x}^{\gamma_x}(x) = \left\langle |F^{\gamma_x}(x)|^2 \right\rangle = \Gamma_{F_x^{\gamma_x}}(x,x) = \int W_{f_x}(x',u')\mathrm{d}u \tag{5.119}$$

corresponds to the WD projection of the input signal, described by $\Gamma_{f_x}(x_1,x_2)$, at the plane associated with the angle γ_x.

The set of these projections for angles in a π-interval, $\gamma_x \in [\gamma_0, \gamma_0+\pi]$, is called as Radon–Wigner transform. The WD associated with the x coordinate can be found by applying the inverse Radon transform, or other algorithms often used in tomography, to the one-dimensional Radon–Wigner transform

$$S_x^{\gamma_x}(x) = \int W_{f_x}(x\cos\gamma_x - s^2 u\sin\gamma_x, s^{-2}x\sin\gamma_x + u\cos\gamma_x)\mathrm{d}u. \tag{5.120}$$

Following the same procedure the $W_{f_y}(y, v)$ can be obtained. The product of $W_{f_x}(x, u)$ and $W_{f_y}(y, v)$ corresponds to the WD of the original 2D separable beam.

Note that the Parseval expression, Eq. (5.114), can also be used to test the field separability. In fact, only fields separable in Cartesian coordinates satisfy the relation

$$\int S_f^{\gamma_x,\gamma_y}(\mathbf{r})\mathrm{d}y \int S_f^{\gamma_x,\gamma_y}(\mathbf{r})\mathrm{d}x = S_f^{\gamma_x,\gamma_y}(\mathbf{r}) \int S_f^{\gamma_x,\gamma_y}(\mathbf{r})\mathrm{d}x\mathrm{d}y. \tag{5.121}$$

Since the only requirement for the ranges of both angles γ_x and γ_y is to cover a π-interval (not necessarily independently), the optical setup performing the symmetric ($\gamma_x = \gamma_y = \gamma$) or antisymmetric ($\gamma_x = -\gamma_y = \gamma$) fractional FT can be applied for the experimental realization of this phase-space tomography method. In particular, we can use an optical setup comprising four convergent cylindrical lenses suitable for the measurement of the antisymmetric fractional FT power spectra, discussed in [39, 65]. Using the programmable setup for the WD projections acquisition (see [65]) both antisymmetric and symmetric fractional FT power spectra can be measured for the WD recovery.

As an example, let us consider the reconstruction of the WD of the coherent beam given by the complex field amplitude proportional to

$$\mathrm{HG}_{0,2}\left(\mathbf{r}; w_x, w_y\right) + 5\mathrm{HG}_{1,2}\left(\mathbf{r}; w_x, w_y\right), \tag{5.122}$$

with $w_x = w_y = 0.73$ mm, whose intensity distribution and phase are displayed in Fig. 5.6. Its symmetric fractional FT power spectra, i.e. the intensity distributions at the output of the setup, for several angles and the corresponding profiles obtained by their integration in the x or y axes are shown in Fig. 5.7. The x-projected profiles for the angles $\gamma \in [\pi/2, 3\pi/2]$ are gathered in the 2D function $S_{f_x}^{\gamma}(x)$ while the collection of the y-projected profiles yields $S_{f_y}^{\gamma}(y)$. These functions are the Radon–Wigner transform of the x and y components of the separable beam, respectively. They are displayed in Fig. 5.8(a), (d). The WDs of such 1D signals are reconstructed by applying the inverse Radon transform [see Fig. 5.8(b)–(c) and (e)–(f)]. The quality of the reconstruction depends on both the number of used projections and the signal complexity, apart from the accuracy of the optical setup. In Fig. 5.8 the WDs of the x and y parts of the beam reconstructed from 10 to 20 projections are shown. It is easy to see that the quality of the reconstruction significantly increases with the number of used projections. It proves the need of a setup for fast generation and acquisition of WD projections, for example the one mentioned in [65, 85].

Notice that the WD projections associated with the antisymmetric fractional FT can be also used for complete characterization of the rotationally symmetric beams defined by the Eqs. (5.91) and (5.101) as it follows from the approach discussed in [6].

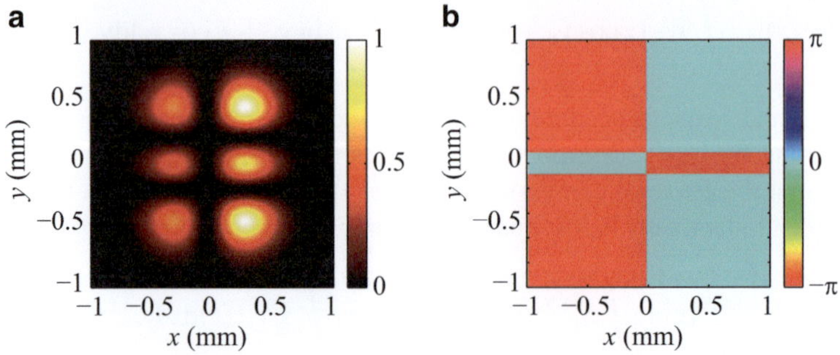

Fig. 5.6 Intensity (**a**) and phase (**b**) distributions of the beam $HG_{0,2}(\mathbf{r}; w_x, w_y) + 5HG_{1,2}(\mathbf{r}; w_x, w_y)$ with $w_x = w_y = 0.73$ mm

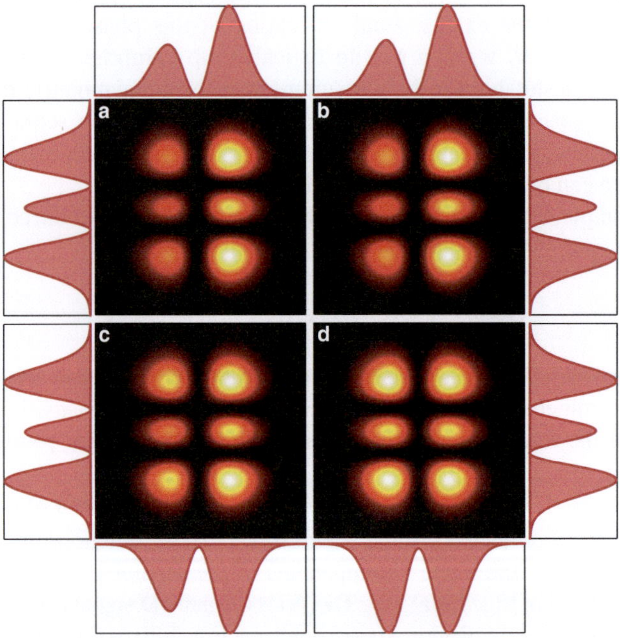

Fig. 5.7 Symmetric fractional power spectra for the beam $HG_{0,2}(\mathbf{r}; w_x, w_y) + 5HG_{1,2}(\mathbf{r}; w_x, w_y)$, with $w_x = w_y = 0.73$ mm for the angle $\gamma = 0°$ (**a**), $\gamma = 30°$ (**b**), $\gamma = 60°$ (**c**), and $\gamma = 90°$ (**d**). The profiles obtaining by integrating the power spectra in the x and y directions are presented in the vertical and horizontal 1D plots, respectively

We conclude that the phase space rotators are important tools for the analysis of the paraxial beams. The process of the recovery of beam characteristics can be optimized if an additional information about the beam is provided.

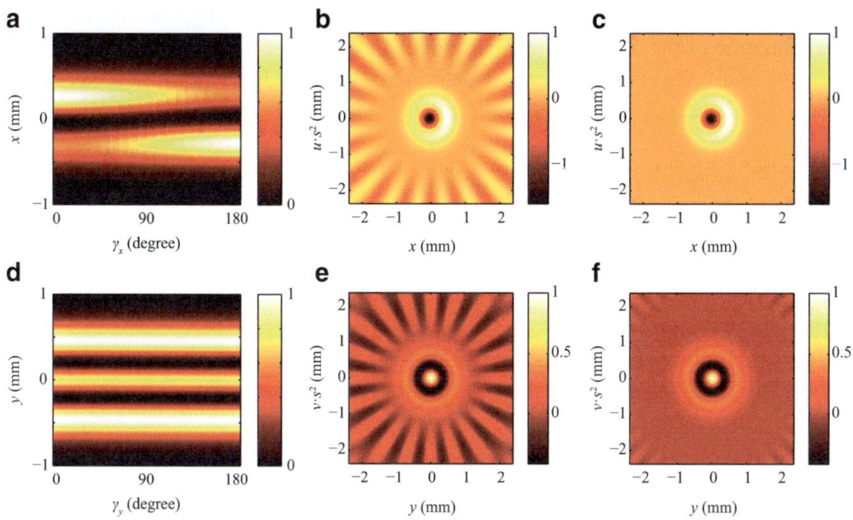

Fig. 5.8 Radon–Wigner transform (**a**) and Wigner distribution reconstructed from 10 (**b**) and 20 (**c**) projections for the x part of beam described in Eq. (5.122), i.e. the 1D signal proportional to $\mathrm{HG}_0(x; w_x) + 5\mathrm{HG}_1(x; w_x)$ with $w_x = 0.73$ mm. Radon–Wigner transform (**d**) and Wigner distribution reconstructed from 10 (**e**) and 20 (**f**) projections for the y part of beam described in Eq. (5.122), i.e. the 1D signal $6\mathrm{HG}_2$ with $w_y = 0.73$ mm

5.6 WD Moments for Beam Characterization

5.6.1 Definition and Measurement of WD Moments

The rather complicated procedure of beam characterization described in the previous section can be simplified if only the global beam properties are of interest. For this purpose the moments of the WD are used. The moment $\mu_{p,q,r,t}$ of the WD, normalized with respect to the beam power $E = \int W(\mathbf{r}, \mathbf{p})\, \mathrm{d}\mathbf{r}\, \mathrm{d}\mathbf{p} = \int I(\mathbf{r})\, \mathrm{d}\mathbf{r}$, is defined by

$$\mu_{p,q,r,t}\, E = \int W(x, y, u, v)\, x^p\, u^q\, y^r\, v^t\, \mathrm{d}\mathbf{r}\, \mathrm{d}\mathbf{p} \tag{5.123}$$
$$= \int \frac{x^p y^r}{(4\pi \mathrm{i})^{q+t}} \left(\frac{\partial}{\partial x_1} - \frac{\partial}{\partial x_2}\right)^q \left(\frac{\partial}{\partial y_1} - \frac{\partial}{\partial y_2}\right)^t \Gamma(x_1, x_2; y_1, y_2)\bigg|_{\substack{x_1=x_2=x \\ y_1=y_2=y}} \mathrm{d}\mathbf{r},$$

where the integers $p, q, r, t \geq 0$. The moments are organized in groups, or orders, according to the sum of its indices: $n = p + q + r + t$. For two-dimensional signals there are four moments of first order, ten moments of second order, twenty moments of third order, etc. In general, there are $N = (1/6)(n+1)(n+2)(n+3)$ moments of n-th order. The more WD moments are known the more complete information of the beam is obtained. As in the case of phase space tomography, the experimental determination of the moments is based on the measurements of the WD projections.

Some of the WD moments can be determined directly from measurements of the intensity distribution in the image plane ($q = t = 0$), $\Gamma(\mathbf{r}, \mathbf{r})$, according to

$$\mu_{p,0,r,0}\, E = \int W(x, y, u, v)\, x^p\, y^r\, \mathrm{d}\mathbf{r}\, \mathrm{d}\mathbf{p} = \int \Gamma(\mathbf{r}, \mathbf{r})\, x^p\, y^r\, \mathrm{d}\mathbf{r}, \tag{5.124}$$

while for the calculation of the other ones, we have to exploit again the affine transformation of the WD in phase-space during beam propagation through first-order optical systems.

For simplicity, here we consider the systems separable with respect to the x and y coordinates, that are described by the ray transformation matrix

$$\mathbf{T} = \begin{bmatrix} \mathbf{A} & \mathbf{B} \\ \mathbf{C} & \mathbf{D} \end{bmatrix} = \begin{bmatrix} a_x & 0 & s^{-1}b_x & 0 \\ 0 & a_y & 0 & s^{-1}b_y \\ sc_x & 0 & d_x & 0 \\ 0 & sc_y & 0 & d_y \end{bmatrix}. \tag{5.125}$$

Taking into account the law of the WD evolution in a first-order optical system, Eq. (5.110), we obtain the relation between the normalized moments of the beam at the output plane of such system, $\mu^{\mathrm{out}}_{p,q,r,t}$, and the normalized moments of the input beam, $\mu^{\mathrm{in}}_{p,q,r,t} = \mu_{p,q,r,t}$, [23, 24]

$$\begin{aligned} \mu^{\mathrm{out}}_{p,q,r,t} E &= \int W_{\mathrm{out}}(x, y, u, v)\, x^p\, u^q\, y^r\, v^t\, \mathrm{d}\mathbf{r}\, \mathrm{d}\mathbf{p} \\ &= \int W_{\mathrm{in}}(d_x x - b_x u, d_y y - b_y v, -c_x x + a_x u, -c_y y + a_y v)\, x^p\, u^q\, y^r\, v^t\, \mathrm{d}\mathbf{r}\, \mathrm{d}\mathbf{p} \\ &= \int W_{\mathrm{in}}(x, y, u, v)(a_x x + b_x u)^p (c_x x + d_x u)^q\, (a_y y + b_y v)^r\, (c_y y + d_y v)^t\, \mathrm{d}\mathbf{r}\, \mathrm{d}\mathbf{p} \\ &= E \sum_{k=0}^{p} \sum_{l=0}^{q} \sum_{m=0}^{r} \sum_{n=0}^{s} \binom{p}{k} \binom{q}{l} \binom{r}{m} \binom{t}{n} a_x^{p-k}\, b_x^k\, c_x^l\, d_x^{q-l} \\ &\quad \times a_y^{r-m}\, b_y^m\, c_y^n\, d_y^{s-n}\, \mu_{p-k+l,q-l+k,r-m+n,t-n+m}. \end{aligned} \tag{5.126}$$

In the particular case of intensity moments, i.e., $q = t = 0$, we have

$$\mu^{\mathrm{out}}_{p,0,r,0} = \sum_{k=0}^{p} \sum_{m=0}^{r} \binom{p}{k} \binom{r}{m} a_x^{p-k}\, b_x^k\, a_y^{r-m}\, b_y^m\, \mu_{p-k,k,r-m,m}. \tag{5.127}$$

It is easy to see that for the Fourier conjugated system the relation between the input and output moments is simplified as $\mu^{\mathrm{out}}_{p,0,r,0} = b_x^p b_y^t \mu_{0,p,0,r}$, where $b_{x,y} = s^2$. In order to obtain all moments of n-th order ($\mu_{p,q,r,t}$), using the Eq. (5.126), the intensity distributions at the output planes of N_s different first-order optical systems have to be measured. It has been shown in [23, 24] that the number of the required systems is $N_s = (1/4)(n+2)^2$ for even n, and $N_s = (1/4)(n+3)(n+1)$ for odd n. Among them, only $n+1$ systems can be isotropic, i.e., $a_x = a_y$ and $b_x = b_y$. Moreover, two systems with equal or proportional parameters $a_{x,y}$ and $b_{x,y}$ cannot be considered different since they yield the same expression, Eq. (5.127).

For the determination of the first-order moments we may use two intensity measurements in an isotropic (rotationally invariant) system corresponding, for example, to the propagation in homogeneous medium. Calculating the first-order intensity moments at input plane, $\mu_{1,0,0,0}$ and $\mu_{0,0,1,0}$, as well as in the output plane, $\mu^{\text{out}}_{1,0,0,0}$ and $\mu^{\text{out}}_{0,0,1,0}$, the other two input moments $\mu_{0,1,0,0}$ and $\mu_{0,0,0,1}$ are obtained from the following equations:

$$\mu^{\text{out}}_{1,0,0,0} = a_x \mu_{1,0,0,0} + b_x \mu_{0,1,0,0}, \tag{5.128}$$

$$\mu^{\text{out}}_{0,0,1,0} = a_y \mu_{0,0,1,0} + b_y \mu_{0,0,0,1}. \tag{5.129}$$

The independence of such equations allows using the isotropic systems.

The determination of the second-order moments requires measurements of four intensity distributions, one of which has to be obtained using anisotropic system, for example, one which contains a cylindrical lens. From the measurements of the intensity distribution at the input plane, three moments $\mu_{2,0,0,0}$, $\mu_{0,0,2,0}$, and $\mu_{1,0,1,0}$ can be calculated. For the determination of the other second-order moments, we use the expressions for beam evolution during its propagation through a separable system, Eq. (5.125):

$$\mu^{\text{out}}_{2,0,0,0} = a_x^2 \mu_{2,0,0,0} + 2a_x b_x \mu_{1,1,0,0} + b_x^2 \mu_{0,2,0,0}, \tag{5.130}$$

$$\mu^{\text{out}}_{1,0,1,0} = a_x a_y \mu_{1,0,1,0} + a_x b_y \mu_{1,0,0,1} + a_y b_x \mu_{0,1,1,0} + b_x b_y \mu_{0,1,0,1} \tag{5.131}$$

$$\mu^{\text{out}}_{0,0,2,0} = a_y^2 \mu_{0,0,2,0} + 2a_y b_y \mu_{0,0,1,1} + b_y^2 \mu_{0,0,0,2}. \tag{5.132}$$

The moments $\mu_{1,1,0,0}$, $\mu_{0,2,0,0}$, $\mu_{0,0,1,1}$, and $\mu_{0,0,0,2}$ can be determined by measuring the intensity moments $\mu^{\text{out}}_{2,0,0,0}$ and $\mu^{\text{out}}_{0,0,2,0}$ at the output planes of other two systems which might be isotropic, see Eqs. (5.130) and (5.132). The other three input moments $\mu_{1,0,0,1}$, $\mu_{0,1,1,0}$, and $\mu_{0,1,0,1}$ are obtained from the intensity moments $\mu^{\text{out}}_{1,0,1,0}$ at the output planes of three different systems, Eq. (5.131), one of which has to be astigmatic. Indeed, by using the isotropic systems only the sum of the moments $\mu_{1,0,0,1} + \mu_{0,1,1,0}$ can be recovered.

Different optical schemes for the determination of the first- and the second-order moments have been proposed (see, for instance, [46, 69, 77, 88]). For the calculation of moments of arbitrary order, the system used for phase space tomography can be applied.

5.6.2 Physical Meaning of the Low-Order Moments

5.6.2.1 First-Order Moments

The normalized first-order moments can be shortly described by the four-dimensional column vector $\mathbf{m} = [\mathbf{m}_\mathbf{r}^\mathrm{t}, \mathbf{m}_\mathbf{p}^\mathrm{t}] = [\mu_{1,0,0,0}, \mu_{0,0,1,0}, \mu_{0,1,0,0}, \mu_{0,0,0,1}]^\mathrm{t}$, defined by

$$\mathbf{m}^\mathrm{t} = \frac{1}{E}\int [\mathbf{r}^\mathrm{t}, \mathbf{p}^\mathrm{t}]\, W(\mathbf{r}, \mathbf{p})\, \mathrm{d}\mathbf{r}\, \mathrm{d}\mathbf{p}. \tag{5.133}$$

Taking into account the evolution of the WD during the beam propagation through the first-order optical system, described by the ray transformation matrix $\mathbf{T}$, it is easy to obtain [13] the propagation law for the first-order moments as $\mathbf{m}^{\mathrm{out}} = \tilde{\mathbf{T}}\mathbf{m}$, where the antidiagonal submatrices of $\mathbf{T}$ and $\tilde{\mathbf{T}}$ are related as $\lambda_0 \mathbf{B} = \tilde{\mathbf{B}}$ and $\lambda_0^{-1}\mathbf{C} = \tilde{\mathbf{C}}$ while the diagonal ones are equal: $\mathbf{A} = \tilde{\mathbf{A}}$ and $\mathbf{D} = \tilde{\mathbf{D}}$. Note that the same equation governs the paraxial ray propagation (see Eq. (5.56)). Therefore, the beam description by the first-order moments is equivalent to its presentation by a single ray whose position coincides with the beam center $(\mu_{1,0,0,0}, \mu_{0,0,1,0})$ and direction is given by $(\mu_{0,1,0.0}, \mu_{0,0,0,1})$. This interpretation is clearly seen if we consider a coherent beam whose moments are expressed through its complex field amplitude $f(\mathbf{r}) = a(\mathbf{r}) \exp[\mathrm{i}\varphi(\mathbf{r})]$ as

$$\mathbf{m}_\mathbf{r}^\mathrm{t} = \frac{1}{E}\int \mathbf{r}^\mathrm{t} f^*(\mathbf{r}) f(\mathbf{r})\, \mathrm{d}\mathbf{r} = \frac{1}{E}\int \mathbf{r}^\mathrm{t} I(\mathbf{r})\, \mathrm{d}\mathbf{r}, \tag{5.134}$$

$$\mathbf{m}_\mathbf{p}^\mathrm{t} = \frac{\mathrm{i}}{2kE}\int \left[f(\mathbf{r})\nabla_\mathbf{r}^\mathrm{t} f^*(\mathbf{r}) - f^*(\mathbf{r})\, \nabla_\mathbf{r}^\mathrm{t} f(\mathbf{r})\right] \mathrm{d}\mathbf{r} = \frac{1}{kE}\int I(\mathbf{r})\nabla_\mathbf{r}^\mathrm{t}\varphi(\mathbf{r})\mathrm{d}\mathbf{r}, \tag{5.135}$$

where $\nabla_\mathbf{r}^\mathrm{t} = [\partial/\partial x, \partial/\partial y]$. Thus the vector $\mathbf{m}_\mathbf{p}$ is proportional to the mean intensity-weighted phase gradient of the beam.

Following the discussion of ray propagation in the first-order optical system (see Sects. 5.2 and 5.3) we can analogously define the trajectory of the beam centroid. In a homogeneous medium it corresponds to a straight line given by $\mathbf{m}_\mathbf{r}^{\mathrm{out}}(z) = \mathbf{m}_\mathbf{r} + \lambda z\, \mathbf{m}_\mathbf{p}$. While during the beam propagation in a selfoc medium, described by the symmetric fractional FT ($\gamma_x = \gamma_y = \gamma$), $\mathbf{m}_\mathbf{r}^{\mathrm{out}}(\gamma) = \mathbf{m}_\mathbf{r} \cos\gamma + s^2 \mathbf{m}_\mathbf{p} \sin\gamma$, it can also follow the plane periodic or spiral trajectory depending on the initial conditions as it is displayed in Fig. 5.2.

For the description of beam propagation in the isotropic first-order systems the beam centroid is commonly used as origin of coordinates in the XY plane.

5.6.2.2 Second-Order WD Moments

According to the recommendations of an International Organization for Standardization standard [61], the second-order moments of the WD form a basis for global beam characterization. The ten second-order moments can be written in the form of a 4×4 real symmetric matrix $\tilde{\mathbf{M}}$ defined by

$$\tilde{\mathbf{M}} = \frac{1}{E}\int \begin{bmatrix}\mathbf{r}\\ \mathbf{p}\end{bmatrix}[\mathbf{r}^{\mathrm{t}}, \mathbf{p}^{\mathrm{t}}]\, W(\mathbf{r},\mathbf{p})\, \mathrm{d}\mathbf{r}\, \mathrm{d}\mathbf{p} = \begin{bmatrix}\mu_{2,0,0,0} & \mu_{1,0,1,0} & \mu_{1,1,0,0} & \mu_{1,0,0,1}\\ \mu_{1,0,1,0} & \mu_{0,0,2,0} & \mu_{0,1,1,0} & \mu_{0,0,1,1}\\ \mu_{1,1,0,0} & \mu_{0,1,1,0} & \mu_{0,2,0,0} & \mu_{0,1,0,1}\\ \mu_{1,0,0,1} & \mu_{0,0,1,1} & \mu_{0,1,0,1} & \mu_{0,0,0,2}\end{bmatrix}. \tag{5.136}$$

Since the value of the moment depends on the choice of the coordinate origin then the *central* second-order moments defined as

$$\mathbf{M} = \frac{1}{E}\int \begin{bmatrix}\mathbf{r}-\mathbf{m}_{\mathbf{r}}\\ \mathbf{p}-\mathbf{m}_{\mathbf{p}}\end{bmatrix}[(\mathbf{r}-\mathbf{m}_{\mathbf{p}})^{\mathrm{t}}, (\mathbf{p}-\mathbf{m}_{\mathbf{p}})^{\mathrm{t}}]\, W(\mathbf{r},\mathbf{p})\, \mathrm{d}\mathbf{r}\mathrm{d}\mathbf{p} = \tilde{\mathbf{M}} - \mathbf{m}\mathbf{m}^{\mathrm{t}} = \begin{bmatrix}\mathbf{M}_{\mathbf{rr}} & \mathbf{M}_{\mathbf{rp}}\\ \mathbf{M}_{\mathbf{rp}}{}^{\mathrm{t}} & \mathbf{M}_{\mathbf{pp}}\end{bmatrix} \tag{5.137}$$

are used for beam characterization. The first-order moments in such a coordinate system are vanished. Moreover, the evolution of the second-order moments during beam propagation through the first-order optical system, described by the ray transformation matrix $\mathbf{T}$, follows the simple law [13]

$$\tilde{\mathbf{M}}^{\mathrm{out}} = \tilde{\mathbf{T}}\tilde{\mathbf{M}}\tilde{\mathbf{T}}^{\mathrm{t}}, \tag{5.138}$$

if they are calculated with respect to the system axes. A similar expression can be written for the central moments

$$\mathbf{M}^{\mathrm{out}} = \tilde{\mathbf{T}}\tilde{\mathbf{M}}\tilde{\mathbf{T}}^{\mathrm{t}} - \tilde{\mathbf{T}}\mathbf{m}\left(\tilde{\mathbf{T}}\mathbf{m}\right)^{\mathrm{t}} = \tilde{\mathbf{T}}\left(\tilde{\mathbf{M}} - \mathbf{m}\mathbf{m}^{\mathrm{t}}\right)\tilde{\mathbf{T}}^{\mathrm{t}} = \tilde{\mathbf{T}}\mathbf{M}\tilde{\mathbf{T}}^{\mathrm{t}}. \tag{5.139}$$

Taking into account that $\det\tilde{\mathbf{T}} = 1$, the determinant of the moment matrix remains invariant during beam propagation.

The moments m_{xx}, m_{yy}, and m_{xy} provide information about the beam *spatial width* and orientation of the principal axes of its intensity distribution. As it was suggested by M. Teague [98], the second-order moments define an ellipse corresponding to the effective area of the beam, which can be parametrized by: Its long semi-axis a, its short semi-axis b, and its orientation ϕ, via

$$a^2 = 2\left[m_{xx} + m_{yy} + \sqrt{(m_{xx} - m_{yy})^2 + 4m_{xy}^2}\right], \tag{5.140}$$

$$b^2 = 2\left[m_{xx} + m_{yy} - \sqrt{(m_{xx} - m_{yy})^2 + 4m_{xy}^2}\right], \tag{5.141}$$

$$\tan(2\phi) = \frac{2m_{xy}}{m_{xx} - m_{yy}}. \tag{5.142}$$

Here ϕ is an angle between the x-axis of the laboratory system and the beam principal axis closer to it. The beam effective area $S_{\mathbf{r}}$ is proportional to the product ab. Thus the first-order moments define the coordinate origin, while the second-order moments composing $\mathbf{M_{rr}}$ determine the direction of the axes. Usually the axes of the coordinate system are chosen such that $m_{xy} = 0$. Correspondingly, the moments m_{uu}, m_{vv}, and m_{uv} describe the beam divergence or *angular cone* in which the light is radiating. Taking into account the phase-space rotation properties of the FT, these moments define the parameters of the ellipse roughly describing the beam intensity distribution in the Fourier conjugated plane or in far field. In Fig. 5.9 we show the position of centroid and the ellipse associated with the beams given as a superposition of two Laguerre–Gaussian modes $\mathrm{LG}_{2,0} + \mathrm{LG}_{4,0}$ with $4w_x/3 = 4w_y/5 = s = 0.73$ mm and two Hermite Gaussian modes $\mathrm{HG}_{0,2} + 5\mathrm{HG}_{1,2}$ with $w_x = w_y = s = 0.73$ mm at two Fourier conjugated planes. Here the scaling of the FT equals s. We underline that the first beam belongs to the class of spiral beams, whose intensity distribution does not change apart from rotation during the propagation through the system described by the fractional FT with scaling parameter s. The centroid of this beam coincides with the coordinate origin and does not change during beam propagation. The centroid of the second beam varies its position during beam propagation: its x coordinate is not zero in Fig. 5.9c, while in the Fourier conjugated plane it coincides with coordinate origin.

Instead of beam characterization by its effective area in the XY (or UV) planes, the beam propagation factor or beam quality factor parameter $M_x^2 = 4\pi\sqrt{m_{xx}m_{uu}}$ (or analogously $M_y^2 = 4\pi\sqrt{m_{yy}m_{vv}}$) is defined at the beam waist. The $M_{x,y}^2$ factor allows predicting the evolution of the beam effective width during its propagation and, in particular, determines the smallest spot size to which the beam can be focused. It holds that $M_{x,y}^2 \geqq 1$ and only for a Gaussian fundamental mode $M_{x,y}^2 = 1$.

The moments associated with the mixed spatial-frequency moments, $\mathbf{M_{rp}}$, m_{xu} and m_{yv} determine the beam *curvature* and the longitudinal projection of the orbital angular momentum carried by the beam. Indeed, the quadratic phase modulation associated with the lens ray transformation matrix $\left[\mathbf{I}, \mathbf{0}; -\tilde{\mathbf{L}}, \mathbf{I}\right]$ produces the following modification of the moment matrix (5.137)

$$\mathbf{M}^{\mathrm{out}} = \begin{bmatrix} \mathbf{M_{rr}} & -\mathbf{M_{rr}}\tilde{\mathbf{L}} + \mathbf{M_{rp}} \\ -\tilde{\mathbf{L}}\mathbf{M_{rr}} + \mathbf{M}_{\mathbf{rp}}^{\mathrm{t}} & \tilde{\mathbf{L}}\mathbf{M_{rr}}\tilde{\mathbf{L}} - \tilde{\mathbf{L}}\mathbf{M_{rp}} - \mathbf{M}_{\mathbf{rp}}^{\mathrm{t}}\tilde{\mathbf{L}} + \mathbf{M_{pp}} \end{bmatrix}. \tag{5.143}$$

Suppose that $\mathbf{M_{rp}} = 0$ and the principal axes of the beam and the lens coincide with the laboratory axes ($\mathbf{M_{rr}}$ and $\tilde{\mathbf{L}} = \left[\tilde{l}_{xx}, 0; 0, \tilde{l}_{yy}\right]$ are diagonal matrices) then $\tilde{l}_{xx} = -m_{xu}^{\mathrm{out}}/m_{xx}$ and $\tilde{l}_{yy} = -m_{yv}^{\mathrm{out}}/m_{yy}$. In the general case, the beam curvature can be found from the knowledge of the beam second order moments as [13]

$$\tilde{\mathbf{L}} = -\mathbf{M}_{\mathbf{rr}}^{-1}\left[\mathbf{M_{rp}} + \sigma\frac{\mathbf{Tr}\left(\mathbf{M_{rr}}\sigma\mathbf{M_{rp}}\right)}{\mathbf{Tr}\left(\mathbf{M_{rr}}\right)}\right], \tag{5.144}$$

where

$$\boldsymbol{\sigma} = \begin{bmatrix} 0 & 1 \\ -1 & 0 \end{bmatrix}. \tag{5.145}$$

The mixed second-order moments are also used for the estimation of the longitudinal projection of the orbital angular momentum on the propagation direction z (OAM), which is proportional to the moment difference $(m_{xv} - m_{yu})$, while the global beam vorticity [3], as we will see further, is closely related to the *twist* parameter [22] (see also [26, Sect. 2.8.2])

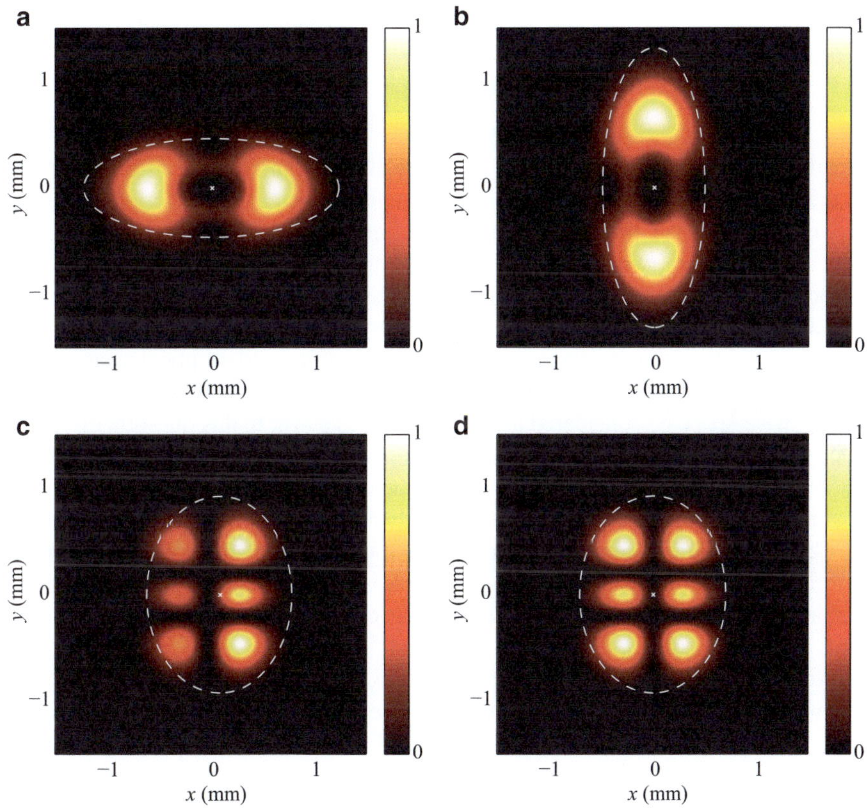

Fig. 5.9 Intensity (**a**) and power spectrum (**b**) of the beam $LG_{2,0} + LG_{4,0}$ with $4w_x/3 = 4w_y/5 = 0.73$ mm. Intensity (**c**) and power spectrum (**d**) of the beam $HG_{0,2} + 5HG_{1,2}$ with $w_x = w_y = 0.73$ mm. The *white cross* defines the beam centroid while the *stripped white ellipse* defines its *effective size* defined by the spatial and angular a, b, and ϕ parameters [see Eqs. (5.140)–(5.142)]

$$T = -\frac{\mathbf{Tr}\left(\mathbf{M_{rr}}\sigma\mathbf{M_{rp}}\right)}{\sqrt{\det\mathbf{M_{rr}}}}. \tag{5.146}$$

These parameters are important for the characterization of vortex or singular beams, which are widely applied in metrology, free-space optical communication and for microparticle manipulation.

Although the higher-order WD moments provide more detailed information about beam's *symmetry* and its *sharpness* [44, 45] they have a little use in practice. Their measurements require the acquisition of a relatively big number of the WD projections, which can be applied for the reconstruction of the entire WD using phase-space tomography methods.

5.6.3 Beam Classification and Comparison Based on the Second-Order Moments

Apart from the valuable information about the global beam characteristics discussed in the previous section (effective size, divergence, principal axes, etc.), the second-order moments also provide the basis for beam classification and comparison. When we compare two beams, the question about how different they are has to be answered. However, as we have seen, the partially coherent paraxial beams are described by 4D functions, for example the WD. The WD suffers affine transformations during beam propagation through a first-order optical system. Can two beams described by the same WD apart from its scaling, sharing, and rotation in the phase-space be considered as different? The answer to this question of course depends on the applications where the compared beams are involved. If we are interested in their focusing properties, then the scaling is certainly important. If the spatial structure of the beam is under study then the scaling can be ignored, but the phase-space rotations cannot. So our first task is to represent the WD of the beam in a normalized form suitable for the comparison with other ones. To find these canonical coordinates in phase space, we will follow the method considered in [26, Sect. 2.8.1] and [10, 25, 96].

According to Williamson's theorem [13, 96, 105] for any positive-definite real symmetric 4×4 matrix, in particular for the second-order moment matrix $\mathbf{M}$, there exists a real symplectic matrix $\tilde{\mathbf{T}}_c$ such that $\mathbf{M}^c = \tilde{\mathbf{T}}_c\mathbf{M}\tilde{\mathbf{T}}_c^{\mathrm{t}}$ takes a canonical (diagonal) form:

$$\mathbf{M}^c = \begin{bmatrix} s^2\boldsymbol{\Lambda} & \mathbf{0} \\ \mathbf{0} & s^{-2}\boldsymbol{\Lambda} \end{bmatrix} \quad \text{with} \quad \boldsymbol{\Lambda} = \begin{bmatrix} \Lambda_x & 0 \\ 0 & \Lambda_y \end{bmatrix} \quad \text{and} \quad \Lambda_x \geq \Lambda_y > 0. \tag{5.147}$$

The dimensionless parameters $\Lambda_{x,y} = \frac{1}{2}(Q_\circ \pm Q)$ are defined by the two invariants I_1 and I_2 or ($Q_\circ^2 = I_2 + 2I_1$ and $Q^2 = I_2 - 2I_1$) of the moment matrix during beam propagation through a first-order optical system [20, 87]:

$$I_1 = \sqrt{\det \mathbf{M}},$$
$$I_2 = -\mathbf{Tr}\left[(\mathbf{MJ})^2\right] = 2\mathbf{Tr}\left(\mathbf{M_{rr}M_{pp}} - \mathbf{M_{rp}^2}\right), \tag{5.148}$$

where s is a dimension factor. Since in the canonical form the second-order matrix is diagonal, the effective volume of the WD (defined as $\Lambda_x\Lambda_y = \sqrt{\det \mathbf{M}}$) may serve as a measure for the beam information capacity. The parameter $\sqrt[4]{\det \mathbf{M}}$ is proportional to the *global beam quality parameter* [46] or effective beam propagation ratio M^2.

The canonical diagonalization of the momentum matrix also allows separating two-dimensional signals into two classes [96]: intrinsically isotropic ($Q = 0$) and anisotropic ($Q \neq 0$). The value of the parameter Q defines the intrinsic beam astigmatism that allows having a deeper look on the vortex properties of the optical beams. Thus a beam with non-zero projection of its orbital angular momentum, but with $Q = 0$, cannot be considered a vortex beam [7, 30] Moreover, it cannot be transformed into it by any LCTs. On the other hand, an intrinsically anisotropic beam with zero OAM projection may not be a vortex but may be transformed into it by LCT. In particular, the non-vortex intrinsically anisotropic HG beam, for example $\mathrm{HG}_{1,2}(\mathbf{r}; w)$, is transformed into the vortex one, the LG beam $\mathrm{LG}_{1,2}(\mathbf{r}; w)$, using the gyrator transformation (see, for example, [26, 83]).

Apart from the beam classification, the knowledge of the canonical form of the moment matrix allows identifying the LCT described by the ray transformation matrix $\tilde{\mathbf{T}}_c^{-1}$, which transforms the beam into this canonical state. The comparison of the WDs or the WD projections of the beams being in the canonical state provides the more objective information about their similarity.

We also may be interested in the analysis of the LCT which brings the beam into the canonical state. Thus the modified Iwasawa decomposition of the $\tilde{\mathbf{T}}_c^{-1}$ [25] allows disclosing the corresponding affine transformation of the phase space. Neglecting the scaling and phase modulation operations, we might map the intrinsically anisotropic beam on the orbital angular Poincarè sphere or radius Q in order to indicate its vortex features. The details of this analysis can be found in [10, 25, 37].

5.6.4 *LCTs for Calculation of Second-Order Moments of a Beam Expressed as Mode Series*

The beam expression given as a superposition of the orthonormal modes is often used for beam design and analysis. For example, the MI can be expanded into orthonormal HG modes as [5, 35]

$$\Gamma\left(\mathbf{r}_1, \mathbf{r}_2\right) = \sum_{m,n}\sum_{m',n'} \left\langle a_{m,n}a^*_{m',n'}\right\rangle \mathrm{HG}_{m,n}\left(\mathbf{r}_1; w\right)\mathrm{HG}_{m',n'}\left(\mathbf{r}_2; w\right), \tag{5.149}$$

with $\left\langle a_{m,n}a^{*}_{m',n'}\right\rangle$ the correlation coefficients where we assume that $\sum_{m,n}\left\langle |a_{m,n}|^2\right\rangle = 1$. Using the moment definition and some recurrence relations for the HG functions (see [38] for further details) we derive the expressions for the first-order moments:

$$w^{-1}m_x + \mathrm{i}wm_u = \sum_{m,n} \sqrt{m+1}\left\langle a_{m+1,n}a^{*}_{m,n}\right\rangle, \tag{5.150}$$

$$w^{-1}m_y + \mathrm{i}wm_v = \sum_{m,n} \sqrt{n+1}\left\langle a_{m,n+1}a^{*}_{m,n}\right\rangle. \tag{5.151}$$

We underline that the centroids in both, *XY* and *UV*, planes are zero (i.e., coincide with the origin of coordinates) if in the HG mode decomposition there are no combinations of modes with consecutive indices.

The second-order moment matrix can be found analogously [29, 38]:

$$\tilde{\mathbf{M}} = \frac{1}{2\pi}\begin{bmatrix} w^2\left(K_x + L_x'\right) & w^2(M+N)' & -L_x'' & (M-N)'' \\ w^2(M+N)' & w^2\left(K_y + L_y'\right) & -(M+N)'' & -L_y'' \\ -L_x'' & -(M+N)'' & w^{-2}\left(K_x - L_x'\right) & w^{-2}(M-N)' \\ (M-N)'' & -L_y'' & w^{-2}(M-N)' & w^{-2}\left(K_y - L_y'\right) \end{bmatrix}, \tag{5.152}$$

where, following the notation of [29],

$$K_x = \frac{1}{2}\sum_{m,n}(2m+1)\left\langle |a_{m,n}|^2\right\rangle, \quad K_y = \frac{1}{2}\sum_{m,n}(2n+1)\left\langle |a_{m,n}|^2\right\rangle, \tag{5.153}$$

$$L_x = \sum_{m,n}\sqrt{(m+1)(m+2)}\left\langle a_{m,n}a^{*}_{m+2,n}\right\rangle, \quad L_y = \sum_{m,n}\sqrt{(n+1)(n+2)}\left\langle a_{m,n}a^{*}_{m,n+2}\right\rangle, \tag{5.154}$$

$$M = \sum_{m,n}\left\langle a_{m,n+1}a^{*}_{m+1,n}\right\rangle\sqrt{(m+1)(n+1)}, \quad N = \sum_{m,n}\left\langle a_{m,n}a^{*}_{m+1,n+1}\right\rangle\sqrt{(m+1)(n+1)}, \tag{5.155}$$

$A' \equiv \Re(A)$, and $A'' \equiv \Im(A)$. The matrix for central-order moments can be obtained from Eq. (5.152) by subtracting the matrix $\mathbf{mm}^{\mathrm{t}}$ constructed from the first-order moments.

Let us now find the second-order moment matrix for the beam whose MI is expanded into series of LG modes, $\mathrm{LG}_{m,n}(\mathbf{r}; w)$, as

$$\Gamma(\mathbf{r}_1, \mathbf{r}_2) = \sum_{m,n}\sum_{m',n'}\left\langle a_{m,n}a^{*}_{m',n'}\right\rangle \mathrm{LG}_{m,n}(\mathbf{r}_1; w)\left[\mathrm{LG}_{m',n'}(\mathbf{r}_2; w)\right]^{*}. \tag{5.156}$$

Notice that the LG mode decomposition is used for the generation of spiral beams [1]. We recall that a HG mode is converted into an LG one under the gyrator transform $\mathcal{G}^{\pi/4}[\cdot]$ (see [26, 84]), i.e.,

$$\mathscr{G}^{\pi/4}\left[\mathrm{HG}_{m,n}\left(\mathbf{r};w\right)\right] = \mathrm{LG}_{m,n}\left(\mathbf{r};w_x,w_y\right). \tag{5.157}$$

In order to find the moments expression for the beam described by the Eq. (5.156) we first consider a *test* beam whose MI is expanded in HG modes with the same correlation coefficients $\left\langle a_{m,n}a^*_{m',n'}\right\rangle$ as in the problem beam, Eq. (5.156):

$$\hat{\Gamma}\left(\mathbf{r}_1,\mathbf{r}_2\right) = \sum_{m,n}\sum_{m',n'}\left\langle a_{m,n}a^*_{m',n'}\right\rangle \mathrm{HG}_{m,n}\left(\mathbf{r}_1;w\right)\mathrm{HG}_{m',n'}\left(\mathbf{r}_2;w\right). \tag{5.158}$$

Since the problem beam is obtained from $\hat{\Gamma}$ applying the gyrator transform $\mathscr{G}^{\pi/4}$, its moments are related to the moments of the test beam according to the propagation equation (5.138) with $\tilde{\mathbf{T}}$ corresponding to the gyrator transform: $\tilde{\mathbf{A}} = \tilde{\mathbf{D}} = 2^{-1/2}\mathbf{I}$ and $w^{-2}\tilde{\mathbf{B}} = -w^2\tilde{\mathbf{C}} = 2^{-1/2}\mathbf{J}$, where $\mathbf{J} = [0,1;1,0]$.

Using this simple method we derive the formulas for the first-order moments of a signal expanded in LG modes,

$$\sqrt{2}m_x = w\sum_{m,n}\left(\sqrt{m+1}\,\Re\left\langle a_{m+1,n}a^*_{m,n}\right\rangle + \sqrt{n+1}\,\Im\left\langle a_{m,n+1}a^*_{m,n}\right\rangle\right), \tag{5.159}$$

$$\sqrt{2}m_y = w\sum_{m,n}\left(\sqrt{n+1}\,\Re\left\langle a_{m,n+1}a^*_{m,n}\right\rangle + \sqrt{m+1}\,\Im\left\langle a_{m+1,n}a^*_{m,n}\right\rangle\right), \tag{5.160}$$

$$\sqrt{2}m_u = w^{-1}\sum_{m,n}\left(\sqrt{m+1}\,\Im\left\langle a_{m+1,n}a^*_{m,n}\right\rangle - \sqrt{n+1}\,\Re\left\langle a_{m,n+1}a^*_{m,n}\right\rangle\right), \tag{5.161}$$

$$\sqrt{2}m_v = w^{-1}\sum_{m,n}\left(\sqrt{n+1}\,\Im\left\langle a_{m,n+1}a^*_{m,n}\right\rangle - \sqrt{m+1}\,\Re\left\langle a_{m+1,n}a^*_{m,n}\right\rangle\right), \tag{5.162}$$

as well as for the second-order moments,

$$w^{-2}\mu_{2,0,0,0} = \left(K_x + L'_x + K_y - L'_y\right)/2 - \left(N'' - M''\right), \tag{5.163}$$

$$w^{-2}\mu_{0,0,2,0} = \left(K_y + L'_y + K_x - L'_x\right)/2 - \left(N'' + M''\right), \tag{5.164}$$

$$w^{2}\mu_{0,2,0,0} = \left(K_x - L'_x + K_y + L'_y\right)/2 + \left(N'' + M''\right), \tag{5.165}$$

$$w^{2}\mu_{0,0,0,2} = \left(K_y - L'_y + K_x + L'_x\right)/2 + \left(N'' - M''\right), \tag{5.166}$$

$$\mu_{1,1,0,0} = \left(L''_y - L''_x\right)/2 - N', \tag{5.167}$$

$$\mu_{0,0,1,1} = \left(L''_x - L''_y\right)/2 - N', \tag{5.168}$$

$$w^{-2}\mu_{1,0,1,0} = M' - \frac{1}{2}\left(L''_x + L''_y\right), \tag{5.169}$$

$$w^{2}\mu_{0,1,0,1} = M' + \frac{1}{2}\left(L''_x + L''_y\right) \tag{5.170}$$

$$\mu_{0,1,1,0} = -\frac{1}{2}\left(K_y - K_x + L'_x + L'_y\right), \tag{5.171}$$

$$\mu_{1,0,0,1} = -\frac{1}{2}\left(K_x - K_y + L'_x + L'_y\right), \tag{5.172}$$

where $K_{x,y}$, $L_{x,y}$, M, and N are defined in Eqs. (5.153)–(5.155). Notice that the problem and test beams have the same effective volume of the WD. The beam decomposition into the series of the LG modes is useful for beam synthesis and analysis. In particular, the OAM of the beam expressed by Eq. (5.156) is given as a weighted sum of the topological charges, i.e. index differences $(n-m)$, of all modes:

$$\mathrm{OAM}^{LG} = \mu_{1,0,0,1} - \mu_{0,1,1,0} = \sum_{m,n}\left\langle |a_{m,n}|^2\right\rangle (n-m). \tag{5.173}$$

Similarly the moments of the beam expressed as the HLG mode superposition can be obtained [38].

5.7 LCTs for Pulse Manipulation and Characterization

In the previous sections the propagation of monochromatic scalar paraxial beams through first-order optical systems has been considered. In this section we show that the LCTs also describe the optical wave propagation in dispersive media and play an important role in the manipulation and characterization of ultrashort pulses.

5.7.1 Quasi-monochromatic Plane Waves in Dispersive Media

In Sect. 5.2 we have considered the solution of the wave equation, Eq. (5.1) in nondispersive medium, where polarization $\mathbf{P} = \mathbf{0}$ and therefore the electric displacement $\mathbf{D} = \varepsilon\mathbf{E}$, where ε is the electric permittivity. Here we will study the wave propagation through the isotropic homogeneous dispersive media described by the following wave equation

$$\nabla\mathbf{E} - \frac{1}{c^2}\frac{\partial^2\mathbf{E}}{\partial t^2} - \mu_0\frac{\partial^2\mathbf{P}}{\partial t^2} = 0, \tag{5.174}$$

where $\mathbf{P}(\boldsymbol{r},t) = \int_0^\infty h(t')\mathbf{E}(\boldsymbol{r},t-t')\mathrm{d}t'$, h is the medium impulse response and $\mathbf{D} = \varepsilon\mathbf{E}+\mathbf{P}$. Let us consider a plane scalar quasi-monochromatic wave propagating in the z direction in the medium with refractive index $n(\omega)$ depending on angular temporal frequencyω. Then its electric field $\mathbf{E}$ (and the polarization $\mathbf{P}$) can be represented by a scalar function of two variables z and t: $E(z,t) = f(z,t)\exp\left[\mathrm{i}\left(\omega_0 t - k_0 z\right)\right]$, where the exponential term describes the rapid changes of the electric field. The ω_0 is a central angular frequency of the wave spectrum and $k_0 = \omega_0/c$. Assuming that f changes slowly with time we can express it by three first terms of the Taylor series at the point t

$$f(z, t - t') \approx f(z, t) - t' \frac{\partial f}{\partial t} + (t')^2 \frac{1}{2} \frac{\partial^2 f}{\partial t^2} \tag{5.175}$$

which yields

$$P(z, t) = \left[\chi(\omega) f - \mathrm{i} \frac{\partial \chi}{\partial \omega} \frac{\partial f}{\partial t} - \frac{1}{2} \frac{\partial^2 \chi}{\partial \omega^2} \frac{\partial^2 f}{\partial t^2} \right] \exp\left[\mathrm{i}\left(\omega_0 t - k_0 z\right)\right], \tag{5.176}$$

where $\chi(\omega) = \int_0^\infty h(t') \exp(-\mathrm{i}\boldsymbol{\omega} t')\, \mathrm{d}t'$ is electric susceptibility related to the refractive index via $n(\omega) = \sqrt{1 + \chi(\omega)}$. Using these approximations the wave equation, Eq. (5.174) is transformed to

$$\frac{\partial f}{\partial z} + \frac{1}{\upsilon} \frac{\partial f}{\partial t} - \frac{\mathrm{i}}{2} g \frac{\partial^2 f}{\partial t^2} + \frac{\mathrm{i}}{2} \left[\frac{\partial^2 f}{\partial z^2} - \frac{1}{\upsilon^2} \frac{\partial^2 f}{\partial t^2} \right] = 0, \tag{5.177}$$

being $\upsilon = (\partial k / \partial \omega)^{-1}$ the group velocity, $g = \partial^2 k / \partial \omega^2$ the parameter describing the group velocity dispersion and $k = k_0 n(\omega)$. The last term $[\cdot]$ in the left side of the equation is smaller than the rest. Notice that $\frac{\partial^2 f}{\partial z^2} - \frac{1}{\upsilon^2} \frac{\partial^2 f}{\partial t^2} = 0$ corresponds to the wave equation for f moving in z with velocity $\upsilon : f(z, t) = f_i(t - z/\upsilon)$, where f_i is defined by the initial condition. The other terms in the Eq. (5.177) are related to the derivatives of the complex field amplitude and therefore larger than $[\cdot]$ which will be further neglected. Thus the evolution of $f(z, t)$ during plane wave propagation in the dispersive medium in the second approximation of the dispersive theory is described by equation

$$\frac{\partial f}{\partial z} + \frac{1}{\upsilon} \frac{\partial f}{\partial t} - \frac{\mathrm{i}}{2} g \frac{\partial^2 f}{\partial t^2} = 0.$$

It can be rewritten in the alternative form for the temporal Fourier transform of the envelop $f(z, \omega)$ as

$$\frac{\partial \bar{f}}{\partial z} - \mathrm{i} \left[\frac{\omega}{\upsilon} + \frac{\omega^2 g}{2} \right] \bar{f} = 0.$$

which has an easy solution

$$\bar{f}(z, \omega) = \bar{f}(0, \omega) \exp\left[-\mathrm{i} \left(\frac{\omega}{\upsilon} + \frac{\omega^2 g}{2} \right) z \right] \tag{5.178}$$

Applying the inverse FT we obtain that the complex amplitude after the propagation a distance $z, f_o(t_o) = f(z, t - z/\upsilon)$ is expressed as a Fresnel transform of the complex field amplitude at the plane $z = 0, f_i(t_i)$,

$$f_o(t_o) = \frac{1}{\sqrt{\mathrm{i} 2\pi g z}} \int f_i(t_i) \exp\left[\mathrm{i} \frac{(t_o - t_i)^2}{2 g z} \right] \mathrm{d}t_i. \tag{5.179}$$

As well as in ray optics we may introduce *temporal ray* [100] defined by the vector $\mathbf{u} = [t, \omega]^t$ measured with respect to center of the wave packet z/υ and angular frequency ω measured with respect to the spectrum center ω_0. These variables compose the so-called *chronocyclic phase space*. The propagation of such rays in the dispersive media is described by the 2×2 matrix Ω

$$\begin{pmatrix} t_o \\ \omega_o \end{pmatrix} = \Omega \begin{pmatrix} t_i \\ \omega_i \end{pmatrix}, \tag{5.180}$$

where

$$\Omega = \begin{bmatrix} 1 & gz \\ 0 & 1 \end{bmatrix}. \tag{5.181}$$

As we will see below this analogy with spatial ray optics is useful for the description of devices applied for pulse manipulation.

5.7.2 Pulse Description

A pulse of light is an optical field of finite time duration which is described by a complex function $E(t) = f(t) \exp(\mathrm{i}\omega_0 t) = |f(t)| \exp(\mathrm{i}\omega_0 t + \mathrm{i}\varphi(t))$, where $f(t)$ is the complex envelop and ω_0 is a central angular frequency. Since we analyze only the temporal pulse behavior the spatial coordinates have been suppressed. The associated intensity distribution is given by $I(t) = |f(t)|^2$. Pulses can also be described in the Fourier domain by its spectrum $\bar{E}(\omega) = (2\pi)^{-1/2} \int E(t) \exp(-\mathrm{i}\omega t)\, \mathrm{d}t = |E(\omega)| \exp[\mathrm{i}\phi(\omega]$ and the so-called spectral intensity $S(\omega) = \left|\bar{E}(\omega)\right|^2$. According to the Fourier uncertainty relation (see [79, 97]) the narrower the spectral distribution is, the larger is the temporal pulse envelop and vice versa. For example, a picosecond pulse has spectral width around 500 GHz, while for a femtosecond pulse it is more than 500 THz.

There exists a wide list of applications of ultrashort pulses, such as chemistry, ultrafast microscopy, telecommunication, to name a few, which requires the development of techniques for pulse generation, manipulation, and characterization. Indeed, for the study of dynamics of ultrafast processes, the amplitude and phase of the complex envelope $f(t)$ have to be known. It is a nontrivial problem because the time response of the fastest available photodetector is significantly larger, $\sim$100 ps, than the pulse duration. Another problem is related to pulse manipulation since the propagation through conventional optical elements such as lenses yields pulse broadening due to dispersion. Different techniques of ultrafast metrology, many of which are based on the application of optical elements whose action is described by the LCTs, help to overcome these obstacles.

We also note that the pulse description by $E(t)$ is not sufficient to specify the character of an ensemble of pulses which are usually slightly different one from another. By analogy with the spatially partially coherent light, the temporal two-point correlation function $\Gamma(t_1, t_2) = \langle E(t_1)E^*(t_2)\rangle$ is used for specification of the pulse ensemble characteristics. Instead the correlation function we may use the Wigner distribution defined by

$$W(t,\omega) = (2\pi)^{-1/2} \int \Gamma(t + \tau/2, t - \tau/2) \exp(-\mathrm{i}\omega\tau)\,\mathrm{d}\tau \tag{5.182}$$

which provides the representation of the pulse in chronocyclic phase space comprised by the time and frequency coordinates [27, 43].

5.7.3 Spectral and Temporal Chirp Filters

From Eq. (5.179) follows that the propagation through a dispersive media changes the pulse shape. Similarly to the evolution of the angular spectrum during beam propagation in homogeneous media, see Sect. 5.2.5.2, it has been shown that the transformation of the pulse spectrum corresponds to quadratic chirp modulation (5.178), which corresponds to the LCT parametrized by the transfer matrix, Eq. (5.180).

There are other optical elements, known as chirp filters, which produce phase modulation of the spectrum. They are based on different phenomena: angular dispersion (gratings, holograms, prisms), interferometric dispersion (stratified media, Bragg gratings, resonators), polarization dispersion (wavelength dependence of the anisotropic properties of materials), and nonlinear dispersion (wavelength dependence of the nonlinear optical effects).

Let us consider first the angular dispersion chirp filter. An angular dispersion element, for example, diffraction grating, changes the direction of propagation of the monochromatic components that compose a polychromatic pulse. Thus, each monochromatic component follows a different direction defined by angle, $\theta(\omega)$, measured with respect to the angle of the component corresponding to the central frequency. The spectral phase shift introduced by the angular dispersion chirp filter at the frequency component ω is given by $\phi(\omega) = \omega l \cos[\theta(\omega)]/c$, where l is the pathlength of the central component. Usually, the chirp filter is formed by four identical dispersive elements (see, Fig. 5.10): the first spatially separates the spectral components; the second collimates the directions of the components; and the other two do the same in reversed order to recover the pulse, apart from an additional phase indicated above, where in this case l is the optical pathlength overall the entire system. If $\theta(\omega)$ is sufficiently small, then $\cos[\theta(\omega)] \approx 1 - \theta^2(\omega)/2$. Moreover, the slowly varying function $\theta(\omega)$ can be approximated by two first terms of the Taylor series that yields to the following expression for the beam phase [102]

$$\phi(\omega_0 + \omega) \approx \phi_0 + \frac{l}{c}\omega + \frac{\phi''}{2}\omega^2 \tag{5.183}$$

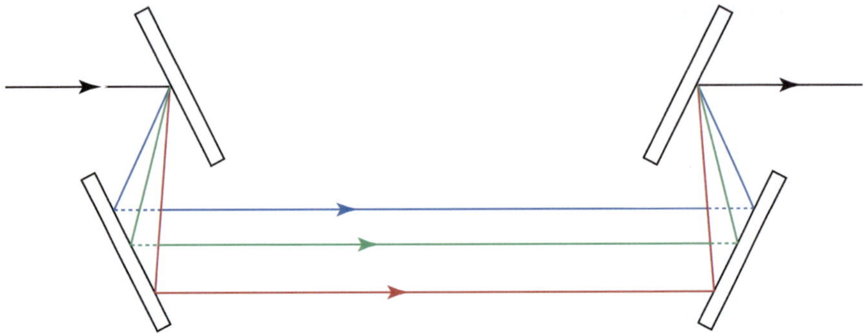

Fig. 5.10 A spectral filter formed by four identical dispersive elements. Polychromatic light entering the system (*black arrow*) is separated in its multiple spectral components (*red*, *green*, and *blue arrows*) to be composed back to polychromatic light with chirp modulated spectrum at the output of the system

where in this approximation $\phi'' = \partial^2\phi/\partial\omega^2\big|_{\omega_0} \approx -l\left(\mathrm{d}\theta/\mathrm{d}\omega|_{\omega_0}\right)^2/c$ and the exact expression depends on the used dispersive element. Note that for this kind of chirp filters, $\phi'' < 0$. Thus we again obtain the LCT transformation described by the matrix

$$\Omega = \begin{bmatrix} 1 & \phi'' \\ 0 & 1 \end{bmatrix}. \tag{5.184}$$

Let us now find the expression for ϕ'' if a diffraction grating is used as dispersive element. The diffraction of a plane wave of angular frequency $\omega_0 + \omega$ on a grating of period d is described by the following equation:

$$\sin\theta_i + \sin\left[\theta_0 + \theta(\omega)\right] = \frac{2\pi c}{\omega d}, \tag{5.185}$$

where θ_i and $\theta_0 + \theta(\omega)$ are the angles between the normal of the grating and the wave vector of the incident and emergent waves. Notice that for the central frequency component $\theta(\omega_0) = 0$. Differentiating this equation we obtain that $\mathrm{d}\theta/\mathrm{d}\omega|_{\omega_0} = -2\pi c/\left(\omega_0^2 d\cos\theta_0\right)$, and thus $\phi'' \approx -lc\left[2\pi/\left(\omega_0^2 d\cos\theta_0\right)\right]^2$.

Another example of chirp filters is a Bragg grating, which is a stratified medium with refractive index gradient parallel to the pulse propagation direction. This kind of filters has frequency-dependent reflectance and transmittance properties, that leads to frequency modulation of the pulse propagating through or reflecting from it. If the grating has a periodic structure with period d, then only the waves with wavelength $\lambda = 2d/m$, where m is an integer, are reflected from it while the rest of spectral components propagate through without any change. If the grating has a pitch varying with position $d(z)$, the waves are reflected from the grating segment

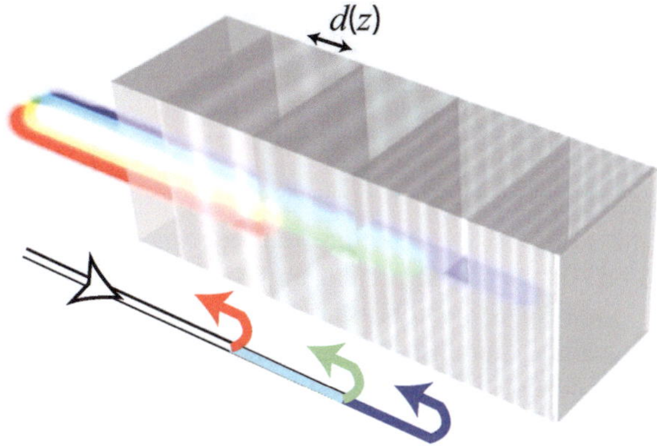

Fig. 5.11 Bragg grating working in reflection mode as a frequency chirp filter. It reflects back the spectral components of a polychromatic pulse from different positions according to the chirped refractive index variation, which yields a spectral-selective behavior

which matches their wavelength. As a consequence, the phase of the reflected waves depends on their wavelengths because they propagate for different distances inside the grating. For instance, in the case of a grating that satisfies $d^{-1}(z) = d_0^{-1} + z/\xi$, as sketched in Fig. 5.11, the wave of frequency ω is reflected at the position which satisfies the relation $d(z_\omega) = \pi mc/\omega$ and, therefore, travels the distance $2z_\omega = 2\omega\xi/(\pi mc) - 2\xi/d_0$ inside the grating. This corresponds to a phase shift $\varphi_\omega = 2z_\omega\omega/c$. The associated spectral phase shift of the pulse has a quadratic form

$$\phi = \frac{2\xi}{cd_0}\omega + \frac{2\xi}{\pi mc^2}\omega^2 \tag{5.186}$$

which means that such Bragg grating can serve as a frequency chirp filter. The parameter ξ, and hence ϕ'', may take positive or negative values.

Another important operation is a temporal quadratic phase modulation

$$f_o(t) = f_i(t)\exp\left(-\mathrm{i}\frac{\varphi'' t^2}{2}\right), \tag{5.187}$$

whose associated LCT is parametrized by a transfer matrix similar to the one describing the action of a lens:

$$\Omega = \begin{bmatrix} 1 & 0 \\ -\varphi'' & 1 \end{bmatrix}, \tag{5.188}$$

where $\varphi'' = \partial^2\varphi/\partial t^2\big|_0$. Therefore, it behaves as a time lens which produces phase changes in time following a parabolic function. Such a *temporal lens* can be experimentally implemented using an electro-optic phase modulator driven by a sinusoidal radio-frequency voltage signal (e.g., based on the Pockels effect) [14, 34, 64], by mixing the original pulse with a chirped pulse in a nonlinear crystal [31], or by means of cross-phase modulation of the original pulse with an intense pump pulse in a nonlinear fiber [72], to name a few.

The combination of quadratic spectral and temporal chirp elements leads to the general one-dimensional LCT described by the 2×2 symplectic matrix

$$\Omega = \begin{bmatrix} a & b \\ c & d \end{bmatrix} \tag{5.189}$$

with $\det \Omega = 1$. Therefore, the equivalent transformations to the ones applied to one-dimensional spatial beams—magnification, Fourier transformation, rotation in phase space—can also be performed for temporal pulses. They are often used for several ultrafast optical signal processing applications, such as waveform magnification [33] based on temporal imaging [64]; waveform diagnosis based on time-to-frequency conversion [17]; elimination of linear distortions [73] based on optical Fourier transformation [62]. Some of these applications are briefly considered below.

5.7.4 Pulse Manipulation and Characterization

One of the most important tasks in ultrafast optics is pulse power amplification. The propagation of a high-peak-power pulse through an optical amplifier leads to undesirable nonlinear effects which can prevent using pulse spectral chirping. Indeed, the application of a spectral chirp filter with parameter ϕ_1'' yields to pulse stretching and, therefore, decreases the power of the pulse peak while maintaining its total energy. This low-peak-power pulse can be successfully amplified increasing the total pulse energy. In order to obtain a pulse with the same duration that the input one, a chirp filter with parameter $\phi_2'' = -\phi_1''$ is applied after the amplifier.

Another application of *temporal lenses* and Fresnel transformation performed by spectral chirp filters is a pulse compression [32]. The system used for this purpose is similar to the beam magnifier. Indeed, based on the matrix formalism it is easy to prove that a system constructed from a quadratic phase modulator with parameter φ'' embedded into two chirp filters with parameters $\phi_{1,2}''$ compresses the pulse by a factor $\tau_o/\tau_i = 1 + \phi_2''\varphi''$, where $\tau_{o,i}$ is the effective width of the output and input pulses, respectively, if the following equation holds

$$\varphi'' = \frac{1}{\phi_1''} + \frac{1}{\phi_2''}. \tag{5.190}$$

By analogy with space optics we may construct chronocyclic phase-space rotator—the fractional Fourier transformer—which is an important tool in pulse metrology. We recall that the temporal intensity distribution of an ultrashort pulse cannot be measured directly, while its spectral intensity can. The temporal intensity distribution of the input pulse can be recovered from the spectral intensity distribution of the pulse at the output of the Fourier transformer (called in this case time–frequency converter). One possible implementation of a time–frequency converter is formed by a temporal lens with parameter φ'' embedded into two identical chirp filters with parameters ϕ'' such that they satisfy the following relation $\varphi''\phi'' = 1$.

For the determination of the pulse correlation function or the pulse phase we can use a phase-space tomography method similar to the one proposed for space optics. Several schemes combining temporal lenses and chirp filters have been proposed for this purpose [27, 43]. As an example we mention one which may serve as a programmable temporal fractional Fourier transformer. Substituting in the system sketched in Fig. 6.1a in [65] the free space intervals by identical frequency chirp filters with parameter ϕ'', and the SLMs, implementing spatial lenses, by electro-optic crystals implementing temporal lenses with parameters $\varphi''_{1,2}(\alpha)$, a programmable chronocyclic phase-space rotator is obtained. The relation between the parameters for the applied elements and the transformation angle α is given by

$$\begin{aligned} \phi''\varphi''_1 &= 1 - (\phi''/s)\cot(\alpha/2), \\ \phi''\varphi''_2 &= 2 - (s/\phi'')\sin\alpha, \end{aligned} \tag{5.191}$$

where s is a constant system parameter. In such case, the spectral intensity of the output pulse corresponds to the squared modulus of the fractional Fourier for angle α of the complex envelop of the input pulse. Varying the temporal lens parameters according to the above equations, a set of WD projections for $\alpha \in [\pi/2, 3\pi/2]$ can be measured. The WD and, therefore, the two-point correlation function of the pulse can be obtained applying the inverse Radon transform to this set.

5.8 Conclusions

In this chapter we have considered different approximations for the fundamental equation of classical optics which lead to the description of the beam or pulse propagation through various elements in the form of the LCTs. In spite of this description is not complete, it is widely used in image formation and signal processing tasks, as well as for optical beam and pulse metrology. The ray transformation matrix formalism of the LCTs is beneficial for optical system design, beam shaping, and phase-space signal analysis. In particular, the transformations of the Wigner distribution of the beam during its propagation through the first-order systems are easily expressed through the ray transformation matrices. Using this approach the tomographic methods for the characterization of the spatial coherence of monochromatic beams and the temporal coherence of optical pulses as well as the systems for the implementation of these methods have been developed. While in this chapter the scalar optical fields have been considered, the results can be easily generalized into the vector case. Moreover, the modified LCTs allow describing the beam propagation in optical systems with loss or gain [58, 76] and in misaligned systems [59].

References

1. E.G. Abramochkin, V.G Volostnikov, Spiral light beams. Phys. Usp. **47**(12), 1177–1203 (2004)
2. E.G. Abramochkin, V.G. Volostnikov, Generalized Gaussian beams. J. Opt. A Pure Appl. Opt. **6**, S157–S161 (2004)
3. G.S. Agarwal, SU(2) structure of the Poincaré sphere for light beams with orbital angular momentum. J. Opt. Soc. Am. A **16**, 2914–2916 (1999)
4. T. Alieva, E. Abramochkin, A. Asenjo-Garcia, E. Razueva, Rotating beams in isotropic optical system. Opt. Express **18**(4), 3568–3573 (2010)
5. T. Alieva, M.J. Bastiaans, Self-affinity in phase space. J. Opt. Soc. Am. A **17**(4), 756–761 (2000)
6. T. Alieva, M.J. Bastiaans, Wigner distribution and fractional Fourier transform for two-dimensional symmetric optical beams. J. Opt. Soc. Am. A **17**(12), 2319–2323 (2000)
7. T. Alieva, M.J. Bastiaans, Evolution of the vortex and the asymmetrical parts of orbital angular momentum in separable first-order optical systems. Opt. Lett. **29**(14), 1587–1589 (2004)
8. T. Alieva, M.J. Bastiaans, Alternative representation of the linear canonical integral transform. Opt. Lett. **30**, 3302–3304 (2005)
9. T. Alieva, M.J. Bastiaans, Orthonormal mode sets for the two-dimensional fractional Fourier transformation. Opt. Lett. **32**(10), 1226–1228 (2007)
10. T. Alieva, M.J. Bastiaans, Phase-space rotations and orbital Stokes parameters. Opt. Lett. **34**(4), 410–412 (2009)
11. L. Allen, M.W. Beijersbergen, R.J.C. Spreeuw, J.P. Woerdman, Orbital angular momentum of light and the transformation of Laguerre-Gaussian laser modes. Phys. Rev. A **45**(11), 8185–8189 (1992)
12. M.A. Alonso, Wigner functions in optics: describing beams as ray bundles and pulses as particle ensembles. Adv. Opt. Photonics **3**(4), 272–365 (2011)
13. Yu. A. Anan'ev, A. Ya. Bekshaev, Theory of intensity moments for arbitrary light beams. Opt. Spectrosc. **76**, 558–568 (1994)
14. J. Azana, N.K. Berger, B. Levit, B. Fischer, Spectro-temporal imaging of optical pulses with a single time lens. IEEE Photonics Technol. Lett. **16**(3), 882–884 (2004)
15. E. Baleine, A. Dogariu, Variable coherence tomography. Opt. Lett. **29**(11), 1233–1235 (2004)
16. M.A. Bandres, J.C. Gutiérrez-Vega, Ince–Gaussian series representation of the two-dimensional fractional Fourier transform. Opt. Lett. **30**(5), 540–542 (2005)
17. W.C. Banyai, A.A. Godil, M.T. Kauffman, D.M. Bloom, Time-to-frequency converter for measuring picosecond optical pulses. Appl. Phys. Lett. **64**(3), 270–272 (1994)
18. M.J. Bastiaans, A frequency-domain treatment of partial coherence. Opt. Acta **24**, 261–274 (1977)
19. M.J. Bastiaans, The Wigner distribution function applied to optical signals and systems. Opt. Commun. **25**, 26–30 (1978)
20. M.J. Bastiaans, Second-order moments of the Wigner distribution function in first-order optical systems. Optik **88**(4), 163–168 (1991)
21. M.J. Bastiaans, ABCD law for partially coherent Gaussian light, propagating through first-order optical systems. Opt. Quantum Electron **24**, 1011–1019 (1992)
22. M.J. Bastiaans, Wigner distribution function applied to twisted Gaussian light propagating in first-order optical systems. J. Opt. Soc. Am. A **17**(12), 2475–2480 (2000)
23. M.J. Bastiaans, T. Alieva, Wigner distribution moments in fractional Fourier transform systems. J. Opt. Soc. Am. A **19**(9), 1763–1773 (2002)
24. M.J. Bastiaans, T. Alieva, Wigner distribution moments measured as intensity moments in separable first-order optical systems. EURASIP J. Appl. Signal Process **10**, 1535–1540 (2005)

25. M.J. Bastiaans, T. Alieva, Signal representation on the angular Poincaré sphere, based on second-order moments. J. Opt. Soc. Am. A **27**(4), 918–927 (2010)
26. M.J. Bastiaans, T. Alieva, The linear canonical transformation: definition and properties, in *Linear Canonical Transforms: Theory and Applications*, Chap. 2 (Springer, Berlin) (2016)
27. M. Beck, M.G. Raymer, I.A. Walmsley, V. Wong, Chronocyclic tomography for measuring the amplitude and phase structure of optical pulses. Opt. Lett. **18**(23), 2041–2043 (1993)
28. M.W. Beijersbergen, L. Allen, H. van der Veen, J.P. Woerdman, Astigmatic laser mode converters and transfer of orbital angular momentum. Opt. Commun. **96**, 123–132 (1993)
29. A. Ya. Bekshaev, Intensity moments of a laser beam formed by superposition of Hermite-Gaussian modes. Fotoelektronika **8**, 22–25 (1999)
30. A.Ya. Bekshaev, M.S. Soskin, M.V. Vasnetsov, Optical vortex symmetry breakdown and decomposition of the orbital angular momentum of the light beams. J. Opt. Soc. Am. A **20**, 1635–1643 (2003)
31. C.V. Bennett, B.H. Kolner, Upconversion time microscope demonstrating 103 × magnification of femtosecond waveforms. Opt. Lett. **24**(11), 783–785 (1999)
32. C.V. Bennett, B.H. Kolner, Principles of parametric temporal imaging, I. System configurations. IEEE J. Quantum Electron. **36**(4), 430–437 (2000)
33. C.V. Bennett, R.P. Scott, B.H. Kolner, Temporal magnification and reversal of 100 Gb/s optical data with an up-conversion time microscope. Appl. Phys. Lett. **65**(20), 2513–2515 (1994)
34. N.K. Berger, B. Levit, S. Atkins, B. Fischer, Time-lens-based spectral analysis of optical pulses by electrooptic phase modulation. Electron. Lett. **36**(19), 1644–1646 (2000)
35. R. Borghi, G. Guattari, L. de la Torre, F. Gori, M. Santarsiero, Evaluation of the spatial coherence of a light beam through transverse intensity measurements. J. Opt. Soc. Am. A **20**(9), 1763–1770 (2003)
36. M. Born, E. Wolf, *Principles of Optics* (Cambridge University Press, Cambridge, 1999)
37. G.F. Calvo, Wigner representation and geometric transformations of optical orbital angular momentum spatial modes. Opt. Lett. **30**, 1207–1209 (2005)
38. A. Cámara, T. Alieva, Second order moments of superpositions of Hermite-Laguerre-Gauss modes, in *PIERS Proceedings*, Cambridge, 5–8 July 2010, pp. 526–530
39. A. Cámara, T. Alieva, J.A. Rodrigo, M.L. Calvo, Phase space tomography reconstruction of the Wigner distribution for optical beams separable in Cartesian coordinates. J. Opt. Soc. Am. A **26**(6), 1301–1306 (2009)
40. A. Cámara, T. Alieva, J.A. Rodrigo, M.L. Calvo, Experimental reconstruction of the mutual intensity based on phase-space tomography, in *Frontiers in Optics/Laser Science XXVIII (FiO/LS)*, OSA, 2012
41. L. Cohen, Time-frequency distributions—A review. Proc. IEEE **77**, 941–981 (1979)
42. S.A. Collins Jr., Lens-system diffraction integral written in terms of matrix optics. J. Opt. Soc. Am. **60**, 1168–1177 (1970)
43. C. Dorrer, I. Kang, Complete temporal characterization of short optical pulses by simplified chronocyclic tomography. Opt. Lett. **28**(16), 1481–1483 (2003)
44. D. Dragoman, Higher-order moments of the Wigner distribution function in first-order optical systems. J. Opt. Soc. Am. A **11**(10), 2643–2646 (1994)
45. D. Dragoman, Applications of the Wigner distribution function in signal processing. EURASIP J. Appl. Signal Process. **2005**, 1520–1534 (2005)
46. B. Eppich, C. Gao, H. Weber, Determination of the ten second order intensity moments. Opt. Laser Technol. **30**(5), 337–340 (1998)
47. J.R. Fienup, Phase-retrieval algorithms for a complicated optical system. Appl. Opt. **32**, 1737–1746 (1993)
48. A.T. Friberg, E. Tervonen, J. Turunen, Interpretation and experimental demonstration of twisted Gaussian Schell-model beams. J. Opt. Soc. Am. A **11**(6), 1818–1826 (1994)
49. D. Furlan, W.C. Soriano, G. Saavedra, Opto-digital tomographic reconstruction of the Wigner distribution function of complex fields. Appl. Opt. **47**(22), E63–E67 (2008)

50. G. Gbur, Simulating fields of arbitrary spatial and temporal coherence. Opt. Express **14**(17), 7567–7578 (2006)
51. G. Gbur, T.D. Visser, *The Structure of Partially Coherent Field*, vol. 55, Chap. 5 (Elsevier, Amsterdam, 2010), pp. 285–341
52. R.W. Gerchberg, W.O. Saxton, A practical algorithm for the determination of phase from image and diffraction plane pictures. Optik **35**, 237–246 (1972)
53. C. Gomez-Reino, GRIN optics and its application in optical connections. Int. J. Optoelectron. **7**(5), 607–680 (1992)
54. J.W. Goodman, *Statistical Optics*, 1st edn. (Wiley-Interscience, London, 2000)
55. J.W. Goodman, *Introduction to Fourier Optics* (Roberts&Company, Englewood, 2005)
56. J.W. Goodman, *Speckle Phenomena in Optics* (Roberts&Company, Englewood, 2007)
57. F. Gori, Collet-Wolf sources and multimode laser. Opt. Commun. **34**, 301–305 (1980)
58. A. Hardy, M. Nazarathy, J. Shamir, Generalized mode propagation in first-order optical systems with loss or gain. J. Opt. Soc. Am. **72**, 1409–1420 (1982)
59. A. Hardy, M. Nazarathy, J. Shamir, Misaligned first-order optics: canonical operator theory. J. Opt. Soc. Am. A **3**, 1360–1369 (1986)
60. J. Healy, D.P. Kelly, U. Gopinathan, J. Sheridan
61. International Organization for Standardization, Lasers and laser-related equipment – test methods for laser beam widths, divergence angles and beam propagation ratios, TC172/SC9, Geneva, Switzerland (1999)
62. T. Jannson, Real-time fourier transformation in dispersive optical fibers. Opt. Lett. **8**(4), 232–234 (1983)
63. H. Kogelnik, T. Li, Laser beams and resonators. Appl. Opt. **5**, 1550–1567 (1966)
64. B.H. Kolner, Space-time duality and the theory of temporal imaging. IEEE J. Quantum Electron. **30**(8), 1951–1963 (1994)
65. A. Kutay, H. Ozaktas, J. Rodrigo, Optical implementations of LCT, in *Linear Canonical Transforms: Theory and Applications*, Chap. 6 (Springer, Berlin)
66. R.K. Luneburg, *Mathematical Theory of Optics* (University of California Press, Berkeley, 1966)
67. L. Mandel, E. Wolf, Spectral coherence and the concept of cross-spectral purity. J. Opt. Soc. Am. **66**, 529–535 (1976)
68. L. Mandel, E. Wolf, *Optical Coherence and Quantum Optics* (Cambridge University Press, Cambridge, 1995)
69. C. Martínez, F. Encinas-Sanz, J. Serna, P.M. Mejías, R. Martínez-Herrero, On the parametric characterization of the transversal spatial structure of laser pulses. Opt. Commun. **139**, 299–305 (1997)
70. D. Mendlovic, H.M. Ozaktas, Fractional Fourier transforms and their optical implementation: I. J. Opt. Soc. Am. A **10**, 1875–1881 (1993)
71. M. Moshinsky, C. Quesne, Linear canonical transformations and their unitary representations. J. Math. Phys. **12**, 1772–1780 (1971)
72. L.K. Mouradian, F. Louradour, V. Messager, A. Barthelemy, C. Froehly, Spectro-temporal imaging of femtosecond events. IEEE J. Quantum Electron. **36**(7), 795–801 (2000)
73. M. Nakazawa, T. Hirooka, F. Futami, S. Watanabe, Ideal distortion-free transmission using optical Fourier transformation and Fourier transform-limited optical pulses. IEEE Photonics Technol. Lett. **16**(4), 1059–1061 (2004)
74. M. Nazarathy, J. Shamir, Fourier optics described by operator algebra. J. Opt. Soc. Am. **70**, 150–159 (1980)
75. M. Nazarathy, J. Shamir, First-order optics—a canonical operator representation: lossless systems. J. Opt. Soc. Am. **72**, 356–364 (1982)
76. M. Nazarathy, J. Shamir, First-order optics—operator representation for systems with loss or gain. J. Opt. Soc. Am. **72**, 1398–1408 (1982)
77. G. Nemes, A.E. Seigman, Measurement of all ten second-order moments of an astigmatic beam by use of rotating simple astigmatic (anamorphic) optics. J. Opt. Soc. Am. A **11**, 2257–2264 (1994)

78. H.M. Ozaktas, D. Mendlovic, Fractional Fourier transforms and their optical implementation: II. J. Opt. Soc. Am. A **10**, 2522–2531 (1993)
79. H.M. Ozaktas, Z. Zalevsky, M.A. Kutay, *The Fractional Fourier Transform with Applications in Optics and Signal Processing* (Wiley, New York, 2001)
80. A.P. Prudnikov, Yu.A. Brychkov, O.I. Marichev (eds.), *Integrals and Series, Volume 1. Elementary Functions* (Gordon and Breach, New York, 1986)
81. M.G. Raymer, M. Beck, D.F. McAlister, Complex wave-field reconstruction using phase-space tomography. Phys. Rev. Lett. **72**(8), 1137–1140 (1994)
82. J.A. Rodrigo, T. Alieva, M.L. Calvo, Optical system design for orthosymplectic transformations in phase space. J. Opt. Soc. Am. A **23**(10), 2494–2500 (2006)
83. J.A. Rodrigo, T. Alieva, M.L. Calvo, Experimental implementation of the gyrator transform. J. Opt. Soc. Am. A **24**(10), 3135–3139 (2007)
84. J.A. Rodrigo, T. Alieva, M.L. Calvo, Gyrator transform: properties and applications. Opt. Express **15**(5), 2190 (2007)
85. J.A. Rodrigo, T. Alieva, M.L. Calvo, Programmable two-dimensional optical fractional Fourier processor. Opt. Express **17**(7), 4976–4983 (2009)
86. A.C. Schell, A technique for the determination of the radiation patterns of a partially coherent aperture. IEEE Trans. Antennas Propag. **AP-15**, 187–188 (1967)
87. J. Serna, R. Martínez-Herrero, P.M. Mejías, Parametric characterization of general partially coherent beams propagating through ABCD optical systems. J. Opt. Soc. Am. A **8**(7), 1094–1098 (1991)
88. J. Serna, F. Encinas-Sanz, G. Nemes, Complete spatial characterization of a pulsed doughnut-type beam by use of spherical optics and a cylindrical lens. Opt. Soc. Am. A **18**, 1726–1733 (2001)
89. J. Shamir, Cylindrical lens described by operator algebra. Appl. Opt. **18**, 4195–4202 (1979)
90. R. Simon, G.S. Agarwal, Wigner representation of Laguerre–Gaussian beams. Opt. Lett. **25**(18), 1313–1315 (2000)
91. R. Simon, N. Mukunda, Twisted Gaussian Schell-model beams. J. Opt. Soc. Am. A **10**(1), 95–109 (1993)
92. R. Simon, K.B. Wolf, Fractional Fourier transforms in two dimensions. J. Opt. Soc. Am. A **17**, 2368–2381 (2000)
93. R. Simon, K.B. Wolf, Structure of the set of paraxial optical systems. J. Opt. Soc. Am. A **17**(2), 342–355 (2000)
94. R. Simon, K.B. Wolf, Structure of the set of paraxial optical systems. J. Opt. Soc. Am. A **17**, 342–355 (2000)
95. R. Simon, E.C.G. Sudarshan, N. Mukunda, Anisotropic Gaussian Schell model beams: passage through optical systems and associated invariants. Phys. Rev. A **31**, 2419–2434 (1985)
96. K. Sundar, N. Mukunda, R. Simon, Coherent-mode decomposition of general anisotropic Gaussian Schell-model beams. J. Opt. Soc. Am. A **12**(3), 560–569 (1995)
97. Tao.
98. M.R. Teague, Image analysis via the general theory of moments. J. Opt. Soc. Am. **70**(8), 920–930 (1980)
99. M.R. Teague, Deterministic phase retrieval: a Green's function solution. J. Opt. Soc. Am. **73**, 1434–1441 (1983)
100. M.E. Testorf, B.M. Hennelly, J. Ojeda-Castañeda (eds.), *Phase-Space Optics: Fundamentals and Applications* (McGraw-Hill, New York, 2009)
101. A. Torre, *Linear Ray and Wave Optics in Phase Space* (Elsevier, Amsterdam, 2005)
102. I.A. Walmsley, C. Dorrer, Characterization of ultrashort electromagnetic pulses. Adv. Opt. Photonics **1**(2), 308–437 (2009)
103. A. Walther, Radiometry and coherence. J. Opt. Soc. Am. **58**, 1256–1259 (1968)
104. E. Wigner, On the quantum correction for thermodynamic equilibrium. Phys. Rev. **40**, 749–759 (1932)

105. J. Williamson, On the algebraic problem concerning the normal forms of linear dynamical systems. Am. J. Math **58**, 141–163 (1936)
106. K.B. Wolf, *Integral Transforms in Science and Engineering*, Chap. 9 (Plenum Press, New York, 1979)
107. K.B. Wolf, *Geometric Optics on Phase Space* (Springer, Berlin, 2004)
108. A.M. Yao, M.J. Padgett, Orbital angular momentum: origins, behavior and applications. Adv. Opt. Photonics **3**, 161–204 (2011)
109. L. Yu, W. Huang, M. Huang, Z. Zhu, X. Zeng, W. Ji, The Laguerre-Gaussian series representation of two-dimensional fractional Fourier transform. J. Phys. A Math. Gen. **31**(46), 9353 (1998)

Chapter 6
Optical Implementation of Linear Canonical Transforms

M. Alper Kutay, Haldun M. Ozaktas, and José A. Rodrigo

Abstract We consider optical implementation of arbitrary one-dimensional and two-dimensional linear canonical and fractional Fourier transforms using lenses and sections of free space. We discuss canonical decompositions, which are generalizations of common Fourier transforming setups. We also look at the implementation of linear canonical transforms based on phase-space rotators.

6.1 Introduction

In this chapter we consider the problem of designing systems for optically implementing linear canonical transforms (LCTs) and fractional Fourier transforms (FRTs). It is well known that an optical Fourier transformer can be realized by a section of free space followed by a lens followed by another section of free space, and also by a lens followed by a section of free space followed by another lens. Another approach is to use a section of quadratic graded-index media. That these approaches can also be used to implement FRTs has been realized in the nineties. One-dimensional systems have been dealt with in [1, 8, 9, 12, 15, 17, 18, 20, 24, 25, 28, 29] and two-dimensional systems have been dealt with in [9, 11, 15, 19, 23, 25, 29, 34, 36], among others. For an overview of the optical implementation of the FRT, see [27].

LCTs can be interpreted as scaled FRTs with additional phase terms. Thus, in principle, if we have an optical FRT system, we can obtain an LCT system with

M.A. Kutay (✉)
The Scientific and Technological Research Council of Turkey, 06100 Kavaklıdere, Ankara, Turkey
e-mail: alper.kutay@tubitak.gov.tr

H.M. Ozaktas
Department of Electrical Engineering, Bilkent University, 06800 Bilkent, Ankara, Turkey
e-mail: haldun@ee.bilkent.edu.tr

J.A. Rodrigo
Universidad Complutense de Madrid, Facultad de Ciencias Físicas, Ciudad Universitaria s/n, Madrid 28040, Spain
e-mail: jarmar@fis.ucm.es

J.J. Healy et al. (eds.), *Linear Canonical Transforms*, Springer Series in Optical Sciences 198, DOI 10.1007/978-1-4939-3028-9_6

some modifications, although handling the scale and phase may not always be convenient. On the other hand, since FRTs are special cases of LCTs, knowing how to realize a desired LCT means we can also realize any FRT easily.

While the design of one-dimensional systems is relatively straightforward, two-dimensional systems bring additional challenges, mostly arising from the fact that the parameters in the two dimensions can be different and this brings a number of constraints with it. We will deal with these challenges and show how all two-dimensional LCTs can be realized [36].

6.2 FRTs and LCTs

Two-dimensional LCT can be defined as:

$$f_o(\mathbf{r}_o) = \mathscr{L}(\mathbf{T})f_i(\mathbf{r}_i) = \int h(\mathbf{r}_o;\mathbf{r}_i)f_i(\mathbf{r}_i)\,d\mathbf{r}_i,$$

$$h(\mathbf{r}_o;\mathbf{r}_i) = (\det i^{-1}\mathbf{L}_{io})^{1/2}\exp\left[i\pi(\mathbf{r}_o^{\mathrm{t}}\mathbf{L}_{oo}\mathbf{r}_o - 2\mathbf{r}_i^{\mathrm{t}}\mathbf{L}_{io}\mathbf{r}_o + \mathbf{r}_i^{\mathrm{t}}\mathbf{L}_{ii}\mathbf{r}_i)\right], \tag{6.1}$$

where we define the column vector $\mathbf{r}$ as $\mathbf{r} = [x, y]^{\mathrm{t}}$. $\mathbf{L}_{ii}$ and $\mathbf{L}_{oo}$ are symmetric 2×2 matrices and $\mathbf{L}_{io}$ is a non-singular 2×2 matrix given by:

$$\mathbf{L}_{ii} \equiv \begin{bmatrix} \ell_{iix} & 0 \\ 0 & \ell_{iiy} \end{bmatrix}, \quad \mathbf{L}_{io} \equiv \begin{bmatrix} \ell_{iox} & 0 \\ 0 & \ell_{ioy} \end{bmatrix}, \quad \mathbf{L}_{oo} \equiv \begin{bmatrix} \ell_{oox} & 0 \\ 0 & \ell_{ooy} \end{bmatrix}, \tag{6.2}$$

where $\ell_{oox}, \ell_{iox}, \ell_{iix}$ and $\ell_{ooy}, \ell_{ioy}, \ell_{iiy}$ are real constants. FRTs, Fresnel transforms, chirp multiplication, and scaling operations are widely used in optics to analyze systems composed of sections of free space and thin lenses. These linear integral transforms belong to the class of LCTs. Any LCT is completely specified by its parameters.

Alternatively, LCTs can be specified by using a transformation matrix:

$$f_o(\mathbf{r}_o) = \mathscr{L}(\mathbf{T})f_i(\mathbf{r}_i) = (\det i\mathbf{B})^{-1/2} \times \int \exp\left[i\pi(\mathbf{r}_o^{\mathrm{t}}\mathbf{D}\mathbf{B}^{-1}\mathbf{r}_o - 2\mathbf{r}_i^{\mathrm{t}}\mathbf{B}^{-1}\mathbf{r}_o + \mathbf{r}_i^{\mathrm{t}}\mathbf{B}^{-1}\mathbf{A}\mathbf{r}_i)\right] f_i(\mathbf{r}_i)\,d\mathbf{r}_i\,, \tag{6.3}$$

The transformation matrix of such a system specified by the parameters $\ell_{oox}, \ell_{iox}, \ell_{iix}$ and $\ell_{ooy}, \ell_{ioy}, \ell_{iiy}$ is

$$\mathbf{T} \equiv \begin{bmatrix} \mathbf{A}\ \mathbf{B} \\ \mathbf{C}\ \mathbf{D} \end{bmatrix} \equiv \begin{bmatrix} A_x & 0 & B_x & 0 \\ 0 & A_y & 0 & B_y \\ C_x & 0 & D_x & 0 \\ 0 & C_y & 0 & D_y \end{bmatrix}$$
$$\equiv \begin{bmatrix} \ell_{iix}/\ell_{iox} & 0 & 1/\ell_{iox} & 0 \\ 0 & \ell_{iiy}/\ell_{ioy} & 0 & 1/\ell_{ioy} \\ -\ell_{iox} + \ell_{oox}\ell_{iix}/\ell_{iox} & 0 & \ell_{oox}/\ell_{iox} & 0 \\ 0 & -\ell_{ioy} + \ell_{ooy}\ell_{iiy}/\ell_{ioy} & 0 & \ell_{ooy}/\ell_{ioy} \end{bmatrix}.$$

with $A_xD_x - B_xC_x = 1$ and $A_yD_y - B_yC_y = 1$ [5, 42].

Propagation in free-space (or a homogeneous medium) and through thin lenses are special forms of LCTs. The transformation matrix for free-space propagation over a distance z and with constant refractive index n can be expressed as

$$\mathbf{T}_{\mathscr{S}}(z) = \begin{bmatrix} 1 & 0 & \frac{\lambda z}{n} & 0 \\ 0 & 1 & 0 & \frac{\lambda z}{n} \\ 0 & 0 & 1 & 0 \\ 0 & 0 & 0 & 1 \end{bmatrix}. \tag{6.4}$$

Similarly, the matrix for a cylindrical lens with focal length f_x along the x direction is

$$\mathbf{T}_{\mathscr{Q}_x}(f_x) = \begin{bmatrix} 1 & 0 & 0 & 0 \\ 0 & 1 & 0 & 0 \\ \frac{-1}{\lambda f_x} & 0 & 1 & 0 \\ 0 & 0 & 0 & 1 \end{bmatrix}, \tag{6.5}$$

and the matrix for a cylindrical lens with focal length f_y along the y direction is

$$\mathbf{T}_{\mathscr{Q}_y}(f_y) = \begin{bmatrix} 1 & 0 & 0 & 0 \\ 0 & 1 & 0 & 0 \\ 0 & 0 & 1 & 0 \\ 0 & \frac{-1}{\lambda f_y} & 0 & 1 \end{bmatrix}. \tag{6.6}$$

More general anamorphic lenses may be represented by a matrix of the form:

$$\mathbf{T}_{\mathscr{Q}_{xy}}(f_x, f_y, f_{xy}) = \begin{bmatrix} 1 & 0 & 0 & 0 \\ 0 & 1 & 0 & 0 \\ \frac{-1}{\lambda f_x} & \frac{-1}{2\lambda f_{xy}} & 1 & 0 \\ \frac{-1}{2\lambda f_{xy}} & \frac{-1}{\lambda f_y} & 0 & 1 \end{bmatrix}. \tag{6.7}$$

The transformation matrix approach has several advantages. First of all, if several systems are cascaded, the overall system matrix can be found by multiplying the corresponding transformation matrices. Second, the transformation matrix corresponds to the ray-matrix in optics [37]. Third, the effect of the system on the Wigner distribution of the input function can be expressed in terms of this transformation matrix. This topic is extensively discussed in [3–7].

The 2D FRT also belongs to the family of LCTs:

$$\begin{aligned} f_o(\mathbf{r}_o) &= \mathscr{F}(\gamma_x, \gamma_y) f_i(\mathbf{r}_i) \\ &= \int A_{\gamma_\mathbf{r}} \exp[i\pi(\mathbf{r}_o^\mathrm{t}\mathbf{C}_\mathrm{t}\mathbf{r}_o - 2\mathbf{r}_o^\mathrm{t}\mathbf{C}_\mathrm{s}\mathbf{r}_i + \mathbf{r}_i^\mathrm{t}\mathbf{C}_\mathrm{t}\mathbf{r}_i)] f_i(\mathbf{r}_i)\, d\mathbf{r}_i, \end{aligned} \tag{6.8}$$

where

$$\mathbf{C}_\mathrm{t} = \begin{bmatrix} \cot\gamma_x & 0 \\ 0 & \cot\gamma_y \end{bmatrix}, \qquad \mathbf{C}_\mathrm{s} = \begin{bmatrix} \csc\gamma_x & 0 \\ 0 & \csc\gamma_y \end{bmatrix},$$

$$A_{\gamma_\mathbf{r}} = A_{\gamma_x} A_{\gamma_y}, \qquad A_{\gamma_x} = \frac{e^{-i(\pi\hat{\gamma}_x/4 - \gamma_x/2)}}{\sqrt{|\sin\gamma_x|}}, \qquad A_{\gamma_y} = \frac{e^{-i(\pi\hat{\gamma}_y/4 - \gamma_y/2)}}{\sqrt{|\sin\gamma_y|}}$$

with $\hat{\gamma}_x = \mathrm{sgn}(\gamma_x)$, $\hat{\gamma}_y = \mathrm{sgn}(\gamma_y)$. γ_x and γ_y are rotational angles of the FRT in the two dimensions, which are related to the fractional orders a_x and a_y through $\gamma_x = a_x\pi/2$ and $\gamma_y = a_y\pi/2$.

The output of a fairly broad class of optical systems can be expressed as the FRT of the input [27]. This is a generalization of the fact that in certain special planes one observes the ordinary Fourier transform. However, when we are dealing with FRTs, the choice of scale and dimensions must always be noted. To be able to handle the scales explicitly, we will modify the definition of the FRT by introducing input and output scale parameters. Also allowing for additional phase factors that may occur at the output, the kernel can be expressed as

$$\begin{aligned} K_{\gamma_x,\gamma_y}(x, y; x', y') = \; & A_{\gamma_x} \exp[i\pi x^2 p_x] \\ & \times \exp\left[i\pi\left(\frac{x^2}{s_2^2}\cot\gamma_x - \frac{2xx'}{s_1 s_2}\csc\gamma_x + \frac{x'^2}{s_1^2}\cot\gamma_x\right)\right] \\ & \times A_{\gamma_y} \exp[i\pi y^2 p_y] \\ & \times \exp\left[i\pi\left(\frac{y^2}{s_2^2}\cot\gamma_y - \frac{2yy'}{s_1 s_2}\csc\gamma_y + \frac{y'^2}{s_1^2}\cot\gamma_y\right)\right]. \end{aligned} \tag{6.9}$$

In this definition, s_1 stands for the input scale parameter, s_2 stands for the output scale parameter, and p_x and p_y are the parameters of the quadratic phase factors. The transformation matrix corresponding to this kernel can be found as

$$\mathbf{T} \equiv \begin{bmatrix} \mathbf{A} & \mathbf{B} \\ \mathbf{C} & \mathbf{D} \end{bmatrix}, \tag{6.10}$$

where

$$\mathbf{A} = \begin{bmatrix} \frac{s_2}{s_1} \cos \gamma_x & 0 \\ 0 & \frac{s_2}{s_1} \cos \gamma_y \end{bmatrix}, \quad \mathbf{C} = \begin{bmatrix} \frac{1}{s_1 s_2} [p_x \cos \gamma_x - \sin \gamma_x] & 0 \\ 0 & \frac{1}{s_1 s_2} [p_y \cos \gamma_y - \sin \gamma_y] \end{bmatrix}, \tag{6.11}$$

$$\mathbf{B} = \begin{bmatrix} s_1 s_2 \sin \gamma_x & 0 \\ 0 & s_1 s_2 \sin \gamma_y \end{bmatrix}, \quad \mathbf{D} = \begin{bmatrix} \frac{s_1}{s_2} \sin \gamma_x (p_x + \cot \gamma_x) & 0 \\ 0 & \frac{s_1}{s_2} \sin \gamma_y (p_y + \cot \gamma_y) \end{bmatrix}. \tag{6.12}$$

It can be deduced from the above equation that any quadratic-phase system can be implemented by appending lenses at the input and output planes of a fractional Fourier transformer [22, 25, 27].

6.3 Canonical Decompositions, Anamorphic Sections of Free Space, and Optical Implementation of LCTs

One way of designing optical implementations of LCTs is to employ the matrix formulation given in (6.3). The LCT matrix can be decomposed into matrices that corresponds to more elementary operations such as free-space propagation, thin lenses, etc.

6.3.1 *One-Dimensional Systems*

We first discuss one-dimensional systems, presenting two decompositions that reduce to familiar optical arrangements for the special case of the Fourier transform.

Canonical Decomposition Type-1

The LCT system matrix $\mathbf{T}$ can be decomposed as

$$\mathbf{T} = \mathbf{T}_{\mathscr{S}}(z_2)\, \mathbf{T}_{\mathscr{Q}_x}(f)\, \mathbf{T}_{\mathscr{S}}(z_1). \tag{6.13}$$

which corresponds to a section of free space of length z_1, followed by a thin lens of focal length f, followed by another section of free space of length z_2, as shown in Fig. 6.1.

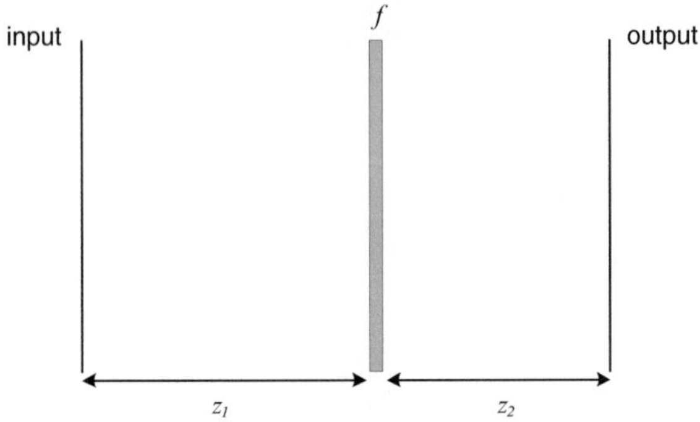

Fig. 6.1 Type-1 system which realizes arbitrary one-dimensional linear canonical transforms [36]

Both the optical system in Fig. 6.1 and the LCT have three parameters. Thus, it is possible to find the system parameters uniquely by solving the above equations. Doing so, the equations for z_1, z_2 and f in terms of ℓ_{oo}, ℓ_{io}, ℓ_{ii} are found as

$$z_1 = \frac{\ell_{io} - \ell_{oo}}{\lambda({\ell_{io}}^2 - \ell_{ii}\ell_{oo})}, \quad z_2 = \frac{\ell_{io} - \ell_{ii}}{\lambda({\ell_{io}}^2 - \ell_{ii}\ell_{oo})}, \quad f = \frac{\ell_{io}}{\lambda({\ell_{io}}^2 - \ell_{ii}\ell_{oo})}. \tag{6.14}$$

Since FRTs are a special case of LCTs, it is possible to implement one-dimensional FRT of any desired order by using this optical setup. The scale parameters s_1 and s_2 may be specified by the designer and the additional phase factors p_x and p_y may be made equal to zero. Letting $\ell_{oo} = \cot\gamma/s_2^2$, $\ell_{ii} = \cot\gamma/s_1^2$ and $\ell_{io} = \csc\gamma/s_1 s_2$, one recovers Lohmann's type-1 fractional Fourier transforming system [15]. In this case, the system parameters are found as

$$z_1 = \frac{(s_1 s_2 - s_1^2 \cos\gamma)}{\lambda \sin\gamma}, \qquad z_2 = \frac{(s_1 s_2 - s_2^2 \cos\gamma)}{\lambda \sin\gamma}, \qquad f = \frac{s_1 s_2}{\lambda \sin\gamma}. \tag{6.15}$$

Since the additional phase factors are set to zero, they do not appear in the equations. However, if one wishes to set p_x and p_y to a value other than zero, it is again possible by setting $\ell_{oo} = p_x \cot\gamma/s_2^2$ and substituting it in Eq. (6.14).

Canonical Decomposition Type-2

In this case, instead of one lens and two sections of free space, we have two lenses separated by a single section of free space, as shown in Fig. 6.2. Again, the parameters z, f_1 and f_2 can be solved similar to that for the Type-1 decomposition:

$$z = \frac{1}{\lambda \ell_{io}}, \qquad f_1 = \frac{1}{\lambda(\ell_{io} - \ell_{ii})}, \qquad f_2 = \frac{1}{\lambda(\ell_{io} - \ell_{oo})}. \tag{6.16}$$

Fig. 6.2 Type-2 system which realizes arbitrary one-dimensional linear canonical transforms [36]

If $\ell_{oo} = \cot\gamma/s_2^2$, $\ell_{ii} = \cot\gamma/s_1^2$ and $\ell_{io} = \csc\gamma/s_1 s_2$ are substituted in these equations, the parameters required to obtain a FRT can be found. The designer can again specify the scale parameters and zero phase factor at the output to find:

$$z = \frac{s_1 s_2 \sin\gamma}{\lambda}, \qquad f_1 = \frac{s_1^2 s_2 \sin\gamma}{s_1 - s_2\cos\gamma}, \qquad f_2 = \frac{s_1 s_2^2 \sin\gamma}{s_2 - s_1\cos\gamma}. \tag{6.17}$$

Equations (6.14) and (6.16) give the expressions for the system parameters of type-1 and type-2 canonical systems. But for some values of ℓ_{oo}, ℓ_{io}, ℓ_{ii}, the lengths of the free space sections required may turn out to be negative, which is not physically realizable. This constraint will restrict the range of LCTs that can be realized with the suggested setups. However, in Sect. 6.3.3, this constraint is removed by employing an optical setup that simulates anamorphic and negative valued sections of free space. This system is designed in such a way that its effect is equivalent to propagation in free space with different (and possibly negative) distances along the two dimensions.

6.3.2 Two-Dimensional Systems

Now we turn our attention to two-dimensional systems. We first present an elementary result which allows us to analyze two-dimensional systems as two one-dimensional systems, which makes the analysis of two-dimensional systems remarkably easier. We write the output of the system in terms of its input as follows:

$$f_o(\mathbf{r}_o) = \int h(\mathbf{r}_o, \mathbf{r}_i) f_i(\mathbf{r}_i)\, d\mathbf{r}_i.$$

If the kernel $h(\mathbf{r}_o, \mathbf{r}_i)$ is separable, that is, $h(\mathbf{r}_o, \mathbf{r}_i) = h_x(x_o, x_i)\, h_y(y_o, y_i)$, then the response in the x direction is the result of the one-dimensional transform

$$f_x(x_o, y_i) = \int h_x(x_o, x_i) f(x_i, y_i)\, dx_i, \tag{6.18}$$

and similar in the y direction. Moreover if the function is also separable, that is, if $f(\mathbf{r}) = f_x(x) f_y(y)$, the overall response of the system is

$$f_o(\mathbf{r}) = f_{ox}(x) f_{oy}(y),$$

where

$$f_{ox}(x) = \int h_x(x, x_i) f_{ix}(x_i)\, dx_i,$$

$$f_{oy}(y) = \int h_y(y, y_i) f_{iy}(y_i)\, dy_i.$$

This result has a nice interpretation in optics which makes the analysis of two-dimensional systems easier. For example, in order to design an optical setup that realizes imaging in the x direction and Fourier transforming in the y direction, one can design two one-dimensional systems that realize the given transformations. When these two systems are merged, the overall effect of the system is imaging in the x direction and Fourier transforming in the y direction. Similarly, if we have a system that realizes a FRT with rotational angle γ_x in the x direction and another system which realizes a FRT with rotational angle γ_y in the y direction, then these two optical setups will together implement a two-dimensional FRT with the rotational angles γ_x and γ_y. So the problem of designing a two-dimensional fractional Fourier transformer reduces to the problem of designing two one-dimensional fractional Fourier transformers.

Canonical Decomposition Type-1

According to the above result, the x and y directions can be considered independent of each other, since the kernel given in Eq. (6.1) or Eq. (6.3) is separable. Hence if two optical setups realizing one-dimensional LCTs are put together, one can implement the desired two-dimensional FRT. The suggested optical system is shown in Fig. 6.3 and employs the following parameters:

$$z_{1x} = \frac{\ell_{iox} - \ell_{oox}}{\lambda(\ell_{iox}^2 - \ell_{iix}\ell_{oox})}, \quad z_{2x} = \frac{\ell_{iox} - \ell_{iix}}{\lambda(\ell_{iox}^2 - \ell_{iix}\ell_{oox})}, \quad f_x = \frac{\ell_{iox}}{\lambda(\ell_{iox}^2 - \ell_{iix}\ell_{oox})}, \tag{6.19}$$

$$z_{1y} = \frac{\ell_{ioy} - \ell_{ooy}}{\lambda(\ell_{ioy}^2 - \ell_{iiy}\ell_{ooy})}, \quad z_{2y} = \frac{\ell_{ioy} - \ell_{iiy}}{\lambda(\ell_{ioy}^2 - \ell_{iiy}\ell_{ooy})}, \quad f_y = \frac{\ell_{ioy}}{\lambda(\ell_{ioy}^2 - \ell_{iiy}\ell_{ooy})}. \tag{6.20}$$

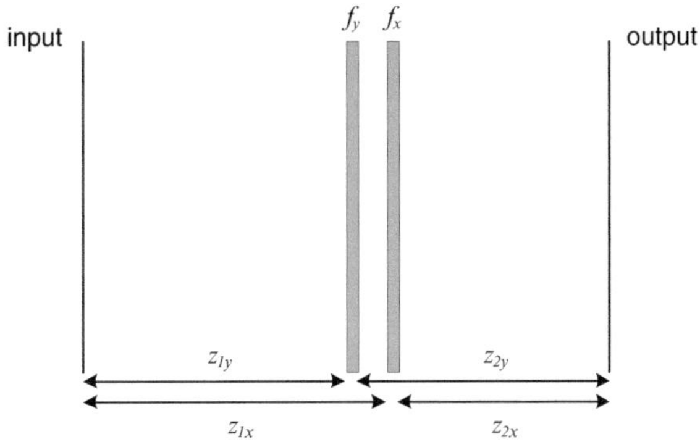

Fig. 6.3 Type-1 system that realizes arbitrary two-dimensional linear canonical transforms [36]

Arbitrary two-dimensional fractional Fourier transforming systems can be obtained as a special case by using:

$$\ell_{oox} = \cot \gamma_x / s_2^2, \qquad \ell_{iix} = \cot \gamma_x / s_1^2, \qquad \ell_{iox} = \csc \gamma_x / s_1 s_2, \tag{6.21}$$

$$\ell_{ooy} = \cot \gamma_y / s_2^2, \qquad \ell_{iiy} = \cot \gamma_y / s_1^2, \qquad \ell_{ioy} = \csc \gamma_y / s_1 s_2. \tag{6.22}$$

When these equations are substituted into (6.19) and (6.20), the parameters of the fractional Fourier transforming optical system can be found.

We saw that the derivations of the required system parameters can be carried out by treating x and y independently. However, $z_{1x} + z_{2x} = z_x = z_{1y} + z_{2y} = z_y$ should always be satisfied so that the actions in the x and y dimensions meet at a single output plane. Another constraint that needs to be satisfied is the positivity of the lengths of the free space sections. z_{1x},z_{1y},z_{2x},z_{2y} should always be positive. These two constraints restrict the set of LCTs that can be implemented. As before, this restriction can be dealt with by simulating anamorphic sections of free space which provides us a propagation distance of z_x in the x direction and a distance of z_y in the y direction where z_x and z_y may take negative values. By removing the restriction that the propagation distance in the two dimensions has to be equal and positive, all LCTs can be realized. This problem is solved in Sect. 6.3.3.

Canonical Decomposition Type-2

Two type-2 systems can also realize arbitrary two-dimensional LCTs, by using the parameters

$$z_x = \frac{1}{\lambda \ell_{iox}}, \qquad f_{1x} = \frac{1}{\lambda(\ell_{iox} - \ell_{iix})}, \qquad f_{2x} = \frac{1}{\lambda(\ell_{iox} - \ell_{oox})}, \tag{6.23}$$

Fig. 6.4 Type-2 system that realizes arbitrary two-dimensional linear canonical transforms [36]

$$z_y = \frac{1}{\lambda \ell_{ioy}}, \qquad f_{1y} = \frac{1}{\lambda(\ell_{ioy} - \ell_{iiy})}, \qquad f_{2y} = \frac{1}{\lambda(\ell_{ioy} - \ell_{ooy})}. \tag{6.24}$$

As before, if Eqs. (6.21) and (6.22) are substituted in (6.23) and (6.24), the design parameters for the FRT can be obtained.

In the optical setup in Fig. 6.4, we have the constraint $z_x = z_y = z$, which is even more restrictive than with type-1 systems. Again z_x and z_y cannot be negative. In order to overcome these difficulties, in the following section, we show how to simulate anamorphic sections of free space with physically realizable components.

6.3.3 *Simulation of Anamorphic Sections of Free Space*

While designing optical setups that implement one-dimensional LCTs, we treated the lengths of the sections of free space as free parameters. But some LCTs specified by the parameters $\ell_{oo}, \ell_{ii}, \ell_{io}$, turned out to require the use of free space sections with negative length. This problem is again encountered in the optical setups realizing two-dimensional LCTs. Besides, two-dimensional optical systems may require different propagation distances in the x and y directions. In order to implement all possible one-dimensional and two-dimensional LCTs, we will design a physically realizable optical system simulating the required, but physically unrealizable free space sections.

The optical system in Fig. 6.5 is composed of a Fourier block, an anamorphic lens and an inverse Fourier block. It can simulate two-dimensional anamorphic sections of free space with propagation distance z_x in the x direction and z_y in the y direction. When the analysis of the system in Fig. 6.5 is carried out, the relation between the input light distribution $f_i(x, y)$ and the output light distribution $f_o(x, y)$ is found as

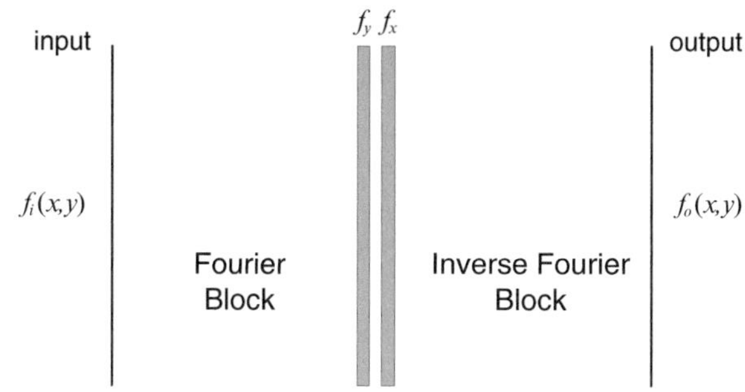

Fig. 6.5 Optical system that simulates anamorphic free space propagation [36]

$$f_o(x,y) = C \iint \exp[i\pi(x-x_i)^2/\lambda z_x + (y-y_i)^2/\lambda z_y] f_i(x_i,y_i)\,dx_i\,dy_i, \tag{6.25}$$

where

$$z_x = \frac{s^4}{\lambda^2 f_x}, \qquad z_y = \frac{s^4}{\lambda^2 f_y}. \tag{6.26}$$

and where s is the scale of the Fourier and inverse Fourier blocks. f_x and f_y can take any real value including negative ones. Thus it is possible to obtain any combination of z_x and z_y by using the optical setup in Fig. 6.5. The anamorphic lens which is used to control z_x and z_y may be composed of two orthogonally situated cylindrical thin lenses with different focal lengths. The Fourier block and inverse Fourier block are 2-f systems with a spherical lens between two sections of free space. Thus, simulating an anamorphic section of free space requires 2 cylindrical and two spherical lenses.

The system in Fig. 6.5 can also be adapted for the one-dimensional case, allowing us to simulate propagation with negative distances. When the required free space sections in the type-1 and type-2 implementations are realized by the optical setup in Fig. 6.5, the optical implementation of all separable LCTs can be realized.

Specializing to the FRT, it is possible to implement all combinations of orders if we can replace the free space sections with sections of anamorphic free space, if need be. All combinations of orders a_x and a_y can be implemented with full control on the scale parameters s_1, s_2 and the phase factors p_x, p_y, the latter which we can set to zero if desired.

6.4 Iwasawa Decomposition, Phase-Space Rotators, and Optical Implementation of LCTs

The modified Iwasawa decomposition [38, 41, 43] states that any ray transformation matrix $\mathbf{T}$ can be written as the product

$$\mathbf{T} = \begin{bmatrix} \mathbf{A} & \mathbf{B} \\ \mathbf{C} & \mathbf{D} \end{bmatrix} = \begin{bmatrix} \mathbf{I} & \mathbf{0} \\ -\mathbf{G} & \mathbf{I} \end{bmatrix} \begin{bmatrix} \mathbf{S} & \mathbf{0} \\ \mathbf{0} & \mathbf{S}^{-1} \end{bmatrix} \begin{bmatrix} \mathbf{X} & \mathbf{Y} \\ -\mathbf{Y} & \mathbf{X} \end{bmatrix} = \mathbf{T}_L \mathbf{T}_S \mathbf{T}_O, \tag{6.27}$$

where

$$\begin{aligned} \mathbf{G} &= -\left(\mathbf{C}\mathbf{A}^t + \mathbf{D}\mathbf{B}^t\right)\left(\mathbf{A}\mathbf{A}^t + \mathbf{B}\mathbf{B}^t\right)^{-1} &&= \mathbf{G}^t, \\ \mathbf{S} &= \left(\mathbf{A}\mathbf{A}^t + \mathbf{B}\mathbf{B}^t\right)^{1/2} &&= \mathbf{S}^t, \\ \mathbf{X} + i\mathbf{Y} &= \left(\mathbf{A}\mathbf{A}^t + \mathbf{B}\mathbf{B}^t\right)^{-1/2}\left(\mathbf{A} + i\mathbf{B}\right) &&= \left(\mathbf{X}^t - i\mathbf{Y}^t\right)^{-1}. \end{aligned} \tag{6.28}$$

The first matrix $\mathbf{T}_L$ corresponds to an anamorphic quadratic-phase modulation, which can be realized with a generalized lens. $\mathbf{T}_S$ is a scaling operation, which corresponds to optical magnification or demagnification. The last one, $\mathbf{T}_O$, is an ortho-symplectic matrix (both orthogonal $\left(\mathbf{T}_O^t = \mathbf{T}_O^{-1}\right)$ and symplectic) [39, 40, 43]. The key to implementing an arbitrary LCT by using the Iwasawa decomposition above is the ortho-symplectic matrix, which corresponds to an optical phase-space rotator. If we know how to realize optical phase-space rotators, we can implement any desired LCT.

The design of an arbitrary phase-space rotator is significantly simplified by using the FRT. Indeed, any phase-space rotator can be written as an FRT, $\mathscr{F}(\gamma_x, \gamma_y)$, embedded between two ordinary image rotators: $\mathscr{R}(\beta)\,\mathscr{F}(\gamma_x, \gamma_y)\,\mathscr{R}(\alpha)$ [30]. Thus, ultimately, the design of arbitrary LCTs boils down to our ability to design arbitrary FRTs.

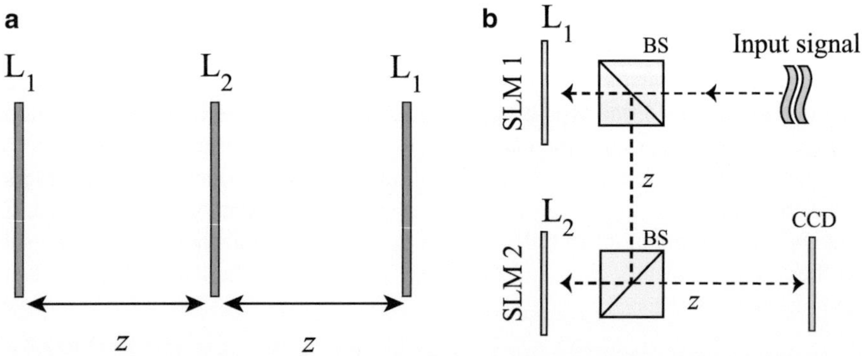

Fig. 6.6 (**a**) Optical system for the FRT using three generalized lenses separated by distance z. (**b**) Experimental implementation of a programmable optical FRT setup: two reflective phase-only SLMs are used to realize the generalized lenses $\mathbf{L}_1$ and $\mathbf{L}_2$. The output signal is registered by a CCD camera in real time [33]

Here we consider a flexible optical setup for the FRT that is suitable for use in many applications. In this setup, a change of the fractional angle γ_x or γ_y does not lead to an additional scaling and/or phase factor, that occurs in other proposed systems [16, 21, 35]. Specifically, this FRT system consists of three generalized lenses with a fixed distance z between them, as shown in Fig. 6.6a. The first and the last lens are identical ($\mathbf{L}_3 = \mathbf{L}_1$). Each generalized lens $\mathbf{L}_j$ ($j = 1, 2$) is an assembled set of two crossed cylindrical lenses, active in the two orthogonal directions x and y, with phase modulation functions $\exp[-i\pi g_x^{(j)} x^2/\lambda]$ and $\exp[-i\pi g_y^{(j)} y^2/\lambda]$, respectively, where we still have the possibility to choose a proper normalization parameter s. The lens powers $g_x^{(j)}$ and $g_y^{(j)}$ are given by [31]

$$
\begin{aligned}
g_x^{(1)}\, z &= 1 - (\lambda z/s)\, \cot(\gamma_x/2), \\
g_y^{(1)}\, z &= 1 - (\lambda z/s)\, \cot(\gamma_y/2), \\
g_x^{(2)}\, z &= 2 - (s/\lambda z)\, \sin\gamma_x, \\
g_y^{(2)}\, z &= 2 - (s/\lambda z)\, \sin\gamma_y.
\end{aligned}
\tag{6.29}
$$

The multiplication of the matrices corresponding to the constituent optical elements yields the FRT transformation matrix [31]. The cylindrical lenses are oriented such that $\phi_1^{(1,2)} = 0$ and $\phi_2^{(1,2)} = \pi/2$, where the angles are measured in the counterclockwise direction and $\phi = 0$ corresponds to the y axis. Using the matrix formalism it is easy to prove that the matrix of the composite system corresponds to the separable phase space rotator and therefore the relation of the complex field amplitudes at the input $f_i(\mathbf{r}_i)$ and output $f_o(\mathbf{r}_o) = \mathscr{F}(\gamma_x, \gamma_y) f_i(\mathbf{r}_i)$ planes are given by the separable FRT.

If we choose the normalization parameter as $s = 2\lambda z$, the lens powers are given by $g_x^{(1)} z = 1 - \cot(\gamma_x/2)/2$, $g_y^{(1)} z = 1 - \cot(\gamma_y/2)/2$, $g_x^{(2)} z = 2 - 2\sin\gamma_x$ and $g_y^{(2)} z = 2 - 2\sin\gamma_y$. Although γ_x or γ_y may take any value in the interval $(0, 2\pi)$, we use the interval $[\pi/2, 3\pi/2]$ because it corresponds to convergent lenses. This interval will be sufficient in most applications. Nevertheless, the entire interval $(0, 2\pi)$ can be covered, if necessary, thanks to the relation $F^{\gamma_x+\pi,\gamma_y+\pi}(\mathbf{r}) = F^{\gamma_x,\gamma_y}(-\mathbf{r})$.

The phase-space rotator $\mathscr{R}(-\alpha)\,\mathscr{F}(\gamma_x, \gamma_y)\,\mathscr{R}(\alpha)$ can be easily realized by rotating the above FRT system by an angle α around the optical axis [30]. In other words, the cylindrical lenses are now oriented according to the angles $\phi_1^{(1,2)} = \alpha$ and $\phi_2^{(1,2)} = \alpha + \pi/2$. Thus, the phase modulation function associated with each generalized lens $\mathbf{L}_j$ ($j = 1, 2$) takes the form

$$
\begin{aligned}
\Psi^{(j)}(x, y) = \exp&\left[-i\pi \frac{g_x^{(j)}}{\lambda}\, (x\cos\alpha - y\sin\alpha)^2\right] \\
\times \exp&\left[-i\pi \frac{g_y^{(j)}}{\lambda}\, (y\cos\alpha + x\sin\alpha)^2\right].
\end{aligned}
\tag{6.30}
$$

This optical configuration permits us to perform various attractive operations. For example, for $\alpha = 0$ we recover the basic FRT setup, whereas for $\alpha = \pi/4$ and $\gamma_x = -\gamma_y = \gamma$, the gyrator operation $\mathscr{R}(-\pi/4)\,\mathscr{F}(\gamma, -\gamma)\,\mathscr{R}(\pi/4)$ is obtained.

One way of implementing a generalized lens is to use a programmable SLM. This type of digital lens implementation allows one to modify the transformation angles $\alpha, \gamma_x, \gamma_y$ in real time. The corresponding optical setup is shown in Fig. 6.6b, where two reflective phase-only SLMs are used for the generalized lens implementation. Note that the third generalized lens is not required here because it only modulates the phase of the output beam, which will be recorded as an intensity image by a CCD camera. The feasibility of such a programmable setup has been demonstrated experimentally [33].

We note that for the special case $\gamma_x = -\gamma_y = \gamma$, the corresponding setup can also be built using glass cylindrical lenses (of fixed power) instead of digital lenses. This subclass of phase-space rotators include the gyrator and the antisymmetric FRT. In such a case, the generalized lens is an assembled set of two identical convergent cylindrical lenses, which are in contact with each other. The distance z between the generalized lenses $\mathbf{L}_j$ is fixed and the lens powers are set according to $g_x^{(j)} = j/z$ and $g_y^{(j)} = j/z$. Note that the first and last generalized lens are identical. While the transverse axes of the cylindrical lenses form angles $\phi_1^{(j)} = \varphi^{(j)} + \alpha + \pi/4$ and $\phi_2^{(j)} = -\varphi^{(j)} + \alpha - \pi/4$ with the y axis, note that the two cylindrical lenses cross at an angle $\phi_1^{(j)} - \phi_2^{(j)} = 2\varphi^{(j)} + \pi/2$. The angles $\varphi^{(1,2)}$ follow from $\sin 2\varphi^{(1)} = (\lambda z/s)\cot(\gamma/2)$ and $2\sin 2\varphi^{(2)} = (s/\lambda z)\sin\gamma$, where s is the normalization parameter. Because of the requirement $|\cot(\gamma/2)| \leq 1$, we conclude that the angle interval $\gamma \in [\pi/2, 3\pi/2]$ is covered if $\lambda z/s = 1$. This scheme (with normalization parameter $s = \lambda z$) has been used for the experimental realization of the gyrator (when $\alpha = 0$) reported in [32] and the antisymmetric fractional FT (when $\alpha = \pi/4$) reported in [10].

6.5 Conclusion

We reviewed some methods for optical implementation of one-dimensional and two-dimensional fractional Fourier transforms (FRTs) and linear canonical transforms (LCTs).

The systems we discussed are good for realizing arbitrary LCTs, which are a more general class of transforms than FRT. Thus, they can be specialized to obtain FRTs with desired orders and parameters as well.

We considered two main groups of approaches. The first is based on canonical decompositions and involves anamorphic sections of free space. The second is based on the modified Iwasawa decomposition and involves phase-space rotators.

LCTs represent a fairly general and important class of optical systems. Thus, their optical implementation is of interest for a variety of optical signal and image processing systems. In particular, these systems can be used for optical implementations of filtering in fractional Fourier or LCT domains [2, 13, 14, 26] and for optical mode converters [32, 33].

Acknowledgements H.M. Ozaktas acknowledges partial support of the Turkish Academy of Sciences.

References

1. T. Alieva, V. Lopez, F. Agullo Lopez, L.B. Almeida, The fractional Fourier transform in optical propagation problems. J. Mod. Opt. **41**, 1037–1044 (1994)
2. B. Barshan, M.A. Kutay, H.M. Ozaktas, Optimal filtering with linear canonical transformations. Opt. Commun. **135**, 32–36 (1997)
3. M.J. Bastiaans, The Wigner distribution function applied to optical signals and systems. Opt. Commun. **25**, 26–30 (1978)
4. M.J. Bastiaans, The Wigner distribution function and Hamilton's characteristics of a geometric-optical system. Opt. Commun. **30**(3), 321–326 (1979)
5. M.J. Bastiaans, Wigner distribution function and its application to first-order optics. J. Opt. Soc. Am. **69**, 1710–1716 (1979)
6. M.J. Bastiaans, Propagation laws for the second-order moments of the Wigner distribution function in first-order optical systems. Optik **82**, 173–181 (1989)
7. M.J. Bastiaans, Second-order moments of the Wigner distribution function in first-order optical systems. Optik **88**, 163–168 (1991)
8. L.M. Bernardo, O.D.D. Soares, Fractional Fourier transforms and imaging. J. Opt. Soc. Am. A. **11**, 2622–2626 (1994)
9. L.M. Bernardo, O.D.D. Soares, Fractional Fourier transforms and optical systems. Opt. Commun. **110**, 517–522 (1994)
10. A. Cámara, T. Alieva, J.A. Rodrigo, M.L. Calvo, Phase space tomography reconstruction of the Wigner distribution for optical beams separable in Cartesian coordinates. J. Opt. Soc. Am. A **26**(6), 1301–1306 (2009)
11. M.F. Erden, H.M. Ozaktas, A. Sahin, D. Mendlovic, Design of dynamically adjustable anamorphic fractional Fourier transformer. Opt. Commun. **136**, 52–60 (1997)
12. J. García, R.G. Dorsch, A.W. Lohmann, C. Ferreira, Z. Zalevsky, Flexible optical implementation of fractional Fourier transform processors. Applications to correlation and filtering. Opt. Commun. **133**, 393–400 (1997)
13. M.A. Kutay, H.M. Ozaktas, Optimal image restoration with the fractional Fourier transform. J. Opt. Soc. Am. A. **15**, 825–833 (1998)
14. M.A. Kutay, H.M. Ozaktas, O. Arıkan, L. Onural, Optimal filtering in fractional Fourier domains. IEEE Trans. Signal Process. **45**, 1129–1143 (1997)
15. A.W. Lohmann, Image rotation, Wigner rotation, and the fractional order Fourier transform. J. Opt. Soc. Am. A **10**, 2181–2186 (1993)
16. A.A. Malyutin, Tunable Fourier transformer of the fractional order. Quantum Electron. **36**, 79–83 (2006)
17. D. Mendlovic, H.M. Ozaktas, Fractional Fourier transforms and their optical implementation: I. J. Opt. Soc. Am. A **10**(9), 1875–1881 (1993)
18. D. Mendlovic, H.M. Ozaktas, A.W. Lohmann, Graded-index fibers, wigner-distribution functions, and the fractional Fourier transform. Appl. Opt. **33**, 6188–6193 (1994)
19. D. Mendlovic, Y. Bitran, R. G. Dorsch, C. Ferreira, J. Garcia, H.M. Ozaktas, Anamorphic fractional Fourier transform: optical implementation and applications. Appl. Opt. **34**, 7451–7456 (1995)

20. D. Mendlovic, H.M. Ozaktas, A.W. Lohmann, Fractional correlation. Appl. Opt. **34**(2), 303–309 (1995)
21. I. Moreno, C. Ferreira, M.M. Sánchez-López, Ray matrix analysis of anamorphic fractional Fourier systems. J. Opt. A Pure Appl. Opt. **8**(5), 427–435 (2006)
22. H.M. Ozaktas, M.F. Erden, Relationships among ray optical, gaussian beam, and fractional Fourier transform descriptions of first-order optical systems. Opt. Commun. **143**, 75–86 (1997)
23. H.M. Ozaktas, D. Mendlovic, Fourier transforms of fractional order and their optical interpretation. Opt. Commun. **101**, 163–169 (1993)
24. H.M. Ozaktas, D. Mendlovic, Fractional Fourier transforms and their optical implementation, II. J. Opt. Soc. Am. A **10**(12), 2522–2531 (1993)
25. H.M. Ozaktas, D. Mendlovic, Fractional Fourier optics. J. Opt. Soc. Am. A **12**, 743–751 (1995)
26. H.M. Ozaktas, B. Barshan, D. Mendlovic, L. Onural, Convolution, filtering, and multiplexing in fractional Fourier domains and their relation to chirp and wavelet transforms. J. Opt. Soc. Am. A **11**, 547–559 (1994)
27. H.M. Ozaktas, Z. Zalevsky, M.A. Kutay, *The Fractional Fourier Transform with Applications in Optics and Signal Processing* (Wiley, New York, 2001)
28. P. Pellat-Finet, Fresnel diffraction and fractional-order Fourier transform. Opt. Lett. **19**(18), 1388 (1994)
29. P. Pellat-Finet, G. Bonnet, Fractional order Fourier transform and Fourier optics. Opt. Commun. **111**, 141–154 (1994)
30. J.A. Rodrigo, T. Alieva, M.J. Bastiaans, *Phase-space rotators and their applications in optics*, chapter, in Optical and Digital Image Processing: Fundamentals and Applications (Wiley-VCH, Weinheim, 2011)
31. J.A. Rodrigo, T. Alieva, M.L. Calvo, Optical system design for ortho-symplectic transformations in phase space. J. Opt. Soc. Am. A **23**, 2494–2500 (2006)
32. J.A. Rodrigo, T. Alieva, M.L. Calvo, Experimental implementation of the gyrator transform. J. Opt. Soc. Am. A **24**(10), 3135–3139 (2007)
33. J.A. Rodrigo, T. Alieva, M.L. Calvo, Programmable two-dimensional optical fractional Fourier processor. Opt. Express **17**(7), 4976–4983 (2009)
34. A. Sahin, H.M. Ozaktas, D. Mendlovic, Optical implementation of the two-dimensional fractional Fourier transform with different orders in the two dimensions. Opt. Commun. **120**, 134–138 (1995)
35. A. Sahin, M.A. Kutay, H.M. Ozaktas, Nonseparable two-dimensional fractional Fourier transform. Appl. Opt. **37**(23), 5444–5453 (1998)
36. A. Sahin, H.M. Ozaktas, D. Mendlovic, Optical implementations of two-dimensional fractional Fourier transforms and linear canonical transforms with arbitrary parameters. Appl. Opt. **37**(11), 2130–2141 (1998)
37. B.E.A. Saleh, M.C. Teich, *Fundamentals of Photonics* (Wiley, New York, 1991)
38. R. Simon, N. Mukunda, Iwasawa decomposition in first-order optics: universal treatment of shape-invariant propagation for coherent and partially coherent beams. J. Opt. Soc. Am. A **15**(8), 2146–2155 (1998)
39. R. Simon, K.B. Wolf, Fractional Fourier transforms in two dimensions. J. Opt. Soc. Am. A **17**, 2368–2381 (2000)
40. R. Simon, K.B. Wolf, Structure of the set of paraxial optical systems. J. Opt. Soc. Am. A **17**, 342–355 (2000)
41. K. Sundar, N. Mukunda, R. Simon, Coherent-mode decomposition of general anisotropic Gaussian schell-model beams. J. Opt. Soc. Am. A **12**(3), 560–569 (1995)
42. K.B. Wolf, *Integral Transforms in Science and Engineering* (Plenum Press, New York, 1979)
43. K.B. Wolf. *Geometric Optics on Phase Space* (Springer, Berlin, 2004)

Part II
Discretization and Computation

Chapter 7
Linear Canonical Domains and Degrees of Freedom of Signals and Systems

Figen S. Oktem and Haldun M. Ozaktas

Abstract We discuss the relationships between linear canonical transform (LCT) domains, fractional Fourier transform (FRT) domains, and the space-frequency plane. In particular, we show that LCT domains correspond to scaled fractional Fourier domains and thus to scaled oblique axes in the space-frequency plane. This allows LCT domains to be labeled and monotonically ordered by the corresponding fractional order parameter and provides a more transparent view of the evolution of light through an optical system modeled by LCTs. We then study the number of degrees of freedom of optical systems and signals based on these concepts. We first discuss the bicanonical width product (BWP), which is the number of degrees of freedom of LCT-limited signals. The BWP generalizes the space-bandwidth product and often provides a tighter measure of the actual number of degrees of freedom of signals. We illustrate the usefulness of the notion of BWP in two applications: efficient signal representation and efficient system simulation. In the first application we provide a sub-Nyquist sampling approach to represent and reconstruct signals with arbitrary space-frequency support. In the second application we provide a fast discrete LCT (DLCT) computation method which can accurately compute a (continuous) LCT with the minimum number of samples given by the BWP. Finally, we focus on the degrees of freedom of first-order optical systems with multiple apertures. We show how to explicitly quantify the degrees of freedom of such systems, state conditions for lossless transfer through the system and analyze the effects of lossy transfer.

F.S. Oktem (✉)
Department of Electrical and Electronics Engineering, Middle East Technical University, 06800 Cankaya, Ankara, Turkey
e-mail: figeno@metu.edu.tr

H.M. Ozaktas
Department of Electrical Engineering, Bilkent University, 06800 Bilkent, Ankara, Turkey
e-mail: haldun@ee.bilkent.edu.tr

J.J. Healy et al. (eds.), *Linear Canonical Transforms*, Springer Series in Optical Sciences 198, DOI 10.1007/978-1-4939-3028-9_7

7.1 Introduction

Optical systems involving thin lenses, sections of free space in the Fresnel approximation, sections of quadratic graded-index media, and arbitrary combinations of any number of these are referred to as first-order optical systems or quadratic-phase systems [1–5]. Mathematically, such systems can be modeled as linear canonical transforms (LCTs), which form a three-parameter family of integral transforms [5, 6]. The LCT family includes the Fourier and fractional Fourier transforms (FRTs), coordinate scaling, chirp multiplication, and convolution operations as its special cases.

One of the most important concepts in Fourier analysis is the concept of the frequency (or Fourier) domain. This domain is understood to be a space where the frequency representation of the signal lives. Likewise, *fractional Fourier domains* are well understood to correspond to oblique axes in the space-frequency plane (phase space) [5, 7]. By analogy with this concept, the term *linear canonical domain* has been used in several papers to refer to the domain of the LCT representation of a signal [8–16]. Because LCTs are characterized by three independent parameters, LCT domains populate a three-parameter space, which makes them hard to visualize. In this chapter, we discuss the relationships between LCT domains, FRT domains, and the space-frequency plane. In particular, we show that each LCT domain corresponds to a scaled FRT domain, and thus to a scaled oblique axis in the space-frequency plane. Based on this many-to-one association of LCTs with FRTs, LCT domains can be labeled and monotonically ordered by the corresponding fractional order parameter, instead of their usual three parameters which do not directly lend to a natural ordering. This provides a more transparent view of the evolution of light through an optical system modeled by LCTs.

Another important concept is the number of degrees of freedom. For simplicity we focus on one-dimensional signals and systems, though most of our results can be generalized to higher dimensions in a straightforward manner. We first discuss the bicanonical width product, which is the number of degrees of freedom of LCT-limited signals. The conventional space-bandwidth product is of fundamental importance in signal processing and information optics because of its interpretation as the number of degrees of freedom of space- and band-limited signals [5, 17–32]. If, instead, a set of signals is highly confined to finite intervals in two arbitrary LCT domains, the space-frequency (phase space) support is a parallelogram. The number of degrees of freedom of this set of signals is given by the area of this parallelogram, which is equal to the BWP, which is usually smaller than the conventional space-bandwidth product. The BWP, which is a generalization of the space-bandwidth product, often provides a tighter measure of the actual number of degrees of freedom, and allows us to represent and process signals with fewer samples.

We illustrate the usefulness of the bicanonical width product in two applications: efficient signal representation and efficient system simulation. First, we show how to represent and reconstruct signals with arbitrary time- or space-frequency support, using fewer samples than required by the classical Shannon–Nyquist sampling theorem. Although the classical approach is optimal for band-limited signals, it is in general suboptimal for representing signals with a known space-frequency support. Based on the LCT sampling theorem, we provide a sub-Nyquist approach to represent signals with arbitrary space-frequency support. This approach geometrically amounts to enclosing the support with the smallest possible parallelogram, as opposed to enclosing it with a rectangle as in the classical approach. The number of samples required for reconstruction is given by the BWP, which is smaller than the number of samples required by the classical approach.

As a second application, we provide a fast discrete LCT (DLCT) computation method which can accurately compute a (continuous) LCT with the minimum number of samples given by the bicanonical width product. Hence the bicanonical width product is also a key parameter in fast discrete computation of LCTs, and hence in efficient and accurate simulation of optical systems.

Lastly, we focus on the degrees of freedom of apertured optical systems, which here refers to systems consisting of an arbitrary sequence of thin lenses and apertures separated by sections of free space. We define the space-frequency window (phase-space window) and show how it can be explicitly determined for such a system. Once the space-frequency window of the system is determined, the area of the window gives the maximum number of degrees of freedom that can be supported by the system. More significantly, it specifies which signals can pass through the system without information loss; the signal will pass losslessly if and only if the space-frequency support of the signal lies completely within this window. When it does not, the parts that lie within the window pass and the parts that lie outside of the window are blocked, a result which is valid to a good degree of approximation for most systems of practical interest. These intuitive results provide insight and guidance into the behavior and design of systems involving multiple apertures and can help minimize information loss.

In the next section, some preliminary material will be reviewed. In Sect. 7.3 we establish the relationships between LCT domains, FRT domains, and the space-frequency plane [33, 34]. The relationships between the space-frequency support, the bicanonical width product, and the number of degrees of freedom of signals is the subject of Sect. 7.4 [33–35]. We then provide a sub-Nyquist approach to represent signals with arbitrary space-frequency support in Sect. 7.5, which requires the number of samples to be equal to the bicanonical width product [33, 34, 36]. In Sect. 7.6 we review a fast DLCT computation method that works with this minimum number of samples [35]. Section 7.7 discusses how to explicitly quantify the degrees of freedom of optical systems with apertures and analyzes lossless and lossy transfer through them [33, 37]. We conclude in Sect. 7.8.

While in this chapter we usually refer to the independent variable in our signals as "space" and speak of the "space-frequency" plane due to the development of many of these concepts in an optical context, virtually all of our results are also valid when the independent variable is "time" or when we speak of the "time-frequency" plane.

7.2 Background

In this section we review some preliminary material that will be used throughout the chapter. This includes the definition and properties of LCTs and FRTs, the Iwasawa decomposition, space-frequency distributions, and the LCT sampling theorem.

7.2.1 Linear Canonical Transforms

Optical systems involving thin lenses, sections of free space in the Fresnel approximation, sections of quadratic graded-index media, and arbitrary combinations of any number of these are referred to as first-order optical systems or quadratic-phase systems. Mathematically, such systems can be modeled as LCTs. The output light field $f_{\mathbf{T}}(u)$ of a quadratic-phase system is related to its input field $f(u)$ through [5, 6]

$$f_{\mathbf{T}}(u) \equiv (\mathcal{C}_{\mathbf{T}}f)(u) \equiv \int_{-\infty}^{\infty} C_{\mathbf{T}}(u,u')f(u')\,du', \tag{7.1}$$

$$C_{\mathbf{T}}(u,u') \equiv \sqrt{\frac{1}{B}}\,e^{-i\pi/4}e^{i\pi\left(\frac{D}{B}u^2-2\frac{1}{B}uu'+\frac{A}{B}u'^2\right)},$$

for $B \neq 0$, where $\mathcal{C}_{\mathbf{T}}$ is the unitary LCT operator with parameter matrix $\mathbf{T} = [A\ B;\ C\ D]$ with $AD - BC = 1$. In the trivial case $B = 0$, the LCT is defined simply as $f_{\mathbf{T}}(u) \equiv \sqrt{D}\exp[i\pi CDu^2]f(Du)$. Sometimes the three real parameters $\alpha = D/B$, $\beta = 1/B$, $\gamma = A/B$ are used instead of the unit-determinant matrix $\mathbf{T}$ whose elements are A, B, C, D. (One of the four matrix parameters is redundant because of the unit-determinant condition.) These two sets of parameters are equivalent and either set of parameters can be obtained from the other [5, 6]:

$$\mathbf{T} = \begin{bmatrix} A & B \\ C & D \end{bmatrix} = \begin{bmatrix} \gamma/\beta & 1/\beta \\ -\beta+\alpha\gamma/\beta & \alpha/\beta \end{bmatrix}. \tag{7.2}$$

The transform matrix $\mathbf{T}$ is useful in the analysis of optical systems because if several systems are cascaded, the overall system matrix can be found by multiplying the corresponding matrices.

The Fourier transform, FRT (propagation through quadratic graded-index media), coordinate scaling (imaging), chirp multiplication (passage through a thin lens), and chirp convolution (Fresnel propagation in free space) are some of the special cases of LCTs.

The ath-order FRT [5] of a function $f(u)$, denoted by $f_a(u)$, can be defined as

$$f_a(u) \equiv (\mathcal{F}^a f)(u) \equiv \int_{-\infty}^{\infty} K_a(u,u')f(u')\,du', \tag{7.3}$$

$$K_a(u,u') \equiv A_\phi \, e^{i\pi\left(\cot\phi u^2 - 2\csc\phi uu' + \cot\phi u'^2\right)},$$

$$A_\phi = \sqrt{1 - i\cot\phi}, \quad \phi = a\pi/2$$

when $a \neq 2k$, and $K_a(u,u') = \delta(u - u')$ when $a = 4k$, and $K_a(u,u') = \delta(u + u')$ when $a = 4k + 2$, where k is an integer. The FRT operator $\mathcal{F}^a$ is additive in index: $\mathcal{F}^{a_2}\mathcal{F}^{a_1} = \mathcal{F}^{a_2+a_1}$ and reduces to the Fourier transform (FT) and identity operators for $a = 1$ and $a = 0$, respectively. The FRT is a special case of the LCT with parameter matrix

$$\mathbf{F}^a = \begin{bmatrix} \cos(a\pi/2) & \sin(a\pi/2) \\ -\sin(a\pi/2) & \cos(a\pi/2) \end{bmatrix}, \tag{7.4}$$

differing only by an inconsequential factor: $\mathcal{C}_{\mathbf{F}^a} f(u) = e^{-ia\pi/4}\mathcal{F}^a f(u)$ [5, 6].

Other than the FRT, another special case of the LCT is multiplication with a chirp function of the form $\exp[-i\pi q u^2]$, which corresponds to a thin lens in optics. The corresponding LCT matrix is given by

$$\mathbf{Q}_q = \begin{bmatrix} 1 & 0 \\ -q & 1 \end{bmatrix}. \tag{7.5}$$

Yet another special case is convolution with a chirp function of the form $e^{-i\pi/4}\sqrt{1/r}\exp[i\pi u^2/r]$, which is equivalent to propagation through a section of free space in the Fresnel approximation. The corresponding LCT matrix is given by

$$\mathbf{R}_r = \begin{bmatrix} 1 & r \\ 0 & 1 \end{bmatrix}. \tag{7.6}$$

The last special case we consider is the scaling operation, which maps a function $f(u)$ into $\sqrt{1/M} f(u/M)$ with $M > 0$. This is often used to model optical imaging. The transformation matrix is

$$\mathbf{M}_M = \begin{bmatrix} M & 0 \\ 0 & 1/M \end{bmatrix}. \tag{7.7}$$

7.2.2 *Iwasawa Decomposition*

An arbitrary LCT can be decomposed into an FRT followed by scaling followed by chirp multiplication [5, 34]:

$$\mathbf{T} = \begin{bmatrix} A & B \\ C & D \end{bmatrix} = \begin{bmatrix} 1 & 0 \\ -q & 1 \end{bmatrix} \begin{bmatrix} M & 0 \\ 0 & \frac{1}{M} \end{bmatrix} \begin{bmatrix} \cos\phi & \sin\phi \\ -\sin\phi & \cos\phi. \end{bmatrix} \tag{7.8}$$

The three matrices, respectively, correspond to the transformation matrices of chirp multiplication with parameter q (multiplication by $\exp(-i\pi q u^2)$), coordinate scaling with factor $M > 0$ (mapping of $f(u)$ into $\sqrt{1/M}f(u/M)$), and ath order FRT with $\phi = a\pi/2$ (transformation of $f(u)$ into $f_a(u)$). The decomposition can be written more explicitly in terms of the LCT and FRT domain representations of the signal as

$$f_{\mathbf{T}}(u) = \exp\left(-i\pi q u^2\right)\sqrt{\frac{1}{M}}\, f_a\left(\frac{u}{M}\right). \tag{7.9}$$

This decomposition is a special case of the Iwasawa decomposition [38–40]. (For a discussion of the implications of this decomposition to the propagation of light through first-order optical systems, see [34, 41]. For a discussion of the implications for sampling optical fields, see [42, 43].) By appropriately choosing the three parameters a, M, q, the above equality can be satisfied for any $\mathbf{T} = [A\ B;\ C\ D]$ matrix. Solving for a, M, q in (7.8), we obtain the decomposition parameters in terms of the matrix entries A, B, C, D:

$$a = \begin{cases} \frac{2}{\pi}\arctan\left(\frac{B}{A}\right), & \text{if } A \geq 0 \\ \frac{2}{\pi}\arctan\left(\frac{B}{A}\right) + 2, & \text{if } A < 0 \end{cases} \tag{7.10}$$

$$M = \sqrt{A^2 + B^2}, \tag{7.11}$$

$$q = \begin{cases} -\frac{C}{A} - \frac{B/A}{A^2+B^2}, & \text{if } A \neq 0 \\ -\frac{D}{B}, & \text{if } A = 0. \end{cases} \tag{7.12}$$

The range of the arctangent lies in $(-\pi/2, \pi/2]$.

7.2.3 *Space-Frequency Distributions*

The Wigner distribution (WD) $W_f(u, \mu)$ of a signal $f(u)$ is a space-frequency (phase-space) distribution that gives the distribution of signal energy over space and frequency, and is defined as [5, 44–46]:

$$W_f(u, \mu) = \int_{-\infty}^{\infty} f(u + u'/2) f^*(u - u'/2) e^{-i2\pi\mu u'}\, du'. \tag{7.13}$$

We refer to the space-frequency region for which the Wigner distribution is considered non-negligible as the space-frequency support of the signal, with the area of this region giving the number of degrees of freedom [5]. A large percentage of the signal energy is confined to the space-frequency support.

All quadratic-phase systems result in an area-preserving geometric transformation in the u-μ plane. Explicitly, the WD of $f_{\mathbf{T}}(u)$ can be related to the WD of $f(u)$ by a linear distortion [5]:

$$W_{f_{\mathbf{T}}}(u, \mu) = W_f(Du - B\mu, -Cu + A\mu). \tag{7.14}$$

The Jacobian of this coordinate transformation is equal to the determinant of the matrix $\mathbf{T}$, which is unity. Therefore, this coordinate transformation will geometrically distort the support region of the WD but the support area (hence the number of degrees of freedom) will remain unchanged.

7.2.4 LCT Sampling Theorem

Just as the LCT is a generalization of the Fourier transform, the LCT sampling theorem [9, 47, 48] is an extension of the classical sampling theorem. According to the LCT sampling theorem, if a function $f(u)$ has an LCT with parameter $\mathbf{T}$ which has a compact support such that $f_{\mathbf{T}}(u)$ is zero outside the interval $[-\Delta u_{\mathbf{T}}/2, \Delta u_{\mathbf{T}}/2]$, then the function $f(u)$ can be reconstructed from its samples taken at intervals $\delta u \leq 1/(|\beta|\Delta u_{\mathbf{T}})$. The reconstruction formula, which we will refer to as the LCT interpolation formula, is given by

$$f(u) = \delta u\,|\beta|\,\Delta u_{\mathbf{T}}\,e^{-i\pi\gamma u^2} \sum_{n=-\infty}^{\infty} f(n\,\delta u)\,e^{i\pi\gamma(n\,\delta u)^2}\,\mathrm{sinc}(\beta\,\Delta u_{\mathbf{T}}(u - n\,\delta u)). \tag{7.15}$$

This reduces to the classical sampling theorem, and to the FRT sampling theorem [49–54] when the parameter matrix $\mathbf{T}$ is replaced with the associated matrices of the FT and FRT operations.

The background material presented in this section employs dimensionless variables and parameters, for simplicity and purity. We assume that a dimensional normalization has been performed and that the coordinates appearing in the definitions of the FRT, LCT, Wigner distribution, etc., are all dimensionless quantities [5]. In Sect. 7.7, however, we will prefer to employ variables with real physical dimensions. There, we will present dimensional counterparts of the background material that we will need. The reader will be able to employ these to obtain dimensional counterparts of other results in this chapter, should the need arise.

7.3 LCT Domains

Because LCTs are characterized by three independent parameters, LCT domains populate a three-parameter space, which makes them hard to visualize. In this section, we discuss the relationships between LCT domains, FRT domains, and the space-frequency plane. In particular, we show that each LCT domain corresponds to a scaled FRT domain, and thus to a scaled oblique axis in the space-frequency plane. This provides a more transparent view of the evolution of light through an optical system modeled by LCTs.

7.3.1 Relationship of LCT Domains to the Space-Frequency Plane

One of the most important concepts in Fourier analysis is the concept of the frequency (or Fourier) domain. This domain is understood to be a space where the frequency representation of the signal lives. Likewise, *fractional Fourier domains* are well understood to correspond to oblique axes in the space-frequency plane (phase space) [5, 7], since the FRT has the effect of rotating the space-frequency (phase space) representation of a signal. More explicitly, the effect of ath-order fractional Fourier transformation on the Wigner distribution of a signal is to rotate the Wigner distribution by an angle $\phi = a\pi/2$ [7, 55, 56]:

$$W_{f_a}(u,\mu) = W_f(u\cos\phi - \mu\sin\phi, u\sin\phi + \mu\cos\phi). \tag{7.16}$$

The Radon transform operator $\mathcal{RDN}_\phi$, which takes the integral projection of the Wigner distribution of $f(u)$ onto an axis making an angle ϕ with the u axis, can be used to restate this property in the following manner [5]:

$$\{\mathcal{RDN}_\phi[W_f(u,\mu)]\}(u_a) = |f_a(u_a)|^2, \tag{7.17}$$

where u_a denotes the axis making angle $\phi = a\pi/2$ with the u axis. That is, projection of the Wigner distribution of $f(u)$ onto the u_a axis gives $|f_a(u_a)|^2$, the squared magnitude of the ath order FRT of the function. Hence, the projection axis u_a can be referred to as the ath-order fractional Fourier domain (see Fig. 7.1) [7, 55]. The space and frequency domains are merely special cases of the continuum of fractional Fourier domains.

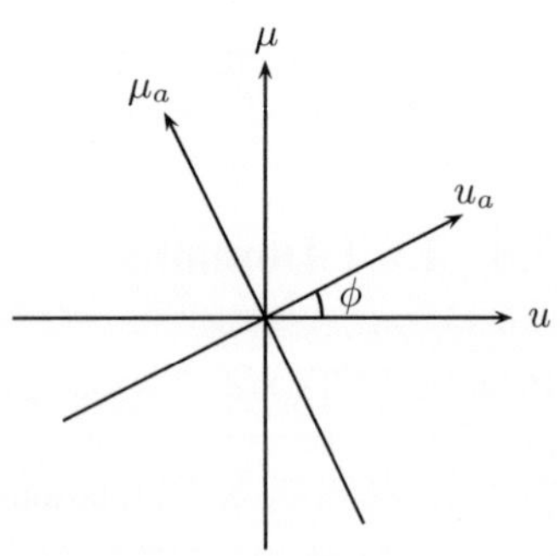

Fig. 7.1 The ath-order fractional Fourier domain [34]

Fractional Fourier domains are recognized as oblique axes in the space-frequency plane [5–7]. By analogy with this concept, the term *linear canonical domain* has been used in several papers to refer to the domain of the LCT representation of a signal [8–16]. However, it is not immediately obvious from these works where these LCT domains exist and how they are related to the space-frequency plane; in other words, while the effect of an LCT on the space-frequency representation of a signal is well understood as a linear geometrical distortion, it is not immediate how members of the three-parameter family of LCT domains are related to the space-frequency plane, or how we should visualize them. LCTs are characterized by three independent parameters, and hence LCT domains populate a three-parameter space, which makes them hard to visualize. Below, we explicitly relate LCT domains to the space-frequency plane [33, 34]. We show that *each LCT domain corresponds to a scaled FRT domain, and thus to a scaled oblique axis in the space-frequency plane*. Based on this many-to-one association of LCTs with FRTs, LCT domains can be labeled and monotonically ordered by an associated fractional order parameter, instead of their usual three parameters which do not directly lend to a natural ordering.

We use the Iwasawa decomposition to relate the members of the three-parameter family of LCT domains to the space-frequency plane. As given in (7.9), any arbitrary LCT can be expressed as a chirp multiplied and scaled FRT. Thus, in order to compute the LCT of a signal, we can first compute the ath-order FRT of the signal, which transforms the signal to the ath-order fractional Fourier domain. Secondly, we scale the transformed signal. Because scaling is a relatively trivial operation, we need not interpret it as changing the domain of the signal, but merely a scaling of the coordinate axis in the same domain. Finally, we multiply the resulting signal with a chirp to obtain the LCT. Multiplication with a function is not considered an operation which transforms a signal to another domain, but which alters the signal in the same domain. (For instance, when we multiply the Fourier transform of a function with a mask, the result is considered to remain in the frequency domain.) Therefore, only the FRT part of the LCT operation corresponds to a genuine domain change, and the linear canonical transformed signal essentially lives in a scaled fractional Fourier domain. In other words, *LCT domains are essentially equivalent to scaled fractional Fourier domains.* That is, despite their three parameters, LCT domains do not constitute a richer family of domains than FRT domains. By using the well-known relationship of FRT domains to the space-frequency plane, we can state a similar relationship for the LCT domains as follows:

> In the space-frequency plane, the LCT domain $u_{\mathbf{T}}$ with parameter matrix $\mathbf{T} = [A\ B; C\ D]$ corresponds to a scaled oblique axis making angle $\arctan(B/A)$ with the u axis (or equivalently, having slope B/A), and scaled with the parameter M where $M = \sqrt{A^2 + B^2}$ [33, 34].

Note that any LCT domain is completely characterized by the two parameters A and B (or equivalently by a and M, or by γ and β) instead of all three of its parameters.

We also note that the relation in (7.17) can be rewritten for the LCT of a signal as

$$\frac{1}{M}\{\mathcal{RDN}_\phi[W_f(u,\mu)]\}\left(\frac{u}{M}\right) = |f_{\mathbf{T}}(u)|^2, \tag{7.18}$$

by using (7.17) and (7.9) (with $s = 1$). This is another way of interpreting scaled oblique axes in the space-frequency plane as the LCT domain with parameter $\mathbf{T}$.

7.3.2 Essentially Equivalent Domains

Observe that LCTs with the same value of B/A (or equivalently the same value of γ) will have the same value of a in the decomposition in (7.10), and therefore will be associated with the same FRT domain. We refer to such LCT domains as well as their associated FRT domain as *essentially equivalent domains* [33, 34]. Note that if a signal has a compact support in a certain LCT domain, then the signal will also have compact support in all essentially equivalent domains.

The concept of essentially equivalent domains we introduce allows many earlier observations and results to be seen in a new light, making them almost obvious or more transparent. For instance, it has been stated that if a particular LCT of a signal is bandlimited, then another LCT of the signal cannot be bandlimited unless $B_1/A_1 = B_2/A_2$ [10]. Since we recognize the domains associated with two LCTs satisfying this relation to be essentially equivalent, this result becomes obvious. Although we will not further elaborate, other results regarding the compactness/bandlimitedness of different LCTs of a signal [57] can be likewise easily understood in terms of the concept of essentially equivalent domains. As a final example, we consider the LCT sampling theorem, according to which if the LCT of a signal has finite extent $\Delta u_{\mathbf{T}}$, then we should sample it with spacing $\Delta u \leq |B|/\Delta u_{\mathbf{T}}$. Such a sampling scheme collapses when $B = 0$. It is easy to understand why if we note that $B = 0$ implies that the LCT domain in question is essentially equivalent to the $a = 0$th FRT domain; that is, the domain in which the signal is specified to have finite extent is essentially equivalent to the domain in which we are attempting to sample the signal.

7.3.3 Optical Interpretation

Let us now optically interpret the equivalence of LCT domains to FRT domains. Consider a signal that passes through an arbitrary quadratic-phase system. Since the light field at any plane within the system is related to the input field through an LCT, the signal will incrementally be transformed through different LCT domains. Because the three parameters of the consequential LCT domains are not sequenced, it is not easy to give any interpretation or visualize the nature of the transformation of the optical field. However, if we think of the LCT domains as being equivalent to scaled FRT domains, it becomes possible to interpret every location along the propagation axis as an FRT domain of specific order, which is equivalent to an oblique axis in the space-frequency plane. Moreover, it has been shown that if we take the fractional order a to be equal to zero at the input of the system, then a monotonically increases as a function of the distance along the optical axis [41, 58]. In other words, propagation through a quadratic-phase system can be understood as passage through a continuum of scaled FRT domains of monotonically increasing order, instead of passage through an unsequenced plethora of LCT domains [34].

To see that the FRT parameter a is monotonically increasing along the z axis, observe from Eq. (7.10) that $a \propto \arctan(B/A)$, so that a increases with B/A. Passage through a lens involves multiplication with the matrix given in Eq. (7.5) which does not change B/A. Passage through an incremental section of free space involves multiplication with the matrix given in Eq. (7.6) which always results in a positive increment in a. This is because r is proportional to the distance of propagation, and the derivative of the new value of B/A with respect to r is always positive, which implies that B/A always increases with r. A similar argument is possible for quadratic graded-index media. A more precise development may be found in [41, 58].

Therefore, the distribution of light is continually fractional Fourier transformed through scaled fractional Fourier domains of increasing order, which we know are oblique axes in the space-frequency plane. This understanding of quadratic-phase systems yields much more insight into the nature of how light is transformed as it propagates through such a system, as opposed to thinking of it in terms of going through a series of unsequenced LCT domains whose whereabouts we cannot visualize. For example, based on this understanding we show in Sect. 7.7 how to explicitly quantify the degrees of freedom of optical systems with apertures and give conditions for lossless transfer.

7.4 Degrees of Freedom of Signals and the Bicanonical Width Product

The conventional space-bandwidth product is of fundamental importance in signal processing and information optics because of its interpretation as the number of degrees of freedom of space- and band-limited signals [5, 17–32]. In this section, we discuss the *bicanonical width product (BWP)*, which is the number of degrees of freedom of LCT-limited signals. The bicanonical width product generalizes the space-bandwidth product and often provides a tighter measure of the actual number of degrees of freedom of signals [33–35].

7.4.1 Space-Bandwidth Product: Degrees of Freedom of Space- and Band-Limited Signals

Consider a family of signals whose members are approximately confined to an interval of length Δu in the space domain and to an interval of length $\Delta\mu$ in the frequency domain in the sense that a large percentage of the signal energy is confined to these intervals. The space-bandwidth product N is then defined [5, 28] as

$$N \equiv \Delta u \Delta\mu, \tag{7.19}$$

and is always greater than or equal to unity because of the uncertainty relation.

The notion of space-bandwidth product, as degrees of freedom of space- and band-limited signals, can be established in a number of different ways. Here we provide two constructions: one based on Fourier sampling theorem, another based on space-frequency analysis.

7.4.1.1 Construction Based on Fourier Sampling Theorem

The conventional space-bandwidth product is the minimum number of samples required to uniquely identify a signal out of all possible signals whose energies are approximately confined to space and frequency intervals of length Δu and $\Delta\mu$. This argument is based on the Shannon–Nyquist sampling theorem, which requires that the spacing between samples (in the space domain) not be greater than $\delta u = 1/\Delta\mu$, so that the minimum number of samples over the space extent Δu is given by $\Delta u/\delta u = \Delta u\Delta\mu$. Alternatively, if we sample the signal in the frequency domain, the spacing between samples should not be greater than $\delta\mu = 1/\Delta u$, so that the minimum number of samples over the frequency extent $\Delta\mu$ is given by $\Delta\mu/\delta\mu = \Delta u\Delta\mu$. The minimum number of samples needed to fully characterize an approximately space- and band-limited signal can be interpreted as the number of degrees of freedom of the set of signals. This number of samples turns out to be the same whether counted in the space or frequency domain, and is given by the space-bandwidth product.

7.4.1.2 Construction Based on Space-Frequency Analysis

Another line of development involves space-frequency analysis. When the approximate space and frequency extents are specified as above, this amounts to assuming that most of the energy of the signal is confined to a $\Delta u \times \Delta\mu$ rectangular region in the space-frequency plane, perpendicular to the space-frequency axes (Fig. 7.2). In this case, the area of this rectangular region, which gives the number of degrees of freedom, is equal to the space-bandwidth product.

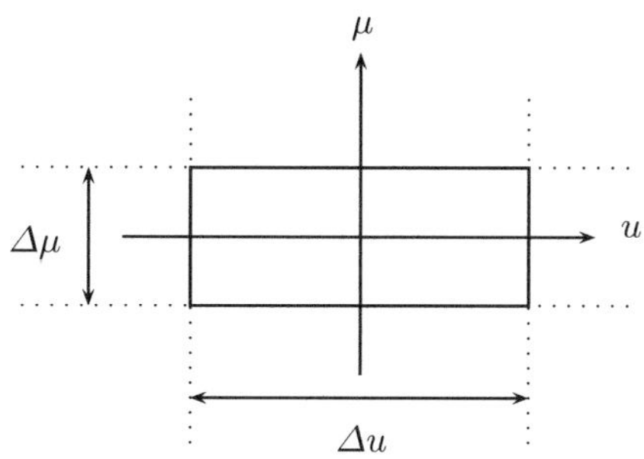

Fig. 7.2 Rectangular space-frequency support with area equal to the space-bandwidth product $\Delta u \Delta\mu$ [34]

More generally, the number of degrees of freedom is given by the area of the space-frequency support (phase space support), regardless of its shape [5, 30]. When the space-frequency support is not a rectangle perpendicular to the axes, the actual number of degrees of freedom will be smaller than the space-bandwidth product of the signal [5, 30].

7.4.1.3 Discussion

The space-bandwidth product is a notion originating from the simultaneous specification of the space and frequency extents. Although this product is commonly seen as an intrinsic property, it is in fact a notion that is specific to the Fourier transform and the frequency domain. It is also possible to specify the extents in other FRT or LCT domains. The set of signals thus specified will in general exhibit a nonrectangular space-frequency support. (For example, we will next show that when two such extents are specified, the support will be a parallelogram [33, 34].) In all cases, the area of the support will correspond to the number of degrees of freedom of the set of signals thus defined. If we insist on characterizing this set of signals with conventional space and frequency extents, the space-bandwidth product will overstate the number of degrees of freedom (see Fig. 7.3).

Obviously, specifying a finite extent in a single LCT domain does not define a family of signals with a finite number of degrees of freedom, just as specifying a finite extent in only one of the conventional space or frequency domains does not. However, specifying finite extents in two distinct LCT domains allows us to define a family of signals with a finite number of degrees of freedom. The number of degrees of freedom will depend on both the specified LCT domains and the extents in those domains.

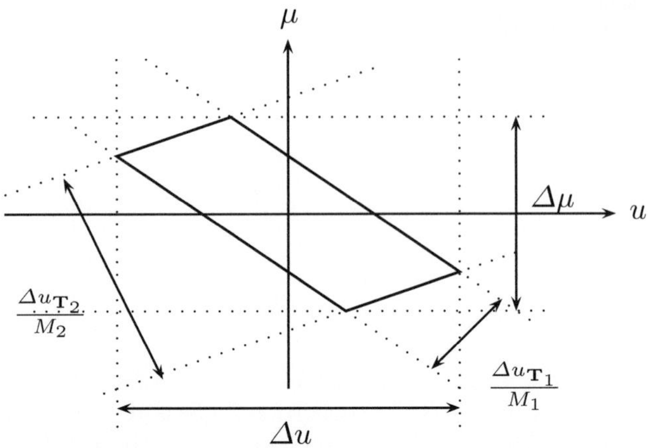

Fig. 7.3 Parallelogram shaped space-frequency support with area equal to the bicanonical width product $\Delta u_{\mathbf{T}_1} \Delta u_{\mathbf{T}_2} |\beta_{1,2}|$, which is smaller than the space-bandwidth product $\Delta u \Delta \mu$ [34]

7.4.2 Bicanonical Width Product: Degrees of Freedom of LCT-Limited Signals

We first define the *space-canonical width product*, which gives the number of degrees of freedom of signals that are approximately confined to a finite interval Δu in the conventional space domain and to a finite interval $\Delta u_{\mathbf{T}}$ in some other LCT domain [33–35]:

$$N \equiv \Delta u \Delta u_{\mathbf{T}} |\beta|. \tag{7.20}$$

This is always greater than or equal to unity because of the uncertainty relation for LCTs [5, 6, 10, 15]. Here $\mathbf{T}$ represents the three parameters of the LCT, where β is one of these three parameters. The space-canonical width product constitutes a generalization of the space-bandwidth product, and reduces to it when the LCT reduces to an ordinary Fourier transform, upon which $\Delta u_{\mathbf{T}}$ reduces to $\Delta \mu$ and $\beta = 1$.

In the above, one of the two domains is chosen to be the conventional space domain. More generally, the two LCT domains can both be arbitrarily chosen. In this case, we use the more general term *bicanonical width product (BWP)* to refer to the product [33–35]

$$N \equiv \Delta u_{\mathbf{T}_1} \Delta u_{\mathbf{T}_2} |\beta_{1,2}|, \tag{7.21}$$

where $\Delta u_{\mathbf{T}_1}$ and $\Delta u_{\mathbf{T}_2}$ are the extents of the signal in two LCT domains and $\beta_{1,2}$ is the parameter of the LCT between these two domains (the LCT which transforms the signal from the first LCT domain to the second). Note that the bicanonical width product is defined with respect to two specific LCT domains.

The notion of bicanonical width product as the degrees of freedom of LCT-limited signals can also be established in two ways [34]: based on the LCT sampling theorem, and based on space-frequency analysis. Before establishing this, we note that if $\beta = \infty$ in (7.20) or $\beta_{1,2} = \infty$ in (7.21), then the product N will not be finite and hence the number of degrees of freedom will not be bounded. This is because when this parameter is infinity (that is, $B = 0$), the two domains are related to each other simply by a scaling or chirp multiplication operation. But as discussed before, domains related by such operations are essentially equivalent. Thus, specification of the extent in two such domains does not constrain the family of signals more than the specification of the extent in only one domain, which, as noted, is not sufficient to make the number of degrees of freedom finite.

7.4.2.1 Construction Based on LCT Sampling Theorem

The space-canonical width product is the minimum number of samples required to uniquely identify a signal out of all possible signals whose energies are approximately confined to a space interval of Δu and a particular LCT interval of Δu_{T}. (This many number of samples can be used to reconstruct the signal.) This argument can be justified by the use of the LCT sampling theorem. According to the LCT sampling theorem, the space-domain sampling interval for a signal that has finite extent Δu_{T} in a particular LCT domain should not be larger than $\delta u = 1/(|\beta|\Delta u_{\mathrm{T}})$. If we sample the space-domain signal at this rate, the total number of samples over the extent Δu will be given by $\Delta u/\delta u = \Delta u \Delta u_{\mathrm{T}}|\beta|$, which is precisely equal to the space-canonical width product. Alternatively, if we sample in the LCT-domain, the sampling interval should not be larger than $\delta u_{\mathrm{T}} = 1/(|\beta|\Delta u)$. Sampling at this rate, the total number of samples over the LCT extent Δu_{T} is given by $\Delta u_{\mathrm{T}}/\delta u_{\mathrm{T}} = \Delta u \Delta u_{\mathrm{T}}|\beta|$, which once again is the space-canonical width product.

The derivation above can be easily replicated for the more general bicanonical width product defined in (7.21). Therefore, *the bicanonical width product can also be interpreted as the minimum number of samples required to uniquely identify a signal out of all possible signals whose energies are approximately confined to finite intervals in two specified LCT domains, and therefore as the number of degrees of freedom of this set of signals [33–35].*

7.4.2.2 Construction Based on Space-Frequency Analysis

Another line of development involves space-frequency analysis. Here we show that when the extents are specified in two LCT domains as above, the space-frequency support becomes a parallelogram (see Fig. 7.4), and the area of this parallelogram, which gives the number of degrees of freedom, is equal to the bicanonical width product [33, 34].

This result follows from the established relationship of LCT domains to the space-frequency plane. Let us consider a set of signals, whose members are approximately confined to the intervals $[-\Delta u_{\mathbf{T}_1}/2, \Delta u_{\mathbf{T}_1}/2]$ and $[-\Delta u_{\mathbf{T}_2}/2, \Delta u_{\mathbf{T}_2}/2]$ in two given LCT domains, $u_{\mathbf{T}_1}$ and $u_{\mathbf{T}_2}$. We want to investigate the space-frequency support of this set of signals. Since LCT domains are equivalent to scaled fractional Fourier domains, each finite interval in an LCT domain will correspond to a scaled interval in the equivalent FRT domain. To see this explicitly, we again refer to (7.9), which implies that if $f_{\mathbf{T}}(u)$ is confined to an interval of length $\Delta u_{\mathbf{T}}$, so is $f_a(u/M)$. Therefore, the extent of $f_a(u)$ in the equivalent ath-order FRT domain is $\Delta u_{\mathbf{T}}/M$. Thus, the set of signals in question is approximately limited to an extent of $\Delta u_{\mathbf{T}_1}/M_1$ in the a_1th order FRT domain, and an extent of $\Delta u_{\mathbf{T}_2}/M_2$ in the a_2th order FRT domain, where a_1, a_2 and M_1, M_2 are related to $\mathbf{T}_1$, $\mathbf{T}_2$ through Eqs. (7.10) and (7.11).

It is well known that if the space-, frequency- or FRT-domain representation of a signal is identically zero (or negligible) outside a certain interval, so is its Wigner distribution [5, 59]. As a direct consequence of this fact, the Wigner distribution of our set of signals is confined to corridors of width $\Delta u_{\mathbf{T}_1}/M_1$ and $\Delta u_{\mathbf{T}_2}/M_2$ in the directions orthogonal to the a_1th order FRT domain u_{a_1}, and the a_2th order FRT domain u_{a_2}, respectively. (With the term corridor we are referring to an infinite strip in the space-frequency plane perpendicular to the oblique u_a axis. The corridor makes an angle $(a+1)\pi/2$ with the u axis (see Fig. 7.5).) Now, if we intersect the two corridors defined by each extent, we obtain a parallelogram, which gives the space-frequency support of the signals (see Fig. 7.4). The area of the parallelogram is equal to the bicanonical width product of the set of signals in question. This result will be formally stated as follows:

Consider a set of signals, whose members are approximately confined to finite extents $\Delta u_{\mathbf{T}_1}$ and $\Delta u_{\mathbf{T}_2}$ in the two LCT domains $u_{\mathbf{T}_1}$ and $u_{\mathbf{T}_2}$, respectively. Let $\beta_{1,2}$ denote the β parameter of the LCT which transforms signals from the first LCT domain to the second. Then, the space-frequency support of these signals is given by a parallelogram defined by these extents (Fig. 7.4), and the area

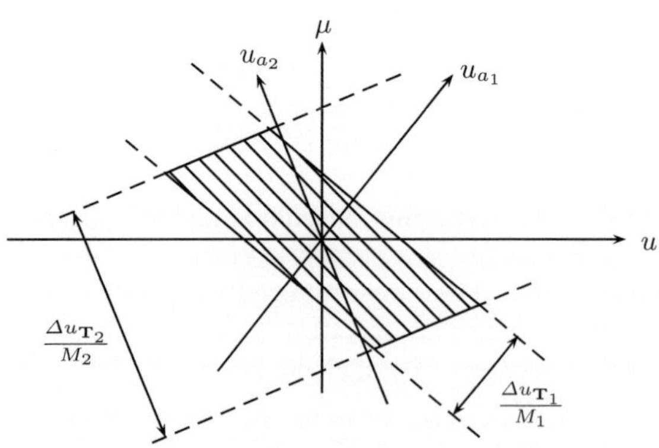

Fig. 7.4 The space-frequency support when finite extents are specified in two LCT domains. The area of the parallelogram is equal to $\Delta u_{\mathbf{T}_1}\Delta u_{\mathbf{T}_2}|\beta_{1,2}|$ [34]

of the parallelogram-shaped support is equal to the bicanonical width product $\Delta u_{\mathbf{T}_1} \Delta u_{\mathbf{T}_2} |\beta_{1,2}|$ *of the set of signals [33, 34].*

Proof. The two heights of the parallelogram defined by the extents $\Delta u_{\mathbf{T}_1}$ and $\Delta u_{\mathbf{T}_2}$, are $\Delta u_{\mathbf{T}_1}/M_1$ and $\Delta u_{\mathbf{T}_2}/M_2$, corresponding to the widths of the corridors. Moreover, the angle between the corridors is $\phi_2 - \phi_1$. Then, the area of the parallelogram is

$$\text{Area} = \frac{\Delta u_{\mathbf{T}_1}}{M_1} \frac{\Delta u_{\mathbf{T}_2}}{M_2} |\csc(\phi_2 - \phi_1)| \tag{7.22}$$

$$= \frac{\Delta u_{\mathbf{T}_1} \Delta u_{\mathbf{T}_2}}{M_1 M_2 |\sin\phi_2 \cos\phi_1 - \cos\phi_2 \sin\phi_1|} \tag{7.23}$$

$$= \frac{\Delta u_{\mathbf{T}_1} \Delta u_{\mathbf{T}_2}}{|A_1 B_2 - B_1 A_2|} \tag{7.24}$$

$$= \Delta u_{\mathbf{T}_1} \Delta u_{\mathbf{T}_2} \frac{|\beta_1 \beta_2|}{|\gamma_1 - \gamma_2|} \tag{7.25}$$

$$= \Delta u_{\mathbf{T}_1} \Delta u_{\mathbf{T}_2} |\beta_{1,2}|, \tag{7.26}$$

where the third and fourth equality follows from (7.8) and (7.2), respectively. The final result can be obtained from the parameter matrix $\mathbf{T}_2 \mathbf{T}_1^{-1}$ which transforms from the first LCT domain to the second domain. □

Since the number of degrees of freedom of a set of signals is given by the area of their space-frequency support, this result provides further justification for interpreting the bicanonical width product as the number of degrees of freedom of LCT-limited signals.

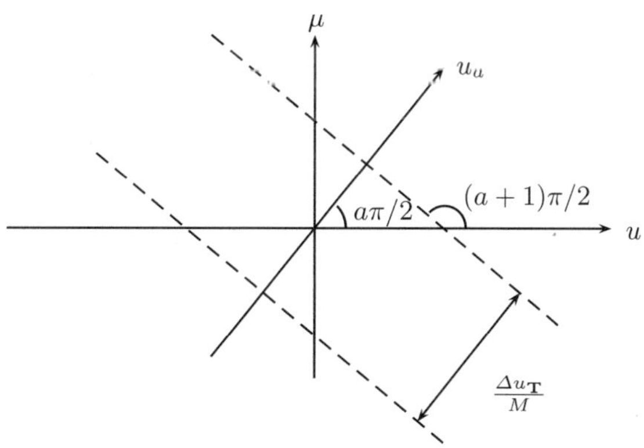

Fig. 7.5 Illustration of a space-frequency corridor [37]

7.4.2.3 Discussion

When confronted with a space-frequency support of arbitrary shape, it is quite common to assume the number of degrees of freedom to be equal to the space-bandwidth product, without regard to the shape of its space-frequency support. In reality, this is a worst-case approach which encloses the arbitrary shape within a rectangle perpendicular to the axes, and overstates the number of degrees of freedom.

The bicanonical width product provides a tighter measure of the number of degrees of freedom than the conventional space-bandwidth product, and allows us to represent and process the signals with a smaller number of samples, since it is possible to enclose the true space-frequency support more tightly with a parallelogram of our choice, as compared to a rectangle perpendicular to the axes, or indeed any rectangle. In applications where the underlying physics involves LCT type integrals (as is the case with many wave propagation problems and optical systems), parallelograms may be excellently, if not perfectly, tailored to the true space-frequency supports of the signals. In the next section, we illustrate how these ideas are useful for representing and reconstructing signals with arbitrary time- or space-frequency support, using fewer samples than required by the Shannon–Nyquist sampling theorem. The developed approach geometrically amounts to enclosing the support with the smallest possible parallelogram, as opposed to enclosing it with a rectangle as in the classical approach.

Another important feature of the bicanonical width product is that it is invariant under linear canonical transformation. The fact that LCTs model an important family of optical systems, makes the bicanonical width product a suitable invariant measure for the number of degrees of freedom of optical signals. On the other hand, the space-bandwidth product, which is the area of the smallest bounding perpendicular rectangle, may change significantly after linear canonical transformation. This has an important implication in DLCT computation as will be discussed in Sect. 7.6. With this computation method, we can accurately compute an LCT with a minimum number of samples given by the bicanonical width product, so that the bicanonical width product is also a key parameter in fast discrete computation of LCTs, and hence in efficient and accurate simulation of optical systems [35].

Given the fundamental importance of the conventional space-bandwidth product in signal processing and information optics, it is not surprising that the bicanonical width product can also play an important role in these areas. In a later section, we discuss how the bicanonical width product is useful for efficiently and accurately simulating optical systems based on an elegant and natural formulation of DLCT computation. Finally we note that the bicanonical width product has been originally introduced in the context of LCTs [33, 35]. However, since the equivalence between FRT and LCT domains has been shown [34], we can also speak of the *bifractional width product* in the context of FRT domains.

7.5 Sub-Nyquist Sampling and Reconstruction of Signals

In this section, we show how to represent and reconstruct signals with arbitrary time- or space-frequency support, using fewer samples than required by the Shannon–Nyquist sampling theorem [33, 36]. The classical Shannon–Nyquist sampling theorem allows us to represent band-limited signals with samples taken at a finite rate. Although the classical approach is optimal for band-limited signals, it is in general suboptimal for representing signals with a known space-frequency support. Application of the classical approach to signals with arbitrarily given space-frequency support amounts to enclosing the support with a rectangle perpendicular to the space and frequency axes. The number of samples is given by the area of the rectangle and equals the space-bandwidth product, which may be considerably larger than the area of the space-frequency support and hence the actual number of degrees of freedom of the signals. When the space-frequency support is not a rectangle perpendicular to the axes, it is possible to represent and reconstruct the signal with fewer samples than implied by the space-bandwidth product. Light fields propagating through optical systems is one example of an application where non-rectangular supports are commonly encountered [37].

The FRT is a generalization of the Fourier transform and the FRT sampling theorem [49–51] is an extension of the classical sampling theorem (while a special case of the LCT sampling theorem). Based on this generalized sampling theorem, here we provide a sub-Nyquist approach to represent signals with arbitrary space-frequency support [33, 34, 36]. This approach reduces to the geometrical problem of finding the smallest parallelogram enclosing the space-frequency support. The area of the parallelogram given by the bicanonical width product is the number of samples needed and the reconstruction is given by an explicit formula. This allows us to represent signals with fewer samples than with the classical approach, since it is possible to enclose the true space-frequency support more tightly with a parallelogram of our choice, than with a rectangle perpendicular to the axes. A Wigner-based approach to related problems has been given in [47, 60].

7.5.1 *Constrained Signal Representation*

Our goal is to determine the minimal sampling rate when the space-frequency support is given and to show how to reconstruct the signal from those samples. First we consider the (constrained) case where the signal needs to be sampled in a specific domain, say the space domain u. In other words, we are not free to choose the domain in which to sample the signal and must sample it in the specified domain. Without loss of generality, suppose the specified domain is the space domain. In the classical approach, the sampling rate in the space domain is determined by the extent in the frequency domain. If we denote this extent by $\Delta\mu$, then the spacing between space-domain samples must not be greater than $\delta u = 1/\Delta\mu$, so that the minimum number of samples over the space extent Δu is given by $\Delta u/(1/\Delta\mu) = \Delta u \Delta\mu$, which is the space-bandwidth product. (The space extent Δu is the projection of the space-frequency support onto the u axis.) This classical approach is geometrically equivalent to enclosing the support with a rectangle perpendicular to the space and frequency axes and having sides of length Δu and $\Delta\mu$. Its area equals $\Delta u \Delta\mu$ and gives the number of samples required for interpolating the continuous signal in the Nyquist–Shannon sense (Fig. 7.6).

For efficient sampling, it is desirable to approach the minimum number of samples possible given by the area of the space-frequency support. When we use the FRT sampling theorem, the sampling rate can be determined by the extent in the FRT domain which minimizes the required number of the samples. Since signals that are extent limited in two FRT domains have parallelogram shaped supports (see Fig. 7.7), determining the optimal value of a is equivalent to the problem of finding the smallest parallelogram enclosing the space-frequency support, under the constraint that two sides of the parallelogram must be perpendicular to the u axis (Fig. 7.6). (This constraint arises because the signal must be sampled specifically in the space domain.) The minimum number of samples needed for reconstruction (based on the FRT sampling theorem) is given by the area of this enclosing parallelogram, which is equal to the bicanonical width product for the two FRT domains orthogonal to the sides of the parallelogram. Reconstruction of the continuous signal is possible through the interpolation formula associated with the FRT sampling theorem, which is a special case of the LCT interpolation formula in (7.15) [33, 34, 36].

This approach is illustrated in Fig. 7.6, where the shaded region shows the space-frequency support. In the classical approach, we would be finding the smallest rectangle perpendicular to the axes that encloses the space-frequency support of the signal. With the proposed approach, we find the smallest enclosing parallelogram with two sides perpendicular to the space axis. Since the FRT includes the ordinary Fourier transform as a special case (and parallelograms include rectangles), the proposed approach will never require more samples than the classical approach. On the other hand, the freedom to optimally choose a can result in a fewer number of samples being necessary [33, 34, 36]. (That is, the area of the fitting parallelogram will be always less than or equal to the area of the fitting rectangle.)

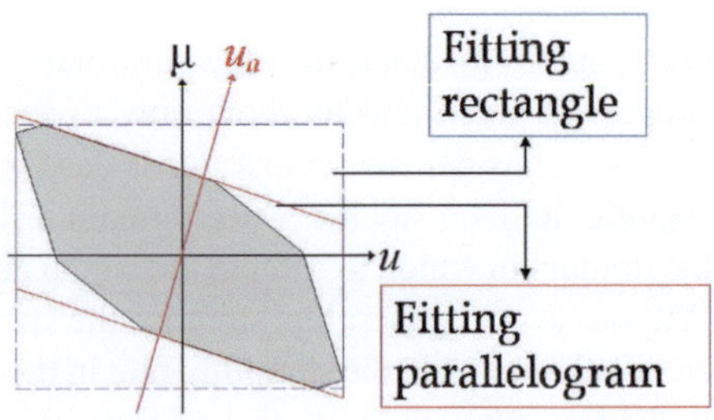

Fig. 7.6 The smallest enclosing parallelogram (*solid*) and rectangle (*dashed*), both under the constraint that two sides be perpendicular to the space axis u. The *shaded region* is the space-frequency support [33, 36]

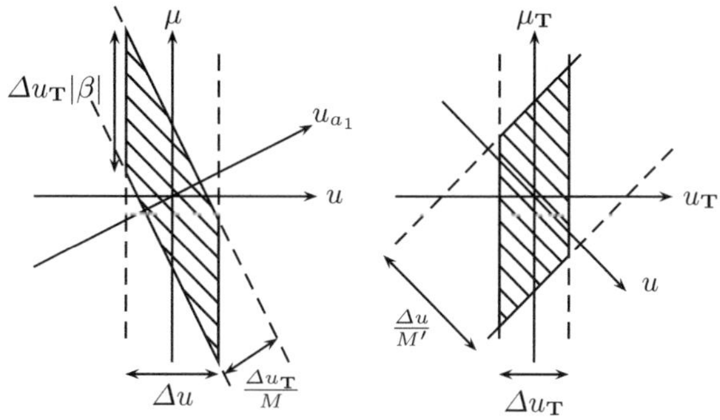

Fig. 7.7 The space-frequency support of $f(u)$ (*left*) and $f_{\mathbf{T}}(u)$ (*right*) for space- and LCT-limited signals. The area of both parallelograms are equal to $\Delta u \Delta u_{\mathbf{T}}|\beta|$ [34]

7.5.2 Unconstrained Signal Representation

In some applications we may have the freedom to process the analog signal prior to sampling and hence to sample the signal at a domain of our choice. In this case the number of samples can be further reduced [33]. This involves computing the FRT of the analog signal prior to sampling. Such computations may involve chirp modulators for time-domain signals and lenses for space-domain signals [5]. After sampling, if necessary we can return back to the original domain in $\sim N \log N$ time since discrete FRTs can be computed in this amount of time [35].

The FRT sampling theorem allows us to work with any two arbitrary domains (the sampling domain and the domain where the extent determines the sampling rate) since any such domains can be related through the FRT; hence, we are free to determine the sampling rate from the extent in any FRT domain of our choice. This allows us to further reduce the number of samples by enclosing the support with an arbitrary parallelogram, instead of a rectangle. This approach reduces to a simple geometrical problem which requires us to find the minimum-area parallelogram enclosing the given space-frequency support. In contrast to the constrained case, having the flexibility of sampling in any FRT domain removes the requirement that the two sides of the parallelogram be perpendicular to the space domain, and allows us to fit an arbitrary parallelogram. The number of samples required is given by the area of the parallelogram, which is equal to the bicanonical width product. The signal can be represented optimally through its samples at either of the two FRT domains that are orthogonal to the sides of the best-fitting parallelogram. Note that this approach gives us two optimal FRT domains in which the signal should be sampled. It is also possible to represent the signal in any *essentially equivalent* LCT domain, with the same sampling efficiency [33, 34].

7.6 Efficient Discrete LCT Computation for System Simulation

We now review how the bicanonical width product is useful for the efficient and accurate simulation of optical systems, based on a natural formulation of discrete LCT (DLCT) computation [35, 61]. It has been recently shown that *if the number of samples N is chosen to be at least equal to the bicanonical width product, the DLCT can be used to obtain a good approximation to the continuous LCT*, limited only by the fundamental fact that a signal cannot have strictly finite extent in more than one domain [35, 61]. The exact relation between the discrete and continuous LCT precisely shows the approximation involved and demonstrates how the approximation improves with increasing N [35]. Because this exact relation generalizes the corresponding relation for Fourier transforms [62], the DLCT defined in [63] approximates the continuous LCT in the same sense that the DFT approximates the continuous Fourier transform, provided the number of samples and the sampling intervals are chosen based on the LCT sampling theorem as specified in [35, 61].

We also note that this DLCT can be efficiently computed in $\mathcal{O}(N \log N)$ time by successively performing a chirp multiplication, a fast Fourier transform (FFT), and a second chirp multiplication, by taking advantage of the simple form of the DLCT [35, 52, 63]. This straightforward fast computation approach does not require sophisticated algorithms or space-frequency support tracking for accurately computing the continuous LCT, as opposed to other LCT computation methods [64–69]. To summarize, a simple fast computation method, a well-defined relationship to the continuous LCT, and unitarity make this definition of the DLCT an important candidate for being a widely accepted definition of the discrete version of the LCT [35].

Note that in order to use any DLCT definition in practice, to approximately compute the samples of the LCT of a continuous signal, it is necessary to know how to choose the number of samples and the sampling intervals, based on some prior information about the signal. The described computation approach (first discussed in [61], and then independently developed in [35]) meets precisely this demand and allows us to accurately compute LCTs with the minimum possible number of samples. In this formulation, the extents of the signal in the input and output LCT domains (the original space domain and the target LCT domain) are assumed to be specified as prior information. This is equivalent to assuming an initial parallelogram-shaped space-frequency support [33, 34, 70]. The minimum number of samples required for accurate computation is then determined from the LCT sampling theorem. This minimum number of samples is equal to the bicanonical width product, which is also the area of the parallelogram support [33, 35]. The DLCT defined in [63] works with this minimum number of samples without requiring any oversampling at the intermediate stages of the computation, in contrast to previously given approaches [66, 67] for the same DLCT. On the other hand, use of the Shannon–Nyquist sampling theorem instead of the LCT sampling theorem, as in [66, 67], leads to problems such as the need to use a greater number of samples, different sampling rates at intermediate stages of the computation, or different numbers of samples at the input and output domains.

This natural DLCT computation method has been revisited in [70], where an interpretation of the method has been given through phase-space diagrams. This allows us to see from yet another perspective how this elegant and accurate LCT computation method [35, 61] works with the minimum number of samples, without requiring interpolation. The DLCT computation presented in [35, 61] and the phase-space illustrations in [70, 71] assume that the extent of the signal is known at the input and output of the system to be simulated. We now discuss how to optimally simulate optical systems by using this DLCT computation method when the space-frequency support of the input signal is specified [36] (rather than its extents in the input and output domains). Different assumptions about the initial space-frequency support have been made in the literature to explore efficient DLCT computation [65–69, 72, 73]. Hence here we explore a unified method that works with any initial support while still ensuring the minimality of the number of samples [36]. The idea is to find the number of samples by fitting a parallelogram to the given space-frequency support, such that two opposing sides are perpendicular to the u axis (the input domain) and the other sides are perpendicular to the oblique axis corresponding to the output LCT domain. The area of the smallest fitting parallelogram gives the number of samples that needs to be used for an accurate DLCT computation [36]. Then the samples of the continuous signal at the output of the optical system can be obtained by sampling the input signal at this rate and then computing its DLCT as described in [35].

This elegant DLCT formulation is mainly achieved through the property that *the bicanonical width product is an invariant measure for the number of degrees of freedom of signals under linear canonical transformation* [33, 34]. To see this, suppose a finite extent has been specified in the space domain and in some other LCT domain. The corresponding space-frequency support is shown in Fig. 7.7a. If we transform to precisely the same LCT domain in which the extent has been specified, the new space-frequency support becomes as shown in Fig. 7.7b. Here M and M' are the scaling parameters associated with the LCT and inverse LCT operations, respectively. Note that in both parts of the figure, the support is bounded by a vertical corridor, perpendicular to the space domain in part a, and to the LCT domain in part b. We are not surprised that the transformed support is again a parallelogram, since the linear geometric distortion imparted by an LCT always maps a parallelogram to another parallelogram. Moreover, the areas of both parallelograms are equal to each other and given by the bicanonical width product $\Delta u \Delta u_{\mathrm{T}} |\beta|$, so that the number of degrees of freedom as measured by the bicanonical width product remains the same after the LCT. This does not surprise us either, since LCTs are known not to change the support area in phase space. The fact that LCTs model an important family of optical systems makes the bicanonical width product a suitable invariant measure for the number of degrees of freedom of optical signals.

On the other hand, the space-bandwidth product, which is the area of the smallest bounding perpendicular rectangle, may change significantly after linear canonical transformation, and quadratic-phase optical systems. (This is the reason why the number of samples must be increased at some intermediate stages of certain previously proposed FRT and LCT algorithms which rely on the space-bandwidth product either as the measure of the number of degrees of freedom or as the minimum number of samples required [65–69, 72, 73]. In contrast, fast computation of LCTs based on the results presented in [35, 61] allows us to work with the same number of samples in both domains without requiring any oversampling. This number of samples is the minimum possible for both domains based on the LCT sampling theorem, and is given by the bicanonical width product [35].) This factor makes the conventional space-bandwidth product undesirable as a measure of the number of degrees of freedom, which we expect to be an intrinsic and conserved quantity under invertible unitary transformations.

The so-called *generalized space-bandwidth product*, which essentially removes the requirement that the rectangular support be perpendicular to the axes, has been proposed [74] as an improvement over the conventional space-bandwidth product. A related approach has also been studied [60]. It has been noted that this entity is invariant under the FRT operation (rotational invariance), but it has also been emphasized that "further research is required in obtaining other forms of generalized space-bandwidth products that are invariant under a more general area preserving space-frequency operations: the symplectic transforms" [74]. The bicanonical width product meets precisely this demand and allows us to compute LCTs with the minimum possible number of samples without requiring any interpolation or oversampling at intermediate stages of the computation [35].

7.7 Degrees of Freedom of Optical Systems

We now discuss how to explicitly quantify the degrees of freedom of first-order optical systems with multiple apertures, and give explicit conditions for lossless transfer [33, 37]. In particular, we answer the following questions about apertured optical systems, which here refers to systems consisting of an arbitrary sequence of thin lenses and apertures separated by sections of free space:

- Given the space-frequency support of an input signal and the parameters of an apertured optical system, will there be any information loss upon passage through the system?
- Which set of signals can pass through a given apertured system without any information loss? In other words, what is the largest space-frequency support that can pass through the system without any information loss?
- What is the maximum number of spatial degrees of freedom that can be supported by a given apertured system?

The *space-frequency support (phase-space support)* of a set of *signals* may be defined as the region in the space-frequency plane (phase space) in which a large percentage of the total energy is confined [5, 30]. The number of degrees of freedom is given by the area of the space-frequency support. We also define the *space-frequency window (phase-space window)* of a *system* [33, 37] as the largest space-frequency support that can pass through the system without any information loss. Here we develop a simple method to find the space-frequency window of a given system in terms of its parameters. Once the space-frequency window of the system is determined, it specifies the set of all signals that can pass through the system without information loss: the optical system preserves the information content of signals whose space-frequency supports lie inside the system window. All we need to do is to compare the space-frequency support of the input signal with the space-frequency window of the system. If the signal support lies completely inside the system window, the signal will pass through the system without any information loss. Otherwise, information loss will occur.

The number of degrees of freedom of the set of signals which can pass through a system can be determined from the area of the space-frequency window of the system. Although the space-frequency window may in general have different shapes [5, 30], it is often assumed to be of rectangular shape with the spatial extent determined by a spatial aperture in the object or image plane, and the frequency extent determined by an aperture in a Fourier plane. (Again, we consider one-dimensional signals and systems for simplicity.) If these apertures are of length Δx and $\Delta\sigma_x$ respectively, then the number of degrees of freedom that can be supported by the system is given by $\Delta x \Delta\sigma_x$. More generally, for space-frequency windows of different (non-rectangular) shapes, the number of degrees of freedom is given by the area of the space-frequency window [33, 37].

Physical systems which carry or process signals always limit their spatial extents and bandwidths to certain finite values. A physical system cannot allow the existence of frequencies outside a certain band because there is always some limit to the resolution that can be supported. Likewise, since all physical events of interest have a beginning and an end, or since all physical systems have a finite extent, the temporal duration or spatial extent of the signals will also be finite. For example, in an optical system the sizes of the lenses will limit both the spatial extent of the images that can be dealt with and their spatial bandwidths. More generally, we may say that they will limit the signal to a certain region in the space-frequency plane. We refer to this region as the space-frequency window of the system. It is these physical limitations that determine the space-frequency support of the signals and thus their degrees of freedom. Just as these may be undesirable physical limitations which limit the performance of the system, they may also be deliberate limitations with the purpose of limiting the set of signals we are dealing with. When a signal previously represented by a system with larger space-frequency window is input into a system with smaller space-frequency window, information loss takes place.

The conventional space-bandwidth product has been of fundamental importance because of its interpretation as the number of degrees of freedom [5, 17–23, 25–28, 30–32, 75, 76]. In most works, the space-bandwidth product, as its name implies, is the product of a spatial extent and a spatial-frequency extent. This implies the assumption of a rectangular space-frequency region. However, the set of input signals may not exhibit a rectangular space-frequency support, and even if they do, this support will not remain rectangular as it propagates through the system [33–35]. Likewise, the space-frequency windows of multi-component optical systems, as we will see in this section, do not in general exhibit rectangular shapes. This possibility and some of its implications were discussed in [30]. In [33, 37] we made concrete the hypothetical concept of a non-rectangular space-frequency window, and showed how it can be actually computed for a broad class of optical systems, as will be discussed here. (To prevent possible confusion, we underline that we are dealing with systems with sequentially cascaded apertures, and not systems with multiple parallel apertures.)

We also note that the phase-space window has been referred to by different names, such as the space-bandwidth product of the system (in short SWY) [30, 77, 78], the system transmission range [77], and the Wigner or space-bandwidth chart of the system [77, 78]. Also, the concept of degrees of freedom can be related to other concepts such as Shannon number and information capacity of an optical system [76], geometrical etendue [79], dimensionality, and so on.

In order to treat systems with real physical parameters, we first revisit some of the background material discussed in Sect. 7.2, and translate them to their dimensional counterparts. We then discuss how to find the phase-space window of an optical system. Next, we treat the cases of lossless and lossy transfer separately, and finally conclude with a discussion of applications.

7.7.1 Scale Parameters and Dimensions

Dimensionless variables and parameters were employed in the previous sections for simplicity and purity (see Sect. 7.2). In this section, we will employ variables with real physical dimensions. For this, we need to revisit a number of earlier definitions and results. When dealing with FRTs, the choice of scale and dimensions must always be noted, as this has an effect on the fractional order observed at a given plane in the system [5, pp. 320–321]. Using x to denote a dimensional variable (with units of length), the ath-order FRT [5] of a function $\hat{f}(x)$, denoted by $\hat{f}_a(x)$, can be defined as

$$\hat{f}_a(x) \equiv (\hat{\mathcal{F}}^a \hat{f})(x) \equiv \int_{-\infty}^{\infty} \hat{K}_a(x, x') \hat{f}(x')\, dx', \tag{7.27}$$

$$\hat{K}_a(x, x') \equiv \frac{A_\phi}{s} e^{i\pi\left(\frac{\cot\phi}{s^2}x^2 - 2\frac{\csc\phi}{s^2}xx' + \frac{\cot\phi}{s^2}x'^2\right)}.$$

Here s is an arbitrary scale parameter with dimensions of length. The scale parameter s serves to convert the dimensional variables x and x' inside the FRT integral to dimensionless form. A hat over a function or kernel shows that it takes dimensional arguments [5, pp. 224–227]. The FRT definition above reduces to the pure mathematical FRT definition with dimensionless arguments if we define the dimensionless variables $u = x/s$ and $u' = x'/s$, or simply if we set $s = 1$ in our measurement unit (meters, etc.). The choice $s = 1$ unit makes the expressions simpler, but we feel that this merely hides the essential distinction between dimensional and dimensionless variables and would actually be a disservice to the reader.

We will denote the LCT of a function $\hat{f}(x)$ with the dimensional parameter matrix $\hat{\mathbf{T}} = [\hat{A}\ \hat{B}; \hat{C}\ \hat{D}]$ as $\hat{f}_{\hat{\mathbf{T}}}(x)$:

$$\hat{f}_{\hat{\mathbf{T}}}(x) \equiv (\hat{C}_{\hat{\mathbf{T}}}\hat{f})(x) \equiv \int_{-\infty}^{\infty} \hat{C}_{\hat{\mathbf{T}}}(x,x')\hat{f}(x')\,dx', \tag{7.28}$$

$$\hat{C}_{\hat{\mathbf{T}}}(x,x') \equiv \sqrt{\frac{1}{\hat{B}}}\, e^{-i\pi/4} e^{i\pi\left(\frac{\hat{D}}{\hat{B}}x^2 - 2\frac{1}{\hat{B}}xx' + \frac{\hat{A}}{\hat{B}}x'^2\right)},$$

for $\hat{B} \neq 0$. A hat over a parameter shows that it is the dimensional counterpart of the same parameter without the hat. Any LCT can be decomposed into a (dimensional) FRT followed by scaling followed by chirp multiplication [5, 37]:

$$\hat{\mathbf{T}} = \begin{bmatrix} \hat{A} & \hat{B} \\ \hat{C} & \hat{D} \end{bmatrix} = \begin{bmatrix} 1 & 0 \\ -\frac{q}{s^2} & 1 \end{bmatrix} \begin{bmatrix} M & 0 \\ 0 & \frac{1}{M} \end{bmatrix} \begin{bmatrix} \cos\phi & s^2\sin\phi \\ -\frac{\sin\phi}{s^2} & \cos\phi \end{bmatrix}. \tag{7.29}$$

The three matrices, respectively, correspond to the transformation matrices of chirp multiplication with parameter q (multiplication by $\exp(-i\pi \frac{q}{s^2}x^2)$), coordinate scaling with factor $M > 0$ (mapping of $\hat{f}(x)$ into $\sqrt{1/M}\hat{f}(x/M)$), and ath order dimensional FRT with $\phi = a\pi/2$ (transformation of $\hat{f}(x)$ into $\hat{f}_a(x)$). The decomposition can be written more explicitly in terms of the LCT and FRT domain representations of the signal $\hat{f}(x)$ as

$$\hat{f}_{\hat{\mathbf{T}}}(x) = \exp\left(-i\pi \frac{q}{s^2}x^2\right)\sqrt{\frac{1}{M}}\,\hat{f}_a\left(\frac{x}{M}\right). \tag{7.30}$$

This is the dimensional version of the Iwasawa decomposition in (7.9).

By appropriately choosing the three parameters a, M, q, the above equality can be satisfied for any $\hat{\mathbf{T}} = [\hat{A}\ \hat{B};\ \hat{C}\ \hat{D}]$ matrix. Solving for a, M, q in (7.8), we obtain the decomposition parameters in terms of the matrix entries $\hat{A}$, $\hat{B}$, $\hat{C}$, $\hat{D}$:

$$a = \begin{cases} \frac{2}{\pi}\arctan\left(\frac{1}{s^2}\frac{\hat{B}}{\hat{A}}\right), & \text{if } \hat{A} \geq 0 \\ \frac{2}{\pi}\arctan\left(\frac{1}{s^2}\frac{\hat{B}}{\hat{A}}\right) + 2, & \text{if } \hat{A} < 0 \end{cases} \tag{7.31}$$

$$M = \sqrt{\hat{A}^2 + (\hat{B}/s^2)^2}, \tag{7.32}$$

$$q = \begin{cases} -s^2 \frac{\hat{C}}{\hat{A}} - \frac{1}{s^2} \frac{\hat{B}/\hat{A}}{\hat{A}^2 + (\hat{B}/s^2)^2}, & \text{if } \hat{A} \neq 0 \\ -s^2 \frac{\hat{D}}{\hat{B}}, & \text{if } \hat{A} = 0. \end{cases} \tag{7.33}$$

The range of the arctangent lies in $(-\pi/2, \pi/2]$.

7.7.2 Phase-Space Window of Optical Systems

We now describe how to find the phase-space window (space-frequency window) of an apertured optical system [33, 37]. Such systems consist of arbitrary concatenations of apertures with quadratic-phase systems (which in turn consist of an arbitrary number of lenses, sections of free space and quadratic graded-index media). Also note that a lens with a finite aperture can be viewed as an ideal lens followed by a finite aperture. Although beyond the scope of the present discussion, these results can be extended to more general systems involving occlusions [80], prisms and gratings [78], and bends and shifts of the optical axis.

Let us first introduce the notation. The input and output planes are defined along the optical axis z at $z = 0$ and $z = d$, where d is the length of the system. If the apertures did not exist, the amplitude distribution at any plane perpendicular to the optical axis could be expressed as an LCT of the input. Hence each z plane corresponds to an LCT domain. Let L denote the total number of apertures in the system. z_j and Δ_j denote the location and extent of the jth aperture in the system, where $j = 1, 2, \ldots, L$. The matrix $\hat{\mathbf{T}}_j$ is used to denote the parameter matrix of the system from the input to the position of the jth aperture; that is, the system lying between 0 and z_j excluding the apertures. The matrix $\hat{\mathbf{T}}_j$ can be readily calculated using the matrices for lenses, sections of free space, quadratic graded-index media, and the concatenation property [5]. The matrix elements of $\hat{\mathbf{T}}_j$ is denoted by $\hat{A}_j$, $\hat{B}_j$, $\hat{C}_j$, $\hat{D}_j$. The associated Iwasawa decomposition parameters is denoted by a_j, M_j, q_j, which can be computed from $\hat{A}_j$, $\hat{B}_j$, $\hat{C}_j$, $\hat{D}_j$ by using the formulas (7.31), (7.32), (7.33). The FRT order in the Iwasawa decomposition begins from 0 at the input of the system, and then monotonically increases as a function of distance [5, 34].

For lossless transfer through the system, the extent of the signal just before each aperture must lie inside the aperture. For simplicity we assume that both the aperture and the signal extents are centered around the origin. Then, the following must be satisfied for $j = 1, 2, \ldots, L$:

$$\Delta x_{\hat{\mathbf{T}}_j} \leq \Delta_j, \tag{7.34}$$

where $\Delta x_{\hat{\mathbf{T}}_j}$ denotes the extent of the signal in the $x_{\hat{\mathbf{T}}_j}$ domain, which corresponds to the LCT domain at the $z = z_j$ plane, where the jth aperture is situated.

As we have showed before, LCT domains are equivalent to scaled FRT domains and thus to scaled oblique axes in the space-frequency plane [34]. Based on this equivalence, each finite interval in an LCT domain will correspond to a scaled interval in the equivalent FRT domain. To see this explicitly in the dimensional case, we refer to (7.30), which implies that if the linear canonical transformed signal $\hat{f}_{\hat{\mathbf{T}}}(x)$ is confined to an interval of length $\Delta x_{\hat{\mathbf{T}}}$, so is $\hat{f}_a(x/M)$. Therefore, the extent of the fractional Fourier transformed signal $\hat{f}_a(x)$ in the equivalent FRT domain is $\Delta x_{\hat{\mathbf{T}}}/M$. Thus, the condition in (7.34) can be reexpressed as

$$\Delta x_{a_j} \leq \Delta_j / M_j, \tag{7.35}$$

where Δx_{a_j} denotes the extent of the signal in the a_jth order (dimensional) FRT domain.

FRT domains are often visualized in the dimensionless space-frequency plane where the coordinates are scaled such that the space and frequency axes are dimensionless. This is achieved by introducing the scaling parameter s and the dimensionless scaled coordinates $u = x/s$ and $\mu = s\sigma_x$. The condition for lossless information transfer then becomes

$$\Delta x_{a_j}/s \leq \Delta_j / M_j s, \tag{7.36}$$

where $\Delta x_{a_j}/s$ denotes the extent of the signal in the a_jth order (dimensionless) FRT domain (along the oblique axis making angle $a_j\pi/2$ with the $u = x/s$ axis). In other words, for every $j = 1, 2, \ldots, L$, the signal must be confined to the normalized aperture extent of $\Delta_j/M_j s$ along the oblique axis with angle $a_j\pi/2$.

It is well known that if the space-, frequency-, or FRT-domain representation of a signal is identically zero (or negligible) outside a certain interval, so is its Wigner distribution [5, 59]. As a direct consequence of this fact, the condition in (7.36) defines a corridor of width $\Delta_j/M_j s$ in the direction orthogonal to the a_jth order FRT domain u_{a_j}. (With the term "corridor" we are referring to an infinite strip in the space-frequency plane perpendicular to the oblique u_{a_j} axis.) The corridor makes an angle $(a_j + 1)\pi/2$ with the $u = x/s$ axis in the dimensionless space-frequency plane (see Fig. 7.5). Now, if we intersect the corridors defined by each aperture, we obtain a bounded region in the space-frequency plane, which has the form of a centrally symmetrical convex polygon (see Fig. 7.8 for $L = 2$ and Fig. 7.9 for $L = 4$). We refer to this convex polygon defined by the normalized aperture extents as the *space-frequency window* of the system [37].

The space-frequency window specifies the set of all signals that can pass through the system without any loss: the optical system preserves the information content of all signals whose supports lie inside the space-frequency window. The area of the space-frequency window gives the number of degrees of freedom that can pass through the system. This is also the minimum number of samples required to faithfully represent an arbitrary signal at the output of the system.

We can summarize the steps for finding the phase-space window (space-frequency window) as follows [37]:

1. Compute the parameter matrix $\hat{\mathbf{T}}_j$ for each aperture $j = 1, 2, \ldots, L$ using the matrices for lenses, sections of free space and quadratic graded-index media, and the concatenation property. Recall that $\hat{\mathbf{T}}_j$ was defined as the parameter matrix of the system lying between the input plane and the location of the jth aperture.
2. Compute the corresponding Iwasawa decomposition parameters a_j and M_j (the fractional order and the magnification) by inserting the matrix entries $\hat{A}_j$, $\hat{B}_j$, $\hat{C}_j$, $\hat{D}_j$ into the formulas (7.31) and (7.32).
3. In the dimensionless space-frequency plane, draw a corridor of width $\Delta_j/M_j s$ making angle $(a_j + 1)\pi/2$ with the x/s axis, for each j (see Fig. 7.9). The corridor is explicitly defined by the following two lines: $y = -\cot(a_j\pi/2)x \pm \frac{\Delta_j}{2M_j s}\csc(a_j\pi/2)$.
4. Intersect the corridors from all apertures to determine the region lying inside all the corridors. This is the phase-space window at the input plane $z = 0$.
5. Scale the horizontal and vertical coordinates by s and $1/s$, respectively, to obtain the phase-space window in the dimensional space-frequency plane x-σ_x.

A few remarks are in order at this point. First, the area of the window and hence the number of degrees of freedom of the system remains the same whether

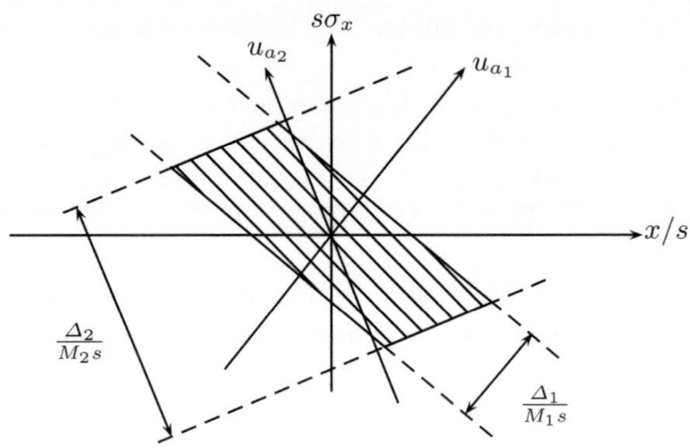

Fig. 7.8 Space-frequency window of a system with two apertures [37]

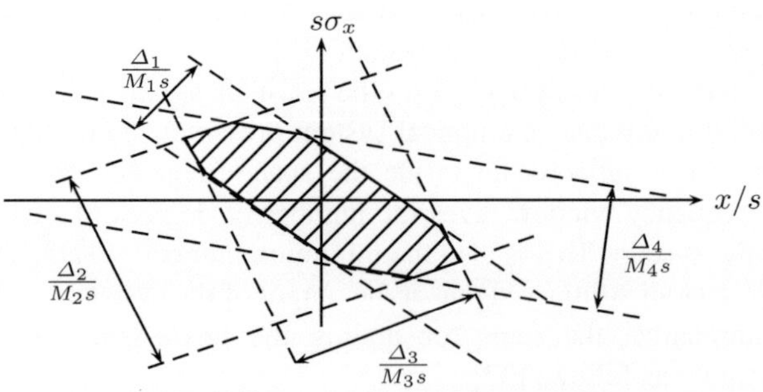

Fig. 7.9 Space-frequency window of a system with four apertures [37]

it is computed in dimensional or dimensionless space. Second, choice of the scale parameter s is arbitrary and the system window in the dimensional space-frequency plane is independent of the choice of s. However, choice of s does affect the value of a as a function of z. Some choices better utilize the range of a (as in Fig. 7.10), whereas poor choices lead to a changing too quickly over a short range of z and then saturating [5, pp. 320–321, 377–378]. One approach is to choose s such that the space and frequency extents in the dimensionless space-frequency plane are comparable to each other. Third, the system window is computed with respect to a chosen reference plane. Above, we compute it with respect to the input plane, so that we can directly compare the input signal support with the system window. The phase-space window at the input plane can be shortly referred to as the *input phase-space window* of the system. If one desires to visualize the system window with respect to a different reference plane, it can be transformed to the new plane using the LCT transformation from the input plane to the new reference plane [37]. (More explicitly, if the ith corner of the system window is expressed as $(x^{(i)}, \sigma_x^{(i)})$, then after LCT transformation with matrix $\hat{\mathbf{T}}$, the new corner will be described by the coordinates $\hat{\mathbf{T}}[x^{(i)}\ \sigma_x^{(i)}]^T$, where T is the transpose operation [33, 65].)

We now illustrate the method on a sample system. Figure 7.10a shows a system consisting of several apertures and lenses, whose aperture sizes and focal lengths are given right above them. The fractional transform order a and the scale parameter M of the system are plotted in Fig. 7.10 as functions of distance z. The emphasis in this paper is on computing a_j and M_j at the aperture locations, since these allow us to determine the system window. However, these quantities can also be computed for all values of z in the system, revealing their continual evolution as we move along the optical axis, as illustrated in Fig. 7.10b, c. We can compute $a(z)$ and $M(z)$ by expressing $\hat{A}, \hat{B}, \hat{C}, \hat{D}$ in terms of z and using them in Eqs. (7.31)–(7.33) [5, 41].

Figure 7.11a and c show the system window at the input plane $z = 0$. This region defines the set of all *input* signals that can pass through the system without any information loss. Input signals whose space-frequency support lies wholly inside this region will not experience any loss. Similarly, Fig. 7.11b and d show the system window at the output plane $z = d$. This region defines the set of all signals that can be observed at the *output* of the system. The region in Fig. 7.11b is just a propagated version of the region in Fig. 7.11a through the entire optical system. This can be obtained by applying the concatenated LCT matrix $\hat{\mathbf{T}}_L \dots \hat{\mathbf{T}}_2\hat{\mathbf{T}}_1$ to the space-frequency window at the input plane, to take into account the linear distortion due to the entire optical system (by multiplying the coordinates of each corner of the window with the LCT matrix, as described before).

Just as the concatenation property of transformation matrices allows us to represent the cumulative action of all optical elements present with a single entity, the system window is an equivalent aperture that appropriately transforms and combines the effects of all individual apertures in different domains, into a single space-frequency aperture [37].

The space-frequency (phase-space) window of the system in Fig. 7.10a is determined only by the 1st, 5th, 7th, and 8th apertures. The other apertures do

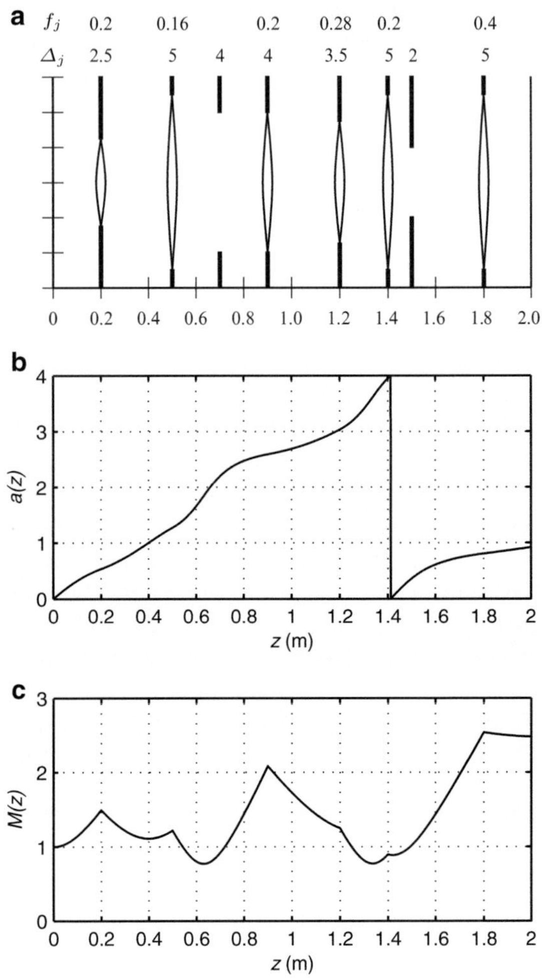

Fig. 7.10 (**a**) An apertured optical system with input plane at $z = 0$ and output plane at $z = 2$ m [37]. The horizontal axis is in meters. The lens focal lengths f_j in meters and the aperture sizes Δ_j in centimeters are given right above them. (**b**) and (**c**) Evolution of $a(z)$ and $M(z)$ as functions of z. $\lambda = 0.5\ \mu$m and $s = 0.3$ mm [5, 41]

not affect or limit the space-frequency window of the system and therefore can be considered as redundant from the system's viewpoint. (Removing the redundant apertures from the system or replacing them with apertures of greater size will have negligible effect on the behavior of the system, for any given input signal.)

It is also worth noting that the information loss caused by an aperture will depend not only on the actual physical size of the aperture, but also the magnification of the signal at that location. If the magnification at the aperture location is small, there will be less or no information loss. For example, although the aperture sizes are the same for the 2nd, 6th, and 8th apertures, only the 8th aperture limits the system window ($M(1.8) \approx 2.5$ whereas $M(0.5) \approx 1.2$ and $M(1.4) \approx 1$). This illustrates that the magnification in the plane of the aperture is as important as the size of the aperture in limiting the system window. If we have some flexibility during the

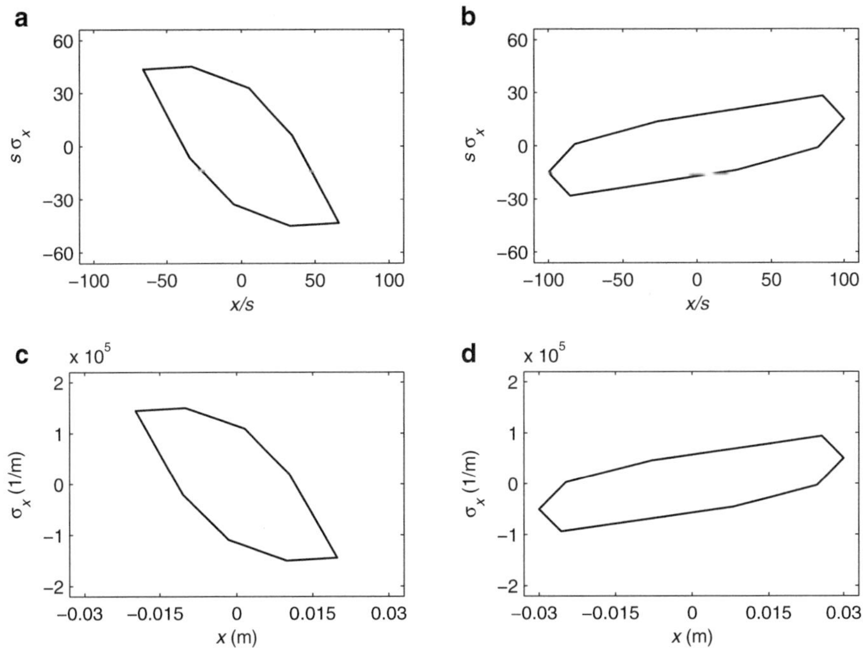

Fig. 7.11 The space-frequency window of the system at the input plane in the dimensionless (**a**) and dimensional spaces (**c**). The space-frequency window of the system at the output plane in the dimensionless (**b**) and dimensional spaces (**d**) [37]

design of the optical system, careful choice of lens and aperture locations can help information losses to be minimized, a process which will be aided by the space-frequency approach and the graphs for $M(z)$ we have discussed.

7.7.3 Necessary and Sufficient Condition for Lossless Transfer

An input signal will pass through the system without any information loss if and only if its space-frequency support is fully contained in the input space-frequency (phase-space) window of the system. That is, if the signal support does not lie completely inside the system window, information loss will occur [37].

Proof. LCT domains correspond to oblique axes in the space-frequency plane. Consider corridors of varying width, orthogonal to such an oblique axis. The extent of the signal in a given LCT domain can be determined from the space-frequency support of the signal, by finding the width of the narrowest orthogonal corridor enclosing the space-frequency support. First, let us consider an input signal whose space-frequency support lies completely inside the input space-frequency window of the system. This guarantees that along any oblique axis in the space-frequency plane corresponding to some LCT domain in which an aperture resides, the extent of the signal will be smaller than the width of the aperture at that LCT domain, and hence the signal will pass through that aperture unhindered. Repeating this for all apertures, the input signal will pass through the whole system without any information loss. (Recall that the space-frequency window is defined by the intersection of the corridors defined by the apertures. If the extent of the signal was larger than the width of the aperture at that LCT domain, the orthogonal corridor enclosing the space-frequency support would have been wider than the corridor defined by the aperture, so that the space-frequency support of the signal could not lie within the space-frequency window of the system.)

Conversely, consider an input signal which passes through the system without any loss. This implies that the signal extent was smaller than the aperture width for each aperture, since otherwise irreversible information loss would occur. Recall that each aperture defines a corridor perpendicular to the LCT domain in which it resides. For any of these LCT domains, the space-frequency support of the signal must lie within this corridor, since if not, the extent of the signal in that domain would not lie within that aperture, leading to information loss and hence a contradiction. Since this argument must be true for all apertures, it follows that the signal space-frequency support must lie inside the region defined by the intersection of the corridors, which is the space-frequency window of the system. This completes the proof. □

A straightforward but lengthy way to determine whether information loss will take place would be to trace the space-frequency support of the signal as it passes through the whole system [33]. When the signal arrives at the first aperture, there will have taken place a linear distortion on the initial space-frequency support of the signal. After this linear distortion, if the extent of the signal in that LCT domain is less than the aperture size, then the signal will pass through this aperture without any information loss. Then another linear distortion will take place as the signal travels to the next aperture. Again, we will determine whether there is any information loss by comparing the extent in this domain to the aperture size. Repeating this procedure throughout the system, we can determine whether the signal passes through the system losslessly. This lengthy way of determining whether there will be information loss is specific to a certain input signal and its support. On the other hand, our method is general in the sense that, once the space-frequency window of the system is determined, it specifies the set of all signals that can pass through the system without information loss. The optical system preserves the information content of all signals whose space-frequency support lies inside the space-frequency window of the system.

7.7.4 Lossy Transfer

If the space-frequency window does not enclose the space-frequency support of the input signal completely, then we would intuitively expect the following: The information contained within the intersection of the space-frequency support of the signal and the space-frequency window of the system will be preserved, and the rest will be lost (Fig. 7.12). This indeed turns out to be approximately true in most cases [37]. In other words, just as a spatial aperture passes certain parts of a signal and blocks the rest, the space-frequency window acts like an aperture in phase-space, passing certain parts and blocking others. In particular, if the set of input signals has a greater number of degrees of freedom than the number of degrees of freedom the system can support, information loss will take place, since a region with larger area can never possibly lie completely within a region with smaller area.

Given an arbitrary space-frequency support at the input, one can obtain the space-frequency support at any position in the system by tracing the support throughout the system [33, 65]. Whenever an aperture narrower than the signal extent is encountered, the outlying parts of the signal will be truncated. The effect of this truncation on the space-frequency support of the signal will be to likewise truncate the regions of the support lying outside the corridor defined by the aperture. If this were the only effect of the aperture in the space-frequency plane, then the statements made above would be exact (rather than being approximate) and the space-frequency support observed at the output could simply be found as follows: (a) Find the intersection of the input space-frequency support and the system space-frequency window at the input plane, (b) Propagate this space-frequency region to the output plane. However, this simple and intuitive result is not exact because each aperture that actually cuts off the outlying parts of the signal will also cause a broadening of the support of the signal along the orthogonal domain, due to the Fourier uncertainty relation.

We now argue that the broadening effects are generally negligible for most real physical signals and systems, so that the simple and intuitive result above is usually valid [37]. The effect of an aperture corresponds to multiplication with a rectangle function. Let Δ_j denote the size of the aperture. Firstly, if the signal extent before the aperture is already smaller than Δ_j, then the windowing operation will affect

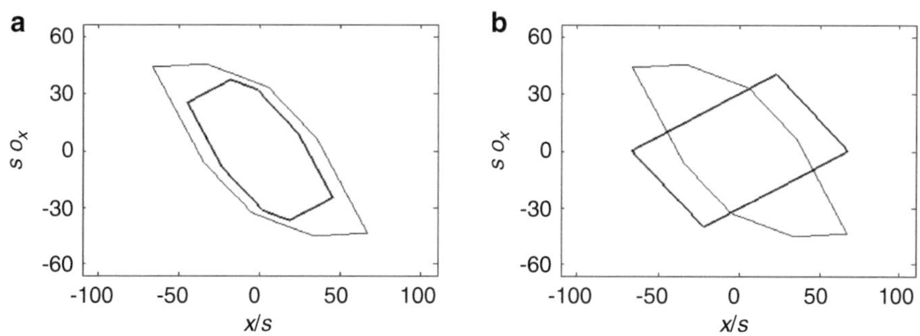

Fig. 7.12 (**a**) The signal support is wholly contained within the system window so there is no loss of information. (**b**) The part of the signal support lying within the system window will pass, and the parts lying outside will be blocked [37]

neither the signal nor its space-frequency support. However, if the signal extent in that domain is larger than Δ_j, then the signal will be truncated and the space-frequency support will also be affected. Because windowing involves multiplication with a rectangle function, it implies convolution of the Wigner distribution of the signal with the Wigner distribution of the rectangle function along the orthogonal direction [44, 59]. (An expression for the Wigner distribution, $\hat{W}_{\text{rect}}(x, \sigma_x)$ of the rectangle function $\text{rect}(x/\Delta_j)$ is known [5, 56] but its exact form is not necessary for our argument.) This operation will cause compaction of the Wigner distribution of the signal to a corridor of width Δ_j. Moreover, convolving the Wigner distribution of the signal with that of the rectangle function along the orthogonal direction will result in broadening of the Wigner distribution by an amount that is comparable with the extent of $\hat{W}_{\text{rect}}$ in that direction. This extent is approximately $1/\Delta_j$, and thus the spread in the orthogonal direction after windowing will be $\sim 1/\Delta_j$ [56].

For simplicity, consider a rectangular region in the space-frequency plane, in which case the space-bandwidth product can be taken as a measure of the number of degrees of freedom. Let us denote the space-bandwidth product as $N = \Delta x \Delta\sigma_x \geq 1$, where Δx ve $\Delta\sigma_x$ denote the spatial and frequency extents. Noting that the apertures can be modeled as rectangle functions, the frequency extent associated with the rectangle function will approximately be the reciprocal of its spatial extent: $1/\Delta_j$. Let us assume that the aperture extent is a fraction κ of the signal extent; that is $\Delta_j = \kappa\Delta x$ where $\kappa < 1$. After the aperture, the new space-domain signal extent will be given by $\Delta x' = \kappa\Delta x$. Moreover, since multiplication in the space-domain implies convolution in the frequency domain, the new extent in the frequency domain will be approximately the sum of the spectral extents of the signal and the aperture. The frequency extent of the signal is $\Delta\sigma_x = N/\Delta x$ and the frequency extent of the window is $\sim 1/\Delta_j = 1/\kappa\Delta x = \Delta\sigma_x/\kappa N$. Then, the new extent in the frequency domain will be $\Delta\sigma_x{}' \approx \Delta\sigma_x + \Delta\sigma_x/\kappa N = \Delta\sigma_x(1 + 1/\kappa N)$. Therefore, the space-bandwidth product of the signal after the aperture will be $\Delta x'\Delta\sigma_x{}' \approx \Delta x\Delta\sigma_x(\kappa + 1/N)$. Here, the first term corresponds to the reduced space-frequency support resulting from the truncation inflicted by the aperture, and the second term corresponds to the increase arising from the broadening in the orthogonal direction. However, if $\kappa \gg 1/N$, or equivalently $N \gg 1/\kappa$, then we can neglect the term $1/N$ in comparison with κ. Thus, we can neglect the broadening effect if $N \gg 1/\kappa$. This condition will hold for most real physical signals and systems. For a physical signal that contains any reasonable amount of information, such as an image, the number of degrees of freedom will be much larger than unity and also much larger than $1/\kappa$, as long as κ is not very close to 0. The case where κ is very close to 0 is not very likely either, since apertures with very small κ truncate nearly all of the signal. For instance, consider a window that allows only 0.1 of the extent of the signal to pass. Even in this case, $N \gg 10$ will be sufficient and most information bearing signals will satisfy this condition easily. Therefore, the broadening effect will be usually negligible when we are dealing with images and other information bearing signals. This in turn means that it is fairly accurate to say that when the space-frequency support of the signal does not wholly lie within the system window, the part that does lie within will pass, and the remaining parts will be lost. On the other hand, this simple result will not hold for some signals that do not exhibit too much spatial structure, such as a laser beam, and the broadening effect must be taken into account.

7.7.5 Discussion and Applications

We considered optical systems consisting of an arbitrary sequence of lenses and apertures separated by arbitrary lengths of free space (or quadratic graded-index media). We defined the space-frequency window (phase-space window) and showed how it can be explicitly determined for such a system. The area of the window gives the maximum number of degrees of freedom that can be supported by the system. More significantly, the window specifies which signals can pass through the system without information loss; we showed that the signal will pass losslessly if and only if the space-frequency support of the signal lies completely within this window. A precondition for lossless passage is of course that the area of the space-frequency support (and thus the number of degrees of freedom) of the set of input signals must be smaller than the area of the space-frequency window (and thus the number of degrees of freedom the system can support). We further saw that when the space-frequency support does not lie completely within the space-frequency window, the parts that lie within the window pass and the parts that lie outside of the window are blocked. While the last result is not exact, we showed that it is valid to a good degree of approximation for many systems of practical interest [33, 37].

These results are very intuitive and provide considerable insight and guidance into the behavior and design of systems involving multiple apertures. They can help designing systems in a manner that minimizes information loss, for instance by ensuring that the magnifications are as small as possible at aperture locations. An advantage of this approach is that it does not require assumptions regarding the input signals during analysis or design, since the concept of a system window is signal-independent.

Being able to determine the space-frequency window as a function of the system parameters as we have shown, and the possibility of tailoring and optimizing it has potential applications in areas including optical superresolution [77, 78, 81–85], holographic imaging [75, 80, 86–89], optical encryption systems [90], analysis and design of recording devices [76, 79], and comparison between different implementations of a particular system [91], where apertured optical systems are involved. The system window approach can yield new perspectives and rigorous approaches for such applications and other previously considered problems in the literature.

A potentially important area of application is optical superresolution and space-bandwidth product adaptation [92]. In this area the goal is to adapt the space-frequency support of the input signal to the space-frequency window of the system based on available a priori information about the signals. In most work in this area, the system window is commonly assumed to be, or approximated as, a rectangular shape or some other simple shape. Being able to precisely calculate the system window for a quite broad class of optical systems will make this superresolution approach more accurate, efficient, and widely applicable.

Moreover, in optical encryption [90], the system window can be used to investigate the optimal design of the encryption system, the most efficient representation of encrypted signals, and determination of the number of degrees of freedom that can be encrypted.

Yet another application of the results of this paper is the selection and optimization of recording devices. The system window at the output plane is of special use for this purpose, since it describes the largest space-frequency support that can be observed at the output of the system. This gives the position, maximum spatial extent, maximum frequency bandwidth (or in general the maximum extent in any LCT domain), and maximum space-bandwidth product (or more generally the bicanonical width product) that can be observed at the output of the system. The spatial extent of the system window should be matched to the location and width of the detector to ensure the recording of the entire output. The number of pixels required can be determined by fitting a rectangle to the system window and computing its area. This gives the number of samples needed to reconstruct any output signal from its Nyquist samples. Once the detector width and the number of pixels are determined, the pixel size is also revealed. Such an approach constitutes a new way of analyzing the optical efficiency of detectors [79]. Moreover, if specifications of the detector are pre-determined by some design limitations (such as limited spatial resolution), then the system can be adapted to work as best as it can with the specified detector [60].

The system window can also be useful in comparing alternative implementations of an optical system. One can choose among different implementations by investigating which implementation supports more degrees of freedom and hence causes less information (or power) loss due to the apertures (by comparing the areas of the system windows). Alternatively, the design goal can be to find the system window that is more compatible with the given detector limitations. Such approaches have been pursued, for example, for comparing different holographic systems [89] and different implementations of optical FRTs [91]. However, these previous approaches are either highly dependent on the input signal considered, involve many simplifications to make the analysis feasible, or yield only limited numerical results.

As a final note, we have mostly used the terms *space-frequency window* or *phase-space window* to distinguish these entities living in the space-frequency plane, from the physical apertures that act on signals in various LCT (or equivalently FRT) domains. However, since we have seen that these windows block or pass the space-frequency support of the signal in a manner very similiar to how apertures block or pass the physical signals, we can also speak of *space-frequency apertures* or *phase-space apertures*.

7.8 Conclusion

We discussed the relationships between LCT domains, FRT domains, and the space-frequency plane. In particular, we showed that LCT domains correspond to scaled fractional Fourier domains and thus to scaled oblique axes in the space-frequency plane. This allows LCT domains to be labeled and monotonically ordered by the corresponding fractional order parameter and provides a more transparent view of the evolution of light through an optical system modeled by LCTs.

We then studied the number of degrees of freedom of optical systems and signals based on these concepts. We first discussed the bicanonical width product, which is the number of degrees of freedom of LCT-limited signals. The bicanonical width product generalizes the space-bandwidth product and often provides a tighter measure of the actual number of degrees of freedom of signals. We illustrated the usefulness of the notion of bicanonical width product in two applications: efficient signal representation and efficient system simulation. In the first application we provided a sub-Nyquist sampling approach to represent and reconstruct signals with arbitrary space-frequency support. This approach geometrically amounts to enclosing the support with the smallest possible parallelogram, as opposed to enclosing it with a rectangle as in the classical approach. In the second application we provided a fast DLCT computation method which can accurately compute a (continuous) LCT with the minimum number of samples given by the bicanonical width product. Thus the bicanonical width product is also a key parameter in fast discrete computation of LCTs, and hence in efficient and accurate simulation of optical systems. Given the fundamental importance of the conventional space-bandwidth product in signal processing and information optics, we believe the bicanonical width product will find other applications in these areas as well.

Finally, we focused on the degrees of freedom of optical systems consisting of an arbitrary sequence of lenses and apertures separated by arbitrary lengths of free space (or quadratic graded-index media). We defined the space-frequency window (phase space window) and showed how it can be explicitly determined for such a system in terms of the system parameters. The area of the window gives the maximum number of degrees of freedom that can be supported by the system. More significantly, the window specifies which signals can pass through the system without information loss; we showed that the signal will pass losslessly if and only if the space-frequency support of the signal lies completely within this window. A precondition for lossless passage is of course that the area of the space-frequency support (and thus the number of degrees of freedom) of the set of input signals must be smaller than the area of the space-frequency window (and thus the number of degrees of freedom the system can support). We further saw that when the space-frequency support does not lie completely within the space-frequency window, the parts that lie within the window pass and the parts that lie outside of the window are blocked. While the last result is not exact, we showed that it is valid to a good degree of approximation for many systems of practical interest.

Thus, just as the concatenation property of transformation matrices allows us to represent the cumulative action of all optical elements with a single entity, the system window is an equivalent aperture that appropriately transforms and combines the effects of all individual apertures in different domains, into a single space-frequency aperture. These results are very intuitive and provide considerable insight and guidance into the behavior and design of systems involving multiple apertures. For example, they can help designing systems in a manner that minimizes information loss, with the advantage that no assumptions regarding the input signals is required, since the system window is a signal-independent entity. We briefly discussed some potential application areas where the system window approach can yield new perspectives. These include optical superresolution, optical encryption, holographic imaging, design and optimization of recording devices, and comparison of alternative implementations of apertured optical systems.

Acknowledgements This chapter is based on [33–37]. H.M. Ozaktas acknowledges partial support of the Turkish Academy of Sciences.

References

1. R.K. Luneburg, *Mathematical Theory of Optics* (University of California Press, Berkeley, 1966)
2. S.A. Collins, Lens-system diffraction integral written in terms of matrix optics. J. Opt. Soc. Am. **60**(9), 1168–1177 (1970)
3. M.J. Bastiaans, Wigner distribution function and its application to first-order optics. J. Opt. Soc. Am. **69**(12), 1710–1716 (1979)
4. M. Nazarathy, J. Shamir, First-order optics—a canonical operator representation: lossless systems. J. Opt. Soc. Am. **72**(3), 356–364 (1982)
5. H.M. Ozaktas, Z. Zalevsky, M.A. Kutay, *The Fractional Fourier Transform with Applications in Optics and Signal Processing* (Wiley, New York, 2001)
6. K.B. Wolf, Construction and properties of canonical transforms, Chap. 9, in *Integral Transforms in Science and Engineering* (Plenum Press, New York, 1979)
7. H.M. Ozaktas, O. Aytur, Fractional Fourier domains. Signal Process. **46**(1), 119–124 (1995)
8. H. Zhao, Q.-W. Ran, J. Ma, L.-Y. Tan, On bandlimited signals associated with linear canonical transform. IEEE Signal Process. Lett. **16**(5), 343–345 (2009)
9. B. Deng, R. Tao, Y. Wang, Convolution theorems for the linear canonical transform and their applications. Sci. China Ser. F Inf. Sci. **49**(5), 592–603 (2006)
10. K.K. Sharma, S.D. Joshi, Uncertainty principle for real signals in the linear canonical transform domains. IEEE Trans. Signal Process. **56**(7), 2677–2683 (2008)
11. K.K. Sharma, S.D. Joshi, Signal separation using linear canonical and fractional Fourier transforms. Opt. Commun. **265**(2), 454–460 (2006)
12. K.K. Sharma, S.D. Joshi, Signal reconstruction from the undersampled signal samples. Opt. Commun. **268**(2), 245–252 (2006)
13. K.K. Sharma, New inequalities for signal spreads in linear canonical transform domains. Signal Process. **90**(3), 880–884 (2010)
14. B.-Z. Li, R. Tao, Y. Wang, New sampling formulae related to linear canonical transform. Signal Process. **87**(5), 983–990 (2007)
15. A. Stern, Uncertainty principles in linear canonical transform domains and some of their implications in optics. J. Opt. Soc. Am. A **25**(3), 647–652 (2008)
16. A. Stern, Sampling of compact signals in offset linear canonical transform domains. Signal Image Video Process. **1**, 359–367 (2007)

17. G. Toraldo di Francia, Resolving power and information. J. Opt. Soc. Am. **45**(7), 497–499 (1955)
18. D. Gabor, Light and information, in *Progress in Optics*, vol. I, Chap. 4, ed. by E. Wolf (Elsevier, Amsterdam, 1961), pp. 109–153
19. G. Toraldo di Francia, Degrees of freedom of an image. J. Opt. Soc. Am. **59**(7), 799–803 (1969)
20. F. Gori, G. Guattari, Effects of coherence on the degrees of freedom of an image. J. Opt. Soc. Am. **61**(1), 36–39 (1971)
21. F. Gori, G. Guattari, Shannon number and degrees of freedom of an image. Opt. Commun. **7**(2), 163–165 (1973)
22. F. Gori, G. Guattari, Degrees of freedom of images from point-like-element pupils. J. Opt. Soc. Am. **64**(4), 453–458 (1974)
23. F. Gori, S. Paolucci, L. Ronchi, Degrees of freedom of an optical image in coherent illumination, in the presence of aberrations. J. Opt. Soc. Am. **65**(5), 495–501 (1975)
24. F. Gori, L. Ronchi, Degrees of freedom for scatterers with circular cross section. J. Opt. Soc. Am. **71**(3), 250–258 (1981)
25. L. Ronchi, F. Gori, Degrees of freedom for spherical scatterers. Opt. Lett. **6**(10), 478–480 (1981)
26. A. Starikov, Effective number of degrees of freedom of partially coherent sources. J. Opt. Soc. Am. **72**(11), 1538–1544 (1982)
27. G. Newsam, R. Barakat, Essential dimension as a well-defined number of degrees of freedom of finite-convolution operators appearing in optics. J. Opt. Soc. Am. A **2**(11), 2040–2045 (1985)
28. A.W. Lohmann, *Optical Information Processing*. Lecture Notes (Optik+Info, Uttenreuth, 1986)
29. F. Gori, Sampling in optics, in *Advanced Topics in Shannon Sampling and Interpolation Theory*, Chap. 2 (Springer, New York, 1993), pp. 37–83
30. A.W. Lohmann, R.G. Dorsch, D. Mendlovic, Z. Zalevsky, C. Ferreira, Space-bandwidth product of optical signals and systems. J. Opt. Soc. Am. A **13**(3), 470–473 (1996)
31. R. Piestun, D.A.B. Miller, Electromagnetic degrees of freedom of an optical system. J. Opt. Soc. Am. A **17**(5), 892–902 (2000)
32. R. Solimene, R. Pierri, Number of degrees of freedom of the radiated field over multiple bounded domains. Opt. Lett. **32**(21), 3113–3115 (2007)
33. F.S. Oktem, Signal representation and recovery under partial information, redundancy, and generalized finite extent constraints, Master's thesis, Bilkent University, 2009
34. F.S. Oktem, H.M. Ozaktas, Equivalence of linear canonical transform domains to fractional Fourier domains and the bicanonical width product: a generalization of the space–bandwidth product. J. Opt. Soc. Am. A **27**(8), 1885–1895 (2010)
35. F.S. Oktem, H.M. Ozaktas, Exact relation between continuous and discrete linear canonical transforms. IEEE Signal Process. Lett. **16**(8), 727–730 (2009)
36. F.S. Oktem, H.M. Ozaktas, Degrees of freedom of optical systems and signals with applications to sampling and system simulation, in *Imaging and Applied Optics Conference*, Optical Society of America, 2013
37. H.M. Ozaktas, F.S. Oktem, Phase-space window and degrees of freedom of optical systems with multiple apertures. J. Opt. Soc. Am. A **30**(4), 682–690 (2013)
38. T. Alieva, M.J. Bastiaans, Alternative representation of the linear canonical integral transform. Opt. Lett. **30**(24), 3302–3304 (2005)
39. M.J. Bastiaans, T. Alieva, Synthesis of an arbitrary ABCD system with fixed lens positions. Opt. Lett. **31**(16), 2414–2416 (2006)
40. J.A. Rodrigo, T. Alieva, M.L. Calvo, Optical system design for orthosymplectic transformations in phase space. J. Opt. Soc. Am. A **23**(10), 2494–2500 (2006)
41. H.M. Ozaktas, M.F. Erden, Relationships among ray optical, Gaussian beam, and fractional Fourier transform descriptions of first-order optical systems. Opt. Commun. **143**(1–3), 75–86 (1997)
42. H. Ozaktas, S. Arık, T. Coşkun, Fundamental structure of Fresnel diffraction: natural sampling grid and the fractional Fourier transform. Opt. Lett. **36**(13), 2524–2526 (2011)

43. H. Ozaktas, S. Arık, T. Coşkun, Fundamental structure of Fresnel diffraction: longitudinal uniformity with respect to fractional Fourier order. Opt. Lett. **37**(1), 103–105 (2012)
44. L. Cohen, *Integral Time-Frequency Analysis* (Prentice-Hall, Englewood Cliffs, 1995)
45. M.J. Bastiaans, Applications of the Wigner distribution function in optics, in *The Wigner Distribution: Theory and Applications in Signal Processing* (Elsevier, Amsterdam, 1997), pp. 375–426
46. G. Forbes, V. Maniko, H. Ozaktas, R. Simon, K. Wolf, Wignerdistributions and phase space in optics. J. Opt. Soc. Am. A **17**(12), 2274–2274 (2000)
47. A. Stern, Sampling of linear canonical transformed signals. Signal Process. **86**(7), 1421–1425 (2006)
48. J.J. Ding, *Research of fractional Fourier transform and linear canonical transform.* Ph.D. thesis, National Taiwan University, Taipei, 2001
49. X.-G. Xia, On bandlimited signals with fractional Fourier transform. IEEE Signal Process. Lett. **3**(3), 72–74 (1996)
50. A. Zayed, On the relationship between the Fourier and fractional Fourier transforms. IEEE Signal Process. Lett. **3**(12), 310–311 (1996)
51. C. Candan, H.M. Ozaktas, Sampling and series expansion theorems for fractional Fourier and other transforms. Signal Process. **83**, 1455–1457 (2003)
52. T. Erseghe, P. Kraniauskas, G. Carioraro, Unified fractional Fourier transform and sampling theorem. IEEE Trans. Signal Process. **47**(12), 3419–3423 (1999)
53. R. Torres, P. Pellat-Finet, Y. Torres, Sampling theorem for fractional bandlimited signals: A self-contained proof application to digital holography. IEEE Signal Process. Lett. **13**(11), 676–679 (2006)
54. R. Tao, B. Deng, W.-Q. Zhang, Y. Wang, Sampling and sampling rate conversion of band limited signals in the fractional Fourier transform domain. IEEE Trans. Signal Process. **56**(1), 158–171 (2008)
55. O. Aytur, H.M. Ozaktas, Non-orthogonal domains in phase space of quantum optics and their relation to fractional Fourier transforms. Opt. Commun. **120**(3–4), 166–170 (1995)
56. H.M. Ozaktas, B. Barshan, D. Mendlovic, L. Onural, Convolution, filtering, and multiplexing in fractional Fourier domains and their relation to chirp and wavelet transforms. J. Opt. Soc. Am. A **11**(2), 547–559 (1994)
57. J.J. Healy, J.T. Sheridan, Cases where the linear canonical transform of a signal has compact support or is band-limited. Opt. Lett. **33**(3), 228–230 (2008)
58. H.M. Ozaktas, D. Mendlovic, Fractional Fourier optics. J. Opt. Soc. Am. A **12**(4), 743–751 (1995)
59. L. Cohen, Time-frequency distributions-a review. Proc. IEEE **77**(7), 941–981 (1989)
60. A. Stern, B. Javidi, Sampling in the light of Wigner distribution. J. Opt. Soc. Am. A **21**(3), 360–366 (2004)
61. A. Stern, Why is the linear canonical transform so little known?, in *AIP Conference Proceedings* (2006), pp. 225–234
62. A. Papoulis, *Signal Analysis* (McGraw-Hill, New York, 1977)
63. S.-C. Pei, J.-J. Ding, Closed-form discrete fractional and affine Fourier transforms. IEEE Trans. Signal Process. **48**(5), 1338–1353 (2000)
64. B.M. Hennelly, J.T. Sheridan, Fast numerical algorithm for the linear canonical transform. J. Opt. Soc. Am. A **22**(5), 928–937 (2005)
65. B.M. Hennelly, J.T. Sheridan, Generalizing, optimizing, and inventing numerical algorithms for the fractional Fourier, Fresnel, and linear canonical transforms. J. Opt. Soc. Am. A **22**(5), 917–927 (2005)
66. J.J. Healy, B.M. Hennelly, J.T. Sheridan, Additional sampling criterion for the linear canonical transform. Opt. Lett. **33**(22), 2599–2601 (2008)
67. J.J. Healy, J.T. Sheridan, Sampling and discretization of the linear canonical transform. Signal Process. **89**, 641–648 (2009)
68. H.M. Ozaktas, A. Koç, I. Sari, M.A. Kutay, Efficient computation of quadratic-phase integrals in optics. Opt. Lett. **31**(1), 35–37 (2006)

69. A. Koc, H.M. Ozaktas, C. Candan, M.A. Kutay, Digital computation of linear canonical transforms. IEEE Trans. Signal Process. **56**(6), 2383–2394 (2008)
70. J.J. Healy, J.T. Sheridan, Reevaluation of the direct method of calculating Fresnel and other linear canonical transforms. Opt. Lett. **35**(7), 947–949 (2010)
71. J.J. Healy, J.T. Sheridan, Fast linear canonical transforms. J. Opt. Soc. Am. A **27**(1), 21–30 (2010)
72. H. Ozaktas, O. Arikan, M. Kutay, G. Bozdagi, Digital computation of the fractional Fourier transform. IEEE Trans. Signal Process. **44**(9), 2141–2150 (1996)
73. A. Koc, H. Ozaktas, L. Hesselink, Fast and accurate computation of two-dimensional non-separable quadratic-phase integrals. J. Opt. Soc. Am. A **27**(6), 1288–1302 (2010)
74. L. Durak, O. Arikan, Short-time Fourier transform: two fundamental properties and an optimal implementation. IEEE Trans. Signal Process. **51**(5), 1231–1242 (2003)
75. A.W. Lohmann, The space-bandwidth product, applied to spatial filtering and holography, Research Paper RJ-438, IBM San Jose Research Laboratory, San Jose, 1967
76. A. Stern, B. Javidi, Shannon number and information capacity of three-dimensional integral imaging. J. Opt. Soc. Am. A **21**(9), 1602–1612 (2004)
77. D. Mendlovic, A. Lohmann, Space–bandwidth product adaptation and its application to superresolution: fundamentals. J. Opt. Soc. Am. A **14**(3), 558–562 (1997)
78. Z. Zalevsky, D. Mendlovic, A. Lohmann, Understanding superresolution in Wigner space. J. Opt. Soc. Am. A **17**(12), 2422–2430 (2000)
79. P. Catrysse, B. Wandell, Optical efficiency of image sensor pixels. J. Opt. Soc. Am. A **19**(8), 1610–1620 (2002)
80. J. Maycock, C. McElhinney, B. Hennelly, T. Naughton, J. McDonald, B. Javidi, Reconstruction of partially occluded objects encoded in three-dimensional scenes by using digital holograms. Appl. Opt. **45**(13), 2975–2985 (2006)
81. D. Mendlovic, A. Lohmann, Z. Zalevsky, Space–bandwidth product adaptation and its application to superresolution: examples. J. Opt. Soc. Am. A **14**(3), 563–567 (1997)
82. K. Wolf, D. Mendlovic, Z. Zalevsky, Generalized Wigner function for the analysis of superresolution systems. Appl. Opt. **37**(20), 4374–4379 (1998)
83. Z. Zalevsky, N. Shamir, D. Mendlovic, Geometrical superresolution in infrared sensor: experimental verification. Opt. Eng. **43**(6), 1401–1406 (2004)
84. Z. Zalevsky, V. Mico, J. Garcia, Nanophotonics for optical super resolution from an information theoretical perspective: a review. J. Nanophotonics **3**(1), 032502–032502 (2009)
85. J. Lindberg, Mathematical concepts of optical superresolution. J. Opt. **14**(8), 083001 (2012)
86. L. Xu, X. Peng, Z. Guo, J. Miao, A. Asundi et al., Imaging analysis of digital holography. Opt. Express **13**(7), 2444–2452 (2005)
87. M. Testorf, A. Lohmann, Holography in phase space. Appl. Opt. **47**(4), A70–A77 (2008)
88. U. Gopinathan, G. Pedrini, B. Javidi, W. Osten, Lensless 3D digital holographic microscopic imaging at vacuum UV wavelength. J. Disp. Technol. **6**(10), 479–483 (2010)
89. D. Claus, D. Iliescu, P. Bryanston-Cross, Quantitative space-bandwidth product analysis in digital holography. Appl. Opt. **50**(34), H116–H127 (2011)
90. B. Hennelly, J. Sheridan, Optical encryption and the space bandwidth product. Opt. Commun. **247**(4), 291–305 (2005)
91. J. Healy, J. Sheridan, Bandwidth, compact support, apertures and the linear canonical transform in ABCD systems, in *Proceedings of the SPIE*, vol. 6994 (2008), p. 69940W
92. Z. Zalevsky, D. Mendlovic, *Optical Superresolution* (Springer, New York, 2004)

Chapter 8
Sampling and Discrete Linear Canonical Transforms

John J. Healy and Haldun M. Ozaktas

Abstract A discrete linear canonical transform would facilitate numerical calculations in many applications in signal processing, scalar wave optics, and nuclear physics. The question is how to define a discrete transform so that it not only approximates the continuous transform well, but also constitutes a discrete transform in its own right, being complete, unitary, etc. The key idea is that the LCT of a discrete signal consists of modulated replicas. Based on that result, it is possible to define a discrete transform that has many desirable properties. This discrete transform is compatible with certain algorithms more than others.

8.1 Introduction

Most of the literature on the LCTs, including many chapters of this book, explicitly or implicitly make use of the continuous transform. There are, however, a number of situations in which it is desirable or necessary to use a discrete transform. Most of these are obvious by analogy with situations in which the fast Fourier transform (FFT) is used to numerically approximate the Fourier transform. In addition, the LCT is of increasing relevance in situations where we wish to model optical systems, including those with inherently discrete components such as spatial light modulators or digital cameras. In this chapter, we will explore the relationship of the LCTs with sampling, and the consequences for everything from the definition of a corresponding discrete transform to how to perform sampling rate changes accurately and efficiently.

The impact of this material should be clear to anyone who has had to comprehend the details of calculations involving the FFT. For many users, the FFT is essentially a black box that performs a Fourier transform on their data. This is close enough

J.J. Healy (✉)
School of Electrical and Electronic Engineering, University College Dublin, Belfield, Dublin 4, Ireland
e-mail: john.healy@ucd.ie

H.M. Ozaktas
Department of Electrical Engineering, Bilkent University, 06800 Bilkent, Ankara, Turkey
e-mail: haldun@ee.bilkent.edu.tr

J.J. Healy et al. (eds.), *Linear Canonical Transforms*, Springer Series in Optical Sciences 198, DOI 10.1007/978-1-4939-3028-9_8

to the truth to make the FFT useful to users with even a primitive understanding of Fourier analysis. Such 'idiot-proofing' arises from the definition of the discrete Fourier transform (DFT), carefully chosen to be unitary, complete and have many of the properties of the Fourier transform of continuous signals. Closer examination reveals subtle differences such as circular convolution. Numerous papers from the past two decades have developed parts of a theory of numerical approximation of the LCTs. This chapter will provide a summary of some of these results, and a discussion of how they fit together to build a complete picture of the relationship between the continuous and discrete LCTs.

First, however, we must introduce some notation and certain key ideas which will crop up repeatedly in this discussion.

8.1.1 Linear Canonical Transform and Notation

This material is covered well in numerous other chapters, most particularly Chap. 2, and the reader is referred there for a broad introduction to the transform. However, there may be some notational differences, and so we have included this brief section for clarity.

Given a function of a single variable, $f(x), f : \mathbb{R} \rightarrow \mathbb{C}$, the LCT of that function for parameter matrix $M = \left(\begin{smallmatrix} a & b \\ c & d \end{smallmatrix}\right)$ is given by

$$F_M(y) = \mathfrak{L}_M\{f(x)\}(y) = \begin{cases} (1/\sqrt{jb}) \int_{-\infty}^{\infty} f(x) \exp\left(\frac{j\pi}{2b} ax^2 - 2xy + dy^2\right) & \text{if} b \neq 0 \\ \sqrt{d} \exp\left(j\pi c d y^2\right) f(dy) & \text{if} b = 0. \end{cases}$$

We will briefly discuss the extension to complex y and complex elements of M later, but unless specified otherwise, these are taken to be real. We will also briefly mention transforms of 2 D signals, but omit the transform definition here. Finally, the $b = 0$ case is trivial, and it is generally assumed that $b \neq 0$ without significant consequence. The elements of M are referred to as the ABCD parameters, and M has unit determinant.

8.1.2 Dirac Delta Function

A key idea in understanding the relationship between discrete and continuous signals and systems is the Dirac delta function. Introduced by Paul Dirac, the delta function is commonly used to model a sampling pulse. It is defined as follows:

$$\int f(x)\delta(x - \tau)\, dx = f(\tau). \tag{8.1}$$

This is called the *sifting* property of the delta function. Note the form of the definition: the delta function is defined by its effect under integration. The following inadequate definition of the delta function can be useful heuristically.

$$\delta(x) = \begin{cases} \infty & \text{if} x = 0 \\ 0 & \text{if} x \neq 0. \end{cases}$$

The delta function may also be viewed as the limit of a series of functions. Consider a rectangle function which has unit area

$$r_L(x) = \begin{cases} 1/L & \text{if} |x| < L/2 \\ 0 & \text{otherwise.} \end{cases}$$

Now consider the limit of this function as $L \to 0$. The function grows narrower and taller, but always has unit area. Thus we can think of the delta function as being a pulse that is infinitely tall, infinitely thin and has unit area.

8.1.3 Time-Frequency Representations

The Fourier transform of a time-varying signal decomposes the signal into a weighted sum of its frequency components, but it doesn't localize them in time. Time-frequency representations of a signal attempt to overcome this limitation. They have proven useful in a variety of applications, but appear here because of their utility in discussing sampling problems associated with the LCTs.

The reader is referred to the introductory chapters of this book for a review of the Wigner–Ville distribution function (WDF). Briefly, a time-varying function $f(t)$ is mapped to a function of time and frequency, $W(t, k)$. Loosely, this function identifies which frequencies are active and how active they are at any given time. Strictly speaking, it is impossible to speak of an instantaneously occurring frequency component—this would breach Heisenberg's uncertainty principle—and so the WDF is referred to as a pseudo-distribution. In an optical context, the time variable is commonly replaced by a spatial variable without repercussion.

A significant simplification of the WDF which retains a lot of useful information is the *phase space diagram* (PSD), see, e.g., [20, 32]. The idea is to pick an arbitrary rectangle enclosing most of the WDF of a given signal. (The WDF must necessarily be of infinite extent, so the whole signal cannot be encapsulated. Hence, we must use nebulous terms like "most". We must choose arbitrary limits, such as those that enclose some proportion, η, of the signal power.) The evolution of this simple geometric shape may then be trapped as the signal passed through an optical system. Such tracking is possible because the effect of the LCT (in this case, implemented as the optical system) is to perform a linear co-ordinate transformation in time-frequency [1]. The resulting parallelograms indicate the width and bandwidth of the signal throughout the system. Figure 8.1 illustrates such a PSD.

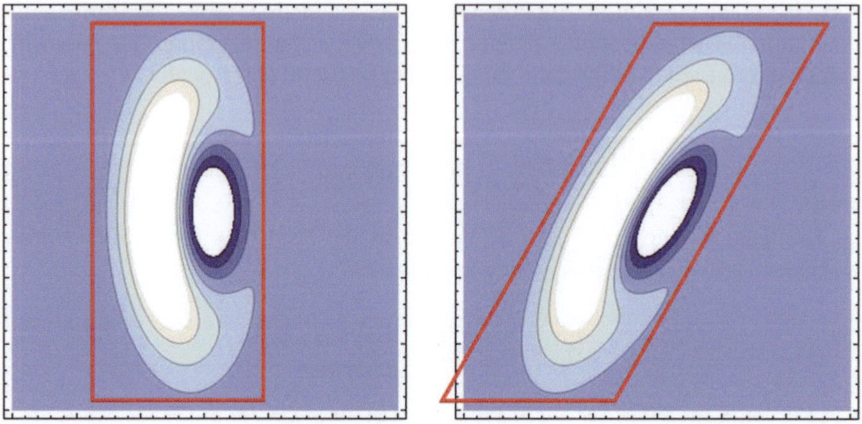

Fig. 8.1 Contour maps of the WDF signal $(3x^2 - 5x + 1)e^{-x^2}$ before (*left*) and after (*right*) free space propagation. The *red boxes* are the PSDs for the same signal

The PSD has been expanded upon by a number of authors [32]. We will encounter some of these at later points in this chapter, but let us mention one of the more relevant examples: Hennelly and Sheridan systematized and applied PSDs to show that by tracking the sampling requirements at each step of an algorithm consisting of a series of LCTs, they could obtain satisfactory results from such algorithms [12]. The flaw in their result is that it does not account for the input to those algorithms being discrete [7]. Shortly, we will see how that may be overcome.

8.2 Sampling

Shannon–Nyquist sampling is so broadly useful and applied that it is natural to attempt to use it in relation to the LCT. However, the underlying assumption—that a signal has finite bandwidth—does not in general lead to the most efficient result, nor is it often the simplest choice. In this section, we will discuss a different assumption, associated with the LCT, which leads to a sampling theorem more general than that of Nyquist. This sampling theorem, which is essential for the theory of a discrete transform, permits more efficient numerical calculations and has physical significance in that it explains higher order diffraction terms in holography [13]. First however, in order to facilitate the reader's comprehension of this more general sampling theory, let us outline the Shannon–Nyquist sampling theorem, which is associated with the Fourier transform (FT). You will recall that the FT is a special case of the LCT, and so Shannon–Nyquist sampling similarly is a special case of the more general theory.

Suppose we have a signal, $f(x)$, which we want to sample. Shannon–Nyquist sampling starts with the assumption that the FT of $f(x)$, $F(k)$, is zero outside of some finite range of k.

$$F(k) = 0, |k| > \Omega.$$

The width of this finite range of support is then directly related to a sampling rate for the signal, the Nyquist rate. Its reciprocal, T, is called the Nyquist period.

$$r_s = 1/T = 2\Omega.$$

A reconstruction filter is specified which allows recovery of the continuous signal, $f(x)$, from its samples, $f(nT)$ taken at not less than that rate.

$$f(x) = \sum_{n=-\infty}^{\infty} f(nT)\frac{\sin((x-nT))}{\pi(x-nT)}.$$

It is illuminating to think about Shannon–Nyquist sampling in the Fourier domain. Given a signal, $f(x)$, we sample it at a rate T. The resulting spectrum of the sampled signal is as follows:

$$F_s(k) = \sum_{p=-\infty}^{\infty} F(k-pT).$$

Hence, the effect of the sampling operation in the Fourier domain is to create periodic copies of the signal. If the signal had finite support prior to sampling, and if the separation between replicas is sufficiently large, we can obtain the original signal from the copies by filtering. We can interpret the equivalent result for the LCT, which we will examine next, in an identical fashion except that we no longer consider the Fourier domain, but the output domain of the LCT.

8.2.1 Uniform Sampling

Prior to the general LCT case, a number of special cases were obtained. First, Franco Gori derived the special case for the Fresnel transform [4], i.e. a theorem that, given a signal that has compact support in some Fresnel plane, determines a sampling rate such that the signal may be recovered from its samples taken at no less than that rate. In the same paper, Gori also proved that a signal with finite bandwidth could not have finite support in any Fourier domain. The equivalent of this latter result for the LCT is more complicated [5] and depends on the ABCD parameters. Results regarding the compactness and bandlimitedness of different LCTs can be easily understood in terms of the concept of essentially equivalent LCT domains [23]. Later, Xia was the first of several authors to derive the corresponding sampling theorem for the fractional Fourier transform [34].

Ding proved that if a particular LCT of a signal is zero outside some range $|x| < \Omega$, then the signal may be sampled at regular intervals of $T = B/\Omega$ [3]. Ding's theorem generalizes those of Shannon and Nyquist, Gori, and Xia in the sense that those theorems refer to special cases of the LCT, namely the Fourier, Fresnel and fractional Fourier transforms, respectively. Ding's proof resembles Gori's, though he does not cite Gori. Ding's theorem was later independently derived by several others, including Stern [29], Deng et al. [2] and Li et al. [17]. Stern's proof uses the Poisson formula for the comb function to derive the LCT of a sampled signal. Deng et al.'s approach was to derive the convolution and multiplication theorems for the LCT, and then calculate the LCT of the product of an arbitrary function and a train of delta functions. Li et al.'s proof is similar to Ding's. All of these proofs use techniques from well-known proofs of the Shannon–Nyquist theorem, and ultimately all of them are equivalent.

Similar to the Shannon–Nyquist theorem, Ding's theorem shows that the effect of sampling a signal on the LCT of that signal is to create an infinite number of shifted copies of the signal, with a regular spacing that depends on the sampling rate. The difference is that each replica is modulated, multiplied by a chirp; hence, the LCT of a discrete signal is referred to as 'chirp-periodic'. Again, as with the Shannon–Nyquist theorem, if the LCT of the continuous signal has finite support we can specify a sampling rate that ensures the replicas do not overlap, and it is possible to extract the continuous signal from among the replicas using an appropriate filter.

We now sketch a proof of the theorem. This proof ignores some constants, but the form of the solution is correct. Consider a signal, $f(x)$. Let us assume that $g(x) = f(x)\exp(j\pi ax^2/2b)$ is bandlimited in the Fourier domain, i.e. $G(k) = 0$ if $|k| > \omega$. According to the Shannon–Nyquist theorem, if we sample $g(x)$ (equivalent to sampling $f(x)$ first) with sample period T, its Fourier transform becomes periodic, i.e.

$$G_s(k) = \sum_{p=-\infty}^{\infty} G(k - p/T).$$

If we scale $G_s(k)$ by a factor b, and substitute $y = bk$, we obtain

$$G_s(y) = \sum_{p=-\infty}^{\infty} G(y - p(b/T)).$$

Finally, we multiply this function by a chirp, yielding the LCT of $f(nT)$,

$$F_{M,s}(y) = \exp(j\pi dy^2/2b) \sum_{p=-\infty}^{\infty} G(y - p(b/T)). \tag{8.2}$$

The parameters of this LCT are given by

$$\begin{pmatrix} a & b \\ c & d \end{pmatrix} = \begin{pmatrix} 1 & 0 \\ d/b & 1 \end{pmatrix} \begin{pmatrix} b & 0 \\ 0 & b^{-1} \end{pmatrix} \begin{pmatrix} 0 & 1 \\ -1 & 0 \end{pmatrix} \begin{pmatrix} 1 & 0 \\ a/b & 1 \end{pmatrix}.$$

In other words, we decomposed the LCT into a series of steps, which are read from right to left as follows: a chirp multiplication, a Fourier transform, magnification or scaling, and a second chirp multiplication. The consequence of Eq. (8.2.1) is that the LCT of a discrete signal is chirp-periodic. The replicas are spaced by b/T, so if the LCT of the original, continuous signal is zero for $|y| > \Omega$, sampling with sample period not less than $T = b/\Omega$, we can filter out the replicas much as in the Fourier case.

The complex parametered case was considered in [19]. By the complex parametered case, we mean not only that the ABCD parameters are complex but so too is the output variable, y. Here, the key result is as follows:

$$\mathfrak{L}_M\{\hat{f}(x))\}(y)$$
$$= \exp\left(\frac{j\pi dy^2}{2b}\right) \sum_{m=-\infty}^{\infty} F_M\left(y - \frac{mB}{T}\right) \exp\left(-j\pi d\left(y - \frac{mB}{T}\right)/2b\right).$$

Consequently, if we sample a function with sampling period T, its complex LCT develops regularly spaced, modulated replicas along a line in the complex plane that depends on the b parameter. The spacing of these replicas also depends on the b parameter, and is given by b/T.

The discussion in this section does not account for noise. This issue has been addressed in some recent papers [28, 38]. Neither does it account for quantisation error, which has also been examined recently [25].

8.2.2 Sampling Rate Conversion

A common problem may be stated as follows. Given the samples of a signal which have been taken at a given rate, how do we obtain the samples of the same signal taken at a different rate without damaging the signal? Such sampling rate conversions are often required to downsample signals for computational savings, demodulation and many other applications. Zhao et al. have proposed methods for sampling rate conversions [36] for signals that are bandlimited in some LCT domain. Their work applies to rational number sampling rate changes. Previously published methods for sampling rate conversion in the Fourier domain and the fractional Fourier domain are special cases of their result.

There are two fundamental operations involved: interpolation and downsampling or decimation. Interpolation was defined as follows:

$$g(n) = \begin{cases} f(n/L) & \text{if } \frac{n}{L} = k, k \in \mathbb{Z} \\ 0 & \text{otherwise.} \end{cases}$$

In the LCT domain, this does not alter the signal. Downsampling is defined as follows:

$$g(n) = x(Qn),$$

where Q is the integer downsampling conversion factor. The consequence of this in the LCT domain is as follows:

$$G_T(y) = \frac{1}{Q}\sum_{k=0}^{Q-1} F_T\left(y - \frac{\text{sgn}\,(b)k}{Q}\right)\exp\left(j2\pi kbd\frac{(\text{sgn}\,(b)y - \pi k)}{(QT)^2}\right).$$

8.2.3 *Nonuniform Sampling*

We may have to deal with nonuniform sampling measurements in various settings, e.g. tomography, or due to factors that impinge on our ability to sample regularly, such as jitter, or dropped packets. A number of papers have addressed this issue for a variety of special cases [18, 27, 31, 33, 35]. For example, in [31], Tao et al. discussed two factors that affect the quality of reconstruction. One was truncation, which we will discuss in the following section on the discrete transform. The other is the situation where we have irregularly or nonuniformly sampled data. In this latter case, four special cases are examined in [31].

The first of the four non-uniform cases considered by Tao *et al.* was that of signals for which the locations of the samples are periodic. The second case considered is a special case of the first. The sample locations are again assumed to be periodic, but the locations in each period are assumed to be further divisible into groups with a common inner period. The next case considered is where we nearly have uniform sampling, but it has been degraded by a finite number of samples being taken at incorrect locations.

Finally, a more general case was considered. Given a signal that is bandlimited in some given LCT domain, such that $\mathfrak{L}_T\{f(x)\}(y) = 0$ for $|y| > \Omega$, then the original signal is uniquely determined by its samples taken at locations t_n if

$$\left|t_n - n\frac{b\pi}{\Omega}\right| \leq D < \frac{b\pi}{4\Omega}.$$

The reconstruction formulae for these cases are omitted here, and may be found in [31].

8.3 Exact Relation Between Discrete and Continuous LCTs

We have seen that the LCT of a discrete signal is chirp-periodic. In this section, we will show how to use that fact to define a well-behaved discrete LCT. This section is largely derived from [6, 22].

The obvious way to fashion a discrete transform is simply to sample the input and output domains. The function to be transformed, $f(x)$, becomes a vector of input samples, the nth element of which is given by $f[n] = f(nT_x)$. Similarly, the transformed function becomes the vector F_M, where the mth element is given by $F_M[m] = \mathfrak{L}_M\{\hat{f}(x))\}(mT_y)$. Applying the same substitutions to the LCT kernel allows us to relate the input and output samples by means of a matrix multiplication.

$$F_M = Wf, \tag{8.3}$$

where W is a square matrix with elements,

$$W[n,m] = K \exp \frac{j\pi}{B}(A(nT_x)^2 + nmT_xT_y + D(mT_y)^2).$$

W is obtained by sampling the kernel of the LCT as described in the preceding paragraph. Here, K is some complex constant, which we determine by requiring the discrete transform to be unitary, which concept we now define.

When discussing discrete transforms, one property we are often interested in is whether or not they are unitary. This is often compared with the continuous property of power conservation, and indeed it resembles it, but it fundamentally determines whether or not the transform is invertible. Given a 1-D linear transformation such as that of Eq. (8.3), then unitarity is the following requirement.

$$WW^\dagger = I_N,$$

where I_N is the identity matrix of dimension N, and the $\dagger$ indicates the Hermitian conjugate.

To our knowledge, the first proposed discrete LCT of this kind was by Pei and Ding in [26]. It has since been found that the sampling rates, T_x and T_y, are critical if we want a genuine discrete transform. The idea is elegant: since the continuous LCT of a discrete signal is chirp-periodic, and hence the continuous LCT of a chirp-periodic signal is discrete, then a true discrete LCT transforms a discrete, chirp-periodic signal into another signal of that kind with the underlying periodically replicated functions being related to each other through a continuous LCT [22, 30]. Oktem and Ozaktas showed that such a relation exists with a discrete transform if the number of samples is determined from the knowledge of the signal's extent in the input and output domains, implying a parallelogram-shaped support in phase space [21]. Figure 8.2 demonstrates this using a series of modified PSDs. Healy and Sheridan proposed that it was necessary to include the location of replicas [10] (which neglects the cross-terms, though this is not critical [11]) and later also incorporated the parallelogram bound [6].

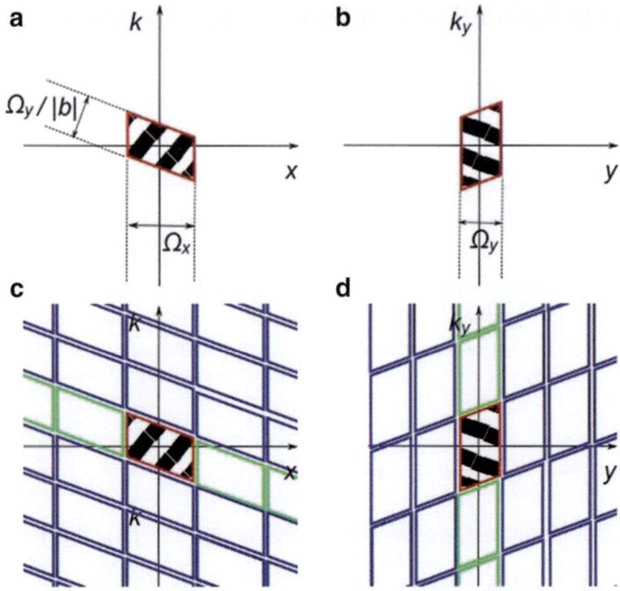

Fig. 8.2 The PSD of a continuous signal that is bandlimited in some LCT domain (**a**) before and (**b**) after that LCT. The PSD of a discrete, chirp-periodic signal (**c**) before and (**d**) after a discrete LCT. The green replicas alone in (**c**) constitute a chirp-periodic signal (and are transformed into a discrete signal, exactly periodic in frequency—see the green replicas in (**d**)), while the blue replicas are a consequence of the signal being discrete

The consequence of this idea of using discrete, chirp-periodic signals is that T_x and T_y should both be chosen using Ding's sampling theorem, i.e.

$$T_x = \frac{\Omega_y}{|b|},$$

and

$$T_y = \frac{\Omega_x}{|b|},$$

where Ω_x and Ω_y are the widths of the signal in the input and output domain, respectively. In practice, the sampling periods are also not independent of one another. Since $\Omega_y = NT_y$,

$$T_x = \frac{N}{|b|}T_y.$$

That is, the three parameters T_x, T_y, and N should be chosen using the preceding three equations together. This choice of sampling rates was first implicitly proposed in [30] (though without any observation of the consequences). Oktem and Ozaktas later independently proposed this rate, and indicated that this choice provides the minimum possible number of samples to approximate the continuous LCT when finite extents in the input and output domains are assumed [22]. Healy and Sheridan interpreted the results in phase space [6], where chirp-periodic signals form a tiling of parallelograms which packs the replicas optimally, indicating that the sampling rate is optimal under these assumptions. Zhao et al. proved that the transform is unitary, and additionally proved that integer multiples of the Ding sampling rate also result in a unitary transform providing the integer is coprime with the number of samples [37].

A key property of the continuous LCT is the fact that it forms a group, in the sense that

$$\mathfrak{L}_{M_2}\{\mathfrak{L}_{M1}\{f(x)\}(y)\}(z) = \mathfrak{L}_{M_2M_1}\{f(x)\}(z).$$

For the special case of the Fourier transform, this reduces to a familiar property. The DFT applied twice is equivalent to time reversal, applied three times is equivalent to the inverse DFT, and applied four times is equivalent to the identity transformation. For larger multiples, the pattern repeats, yielding $n \mod 4$ instances of the DFT. Zhao et al. have recently proven that providing we sample the input and output at a coprime integer multiple of Ding's sampling rate (just as for unitarity), the discrete LCT exhibits the group property similar to the continuous LCT [9, 40].

The literature to date on decomposition based algorithms, e.g. [12, 14, 24] and their two-dimensional and complex extensions [15, 16], has not explicitly used a DLCT, but this discussion may nevertheless have relevance for those algorithms.

The LCT of 2 D signals may often be decomposed into 1D transforms in orthogonal spatial directions, and so has received less attention, but there are cases with independent significance. These 2 D non-separable LCTs have also been shown to be unitary for appropriate sampling rates [39].

8.4 Comparison of Algorithms

We have fulfilled two thirds of our goal: we have a discrete transform and a sampling theorem. However, the discrete LCT is not efficient enough for most purposes. We turn now to algorithms for calculating the discrete transform quickly. This topic is dealt with more comprehensively in Chap. 10 of this book. Here, we focus on the consequences of our new understanding of discrete signals and their LCTs for two such algorithms.

We believe Hennelly and Sheridan were the first to look at systematic ways of accounting for sampling requirements at each stage of LCT algorithms based on decomposition of the matrix of parameters, M, which they described in [12]. Around the same time, Ozaktas et al. demonstrated an Iwasawa-type decomposition with speed and accuracy comparable to the FFT [24]. Healy and Sheridan later reduced many of the consequences of [12] to two parameters, the well-known space-bandwidth product and a new metric they dubbed the space-bandwidth ratio [8]. However, these results are only accurate in certain circumstances where the replicas due to sampling don't substantially alter the calculations.

One interesting result of the analysis in [8] is that fast LCT algorithms of the kind presented in [6] require at worst the same number of samples as two (and perhaps many more) of the more common algorithms. The algorithm of [6] decomposes the discrete transform matrix into smaller matrices iteratively in the style of the FFT, while the others decompose the matrix of parameters, i.e. decompose the discrete transform into a sequence of simpler discrete transforms. However, in practice, the slowest step in algorithms of the second type are Fourier transforms, and commercial FFT software is fast enough to overcome any advantages of algorithms of the first kind have in terms of the number of samples required.

Let us turn to the consequences of discrete, chirp-periodic signals for the direct and spectral methods [7]. First, let us define these algorithms. The direct method is given by the following decomposition:

$$\begin{pmatrix} a & b \\ c & d \end{pmatrix} = \begin{pmatrix} 1 & 0 \\ d/b & 1 \end{pmatrix} \begin{pmatrix} b & 0 \\ 0 & b^{-1} \end{pmatrix} \begin{pmatrix} 0 & 1 \\ -1 & 0 \end{pmatrix} \begin{pmatrix} 1 & 0 \\ a/b & 1 \end{pmatrix}.$$

This is the decomposition we used to prove Ding's sampling theorem in Sect. 8.2.1. The spectral method is given by a different decomposition [8].

$$\begin{pmatrix} a & b \\ c & d \end{pmatrix} = \begin{pmatrix} 1 & 0 \\ c/a & 1 \end{pmatrix} \begin{pmatrix} a & 0 \\ 0 & a^{-1} \end{pmatrix} \begin{pmatrix} 0 & -1 \\ 1 & 0 \end{pmatrix} \begin{pmatrix} 1 & 0 \\ -b/a & 1 \end{pmatrix} \begin{pmatrix} 0 & 1 \\ -1 & 0 \end{pmatrix}.$$

In both cases, the diagonal matrices are scalings (scaling factor given by the top left character), the anti-diagonals are Fourier transforms (top right entry is 1) or inverse Fourier transforms (top right entry is -1), and the remaining triangular matrices are chirp multiplications with parameter given by the lower left entry.

Before we continue, an aside. In the discussion below, we will consider the discrete LCT to transform one discrete, chirp-periodic signal into another. This can be a source of confusion. The data input to such a discrete transform can be thought of as merely a vector of numbers, or as the samples of a continuous signal under examination. Indeed, the means of obtaining our input data may support one of those interpretations. However, for the purposes of PSD-based analyses of discrete signals, it is useful to think of the vector of numbers as being the samples of a single period of a chirp-periodic signal, and hence of the input as a discrete, chirp-periodic signal. It may be helpful for the reader to consider the special case of the DFT, which can be thought of as relating discrete, periodic functions; necessarily so because the input being discrete means the output is periodic, and the output being discrete means the input is periodic.

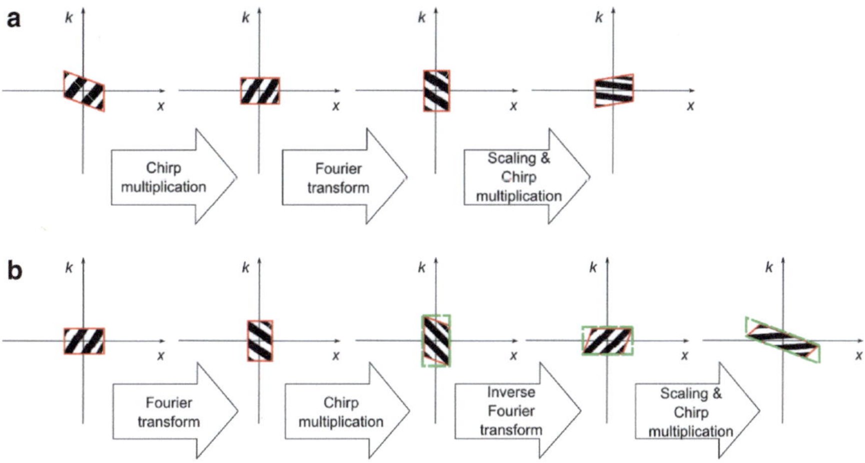

Fig. 8.3 Sequence of PSDs of signal undergoing the stages of (**a**) the direct method algorithm and (**b**) the spectral method. Replicas are not shown for convenience. Note the resampling (*green box*) necessary at the third step of the spectral method

Let us consider the direct method first. Given a signal that is discrete and chirp-periodic, the first operation is a chirp multiplication. It is no coincidence that this operation converts the signal into a simple discrete, periodic signal that is suitable for the second operation—a Fourier transform (numerically implemented using an FFT). The signal remains discrete and periodic after the Fourier transform and the third operation—scaling. Finally, the second chirp multiplication converts the signal to chirp-periodic. This sequence is illustrated in Fig. 8.3.

Now let us consider the spectral method. The first operation is a Fourier transform, so we require the signal to be discrete and periodic in order to use the FFT. (How can we require this? We implicitly impose this form on the signal simply by assuming we can sample the input and output at some rates for which aliasing is acceptable.) The resulting signal is also discrete and periodic. The second operation is a chirp multiplication, which converts the signal to being discrete and chirp-periodic. The third operation is a Fourier transform, for which we require a discrete-periodic signal. This mismatch means that we must oversample the signal at this point. Consequently, the spectral method is inherently less efficient than the direct method.

A second mark against the spectral method is that it requires two FFTs to the direct method's one. Why then, does the spectral method remain useful? The two algorithms make different assumptions about the signal. The direct method assumes the signal is bounded in the output LCT domain, while the spectral method assumes it is bounded in the Fourier domain. For certain domains where the ratio a/b is large, such as Fresnel transforms for short distances ($a = 1$, $b = \lambda z$), the signal is much more efficiently bound in the Fourier domain.

There are many other algorithms that decompose the ABCD matrix, and each of these may be analysed in this fashion.

8.5 Conclusion

While by no means an exhaustive treatment of the issues surrounding numerical approximation of LCTs, this chapter has reviewed a number of key ideas. Central to the topic is Ding's sampling theorem, which shows that the LCT of a discrete signal is 'chirp-periodic', and that a signal may be recovered from its samples if it has finite support in some LCT domain such that the periodic replicas do not overlap. The second really critical idea is that a discrete, chirp-periodic signal may be transformed by certain LCTs (those on the manifolds given by certain multiples of a particular ratio of parameters a/b) into discrete, chirp-periodic signals. Completely equivalently, any discrete, chirp-periodic signal may be transformed by LCT into another discrete, chirp-periodically providing the choice of sampling rate is appropriately constrained. However we frame the discussion, the resulting discrete transform exhibits a number of desirable properties including unitarity and additivity. With these properties, the discrete LCT becomes a stable and powerful tool in its own right.

Acknowledgements H.M. Ozaktas acknowledges partial support of the Turkish Academy of Sciences.

References

1. M.J. Bastiaans, Wigner distribution function and its application to first order optics. J. Opt. Soc. Am. **69**, 1710–1716 (1979)
2. B. Deng, R. Tao, Y. Wang, Convolution theorems for the linear canonical transform and their applications. Sci. China (Ser. F Inf. Sci.) **49**(5), 592–603 (2006)
3. J.J. Ding, Research of fractional Fourier transform and linear canonical transform. Ph.D. thesis, National Taiwan University, Taipei, Taiwan, ROC, 2001
4. F. Gori, Fresnel transform and sampling theorem. Opt. Commun. **35**(5), 293–297 (1981)
5. J.J. Healy, J.T. Sheridan, Cases where the linear canonical transform of a signal has compact support or is band-limited. Opt. Lett. **33**, 228–473 (2008)
6. J.J. Healy, J.T. Sheridan, Fast linear canonical transforms - FLCTs. J. Opt. Soc. Am. A **27**, 21–30 (2010)
7. J.J. Healy, J.T. Sheridan, Reevaluation of the direct method of calculating Fresnel and other linear canonical transforms. Opt. Lett. **35**(7), 947–949 (2010)
8. J.J. Healy, J.T. Sheridan, Spacebandwidth ratio as a means of choosing between Fresnel and other linear canonical transform algorithms. J. Opt. Soc. Am. A **28**(5), 786–790 (2011)
9. J.J. Healy, K.B. Wolf, Discrete canonical transforms that are Hadamard matrices. J. Phys. A Math. Theor. **44**, 265302 (2011)
10. J.J. Healy, B.M. Hennelly, J.T. Sheridan, Additional sampling criterion for the linear canonical transform. Opt. Lett. **33**, 2599–2601 (2008)

11. J.J. Healy, W.T. Rhides, J.T. Sheridan, Cross terms of the Wigner distribution function and aliasing in numerical simulations of paraxial optical systems. Opt. Lett. **35**(8), 1142–1144 (2010)
12. B.M. Hennelly, J.T. Sheridan, Generalizing, optimizing, and inventing numerical algorithms for the fractional Fourier, Fresnel, and linear canonical transforms. J. Opt. Soc. Am. A **22**(5), 917–927 (2005)
13. D.P. Kelly, J.J. Healy, B.M. Hennelly, J.T. Sheridan, Quantifying the 2.5D imaging performance of digital holographic systems. J. Eur. Opt. Soc. RP **6**, 11034 (2011)
14. A. Koç, H.M. Ozaktas, C. Candan, M.A. Kutay, Digital computation of linear canonical transforms. IEEE Trans. Signal Process. **56**(6), 2383–2394 (2008)
15. A. Koç, H.M. Ozaktas, L. Hesselink, Fast and accurate algorithm for the computation of complex linear canonical transforms. J. Opt. Soc. Am. A **27**(9), 1896–1908 (2010)
16. A. Koç, H.M. Ozaktas, L. Hesselink, Fast and accurate computation of two-dimensional non-separable quadratic-phase integrals. J. Opt. Soc. Am. A **27**(6), 1288–1302 (2010)
17. B.Z. Li, R. Tao, Y. Wang, New sampling formulae related to linear canonical transform. Signal Process. **87**(5), 983–990 (2007)
18. C. Li, B. Li, T. Xu, Approximating bandlimited signals associated with the lct domain from nonuniform samples at unknown locations. Signal Process. **92**, 1658–1664 (2012)
19. C. Liu, D. Wang, J.J Healy, B.M. Hennelly, J.T. Sheridan, M.K. Kim, Digital computation of the complex linear canonical transform. J. Opt. Soc. Am. A **28**(7), 1379–1386 (2011)
20. A.W. Lohmann, R.G. Dorsch, D. Mendlovic, Z. Zalevsky, C. Ferreira, Space-bandwidth product of optical signals and systems. J. Opt. Soc. Am. A **13**, 470–473 (1996)
21. F. Oktem, Signal representation and recovery under partial information, redundancy, and generalized finite extent constraints. Master's thesis, Bilkent University, 2009
22. F. Oktem, H.M. Ozaktas, Exact relation between continuous and discrete linear canonical transforms. IEEE Signal Process. Lett. **16**, 727–730 (2009)
23. F.S. Oktem, H.M. Ozaktas, Equivalence of linear canonical transform domains to fractional fourier domains and the bicanonical width product: a generalization of the space–bandwidth product. J. Opt. Soc. Am. A **27**(8), 1885–1895 (2010)
24. H.M. Ozaktas, A. Koç, I. Sari, M.A. Kutay, Efficient computation of quadratic-phase integrals in optics. Opt. Lett. **61**(1), 35–37 (2006)
25. A. Ozcelikkale, H.M. Ozaktas, Beyond nyquist sampling: a cost-based approach. J. Opt. Soc. Am. A **30**(4), 645–655 (2013)
26. S.C. Pei, J.J. Ding, Closed-form discrete fractional and affine Fourier transforms. IEEE Trans. Signal Process. **48**, 1338–1353 (2000)
27. K. Sharma, Approximate signal reconstruction using nonuniform samples in fractional Fourier and linear canonical transform domains. IEEE Trans. Signal Process. **57**, 4573–4578 (2009)
28. D. Song, H. Zhao, Stochastic formulation of (a,b,c,d)-bandlimited signal reconstruction. Circuits Syst. Signal Process. **34**(6), 2053–2064 (2015). doi:10.1007/s00034-014-9932-0
29. A. Stern, Sampling of linear canonical transformed signals. Signal Process. **86**(7), 1421–1425 (2006)
30. A. Stern, Why is the linear canonical transform so little known?, in *Proceedings of the 5th International Workshop on Information Optics*, ed. by G. Cristóbal, B. Javidi, S. Vallmitjana (Springer, Toledo, 2006), pp. 225–234
31. R. Tao, B.Z. Li, Y. Wang, G.K. Aggrey, On sampling of band-limited signals associated with the linear canonical transform. IEEE Trans. Signal Process. **56**, 5454–5464 (2008)
32. M. Testorf, B. Hennelly, J. Ojeda-Castanyeda, *Phase-Space Optics* (McGraw-Hill, New York, 2010)
33. D. Wei, Q. Ran, Y. Li, Reconstruction of band-limited signals from multichannel and periodic nonuniform samples in the linear canonical transform domain. Opt. Commun. **284**, 4307–4315 (2011)
34. X.G. Xia, On bandlimited signals with fractional Fourier transform. IEEE Signal Process. Lett. **3**(3), 72–74 (1996)

35. Xiao, L., Sun, W.: Sampling theorems for signals periodic in the linear canonical transform domain. Opt. Commun. **290**, 14–18 (2013)
36. J. Zhao, R. Tao, Y. Wang, Sampling rate conversion for linear canonical transform. Signal Process. **88**(11), 2825–2832 (2008)
37. L. Zhao, J.J. Healy, J.T. Sheridan, Unitary discrete linear canonical transform: analysis and application. Appl. Opt. **52**(7), C30–C36 (2013)
38. H. Zhao, R. Wang, D. Song, Recovery of bandlimited signals in linear canonical transform domain from noisy samples. Circuits Syst. Signal Process. **33**(6), 1997–2008 (2014)
39. L. Zhao, J.J. Healy, J.T. Sheridan, Two-dimensional nonseparable linear canonical transform: sampling theorem and unitary discretization. J. Opt. Soc. Am. A **31**(12), 2631–2641 (2014)
40. L. Zhao, J.J. Healy, C.-L. Guo, J.T. Sheridan, Additive discrete 1D linear canonical transform. Proc. SPIE **9599**, 95992E (2015)

Chapter 9
Self-imaging and Discrete Paraxial Optics

Markus Testorf and Bryan Hennelly

To the memory of Adolf Lohmann (1926–2013)

Abstract Coherent self-imaging, also known as the Talbot effect, describes Fresnel diffraction of strictly periodic wavefronts. Based on a phase-space interpretation of optical signals and systems we review a number of self-imaging phenomena, including the fractional Talbot effect. Recognizing Fresnel diffraction as merely one particular instance of the wider class of linear canonical transforms allows us to discuss various generalizations of self-imaging. Recognizing the discrete Fresnel transformation as a special case of the fractional Talbot effect leads to a definition of discrete linear canonical transformations, which preserve important properties of the continuous transformation for a subset of paraxial signals and systems. This subset can be used as a framework for practical implementations of discrete LCTs discrete linear canonical transform.

9.1 Introduction

Coherent self-imaging, also known as the Talbot effect, is one of the most intriguing optical phenomena. The Talbot effect describes the periodic revival of a periodic complex amplitude distribution in equidistant diffraction planes along the optical axis. Henry Fox Talbot, also known for his contributions to photography, was the first to describe self-imaging in 1836 [76]. Almost half a century later, Lord Rayleigh offered the first theoretical analysis [64]. It would require more than another half century, however, before the Talbot effect would slowly start to emerge

---

M. Testorf (✉)
Thayer School of Engineering, Dartmouth College, 14 Engineering Drive,
Hanover, NH 03755, USA
e-mail: Markus.Testorf@osamember.org

B. Hennelly
Department of Computer Science, National University of Ireland,
Maynooth, Co. Kildare, Ireland
e-mail: bryanh@cs.nuim.ie

J.J. Healy et al. (eds.), *Linear Canonical Transforms*, Springer Series
in Optical Sciences 198, DOI 10.1007/978-1-4939-3028-9_9

from relative obscurity to become part of the canon of Fourier optics. This can be linked, at least in part, to the invention of the laser and the availability of high power coherent light sources. What had required skillful experimentation in Talbot's time can now easily be observed with low cost optical equipment. This is also the basis for applications of optical self-imaging, including the Talbot interferometer [51, 56], wavefront restoration [26], wavefront sensing [62, 67, 71], and the design of Talbot array generators [11, 48, 82]. The translation of the Talbot effect to the temporal domain introduced the idea of dispersion management via temporal self-imaging [15, 19, 43], and new concepts for clock distribution in fiber optical systems [16]. More recently, the temporal Talbot effect was used to demonstrate time-domain filtering with diffractive optics [41, 42].

While self-imaging claimed its place as a practical tool for optical system development, theoretical investigations of self-imaging phenomena revealed equally fascinating details. This is particularly true for the study of the fractional Talbot effect [89]. For all, but a few cases, the Fresnel diffraction integral has no simple analytic solutions. However, the fractional Talbot effect defines a unique class of problems, for which the diffraction amplitude can be understood as the superposition of a finite number of shifted and modulated copies of the input signal [33, 35, 55]. The number of copies, the lateral shift, and the modulation coefficients depend on the specific diffraction plane, which is evaluated. This makes it possible to identify fractional Talbot planes, for which the discrete samples of periodic signals in the input and output plane are linked by discrete unitary transforms [9, 10].

This link between the continuous Fresnel diffraction integral of periodic functions and a discrete transformation based on the fractional Talbot effect provides the motivation for a more general exploration of the connection between the diffraction of periodic signals and discrete paraxial optics, i.e. formulations of paraxial optics suitable for numeric simulation. To this end, we study self-imaging for more general classes of paraxial systems, also known as linear canonical transforms (LCTs). By comparing the matrix formulation of the fractional Talbot effect with properties of the discrete Fourier transform (DFT) we formulate conditions for constructing discrete versions of LCTs, which preserve many of the properties of the continuous LCT for bandlimited and periodic signals.

Motivated by attempts to formulate a working definition for the discrete fractional Fourier transform [22, 24, 54, 69], much work has been directed toward the development of discrete LCTs (DLCTs). Remarkably, the implementation of DLCTs has remained an open problem, despite much efforts and a variety of proposed solutions. A naive discretization of the generalized Fresnel transformation as a Riemann sum is plagued with inherent problems. This includes the potential loss of unitarity [91], and the problem of defining a proper inverse DLCT [28]. Furthermore, LCTs generally transform bandlimited signals into signals, which are neither compact nor bandlimited [90]. Appropriate measures have to be taken to find approximate solutions, which do not suffer from signal aliasing [38].

There are a number of excellent reviews of existing implementations of DLCTs [37, 57], and it is not or goal to provide a comprehensive survey of DLCTs theory. Instead, we want to demonstrate that the fractional Talbot effect provides a framework for constructing DLCTs. This is based on the insight that continuous, periodic signals can be represented rigorously with discrete signals of finite dimension. The fractional Talbot effect provides the interface between the diffraction of continuous, periodic signals and their discrete representation.

Instrumental to this study of self-imaging discrete paraxial optics will be the phase-space formulation of optical signals and systems. It was shown previously how this can provide a rather intuitive way to study self-imaging phenomena in general [78]. Here we emphasize the utility of phase-space optics to introduce discrete formulations of optical phenomena by investigating the phase space of periodic signals.

9.2 The Phase Space of Paraxial Optics

For our exploration of generalized self-imaging it will be convenient to rely on phase-space optics as a formal representation of coherent optical signals and systems. A systematic introduction to phase-space optics is beyond the scope of our discussion and we refer to the growing literature on the subject [18, 83–85]. Here, we only summarize briefly those relationships, which we will require for our exploration of self-imaging and discrete paraxial optics. We will limit the notation to one-dimensional complex amplitudes. This will allow us to make efficient use of phase-space diagrams, i.e. schematic graphical depictions of phase space.

For paraxial wavefronts we use the Wigner distribution function (WDF) [18] as a suitable phase-space representation. The complex amplitude $u(x)$ of scalar wave optics is transformed into

$$W(x,\nu) = \int_{-\infty}^{\infty} u\left(x + \frac{x'}{2}\right) u^*\left(x - \frac{x'}{2}\right) \exp(-i2\pi\nu x')\, dx' \qquad (9.1)$$

which is a function of space x, as well as of spatial frequency ν. The WDF is a bilinear transformation of $u(x)$. This means, the sum of two signals will be transformed into the respective WDFs, with additional interference or cross terms added. For instance, a signal consisting of two plane waves, $u_{1,2}(x) = \exp(i2\pi\nu_1 x) + \exp(i2\pi\nu_2 x)$, is transformed into

$$W_{1,2}(x,\nu) = \delta(\nu-\nu_1) + \delta(\nu-\nu_2) + 2\cos[2\pi(\nu_1-\nu_2)]\delta\left(\nu - \frac{\nu_1+\nu_2}{2}\right). \qquad (9.2)$$

Apart from the two signal components, which refer to the spatial frequencies of the two plane waves, we encounter an additional frequency component. This frequency is located halfway between the other two frequencies and is modulated by a cosine function. The modulation frequency of the interference term reflects the spacing between the unmodulated frequency terms.

The WDF is symmetric with respect to the two variables x and ν, and we find the alternative definition,

$$W(x,\nu) = \int_{-\infty}^{\infty} \tilde{u}\left(\nu + \frac{\nu'}{2}\right) \tilde{u}^*\left(\nu - \frac{\nu'}{2}\right) \exp\left(i2\pi\nu' x\right) d\nu', \tag{9.3}$$

based on the Fourier transform of the signal

$$\tilde{u}(\nu) = \int_{-\infty}^{\infty} u(x) \exp(-i2\pi\nu x) dx. \tag{9.4}$$

Intensity and power spectrum of the complex signal are calculated as projections of the WDF,

$$\int_{-\infty}^{\infty} W(x,\nu) d\nu = |u(x)|^2, \quad \text{and} \quad \int_{-\infty}^{\infty} W(x,\nu) dx = |\tilde{u}(\nu)|^2, \tag{9.5}$$

and it is possible to recover the original signal as a Fourier transformation of the WDF. The projection theorems highlight the role of the interference term in Eq. (9.2) as a representation of the mutual coherence of the two plane waves. Only with the interference term being present does the projection of the WDF show the interference pattern of two plane waves.

One of the most important properties associated with the phase-space concept is the correspondence between the signal dynamics and the associated geometrical transformation of phase space. The propagation of paraxial optical waves through quadratic phase systems can be calculated as an LCT of the input signal $u_{\text{in}}(x)$ [73],

$$u_{\text{out}}(x) = \frac{1}{\sqrt{iB}} \int_{-\infty}^{\infty} u_{\text{in}}(x') \exp\left[\frac{i\pi}{B}\left(A\, x'^2 - 2xx' + D\, x^2\right)\right] dx'. \tag{9.6}$$

For $B = 0$ the integral expression is replaced by a scaling and a chirping operation,

$$u_{\text{out}}(x) = \frac{1}{\sqrt{A}} u_{\text{in}}(x/A) \exp\left(i\pi \frac{C}{A} x^2\right). \tag{9.7}$$

The corresponding change of the WDF is given by the rules of ABCD-matrix optics (see, for example, [68]). Each point (x, ν) in phase space represents the generalized coordinates of a geometric optical ray, with x representing the position and $\nu = \sin\theta/\lambda$ the direction of propagation determined by the angle θ relative to the optical axis. Thus, paraxial ray tracing corresponds to a linear coordinate transform, where each optical element or system is represented by a 2×2 matrix, with

$$\begin{pmatrix} x \\ \nu \end{pmatrix}_{\text{out}} = \begin{pmatrix} A & B \\ C & D \end{pmatrix} \begin{pmatrix} x \\ \nu \end{pmatrix}_{\text{in}}. \tag{9.8}$$

For paraxial signals and systems the geometrical transform in Eq. (9.8) not only describes the evolution of the phase-space of geometrical optics, but equally the phase-space of scalar wave optics. This means, the WDF of the input signal is identical to the WDF at the output except that the input coordinate $(x, \nu)_{\text{in}}$ is translated to a new coordinate $(Ax + B\nu, Cx + D\nu)_{\text{out}}$ at the output of the paraxial system, or

$$W_{\text{out}}(x, \nu) = W_{\text{in}}(Dx - B\nu, -Cx + A\nu). \tag{9.9}$$

In other words, as long as the system function is given as a quadratic phase system and diffraction at system apertures can be ignored, the WDF follows the rules of paraxial ray optics.

In this scheme, paraxial free space propagation or Fresnel diffraction over a distance z corresponds to a shear of the WDF parallel to the x axis,

$$W_z(x, \nu) = W_0(x - \lambda z\nu, \nu). \tag{9.10}$$

The Fourier dual operation is the modulation with a linear chirp function, i.e. the function of a parabolic lens. For a convex lens of focal length f we find

$$W_L(x, \nu) = W_0(x, \nu + x/(\lambda f)). \tag{9.11}$$

In Fig. 9.1b, c both operations are applied to the generic phase-space volume of rectangular shape in Fig. 9.1a. The corresponding ABCD matrix is

$$S_z(\lambda z) = \begin{pmatrix} 1 & \lambda z \\ 0 & 1 \end{pmatrix}, \text{ and } S_L(-1/\lambda f) = \begin{pmatrix} 1 & 0 \\ -\frac{1}{\lambda f} & 1 \end{pmatrix} \tag{9.12}$$

respectively. Figure 9.1d illustrates the effect of a fractional Fourier transformation, which corresponds to a rotation of phase space by an angle α. The associated system matrix reads

$$S_{F\alpha}(\alpha) = \begin{pmatrix} \cos\alpha & \sin\alpha \\ -\sin\alpha & \cos\alpha \end{pmatrix}. \tag{9.13}$$

The Fourier transformation, Fig. 9.1d, can be interpreted as a fractional Fourier transformation with a rotation angle $\alpha = \pi/2$, and a system matrix $S_{\text{FT}} = S_{F\alpha}(\pi/2)$. Equally important for our discussion is magnification or scaling, in Fig. 9.1f. We may interpret scaling as perfect imaging with a magnification factor $m \neq 1$. Scaling of one of the coordinates is accompanied by the inverse scaling of the conjugate coordinate, i.e.

$$S_s(m) = \begin{pmatrix} m & 0 \\ 0 & 1/m \end{pmatrix}. \tag{9.14}$$

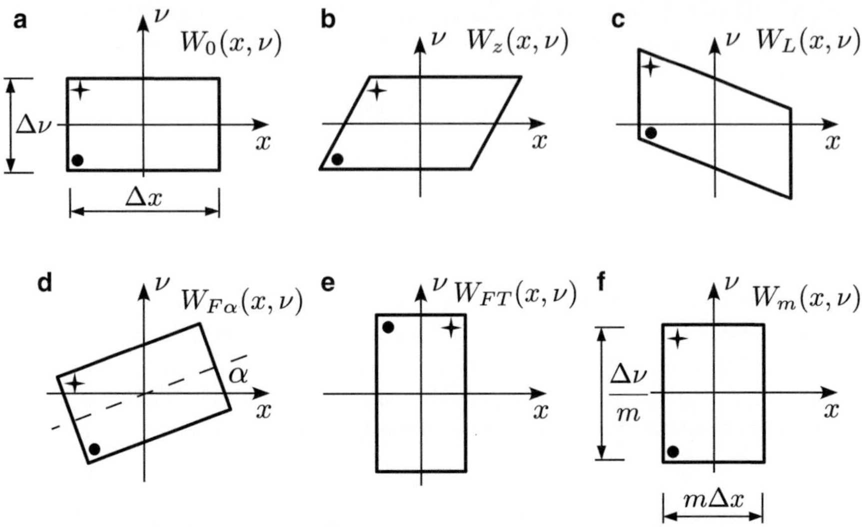

Fig. 9.1 Paraxial optics in phase space: (**a**) generic phase-space distribution of an optical signal; (**b**) diagram of the signal after Fresnel diffraction, and (**c**) after passage through a thin lens (chirping); (**d**) the fractional Fourier transform of the input signal corresponds to a rotation of phase space, where (**e**) the ordinary Fourier transform corresponds to a rotation of $\pi/2$; (**f**) signal scaling magnifies one coordinate of phase space and demagnifies the conjugate coordinate by the same factor

Further rules for the use of phase-space optics govern the modulation and convolution of two signals. For the product of two functions, $u(x) = g(x)\,h(x)$, the corresponding WDFs are convolved with respect to the frequency variable

$$W_u(x,\nu) = \int_{-\infty}^{\infty} W_g(x,\nu')W_h(x,\nu-\nu')d\nu' = W_g(x,\nu) *_\nu W_h(x,\nu). \quad (9.15)$$

The symmetry between x and ν implies that the convolution between the two signals is translated to a convolution between the corresponding WDFs with respect to x.

Important for our discussion will be the phase-space representation of a linear chirp,

$$u_{\mathrm{ch}}(x) = \exp[i2\pi(\alpha x^2 + \beta x + \gamma)], \quad (9.16)$$

and its WDF

$$W_{\mathrm{ch}}(x,\nu) = \delta(\nu - 2\alpha x - \beta). \quad (9.17)$$

This provides us with an alternative interpretation of chirping and Fresnel diffraction in phase space. Fresnel diffraction can be understood as a convolution of the complex amplitude distribution with the point response function of free space,

$$h_{\mathrm{Fr}}(x,z) = \frac{1}{\sqrt{i\lambda z}} \exp\left(\frac{i\pi}{\lambda z}x^2\right). \tag{9.18}$$

This translates into

$$W_h(x,\nu) = \frac{1}{|\lambda z|}\delta\left(\nu - \frac{x}{\lambda z}\right) = \delta(x - \lambda z\nu), \tag{9.19}$$

which is a straight line in phase space. From this we obtain Eq. (9.10) as the convolution in x between the input WDF and $W_h(x,\nu)$. Similarly, convolution in ν of the oblique delta-line with the input WDF corresponds to the chirping operation in Eq. (9.11).

9.3 The Talbot Effect

The classical Talbot effect is best defined as Fresnel diffraction of periodic wavefronts. The phase-space interpretation relies on a generalization of Eq. (9.2). Strictly periodic signals have a discrete spectrum and can be expanded into a Fourier series,

$$u(x) = u(x + x_p) = \sum_{n=-\infty}^{\infty} u_n \exp(i2\pi nx/x_p) \tag{9.20}$$

with

$$u_n = \tilde{u}_s(n/x_p) \tag{9.21}$$

and $\tilde{u}_s(\nu)$ being the Fourier transform of a single period of the signal $u_s(x) = \mathrm{rect}(x/x_p)u(x)$. The WDF of a periodic signal is calculated straightforwardly,

$$W(x,\nu) = \sum_{n=-\infty}^{\infty}\sum_{n'=-\infty}^{\infty} u_n u_{n'}^* \,\delta\left(\nu - \frac{n+n'}{2x_p}\right)\exp\left(i2\pi\frac{n-n'}{x_p}x\right). \tag{9.22}$$

Terms $n = n'$ in this series expansion correspond to the WDFs of individual discrete frequencies in the Fourier series. All other terms $n \neq n'$ are the associated interference terms, which appear equidistantly spaced along the frequency axis at integer multiples of $1/(2x_p)$. Figure 9.2 schematically depicts the WDF, with solid lines representing the WDFs of individual signal harmonics.

Fresnel diffraction corresponds to a horizontal shear of phase space and we recover the phase-space distribution of the input function, in Eq. (9.22), if the shear moves the term at frequency $1/(2x_p)$ by exactly one period x_p. In Fig. 9.2 line S indicates this shear as the horizontal distance between the ν axis and line S. Recalling Eq. (9.10), the condition for self-imaging becomes

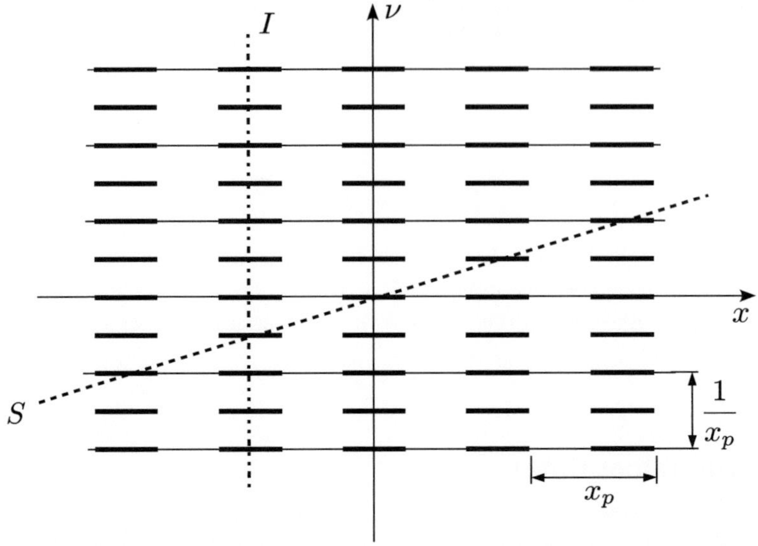

Fig. 9.2 Phase-space diagram of a periodic function. Talbot imaging corresponds to a horizontal shear of the phase-space distribution indicated by *line S*. In contrast, conventional image formation requires all points of phase space along *vertical lines I* to map to the same distribution of *vertical lines* in the phase space representation of the image

$$\lambda z_T \frac{1}{2x_p} = x_p \tag{9.23}$$

or

$$z_T = \frac{2x_p^2}{\lambda} \tag{9.24}$$

which is known as the Talbot distance or self-imaging distance. It is immediately obvious that the same effect can be observed at multiples of the Talbot distance. The phase space interpretation also illustrates the difference between self-imaging and image formation based on a lens system. The point-to-point mapping of conventional imaging would require phase-space points along vertical lines I to map to vertical lines in the phase space of the image. Considering the projection property of the WDF, in Eq. (9.5), only then is the signal of a single image point formed by the signal of a single object point. Self-imaging relaxes this condition and only requires the same phase-space distribution to occur irrespective of how the signal energy of the input is redistributed in the space-spatial frequency plane.

9.4 Self-imaging Beyond Fresnel Diffraction

Self-imaging is not limited to Fresnel diffraction, but can be observed for other first order optical systems as well. In fact, self-imaging has been generalized in more than one way. In each case, the phase-space interpretation either motivates the generalization, or helps us to relate generalized self-imaging to the generic Talbot effect. We highlight three possible ways to generalize self-imaging for paraxial optics.

9.4.1 Spectral Self-imaging

The strong symmetry between signal and spectrum in phase space suggests a form of self-imaging, where the roles of space and spatial frequency are reversed. In this case, the signal is discrete

$$u_{sT}(x) = \sum_{n=-\infty}^{\infty} u_n \delta(x - n\Delta x) \tag{9.25}$$

and the spectrum is periodic. The associated phase-space distribution consists of discrete lines perpendicular to the space axis. We obtain the standard form of Talbot self-imaging after rotating phase space by $\pi/2$, i.e by Fourier transforming $u_{sT}(x)$. The Fourier dual LCT of Fresnel diffraction is a chirp modulation and the optical interpretation of spectral self-imaging is that of a modulated pinhole array. Self-imaging occurs, if the discrete signal is passing a lens with focal length f,

$$u_L(x) = \exp\left(-i\pi \frac{x^2}{\lambda f}\right), \tag{9.26}$$

without any change to its complex amplitude distribution. Similar to finding the Talbot distance for periodic functions, the self-imaging condition for discrete pulse trains, i.e. periodic spectra is determined from the phase space of a point array,

$$W_{sT}(x,\nu) = \sum_{n=-\infty}^{\infty} \sum_{n'=-\infty}^{\infty} u_n u_{n'}^* \exp[i2\pi(n'-n)\Delta x\nu]\, \delta\left(x - \frac{n+n'}{2}\Delta x\right). \tag{9.27}$$

The lens function corresponds to the vertical shear of the phase-space distribution, in Eq. (9.11). To reproduce the WDF of the input pulse train, the vertical translation at $x = x_p/2$ has to be equal to the period of the spectrum $1/\Delta x$,

$$\frac{\Delta x/2}{\lambda f_S} = \frac{1}{\Delta x} \tag{9.28}$$

or

$$f_S = \frac{\Delta x^2}{2\lambda}. \tag{9.29}$$

The spectral Talbot effect was first described for time domain signals [14, 88], and is primarily discussed in the context of fiber optics. Recent work describes the spectral Talbot effect in sampled fiber Bragg gratings with super-periodic structures [92] as well as the demonstration of a comb optical filter [20]. This also includes a scheme to implement the spectral Talbot effect with diffractive optics [23].

In the context of our discussion of self-imaging and discrete Fourier optics, we may also contemplate signals, which are simultaneously discrete and periodic. Here, any combination of Talbot self-imaging and spectral self-imaging results in an image of the input signal.

9.4.2 *Self-imaging of Discrete Chirps*

Another way to interpret Fig. 9.2 and the formation of Talbot images is to recognize that the shear of phase space has to occur parallel to the orientation of the discrete periodically modulated lines, which define the harmonics of the periodic signals. Lines at oblique angles in phase space can be identified as chirp functions [compare Eq. (9.16)] and an array of equidistantly spaced chirp functions may be written as

$$u_{\text{ca}}(x) = \sum_n u_n \exp\left(-i\pi\alpha x^2 + i2\pi n\, x/x_p\right). \tag{9.30}$$

We note that Eq. (9.30) may also be interpreted as the result of multiplying a discrete signal with a chirp function, or as the result of a fractional Fourier transform of either a periodic, or of a discrete signal.

It is now straightforward to relate the chirp array to the standard form of Talbot self-imaging. If we first remove the chirp factor, which is common to all signal harmonics, self-imaging of the remaining periodic signal occurs for multiples of $z_T = 2x_p^2/\lambda$. To restore the original chirp array we only need to re-modulate the result with the chirp function. This means, we observe self-imaging for the chirp array in Eq. (9.30), with any system of the form

$$S_{\text{ca}} = S_L(-\alpha)\, S_z(z_T\lambda)\, S_L(\alpha) = \begin{pmatrix} 1+\alpha z_T\lambda & z_T\lambda \\ -\alpha^2 z_T\lambda & 1-\alpha z_T\lambda \end{pmatrix}. \tag{9.31}$$

9.4.3 Scaled Self-imaging

A further extension of self-imaging is based on relaxing the self-imaging condition and by including cases, where magnified replica of the input signal are recovered. This problem is typically discussed in the context of self-imaging with spherical illumination [25, 56, 78]. Here we reinterpret this configuration as an extension of self-imaging to a larger set of LCTs. We assume a periodic input function, Eq. (9.20), as input signal. A lens, with focal length f, is used for chirping the signal, thus creating a converging (or diverging) wave modulated with a periodic signal. After free space propagation over a distance z a second lens, with focal length $-(f-z)$, is used to remove the chirp in the self-imaging plane resulting in a scaled replica of the input signal.

Given the magnification factor M we can now determine the self-imaging condition as a function of the focal length of the lenses and the propagation distance. In principle, it is again possible to use phase-space diagrams to a obtain the self-imaging condition [78]. Here, we demonstrate how the system matrix can be used for this purpose. The system matrix reads

$$S_{\text{scale}} = S_L\left(\frac{1}{\lambda(f-z)}\right) S_z(\lambda z) S_L\left(-\frac{1}{\lambda f}\right) = \begin{pmatrix} \frac{f-z}{f} & z\lambda \\ 0 & \frac{f}{f-z} \end{pmatrix}. \tag{9.32}$$

This system matrix can be interpreted as free space propagation followed by a scaling operation,

$$S_{\text{scale}} = S_M(M)\, S_z(\lambda z/M), \tag{9.33}$$

with $M = (f-z)/f$. Self-imaging occurs, if the free space propagation is equal to the Talbot length of the input signal,

$$z_{\text{sc}} = M\frac{2x_p^2}{\lambda} = M\, z_T. \tag{9.34}$$

Clearly, we are again at liberty to extend the scaled self-imaging theory to include spectral self-imaging and self-imaging of discrete chirp functions. We also note that typically, in the discussion of Talbot imaging under spherical illumination, the lenses in the input plane and in the exit plane of the system are not part of the system, but part of the propagating signal. This means, the self-imaging condition is actually further relaxed to include situations, where neither the complex amplitude of the input signal, nor of the self-image are strictly periodic.

9.4.4 Self-imaging as an Eigenvalue Problem

The Talbot effect reproduces the coherent amplitude after passing the input signal through the optical system. All input signals, which exhibit self-imaging after passage through a particular LCT can be interpreted as eigenfunctions of the associated generalized Fresnel transformation [58, 59]. This perspective can be used to develop a framework, which includes both self-imaging, and conventional imaging [1]. As we mentioned before, the latter also reproduces the input signal, albeit under more stringent conditions. Translating this notion to phase space, self-imaging corresponds to a self-affine transformation of phase space [6].

One important example of generalized self-imaging, which is conveniently discussed in this framework, is the fractional Fourier transform [3, 58, 59]. The eigenfunctions in this case are Hermite-Gauss polynomials, and the associated phase-space functions are Laguerre–Gauss polynomials, which have the required rotational symmetry to ensure the self-affinity of phase-space rotations.

In this context, it is instructive to consider the explicit construction of self-imaging functions for a subset of fractional Fourier transforms. Most importantly, we recognize that four consecutive Fourier transforms reproduce the original input signal. This corresponds to four $\pi/2$ rotations in phase space. Thus, for any signal to reproduce itself under Fourier transformation, it has to have a fourfold rotational symmetry in phase space. A simple way to construct signals with this symmetry is to add for any signal the results obtained after each of the four Fourier transform steps [49, 50]. For any signal $u(x)$, the associated signal

$$f(x) = u(x) + \tilde{u}(x) + u(-x) + \tilde{u}(-x) \tag{9.35}$$

is equal to its own Fourier transform. Incidentally, $f(x)$ is also self-transform for two and three consecutive Fourier transforms. The phase-space interpretation [29] suggests that the same concept can be generalized to other LCTs [5], and in particular to fractional Fourier transforms, where the order of the transform defines an N-fold symmetry, i.e. the original signal is obtained after N consecutive fractional Fourier transforms of the same order [4, 53].

The analysis of self-imaging as an eigenvalue problem also inspired self-imaging of finite energy signals [60, 61]. In particular, in two dimensions this has motivated a further extension of self-imaging to signals, which form scaled and rotated copies of the input beam. Remarkably, the eigenvalue analysis even provides an interpretation of the Gouy phase and, for two-dimensional signals, of the geometrical phase of Gaussian beams in first order optical systems [7]. For a physical interpretation of the Gouy phase in the context of phase-space optics, we refer to [79].

9.5 The Fractional Talbot Effect

Self-imaging is remarkable in that it identifies instances, where the tedious evaluation of the Fresnel diffraction integral can be bypassed and the diffraction amplitude is known without further effort. In 1965 Winthrop and Worthington investigated Fresnel diffraction at rational fractions of the Talbot distance, i.e. at distances $z_{P,Q} = (P/Q)z_T$, with P and Q being integer numbers [89]. This, again, leads to rather simple expressions for the propagation of periodic signals in free space. The term "Fresnel image" is generally associated with the patterns observed in these distinct diffraction planes. Several alternative formulations of the fractional Talbot effect are available. The most compact notations express the complex amplitude of Fresnel images as the superposition of a finite number of copies of the input signal. Each copy is shifted laterally and multiplied with a complex coefficient, the so-called Talbot coefficient. The number of copies which are superimposed depends on the ratio P/Q.

Apart from an efficient means to analyze Fresnel diffraction the fractional Talbot effect has been employed most prominently for the design of Talbot array illuminators, which split the incoming wavefront into an array of high intensity beams [34, 45, 48]. The fractional Talbot effect was also used for beam splitting in multi-mode waveguide systems [72]. The counterpart of the fractional Talbot effect in the temporal domain has been used to multiply the repetition rate of clock pulses [13]. Both, the analysis and the design of array generators, in fact, is one of the motivations to interpret self-imaging in phase space [77, 81].

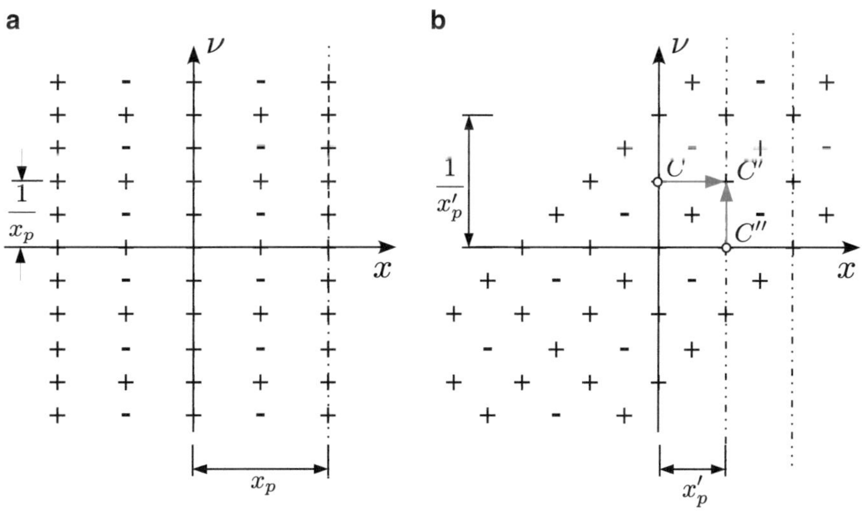

Fig. 9.3 Fresnel diffraction of a comb-function: (**a**) phase-space diagram of a comb function; (**b**) sheared phase-space distribution of the comb function at $z_T/4$. The horizontally sheared delta-comb function can be interpreted as a vertically sheared comb-function of period $x_p/2$

9.5.1 Phase-Space Analysis of the Fractional Talbot Effect

Phase-space optics offers a rather elegant way to understand and derive formal expressions for the fractional Talbot effect [78]. We continue to restrict our attention to periodic signals, this time expressed as a periodic continuation of a single period $u_{xp}(x)$,

$$u(x) = u_{xp}(x) * \sum_{n=-\infty}^{\infty} \delta\left(x - nx_p\right) = u_{xp}(x) * \mathrm{comb}(x, x_p) \tag{9.36}$$

with $\mathrm{comb}(x, x_p)$ defining the comb-function. The diffraction pattern after free space propagation over a distance z can be expressed as an additional convolution with the point spread function of free space,

$$u(x, z) = u(x) * h_{\mathrm{Fr}}(x, z). \tag{9.37}$$

Since the order of computing convolutions can be exchanged, we are at liberty to investigate the Fresnel diffraction of the delta-comb function first. This automatically solves the Fresnel diffraction problem for any periodic function, assuming the second convolution with the groove shape of the input signal is sufficiently straightforward. Fresnel diffraction can be analyzed in phase space by calculating the WDF of the comb-function,

$$W_{\mathrm{comb}}(x, \nu) = \frac{1}{2x_p} \sum_{n=-\infty}^{\infty} \sum_{n'=-\infty}^{\infty} (-1)^{n \cdot n'} \delta\left(x - n\frac{x_p}{2}\right) \delta\left(\nu - n'\frac{1}{2x_p}\right). \tag{9.38}$$

The resulting grid of equidistant delta functions is shown in Fig. 9.3a. We note the alternating sign of the delta functions (symbolically represented by + and – signs), and the interference terms located at points interlaced with the grid of integer multiples of x_p and $1/x_p$. In general, the horizontal shear in phase space, which corresponds to Fresnel propagation, will disperse the delta-functions without yielding any simple solution to the diffraction problem. However, for certain shears, the resulting pattern of delta functions can again be interpreted as the WDF of a comb-function, now sheared in vertical direction. This is the case, when all delta functions are arranged along a discrete set of equidistant vertical lines (Fig. 9.3b). The period $x_p' = x_p/Q$ of the new delta-comb is decreased in x and increased in ν by the same integer factor Q. The corresponding propagation distance is determined from the shear which moves point C at $\nu = 1/x_p$ horizontally by an integer multiple of x_p/Q to point C', or

$$z_{1,2Q} = \frac{1}{2Q} z_T. \tag{9.39}$$

Here we restrict our attention to the case of even Q. For a complete discussion of the fractional Talbot effect, we refer to [8, 12, 35, 78, 81].

We can now understand Fresnel diffraction of a comb-function at fractional Talbot planes as a chirp modulation of a new comb-function with reduced period. The chirp can be deduced from the vertical shear necessary to move point C'' to point C' in Fig. 9.3b. For $z_{1,2Q}$ we find

$$\begin{aligned} \mathrm{comb}(x, x_p; z) &= \mathrm{comb}(x, x_p) * h_{\mathrm{Fr}}(x, z_{1,2Q}) \\ &= \frac{1}{\sqrt{Q}} \exp\left(i\pi \frac{Q}{x_p^2} x^2\right) \sum_{n=-\infty}^{\infty} \delta\left(x - n x_p / Q\right) \end{aligned} \tag{9.40}$$

where the factor $1/\sqrt{Q}$ accounts for the conservation of energy, if the comb-function is compressed by a factor of Q. The sampled chirp function can also be written in terms of the original comb function,

$$\begin{aligned} \mathrm{comb}(x, x_p; z) &= \frac{1}{\sqrt{Q}} \sum_{n=-\infty}^{\infty} \exp\left[i\pi \frac{Q}{x_p^2} \left(n \frac{x_p}{Q} \right)^2 \right] \delta\left(x - n x_p / Q\right) \\ &= \left[\sum_{q=0}^{Q-1} c_q \delta\left(x - q x_p / Q\right) \right] * \mathrm{comb}(x, x_p; 0) \end{aligned} \tag{9.41}$$

with

$$c_q = \frac{1}{\sqrt{Q}} \exp(i\pi q^2 / Q), \tag{9.42}$$

and where we made use of the Q periodicity of the Talbot coefficients c_q. We are finally able to write down the solution of the Fresnel diffraction problem by substituting Eq. (9.42) into Eqs. (9.36) and (9.37),

$$u(x, z) = u'_{xp}(x) * \mathrm{comb}(x, x_p) \tag{9.43}$$

with

$$u'_{xp}(x) = u_{xp}(x) * \sum_{q=0}^{Q-1} c_q \delta\left(x - q x_p / Q\right) = \sum_{q=0}^{Q-1} c_q u_{xp}(x - q x_p / Q). \tag{9.44}$$

Equation (9.44) is a valid expression for all $2Q$ diffraction planes $z_{P,2Q}$, where each plane requires a different set of Talbot coefficients. If P and $2Q$ have common factors, some of the Q Talbot coefficients are typically zero.

9.5.2 Generalizations to Other LCTs

Similar to the Talbot effect, it is possible to generalize the fractional Talbot effect to other LCTs. In particular, the fractional spectral Talbot effect is obtained straightforwardly by exchanging the roles of the coordinate axes in the phase-space analysis [88].

A second case of particular interest for our discussion is the fractional Talbot effect under spherical illumination. We express this problem as an LCT consisting of a chirp modulation and Fresnel diffraction. Matrix optics allows us to find a substitute system, to which Eq. (9.44) can directly be applied,

$$S_{sf} = S_z(\lambda z)\, S_L\left(-\frac{1}{\lambda f}\right) = S_s(M)\, S_L\left(-\frac{M}{\lambda f}\right)\, S_z\left(\frac{z\lambda}{M}\right) \tag{9.45}$$

with $M = 1 - z/f$. This means, spherical illumination followed by Fresnel diffraction can be replaced by Fresnel diffraction followed by chirping and a scaling operation. The fractional Talbot planes are again determined exclusively by the initial propagation step, where $z\lambda/M$ now has to fulfill the condition for a fractional Talbot plane [36]. In principle, the remaining chirp modulation of the Fresnel image can be removed by multiplying the complex amplitude with a second chirp, and ensuring a periodic output signal.

9.5.3 The Matrix Description of the Fractional Talbot Effect

Our analysis of the fractional Talbot effect was based on separating the particular shape of a single period from the analysis of Fresnel diffraction. We can employ the same concept again, if we consider periodic signals, which can be expressed via a sampling expansion

$$u_{xp}(x) = u_i(x/\Delta x) * \sum_{q=0}^{Q-1} u_q \delta(x - q\Delta x), \tag{9.46}$$

with $Q\Delta x = x_p$. For bandlimited functions, the interpolation function is $u_i(x) = \mathrm{sinc}(x)$. Fresnel diffraction can now be analyzed as the fractional Talbot effect of the discrete and periodic signal

$$u^s(x) = \sum_{q=0}^{Q-1} u_q \delta(x - q\Delta x) * \mathrm{comb}(x, x_p). \tag{9.47}$$

At the fractional Talbot distance $z_{1,2Q}$ we find Q copies of the discrete period in Eq. (9.46) shifted by a multiple of the sampling distance. Thus, we expect the

Fresnel diffraction amplitude of the discrete signal to have the same structure as Eq. (9.47), where the coefficients u_q are replaced by a new set of coefficients w_q, with

$$w_q = \sum_{q'=0}^{Q-1} c_{q-q'} u_{q'} . \tag{9.48}$$

The Fresnel diffraction pattern, in this case, is completely characterized by a discrete linear transformation of the Q samples which uniquely describe input and output signal. The matrix $C = c_{q-q'}$ can be applied multiple times, thus providing access to planes $z_{P,2Q}$.

The matrix description of the fractional Talbot effect was first formulated to calculate Fresnel diffraction for staircase profiles and binary optics, with $u_i(x) = \text{rect}(x)$ [9, 10]. However, the discrete formulation of Fresnel diffraction remains valid for any sampling expansion of the form in Eq. (9.46).

It is interesting to interpret this model of Fresnel diffraction in phase space. The sampled signal in Eq. (9.47) corresponds to a WDF of the form of Eq. (9.27). Convolution with the WDF of the comb function with respect to x and sampling in ν results in a WDF of the discrete signal, which is a grid of delta functions, periodic in x and ν. Due to interference terms the WDF will be $2Q$ periodic in both coordinates,

$$\begin{aligned} W^s(x,\nu) &= \sum_{q=0}^{2Q\ 1} \sum_{q'=0}^{2Q\ 1} W_{q,q'} \delta(x - q\Delta x/2)\delta(\nu - q'/2x_p) \\ &*_{x,\nu} \sum_{n=-\infty}^{\infty} \sum_{n'=-\infty}^{\infty} \delta\left(x - nx_p\right) \delta\left(\nu - n'/\Delta x\right), \end{aligned} \tag{9.49}$$

which is schematically depicted in Fig. 9.4. In passing, we note that coefficients $W_{q,q'}$, which are functions of the samples u_q, may serve as a definition for a discrete WDF [21, 65, 66].

The conditions for applying the matrix formulation of the fractional Talbot effect correspond to a horizontal shear of phase space, which moves all points in Fig. 9.4 to an identical grid of δ-points. The first time this happens is represented by line T, i.e. the row with frequency $1/(2x_p)$ is translated in space by $1/(2\Delta x)$. Repeating this operation Q times recovers the input distribution and is identical to the first Talbot distance. It is intuitively clear that any LCT which preserves the sampling grid of the WDF in Eq. (9.49) can be used to establish a linear transformation between the samples of the respective signals. However, recalling the result of the previous section on Fresnel imaging under spherical illumination, we can relax this constraint to LCTs, where all points of the phase-space samples are aligned in discrete vertical rows but dispersed along the frequency axis by the chirp modulation.

The matrix formulation was also promoted as an efficient way to simulate light propagation, not only in free space, but also in fractional Fourier transform systems

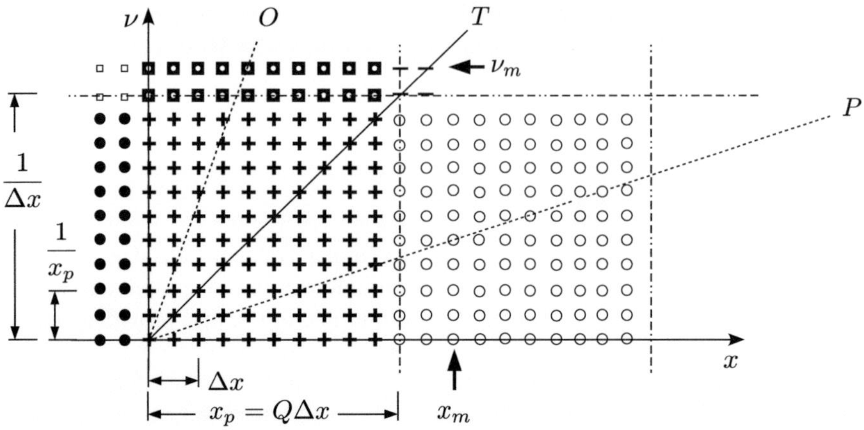

Fig. 9.4 WDF of a discrete, bandlimited, and periodic signal: schematic depiction of the discrete non-zero locations in the space-frequency plane associated with the discrete signal; each period of the WDF in both conjugate coordinates is represented with different symbols. Lines *O*, *T*, *P* refer to the horizontal shear indicating the position of points on the ν axis after Fresnel diffraction. Only line *T* corresponds to a periodic and discrete transfer function of free space

[86] suggesting the possibility to use the matrix method for a more general class of LCTs.

At this point, we may step back and restate the result of our discussion: For a discrete set of diffraction problems, the continuous Fresnel diffraction integral of periodic (and bandlimited) signals is uniquely described by a discrete linear transformation of a finite set of samples. For this to be possible, it is necessary to link the sampling interval to the propagation distance.

This close relationship between the continuous transformation of a periodic function and its discrete counterpart is not unique. In particular, it is a well-known fact of the Fourier transformation, and it is worthwhile to review briefly the properties of the Fourier transform of periodic and sampled signals as a prototype for the development of DLCTs in general.

9.6 Diffraction of Periodic Wavefronts and Discrete Optics

The DFT undoubtedly is the most widely used DLCT. The motivation for the DFT is the desire to calculate the continuous Fourier integral transformation of a complex signal $u(x)$ with a discrete linear transform suited for digital computation. There are at least three ways to introduce the DFT. One possibility is to define the DFT as an orthogonal expansion of discrete signals into a set of suitable basis functions. Relationships between the continuous and the discrete transform are then treated as premeditated coincidences.

Alternatively, it is possible to interpret the DFT as the result of a staircase approximation of the truncated Fourier integral [63],

$$\tilde{u}(\nu) \approx \Delta x \sum_{n=0}^{N-1} u(n\Delta x) \exp(-i2\pi \nu n \Delta x). \tag{9.50}$$

The DFT is found by sampling $\tilde{u}(\nu)$ up to the Nyquist frequency, which is equal to the inverse of the sampling interval Δx. This approximation of the Fourier transform as a Riemann sum, however, deserves a fair amount of critique. It implies that the DFT has to be interpreted as an approximate representation of the continuous signal, where the accuracy of the DFT can always be improved by increasing the integration window, and by decreasing the sampling interval Δx. In addition, it implicitly uses two different interpolation schemes to connect the discrete series representation with the continuous signal. The sampling at discrete distances is connected with the sampling theorem, and the continuous signal is derived from the discrete samples with the help of a trigonometric sinc interpolation. The Riemann sum, in contrast, approximates the signal with the simplest form of a polynomial type interpolation. This inconsistency requires some "handwaving" to justify the DFT, without offering a truly intuitive interpretation of potential computation errors.

We can choose a third option and restrict our attention to strictly periodic and bandlimited functions, $\tilde{u}(\nu) = 0$ for $|\nu| > \nu_0/2$. Applying the sampling theorem to Eq. (9.36) with $\Delta x = 1/\nu_0$, the signal can be re-written as

$$u(x) = \left\{ \sum_{q=0}^{Q-1} u(q\Delta x)\delta(x - q\Delta x) \right\} * \left\{ \sum_{n=-\infty}^{\infty} \delta(x - nx_p) \right\} * u_i(x/\Delta x), \tag{9.51}$$

with $u_i(x) = \operatorname{sinc}(x)$, and $x_p = Q\Delta x$. Equation (9.51) expresses the signal in terms of a single signal period, which is sampled and periodically continued. Finally, the discrete signal is interpolated to recover the continuous signal.

If the Fourier integral transformation is applied to Eq. (9.51), we obtain

$$\begin{aligned} \tilde{u}(\nu) &= \left\{ \sum_{q=0}^{Q-1} u(q\Delta x) \exp(-i2\pi \nu q \Delta x) \right\} \frac{1}{x_p} \sum_{n=-\infty}^{\infty} \delta(\nu - n/x_p)\, \Delta x\, \tilde{u}_i(\Delta x \nu) \\ &= \left\{ \sum_{q=0}^{Q-1} u(q\Delta x) \exp(-i2\pi nq/Q) \right\} \frac{1}{Q} \sum_{n=-\infty}^{\infty} \delta(\nu - n/x_p)\, \tilde{u}_i(\Delta x \nu). \end{aligned} \tag{9.52}$$

The first term in Eq. (9.52) is Q periodic in n and we identify the DFT,

$$\tilde{u}_n = \sum_{q=0}^{Q-1} u_q \exp(-i2\pi nq/Q), \tag{9.53}$$

with $\tilde{u}_n = \tilde{u}(n/x_p)$, $u_q = u(q\Delta x)$, and $n, q = 0, \ldots, Q-1$. The Fourier transformation of the continuous input signal, thus, can be re-written

$$\tilde{u}(\nu) = \frac{1}{Q}\left\{\sum_{q'=0}^{Q-1} \tilde{u}_{q'}\delta(\nu - q'/x_p) * \sum_{n=-\infty}^{\infty} \delta(\nu - nQ/x_p)\right\} \tilde{u}_i(\Delta x\nu). \tag{9.54}$$

Recovering the spectrum of the continuous transform is expressed as a multiplication with the Fourier transform of the interpolation function, in this case $\tilde{u}_i(\nu) = \mathrm{rect}(\nu)$.

Here, the DFT is derived without introducing any type of approximation. Instead, we identify a class of functions, periodic bandlimited signals, for which the continuous integral transform is identical to the discrete transform, which operates on the finite number of discrete samples necessary to represent the signal rigorously. For the case of the Fourier transform no additional interpolation is needed to recover the output of the continuous transform from the discrete output of the discrete transform. Since the DFT is derived as the continuous Fourier transform of generalized functions, we expect all properties of the integral transform to be preserved for the DFT, as long as the discrete function can be interpreted as a sampled periodic and bandlimited signal. We note that Fig. 9.4 again illustrates the associated phase space. The Fourier transform rotates phase space by $\pi/2$ mapping one point of the discrete WDF grid to another point of the grid. This means, the continuous transform of the discrete signal, represented by a train of delta-functions, is identical to the discrete transform of the sample values.

Having based our derivation exclusively on the sampling theorem, we can now use the latter to explore what happens, if we choose more general sets of functions. For instance, if we represent non-periodic functions in a truncated computation window, we force implicit periodicity of the discrete function, which is no longer bandlimited in a rigorous sense. However, it is possible to develop approximate methods, such as windowing, to minimize the truncation effect.

In the remainder of the chapter, we seek definitions of DLCTs, which are inspired by this interpretation of the DFT. The theory of the fractional Talbot effect will allow us to develop a similar perspective for discrete Fresnel diffraction and for LCTs in general. A valid implementation of DLCTs may be identified in the following way: For a certain class of functions, which include periodic and bandlimited functions, the signal can be represented by a finite number of discrete samples. The DLCT of these samples should be identical, both to the sampled output of the LCT of the signal, and to the continuous transform of the sampled input, the latter being represented by a train of δ-functions.

9.7 The Discrete Fresnel Transform

The starting point for our continuing exploration is Eq. (9.37), which describes free space propagation as a convolution of the input signal and the point response of free space. It is often convenient to express this convolution as a product in the frequency domain. In optics, the spatial frequency is linked with the propagation angle of plane wave modes, which has earned this approach the name "angular spectrum method" [17, 31]. The efficiency of the FFT algorithm makes the angular spectrum method attractive for numerical computations as well.

9.7.1 The Discrete Angular Spectrum Method

The discrete implementation of the angular spectrum method is arguably one of the most widely used methods to simulate free space propagation, both for paraxial, and for non-paraxial optical signals [30, 44, 52, 75, 80]. The convolution with the point response of free space is transformed into a modulation of the spectrum with the transfer function of free space,

$$\tilde{u}(\nu, z) = \tilde{u}(x, 0)\, \tilde{h}_{\mathrm{Fr}}(\nu, z) = \tilde{u}(x, 0)\, \exp(-i\pi\lambda z \nu^2), \tag{9.55}$$

where $\tilde{h}_{\mathrm{Fr}}(\nu, z)$ is the Fourier transform of $h_{\mathrm{Fr}}(x, z)$ in Eq. (9.18). The angular spectrum method can be expressed as a system matrix,

$$S_z(\lambda z) = S_{\mathrm{FT}}^{-1} S_L(\lambda z) S_{\mathrm{FT}}. \tag{9.56}$$

For numerical computations, the discrete input is the sampled version of the continuous input signal. True to our theme, we consider strictly periodic and bandlimited signals. Each period x_p of the input signal is represented by Q samples, and we again assume Q to be even. Then, the entire chain of using discrete computation to calculate the Fresnel integral of the continuous signal $u(x)$ can be summarized as

$$u(x, z) = u^{(s)}(x, z) \ast u_i(x/\Delta x), \tag{9.57}$$

where the discrete output of the Fresnel transform reads

$$u^{(s)}(x, z) = \left\{ \sum_{q=0}^{Q-1} u_q \delta(x - q\Delta x) \right\} \ast \sum_{n=-\infty}^{\infty} \delta(x - n x_p) \ast h_{\mathrm{Fr}}(x, z), \tag{9.58}$$

and with the u_q are being the samples of the continues input signal $u(x)$ at $x = q\Delta x$. This is the Fresnel diffraction of the signal in Eq. (9.51). The first term corresponds to the samples of a single signal period, which is replicated periodically by the comb-function. After Fresnel propagation the continuous Fresnel diffraction amplitude can be obtained by convolution with the interpolation function $u_i(x) = \mathrm{sinc}(x)$. In the frequency domain, the spectrum of the signal reads

$$\begin{aligned}\tilde{u}(\nu, z) &= \tilde{u}^{(s)}(\nu, z)\, \Delta x\, \tilde{u}_i(\Delta x \nu) \\ &= \frac{1}{Q}\left\{\sum_{q'=0}^{Q-1} \tilde{u}_{q'}\delta(\nu - q'/x_p) * \sum_{n=-\infty}^{\infty} \delta(\nu - n/\Delta x)\right\} \tilde{h}_{\mathrm{Fr}}(\nu, z)\tilde{u}_i(\Delta x \nu)\end{aligned} \tag{9.59}$$

where coefficients $\tilde{u}_{q'}$ again refer to the DFT of the samples u_q.

The discrete angular spectrum method is based on sampling the transfer function in the frequency domain at discrete frequencies $(q' + Qn)/x_p$. In much of the literature on the subject, conditions are derived to ensure the sampling of the transfer function is done appropriately [30, 52]. However, for the class of signals we are considering here, aliasing due to under-sampling the chirp of the transfer function is not necessarily a fundamental issue. Since the signal is defined as being periodic, any alias term can be interpreted physically as signal energy of one period spreading through diffraction and leaking into the neighboring period. In some sense, we have decided on a physical model, which shares the cyclic boundary conditions of the discrete model.

A second issue for the class of periodic signals is often ignored. This concerns the "out-of-bin" sampling of the transfer function [87]. In general, the samples of the transfer function do not show the same inherent periodicity as the DFT of the discrete samples. This is relevant, because the actual numerical implementation of the angular spectrum method reads

$$\tilde{u}^{\mathrm{asm}}(\nu, z) = \frac{1}{Q}\left\{\tilde{h}_{\mathrm{Fr}}(\nu, z) \sum_{q'=0}^{Q-1} \tilde{u}_{q'}\delta(\nu - q'/x_p) * \sum_{n=-\infty}^{\infty} \delta(\nu - n/\Delta x)\right\} \tilde{u}_i(\Delta x \nu). \tag{9.60}$$

Only the first period of the spectrum is multiplied with the transfer function and this truncated sequence is implicitly continued. We distinguish several cases, for which Eq. (9.60) is equivalent to Eq. (9.59). Most importantly both equations are equivalent, if we are only interested in the continuous output of the DLCT. The bandlimit is restored by the sinc-interpolation. In this case, the frequency domain equivalent is a perfect low pass filter, which truncates everything outside the first period of the spectrum. In other words, the output of the discrete angular spectrum algorithm is always a properly sampled version of the continuous Fresnel diffraction pattern, and the latter, which is obtained via a sinc-interpolation of the output samples, is a truthful representation of the Fresnel diffraction integral of the continuous input function.

The same does not hold for the discretized input of bandlimited and periodic functions. Equation (9.59) samples the transfer function of free space with the entire discrete and periodic spectrum instead of a single period, which is a truncated and replicated period of the spectrum. Only if the sampled transfer function matches the periodicity of the input spectrum can the discrete output of the angular spectrum method be regarded as the Fresnel diffraction pattern of the discrete input.

It is worthwhile to explore this subtle point in some more detail. The sampled transfer function shows Q periodicity when $-\pi\lambda z q'^2/x_p^2 = -\pi\lambda z(q' + Q)^2/x_p^2$. For even Q this is the case at integer multiples of

$$z_s = \frac{1}{2Q}\frac{2x_p^2}{\lambda}. \tag{9.61}$$

We recognize condition Eq. (9.61) as those fractional Talbot planes, which can be related to the matrix description of Fresnel diffraction. In the frequency domain, the matrix formulation of the fractional Talbot effect thus reads

$$\tilde{u}_{q'}(z_s) = \tilde{u}_{q'}\,\tilde{h}_{q'}(z_s) \tag{9.62}$$

with $\tilde{h}_{q'}(z_s) = \tilde{h}_{\mathrm{Fr}}(q'/x_p, z_s) = \exp(-i\pi q'^2/Q)$. This can be translated back into the spatial domain, and the inverse DFT transforms the product into the discrete cyclical convolution of the matrix method of the fractional Talbot effect,

$$u_q(z_s) = \sum_{p=0}^{Q-1} h_{q-p} u_p. \tag{9.63}$$

The matrix elements can be determined as the inverse DFT of $\tilde{h}_p(z_s)$, and with the help of Gaussian sums [70] we find

$$h_q(z_s) = \exp(i\pi/4 + i\pi q^2/Q), \tag{9.64}$$

which reproduces the Talbot coefficients we found in Sect. 9.5.3. The matrix formulation of the fractional Talbot effect was promoted as a possible alternative to simulate Fresnel diffraction numerically [86]. However, in terms of numerical cost it has to compete with the complexity of the FFT algorithm, which can be used for the angular spectrum method. We emphasize, however, that as long as the condition in Eq. (9.61) is fulfilled, all properties we encountered for the fractional Talbot effect can be transferred to the discrete angular spectrum method. In particular, it is possible to interpret Eq. (9.60) not merely as a valid sampling expansion for bandlimited signals, but for any signal, which can be represented in this way. This includes a rectangular interpolation function $u_i(x) = \mathrm{rect}(x)$ as one important example.

We can again employ the phase-space representation of periodic and discrete signals to develop a better appreciation of the "out-of-bin" sampling condition, if the fractional Talbot condition is not met. In Fig. 9.4, line O corresponds to an out-of-bin sampling of the transfer function. The original phase-space samples are moved away from the sampling grid and the horizontal shift of positions into neighboring periods is not at a multiple of the grid period. It is clear that the distorted grid corresponds to additional samples of the delta train representation of the discrete signal. The same is true for line P which in addition moves sampling points beyond the boundaries of their respective period in space corresponding to alias terms of the propagated signal. Only the case of line T and multiples of the associated propagation distance ensure strict periodicity of the sampled transfer function.

The condition in Eq. (9.61), which links the number of samples per period to the propagation distance was previously identified by Voelz and Roggemann as the ideal sampling condition for optical signals, albeit without recognizing its relationship with the fractional Talbot effect [87]. In the same work it is also argued that alternative implementations of discrete Fresnel diffraction are subject to the same discretization condition. In particular, for the so-called direct methods, which sample the point response function $h_{Fr}(x, z)$ instead of the transfer function of free space, we again encounter the issues of sampling the chirp of the response function and to find conditions which ensure periodicity for both the continuous and the discrete transformation. Some additional flexibility can be gained, however, by considering non-periodic functions of finite support.

9.7.2 Signals with Finite Space-Bandwidth Product

While the class of periodic bandlimited signals has allowed us to define cases, for which the discrete Fresnel transform yields the same result as the continuous transformation, we are typically interested to simulate light propagation of signals with finite support. Despite the fact that these signals are not bandlimited in any strict sense we invoke the notion of the space-bandwidth product and assume an effective bandlimit determined by the frequency, for which the signal spectrum drops below the noise level. We can apply the angular spectrum method to this wider class of signals, as long as we appreciate that any use of the DFT assumes an implicit periodicity for both the signal and the spectrum. The implicit periodicity on non-periodic signals of finite length is typically accommodated by zero-padding the signal [47].

In this context, it is again beneficial to consider the propagation problem in phase space [38]. Figure 9.5 schematically shows three cases, for which Fresnel diffraction of a signal with finite space-bandwidth product is simulated. The signal resides entirely within one cell of the implicitly periodic signal space. Figure 9.5a demonstrates that we will always encounter aliasing, if the signal occupies the entire phase-space cell defined by the periodicity of the discrete signal and its sampling distance. A propagation step of any size will inevitably lead to a leakage of the signal into the neighboring cell, which is no longer representing the physical problem correctly. The Nyquist frequency is determined by the maximum frequency

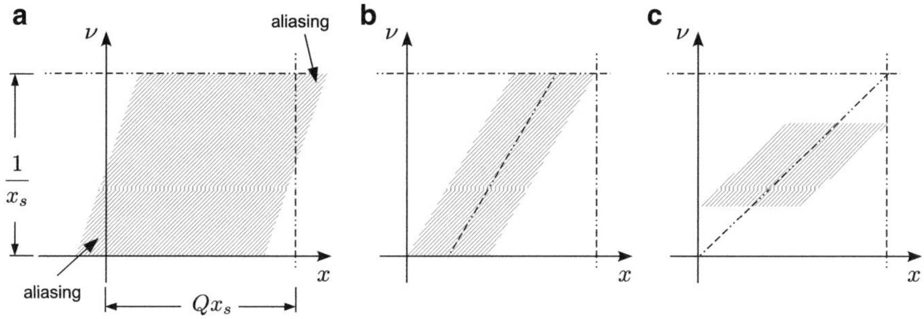

Fig. 9.5 Fresnel propagation of finite space-bandwidth products: (**a**) signal occupying the entire phase-space cell, the inherently periodic discrete Fresnel diffraction. (**b**) Signal which is only zero-padded in the spatial coordinate. (**c**) Signal with zero-padding in both space and frequency; only in this case is alias-free propagation to the fractional Talbot distance z_s possible

of the transfer function inside the unit phase-space cell, combined with the highest frequency of the signal, contrary to some work, which emphasizes exclusively the maximum frequency of the transfer (or response) function (see, for example, reference [52]).

The phase-space diagram also reveals that it is not sufficient to zero-pad the signal along the spatial coordinate, if we want to propagate the signal over the ideal distance z_s. The simulation with ideal sampling conditions will still suffer from aliasing, unless zero-padding is done both in space and frequency. Only for the configuration in Fig. 9.5c does the entire phase-space volume remain contained in a single phase-space cell as we propagate the signal to the distinct fractional Talbot plane z_s. It is easy to show that we need at least a phase-space cell with four times the space-bandwidth product of the signal. In practical applications zero padding is often applied to only one of the two conjugate domains combined with a quadratic phase propagator, which is not ideally sampled (Fig. 9.5b).

It is for non-ideal sampling conditions that the direct method, i.e. the sampling of the response function, is distinct from the angular spectrum method. The former requires large propagation distances and zero-padding in the frequency domain, to avoid aliasing in the spectral domain. The latter is restricted to propagation distances smaller than the ideal condition in Eq. (9.61) and requires zero-padding in the signal domain. The spectral leakage related to out-of-bin sampling of the response or transfer function is equally addressed through sufficient zero-padding [87].

9.7.3 Zooming Algorithms

In particular in the context of digital holography, there is considerable interest to combine discrete Fresnel diffraction with zooming functions, where the image shows parts of the object space with a specific scale. Apart from resampling and interpolation, the implementation of discrete diffraction can be used to control the image scale. The angular spectrum method as introduced in the previous sections maintains the sampling interval.

Zooming can be achieved through implementations of the aforementioned direct simulation method. To this end, Fresnel diffraction is re-written as

$$S_z(\lambda z) = S_L(1/\lambda z) S_s(\lambda z) S_{\mathrm{FT}} S_L(1/\lambda z). \tag{9.65}$$

The corresponding integral transform can be discretized with different sampling intervals in the input and the exit plane. The resulting discrete transformation can be written as [40]

$$u(n\Delta x_2, z) = \exp\left(\frac{i\pi (n\Delta x_2)^2}{\lambda z}\right) \times \sum_{m=0}^{N-1} u(m\Delta x_1) \exp\left(\frac{i\pi (n\Delta x_1)^2}{\lambda z}\right) \exp\left(\frac{-i2\pi nm\Delta x_2 \Delta x_1}{\lambda z}\right). \tag{9.66}$$

The scaled discrete Fourier transformation in this expression reduces to the DFT, if

$$\lambda z = N\Delta x_1 \Delta x_2. \tag{9.67}$$

This provides us with a practical condition to link input and output scale. More interestingly, we find again a relationship between the sampling interval and the propagation distance. In fact, for $\Delta x_1 = \Delta x_2$ we recover the condition in Eq. (9.61) and we can identify Eq. (9.67) as an alternative ideal sampling condition. While neither chirp modulation in Eq. (9.66) corresponds to ideal sampling, the overall process again moves points of the grid in phase space to other points on the same grid and spectral leakage is avoided.

It is worthwhile to highlight a second zooming algorithm in this context. By sandwiching the Fresnel propagation step between two additional chirp modulations [40]

$$S_{\mathrm{zoom}} = S_L(1/\lambda f_2) S_z(\lambda z) S_{\mathrm{FT}} S_L(1/\lambda z) \tag{9.68}$$

it is possible to simulate the propagation of a signal, which can be interpreted as a modulated spherical wave. This, in effect, is equivalent to the discussion of the Talbot effect and the fractional Talbot effect for the case of spherical illumination in Sects. 9.4.3 and 9.5.2, respectively. The first lens of focal length f_1 introduces a convergent or divergent carrier, while the second lens f_2 removes the remaining chirp after a scaled propagation step. Maintaining the number of samples across one period of the scaled wavefront provides the desired zooming functionality.

As we have pointed out in Sect. 9.5.2 the configuration in Eq. (9.68) can be replaced by Fresnel diffraction over a scaled distance, followed by a single chirp modulation and scaling. This, in effect, can be used as the prototype for computing arbitrary LCTs.

9.8 The Fractional Talbot Effect and DLCTs

One of the curios of paraxial optics is the perpetual problem of defining a universally accepted definition of the general DLCTs. Various definitions provide working solutions for estimating LCTs numerically, yet the growing body of literature on the subject also highlights the difficulties of finding a definition, which is both rigorous and practical. Much of the work on DLCTs is motivated by the need to find numerical implementations of the fractional Fourier transformation, yet even for this rather narrow set of LCTs no definition has found the same universal acceptance the DFT enjoys as the discrete counterpart of the continuous Fourier transform.

In part, this may be attributed to the same reasons, which guides discussions about the most suitable implementation of the discrete Fresnel transformation. Rather than establishing conditions, for which the DLCT can be linked rigorously to the continuous transform, many studies assume a priori that any numerical scheme for computing LCTs inevitably has to be an approximation of the continuous transform. However, after exploring the self-imaging phenomenon both for Fresnel diffraction, and for more general classes of LCTs, we may well expect to carry the general concept of constructing the discrete Fresnel diffraction algorithm over to the general class of LCTs. Any LCT can be factorized into a sequence of DFTs, chirp modulations and chirp convolutions. The implicit periodicity of finite discrete signals introduced through the DFT suggests that we can apply a subset of LCTs to bandlimited and periodic signals and obtain a rigorous formulation for the respective DLCT of the finite number of samples. Those LCTs, which belong to this set, can readily be identified, since they again move each point of the phase-space sampling grid in Fig. 9.4 to another point of that sampling grid. Hence, we are limited to chirps modulating either the signal or the spectrum corresponding to the vertical or horizontal shear indicated by line T in Fig. 9.4. The modulation of the spectrum corresponds to multiples of the fractional Talbot plane $z_s = z_{1,2Q}$, while the chirp of the signal corresponds to a parabolic lens with a focal length of integer fractions of z_s. If these conditions are met, the discrete output signal, again, represents a periodic and discrete signal in both phase-space coordinates. We can extend this class of LCTs by accepting scaling transformations of the input and the output signal. This is analogous to the optical Fourier transformation implemented with the help of a 2f-system. The specific focal length of the Fourier transform lens introduces a scaling of the Fraunhofer spectrum, which does not change the essence of the mathematical operation, but which still corresponds to a different LCT.

This framework, however, is unsatisfying for several reasons. First, we are left with a rather limited set of ABCD parameters, far less than necessary to accommodate even the most generic LCTs, namely the fractional Fourier transformation. Even including scaling operations, it is not possible to establish conditions, which move the points of a rotated cartesian grid to points of the same cartesian grid. This, in fact, may explain some of the difficulties of finding a universal definition for DLCTs. Only for a small number of LCTs, which are compatible with the fractional self-imaging condition in space and frequency, is there a unique and rigorous way to accomplish this task.

Second, and perhaps more importantly, LCTs typically transform bandlimited and compact signals into signals, which are neither bandlimited nor compact. This means, we can no longer use a sinc-interpolation to establish a rigorous relationship between the discrete samples and the continuous transform, even where the set of sampling points in phase space maps to itself .

Finally, the mapping of the phase-space grid to itself is a sufficient, but not a necessary condition for establishing rigorous conditions for DLCTs. We recall the possibility to allow for signals carried by a chirp signal. Prior knowledge of the chirp carrier allows us to remove the chirp modulation at any point in the process.

Thus, the remaining task is to find a definition for DLCTs, where we surrender some of the properties we observed with the DFT and the discrete Fresnel transform, but which allows us to establish relationships between the discrete and the continuous signals without loss of rigor. Our particular interest is again directed toward implementations based on self-imaging theory. A suitable factorization substitutes the LCT with a Fresnel diffraction step followed by a chirp modulation and a scaling operation,

$$\begin{pmatrix} A & B \\ C & D \end{pmatrix} = S_s(M)S_L(\gamma)S_z(\beta) = \begin{pmatrix} M & M\beta \\ \frac{\gamma}{M} & \frac{1+\beta\gamma}{M} \end{pmatrix}. \quad (9.69)$$

The magnification M can be treated as a continuous parameter to accommodate the matrix element A. In practical terms, we only need to consider the magnification to relate the discrete signal to physical length or frequency, i.e. as part of the interpolation from the discrete to the continuous representation and vice versa. The sampling distance has to be defined explicitly for the sampling and the interpolation step and the interpolation function has to be changed accordingly. The matrix element $B = \beta M$ is constrained by the conditions we derived for the fractional Talbot effect and numerical Fresnel propagation.

The step which deserves slightly more attention is the chirp modulation, expressed by the matrix element $C = \gamma/M$. Closer inspection reveals that it is not necessary to match the condition for the spectral fractional Talbot effect to select γ. As long the output signal of LCTs is the last step in a line of numerical simulations, we can in fact ignore any constraints on γ. Even if the chirp is under-sampled, the chirp modulation of the discrete signal always corresponds to a diagonal matrix, i.e. a unitary, invertible operation. Since we know the chirp function explicitly, we can always recover the samples of the simulation window as well as in all other implicitly defined periods. By incorporating the chirp function into the interpolation function we can also recover the correct continuous LCTs from under-sampled discrete data.

This is accomplished, by what has become known as a form of generalized sampling, applicable to signals which occupy a diagonal band in phase-space

[39, 46, 74]. Any such signal can be interpreted as a chirped bandlimited signal. Conceptually, the sampling interval does not need to match the overall bandlimited of the signal, but corresponds to the inverse of the local bandwidth. The interpolation of the continuous LCT corresponds to removing the sampled chirp function from the discrete data, then performing a standard sinc-interpolation, and multiplying the continuous chirp with the result of the interpolation. Reversely, the discrete output of this DLCTs can be computed from the continuous output signal of the LCT by sampling the signal and selecting the sampling frequency based on the Nyquist sampling rate of the bandlimited input, which has to be scaled by the magnification factor M.

It is also possible to chain several DLCTs. This can be accomplished by first removing the chirp from the first LCT, and adding this chirp as the first step of the consecutive LCT. The sequence of chirp-Fresnel propagation-chirp-magnification can now be recast into the form in Eq. (9.69). In other words, with each step, we increase the overall Fresnel propagation step and combine the necessary chirping steps as a single chirp at the end of the LCT chain. This perspective of LCTs is unique, since it allows us to deemphasize the difference between different types of LCTs and identify the Fresnel propagation, and thus the fractional Talbot effect, as the core operation concerning DLCTs. One key ingredient for this interpretation is the observation that any LCT of a bandlimited signal can be regarded as a chirped bandlimited signal [39].

Similar to the discrete Fresnel diffraction, for non-periodic signals with finite space-bandwidth product we need to be concerned about aliasing which is introduced by the propagation step. Zero-padding remains necessary to ensure the fractional Talbot condition for the free space propagation step. Again, however, sampling of the chirp modulation function does not require compliance with the sampling theorem. Instead, this issue is shifted to the interpolation function and we are not concerned about parts of the signal's phase-space volume leaking into higher and lower frequency periods of the phase-space grid in Fig. 9.5. The difference between under-sampling the chirp of the propagation step in the frequency domain and the chirp in the signal domain is related to the inherent properties of the DFT, which is needed for any implementation of the Fresnel propagation step. Aliasing artifacts only impact the computation, if the chirp modulation is followed by a DFT operation, which forces implicit periodicity for both the signal and its spectrum. This is the case, for instance, if several DLCTs need to be executed in consecutive order, and which may be addressed in the way outlined above.

9.9 Test Case: The Fractional Fourier Transform

Much of the interest in DLCTs was motivated by the need for efficient and accurate numerical implementations of the fractional Fourier transformation. The fractional Fourier transform can be regarded as an important test case for exploring LCTs beyond Fresnel diffraction. The relationship between the fractional Talbot effect and the calculation of fractional Fourier transforms has been noted by a number of authors, who have highlighted parts of the framework outlined in the preceding section [2, 32, 86].

The fractional Fourier transform corresponds to a rotation of phase space, expressed by a system matrix with $A = D = \cos(\phi)$, and $B = -C = \sin(\phi)$. The rotation angle ϕ defines the order of the fraction Fourier transform, with $\phi = \pi/2$ corresponding to the ordinary Fourier transformation. For the factorization of the LCTs this identifies $M = \cos(\phi)$ as the scaling operation. Consequently, the chirp modulation is given as $\gamma = -\sin(\phi)\cos(\phi)$. The Fresnel propagation step is determined from two conditions. Given the number of samples Q used to compute the DLCT, the window size x_p and sampling interval Δx are linked via $x_p/\Delta x = Q$. In addition, the propagation step in Eq. (9.69) has to be $\beta = \tan(\phi) = x_p \Delta x$. We find

$$\Delta x = \left(\frac{\tan(\phi)}{Q} \right)^{1/2}. \tag{9.70}$$

Figure 9.6 illustrates the phase-space transformation of the Cartesian grid under these conditions. We observe that the rotated grid again aligns grid points in vertical direction. The resulting grid can be interpreted as a chirped Cartesian grid. As expected, this new grid corresponds to the new sampling interval $\Delta x' = M\Delta x$. Removing the chirp, in Fig. 9.7c, leaves us with a Fresnel diffracted signal, where all grid points again coincide with a scaled Cartesian grid.

For the fractional Fourier transformation, there are a number of additional properties, which we may wish to translate into the discrete formulation. Most notably, this is the addition theorem of the respective phase-space rotations, i.e.

$$S_{F\alpha}(\alpha_1 + \alpha_2) = S_{F\alpha}(\alpha_2) S_{F\alpha}(\alpha_1). \tag{9.71}$$

In the spirit of our exploration, Eq. (9.71), $\alpha_1 + \alpha_2$ as well as α_1 have to be limited to cases with in-bin sampling of the Fresnel diffraction.

For signals with finite support and finite bandwidth zero-padding is obligatory to avoid aliasing of the actual signal. This also reduces the impact of out-of bin sampling of the Fresnel diffraction step. As an example, Fig. 9.7 shows the numerical result of applying our procedure to two Hermite-Gauss polynomials, i.e. the eigenfunctions of the fractional Fourier transform. The computation was performed for $\phi = \pi/6$, with $Q = 2^{10}$ samples of the polynomials of order 2 and 3. To compare the input and output of the DLCT, the output vector was divided by the respective eigenvalue. Figure 9.7a was computed with ideal sampling, while for Fig. 9.7b the number of samples were not matched to the propagation distance. The accuracy of both results is virtually identical, which shows that the DLCT based on self-imaging theory provides a sufficient condition. However, for practical implementations, a satisfactory accuracy can always be obtained, if the number of samples can be chosen freely. This returns our discussion of DLCTs to its starting point. Rather than defining necessary conditions for an implementation of DLCTs,

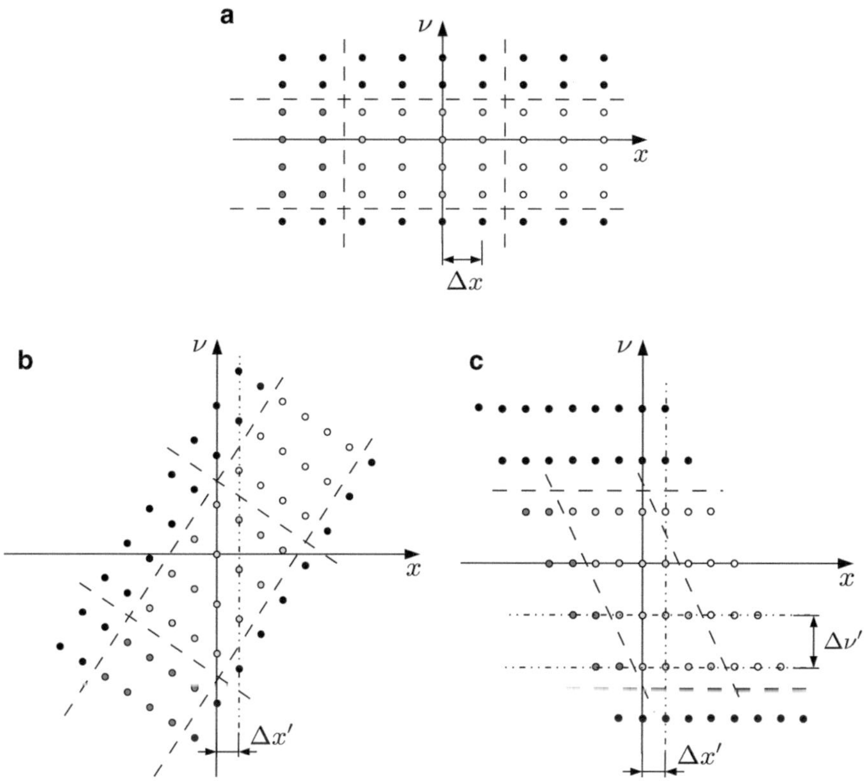

Fig. 9.6 Fractional Fourier transform of the discrete phase space: (**a**) Cartesian grid of the discrete bandlimited and periodic signal. (**b**) Fractional Fourier transformation of the grid in (**a**); rotation angle and sampling are selected to align grid points in vertical direction. (**c**) Cartesian grid after removing the chirp of the distribution in (**b**)

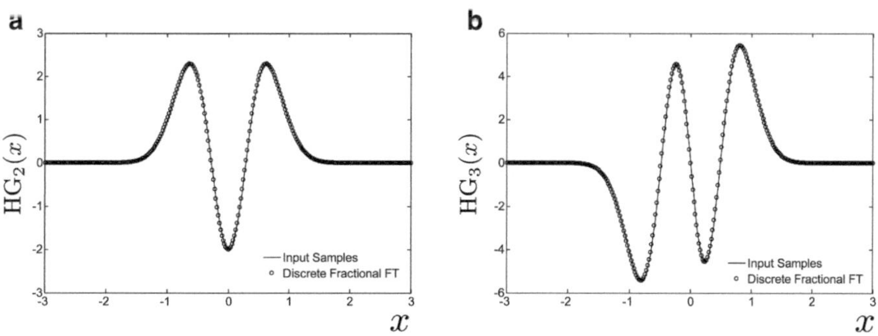

Fig. 9.7 Fractional Fourier transform ($\phi = \pi/6$) of Hermite–Gauss polynomials of order 2 and 3. The *solid line* and *circles* correspond to the input and output signal, respectively

self-imaging defines an ideal case, which allows us to develop a more intuitive understanding of discrete paraxial optics. This includes issues, which may arise, if these ideal conditions cannot be implemented for practical reasons.

9.10 Conclusions

Our exploration of self-imaging and LCTs has revealed some interesting insights into the relationship between continuous periodic signals and discrete representations. Undoubtedly, the theory of self-imaging provides an easily accessible physical approach to explore numerical algorithms and in particular DLCTs. As we have pointed out, we see the most important contribution of self-imaging in providing a framework for thinking about problems involving discrete representations, rather than as a solution to the problem of finding the most suitable and efficient implementation.

One important advantage of restricting our attention to periodic bandlimited functions is a rigorous definition of the number of degrees of freedom. The finite number of discrete samples necessary to represent a single period of the continuous signal contains all information available about the signal and there is no need to resort to heuristic definitions of the space-bandwidth product for estimating the number of independent variables. For paraxial signals and systems this corresponds directly to configurations best explored in terms of self-imaging theory. As we demonstrated, the propagation of these independent variables can be traced most conveniently, if we restrict the set of operations to those compatible with the matrix formulation of the fractional Talbot effect.

We also speculate that apart from the fractional Talbot effect of periodic functions, other self-imaging phenomena, which we mentioned briefly in Sect. 9.4 could be used to model optical signals and system more flexibly. In fact, some alternative methods for defining DLCTs, such as the definition in terms of eigenfunctions [22, 27] may find a more intuitive physical interpretation, if explored in the context of generalized self-imaging.

Instrumental to our exploration was the phase-space formalism of paraxial optics. The phase-space of discrete and periodic signals corresponding to a discrete and double periodic phase space emphasizes a novel approach for obtaining geometrical solutions for discrete optical problems. While closely associated with the properties of phase-space optics in general, it aids our understanding of the increasingly important link between continuous signals and discrete representations. In this sense, we believe, discrete phase-space optics has the potential to assume a role in optical system theory distinct from its continuous counterpart. Our narrowly focused exploration of self-imaging and discrete paraxial optics may serve as an example of how to exploit this theoretical framework.

References

1. T. Alieva, F. Agullo-Lopez, Imaging in first-order optical systems. J. Opt. Soc. Am. A **13**(12), 2375–2380 (1996)
2. T. Alieva, A. Barbé, Fractional Fourier and Radon - Wigner transforms of periodic signals. Signal Process. **69**(2), 183–189 (1998)
3. T. Alieva, A. Barbé, Self-imaging in fractional Fourier transform systems. Opt. Commun. **152**(1–3), 11–15 (1998)
4. T. Alieva, A.M. Barbé, Self-fractional Fourier images. J. Mod. Opt. **46**(1), 83–99 (1999)
5. T. Alieva, M.J. Bastiaans, Powers of transfer matrices determined by means of eigenfunctions. J. Opt. Soc. Am. A **16**(10), 2413–2418 (1999)
6. T. Alieva, M.J. Bastiaans, Self-affinity in phase space. J. Opt. Soc. Am. A **17**(4), 756–761 (2000)
7. T. Alieva, M.J. Bastiaans, Dynamic and geometric phase accumulation by Gaussian-type modes in first-order optical systems. Opt. Lett. **33**(15), 1659–1661 (2008)
8. V. Arrizón, J. Ojeda-Castañeda, Multilevel phase gratings for array illuminators. Appl. Opt. **33**, 5925–5931 (1994)
9. V. Arrizón, J. Ojeda-Castañeda, Fresnel diffraction of substructured gratings: matrix description. Opt. Lett. **20**, 118–120 (1995)
10. V. Arrizón, J.G. Ibarra, J. Ojeda-Castañeda, Matrix formulation of the Fresnel transform of complex transmittance gratings. J. Opt. Soc. Am. A **13**, 2414–2422 (1996)
11. V. Arrizón, E. López-Olazagasti, A. Serrano-Heredia, Talbot array illuminators with optimum compression ratio. Opt. Lett. **21**, 233–235 (1996)
12. V. Arrizón, G. Rojo-Valázquez, J.G. Ibarra, Fractional Talbot effect: compact description. Opt. Rev. **7**, 129–131 (2000)
13. S. Atkins, B. Fischer, All-optical pulse rate multiplication using fractional Talbot effect and field-to-intensity conversion with cross-gain modulation. IEEE Photonics Technol. Lett. **15**(1), 132–134 (2003)
14. J. Azaña, Spectral Talbot phenomena of frequency combs induced by cross-phasemodulation in optical fibers. Opt. Lett. **30**(3), 227–229 (2005)
15. J. Azaña, L.R. Chen, General temporal self-imaging phenomena. J. Opt. Soc. Am. B **20**(7), 1447–1458 (2003)
16. J. Azaña, M.A. Muriel, Technique for multiplying the repetition rates of periodic trains of pulses by means of atemporal self-imaging effect in chirped fiber gratings. Opt. Lett. **24**(23), 1672–1674 (1999)
17. R. Barakat, The caculation of integrals encountered in optical diffraction theory, in *Topics in Applied Physics*, vol. 41, ed. by B.R. Frieden (Springer, Heidelberg, 1980)
18. M. Bastiaans, Application of the Wigner distribution function in optics, in *The Wigner Distribution - Theory and Applications in Signal Processing*, ed. by W. Mecklenbräuker, F. Hlawatsch (Elsevier, Amsterdam, 1997), pp. 375–426
19. P.A. Bélanger: Periodic restoration of pulse trains in a linear dispersive medium. IEEE Photonics Technol. Lett. **1**, 71–72 (1989)
20. N. Berger, B. Levit, B. Fischer, Optical comb filter based on spectral Talbot effect in uniform fibre Bragg gratings. Electron. Lett. **43**(12), 665–667 (2007)
21. K.H. Brenner, A discrete version of the Wigner distribution function, in *Selected Papers on Phase-Space Optics*, ed. by M. Testorf, J. Ojeda-Castañeda, A. Lohmann. Milestone, vol. 181 (SPIE, Bellingham, 2006)
22. C. Candan, M. Kutay, H. Ozaktas, The discrete fractional Fourier transform. IEEE Trans. Signal Process. **48**(5), 1329–1337 (2000)
23. J. Caraquitena, M. Beltrán, R. Llorente, J. Martí, M.A. Muriel, Spectral self-imaging effect by time-domain multilevel phase modulation of a periodic pulse train. Opt. Lett. **36**(6), 858–860 (2011)

24. G. Cariolaro, T. Erseghe, P. Kraniauskas, N. Laurenti, Multiplicity of fractional Fourier transforms and their relationships. IEEE Trans. Signal Process. **48**(1), 227–241 (2000)
25. J.M. Cowley, A.F. Moodie, Fourier images: I - the point source. Proc. Phys. Soc. Sect. B **70**(5), 486 (1957)
26. H. Dammann, G. Groh, M. Kock, Restoration of faulty images of periodic objects by means of self-imaging. Appl. Opt. **10**(6), 1454–1455 (1971)
27. J.J. Ding, S.C. Pei, Eigenfunctions and self-imaging phenomena of the two-dimensional nonseparable linear canonical transform. J. Opt. Soc. Am. A **28**(2), 82–95 (2011)
28. B.Z. Dong, Y. Zhang, B.Y. Gu, G.Z. Yang, Numerical investigation of phase retrieval in a fractional Fourier transform. J. Opt. Soc. Am. A **14**(10), 2709–2714 (1997)
29. D. Dragoman, The Wigner distribution function of self-Fourier functions. J. Mod. Opt. **43**(9), 1933–1938 (1996)
30. J. Garcia, D. Mas, R. Dorsch, Fractional - Fourier - transform calculation through the fast-fourier-transform algorithm. Appl. Opt. **35**, 7013–7018 (1996)
31. J. Goodman, *Introduction to Fourier Optics* (McGraw-Hill, New York, 1969)
32. S. Granieri, O. Trabocchi, E.E. Sicre, Fractional Fourier transform applied to spatial filtering in the Fresnel domain. Opt. Commun. **119**, 275–278 (1995)
33. J.P. Guigay, On Fresnel diffraction by one-dimensional periodic objects, with application to structure determination of phase objects. Opt. Commun. **18**, 677–682 (1971)
34. H. Hamam, Design of Talbot array illuminators. Opt. Commun. **131**, 359–370 (1996)
35. H. Hamam, Simplified linear formulation of Fresnel diffraction. Opt. Commun. **144**, 89–98 (1997)
36. H. Hamam, Design of array illuminators under spherical illumination. Appl. Opt. **37**, 1393–1400 (1998)
37. J.J. Healy, J.T. Sheridan, Fast linear canonical transforms. J. Opt. Soc. Am. A **27**, 21–30 (2010)
38. B.M. Hennelly, J.T. Sheridan, Generalizing, optimizing, and inventing numerical algorithms for the fractional Fourier, Fresnel, and linear canonical transforms. J. Opt. Soc. Am. A **22**(5), 917–927 (2005)
39. B. Hennelly, J.J. Healy, J.T. Sheridan, Sampling in phase space, in *Phase-Space Optics: Fundamentals and Applications*, Chap. 10, ed. by M. Testorf, B. Hennelly, J. Ojeda-Castañeda (McGraw-Hill, New York, 2009), pp. 309–336
40. B.M. Hennelly, D.P. Kelly, D.S. Monaghan, N. Pandey, Zoom algorithms for digital holography, in *Information Optics and Photonics, Algorithms Systems and Applications* (Springer, New York, 2010), pp. 187–204
41. J. Jahns, A.W. Lohmann, Temporal filtering by double diffraction. Appl. Opt. **43**(22), 4339–4344 (2004)
42. J. Jahns, E. ElJoudi, D. Hagedorn, S. Kinne, Talbot interferometer as a time filter. Optik Int. J. Light Electron Opt. **112**(7), 295–298 (2001)
43. T. Jannson, J. Jannson, Temporal self-imaging effect in single-mode fibers. J. Opt. Soc. Am. **71**(11), 1373–1376 (1981)
44. T. Kozacki, Numerical errors of diffraction computing using plane wave spectrum decomposition. Opt. Commun. **281**, 4219–4223 (2008)
45. J.R. Leger, G.J. Swanson, Efficient array illuminator using binary-optics phase plates at fractional Talbot planes. Opt. Lett. **15**, 288–290 (1990)
46. B.Z. Li, R. Tao, Y. Wang, New sampling formulae related to linear canonical transform. Signal Process. **87**(5), 983–990 (2007)
47. J.P. Liu, Controlling the aliasing by zero-padding in the digital calculation of scalar diffraction. J. Opt. Soc. Am. A **29**, 1956–1964 (2012)
48. A.W. Lohmann, An array illuminator based on the Talbot effect. Optik **79**, 41–45 (1988)
49. A.W. Lohmann, D. Mendlovic, Self-Fourier objects and other self-transform objects. J. Opt. Soc. Am. A **9**, 2009–2012 (1992)
50. A.W. Lohmann, D. Mendlovic, Image formation of a self-Fourier object. Appl. Opt. **33**, 153–157 (1994)

51. A. Lohmann, D. Silva, An interferometer based on the Talbot effect. Opt. Commun. **2**, 413–415 (1971)
52. D. Mas, J. Garcia, C. Ferreira, L.M. Bernardo, F. Marinho, Fast algorithms for free-space diffraction patterns calculation. Optics Commun. **164**, 233–245 (1999)
53. D. Mendlovic, H.M. Ozaktas, A.W. Lohmann, Self Fourier functions and fractional Fourier transforms. Optics Commun. **105**, 36–38 (1994)
54. H. Ozaktas, O. Arikan, M. Kutay, G. Bozdagt, Digital computation of the fractional fourier transform. IEEE Trans. Signal Process. **44**, 2141–2150 (1996)
55. K. Patorski, Self-imaging phenomenon, lateral shift of Fresnel images. Opt. Acta **30**, 1255–1258 (1983)
56. K. Patorski, The self-imaging phenomenon and its applications, in *Progress in Optics*, vol. 27, ed. by E. Wolf (Elsevier Science, Amsterdam, 1989), pp. 1–108
57. S.C. Pei, J.J. Ding, Closed-form discrete fractional and affine Fourier transforms. IEEE Trans. Signal Process. **48**(5), 1338–1353 (2000)
58. S.C. Pei, J.J. Ding, Eigenfunctions of the canonical transform and the self-imaging problems in optical system, in *Proceedings of the 2000 IEEE International Conference on Acoustics, Speech, and Signal Processing (ICASSP '00)*, vol. 1 (2000), pp. 73–76
59. S.C. Pei, J.J. Ding, Eigenfunctions of linear canonical transform. IEEE Trans. Signal Process. **50**(1), 11–26 (2002)
60. R. Piestun, J. Shamir, Generalized propagation-invariant wave fields. J. Opt. Soc. Am. A **15**(12), 3039–3044 (1998)
61. R. Piestun, Y.Y. Schechner, J. Shamir, Propagation-invariant wave fields with finite energy. J. Opt. Soc. Am. A **17**(2), 294–303 (2000)
62. D. Podanchuk, V. Kurashov, A. Goloborodko, V. Dan'ko, M. Kotov, N. Goloborodko, Wavefront sensor based on the Talbot effect with the precorrected holographic grating. Appl. Opt. **51**(10), C125–C132 (2012)
63. W.H. Press, S.A. Teukolsky, W.T. Vetterling, B.P. Flannery, *Numerical Recipes in C* (Cambridge University Press, New York, 1992)
64. L. Rayleigh, On copying diffraction gratings, and on some phenomena connected therewith. Philos. Mag. **11**, 196–205 (1881)
65. M. Richman, T. Parks, R. Shenoy, Discrete-time, discrete-frequency time-frequency representations, in *1995 International Conference on Acoustics, Speech, and Signal Processing (ICASSP-95)*, vol. 2 (1995), pp. 1029–1032
66. M. Richman, T. Parks, R. Shenoy, Discrete-time, discrete-frequency, time-frequency analysis. IEEE Trans. Signal Process. **46**(6), 1517–1527 (1998). doi: 10.1109/78.678465
67. N.H. Salama, D. Patrignani, L.D. Pasquale, E.E. Sicre, Wavefront sensor using the talbot effect. Opt. Laser Technol. **31**(4), 269–272 (1999)
68. B.E.A. Saleh, M.C. Teich, *Fundamentals of Photonics* (Wiley, New York, 1991)
69. B. Santhanam, J. McClellan, The discrete rotational Fourier transform. IEEE Trans. Signal Process. **44**(4), 994–998 (1996)
70. M.R. Schroeder, *Number Theory in Science and Communication*, 2nd edn. (Springer, Heidelberg, 1986)
71. C. Siegel, F. Loewenthal, J. Balmer, A wavefront sensor based on the fractional talbot effect. Opt. Commun. **194**, 265–275 (2001)
72. L.B. Soldano, E.C.M. Pennings, Optical multi-mode interference devices based on self-imaging: principles and applications. J. Lightwave Technol. **13**, 615–627 (1995)
73. J. St. Collins, Lens-system diffraction integral written in terms of matrix optics. J. Opt. Soc. Am. **60**(9), 1168–1177 (1970)
74. A. Stern, Sampling of linear canonical transformed signals. Signal Process. **86**, 1421–1425 (2006)
75. E. Sziklas, A. Siegman, Diffraction calculations using fast Fourier transform methods. Proc. IEEE **62**, 410–412 (1974)
76. H.F. Talbot, Facts relating to optical science, No. IV. Philos. Mag. **9**, 401–407 (1836)

77. M. Testorf, Designing Talbot array illuminators with phase-space optics. J. Opt. Soc. Am. A **23**, 187–192 (2006)
78. M. Testorf, Self-imaging in phase space, in *Phase-Space Optics: Fundamentals and Applications*, Chap. 9, ed. by M. Testorf, B. Hennelly, J. Ojeda-Castañeda (McGraw-Hill, New York, 2009), pp. 279–307
79. M. Testorf, The phase-space approach to optical system theory. Opt. Photonics Lett. **6**, 1330001 (2013). doi:10.1142/S1793528813300015. http://www.worldscientific.com/doi/abs/10.1142/S1793528813300015
80. M.E. Testorf, M.A. Fiddy, Simulation of light propagation in planar-integrated free-space optics. Opt. Commun. **176**(4–6), 365–372 (2000)
81. M. Testorf, J. Ojeda-Castañeda, Fractional Talbot effect: analysis in phase space. J. Opt. Soc. Am. A **13**, 119–125 (1996)
82. M. Testorf, V. Arrizón, J. Ojeda-Castañeda, Numerical optimization of phase-only elements based on the fractional Talbot effect. J. Opt. Soc. Am. A **16**, 97–105 (1999)
83. M. Testorf, J. Ojeda-Castañeda, A.W. Lohmann, *Selected Papers on Phase-Space Optics*. SPIE Milestone Series, vol. MS 181 (SPIE, Bellingham, 2006)
84. M. Testorf, B. Hennelly, J. Ojeda-Castañeda, *Phase-Space Optics: Fundamentals and Applications* (McGraw-Hill, New York, 2009)
85. A. Torre, *Linear Ray and Wave Optics in Phase Space* (Elsevier, Amsterdam, 2005)
86. S.B. Tucker, J. Ojeda-Castañeda, W.T. Cathey, Matrix description of near field diffraction and the fractional Fourier transform. J. Opt. Soc. Am. A **16**, 316–322 (1999)
87. D.G. Voelz, M.C. Roggemann, Digital simulation of scalar optical diffraction: revisiting chirp function sampling criteria and consequences. Appl. Opt. **48**, 6132–6142 (2009)
88. C. Wang, J. Azaña, L.R. Chen, Spectral Talbot-like phenomena in one-dimensional photonic bandgapstructures. Opt. Lett. **29**, 1590–1592 (2004)
89. J.T. Winthrop, C.R. Worthington, Theory of Fresnel images I. Plane periodic objects in monochromatic light. J. Opt. Soc. Am. **55**, 373–381 (1965)
90. X.G. Xia, On bandlimited signals with fractional Fourier transform. IEEE Signal Process. Lett. **3**, 72–74 (1996)
91. L. Zhao, J.J. Healy, J.T. Sheridan, Unitary discrete linear canonical transform: analysis and application. Appl. Opt. **52**(7), C30–C36 (2013)
92. X.H. Zou, W. Pan, B. Luo, M.Y. Wang, W.L. Zhang, Spectral Talbot effect in sampled fiber Bragg gratings with super-periodic structures. Opt. Express **15**(14), 8812–8817 (2007)

Chapter 10
Fast Algorithms for Digital Computation of Linear Canonical Transforms

Aykut Koç, Figen S. Oktem, Haldun M. Ozaktas, and M. Alper Kutay

Abstract Fast and accurate algorithms for digital computation of linear canonical transforms (LCTs) are discussed. Direct numerical integration takes $O(N^2)$ time, where N is the number of samples. Designing fast and accurate algorithms that take $O(N \log N)$ time is of importance for practical utilization of LCTs. There are several approaches to designing fast algorithms. One approach is to decompose an arbitrary LCT into blocks, all of which have fast implementations, thus obtaining an overall fast algorithm. Another approach is to define a discrete LCT (DLCT), based on which a fast LCT (FLCT) is derived to efficiently compute LCTs. This strategy is similar to that employed for the Fourier transform, where one defines the discrete Fourier transform (DFT), which is then computed with the fast Fourier transform (FFT). A third, hybrid approach involves a DLCT but employs a decomposition-based method to compute it. Algorithms for two-dimensional and complex parametered LCTs are also discussed.

10.1 Introduction

Linear canonical transforms (LCTs) are commonly referred to as quadratic-phase integrals or quadratic-phase systems in optics [1]. They have also been referred to

A. Koç, (✉)
Intelligent Data Analytics Research Program Department, ASELSAN Research Center, ASELSAN Inc., Ankara, Turkey
e-mail: aykutkoc@aselsan.com.tr

F.S. Oktem
Department of Electrical and Electronics Engineering, Middle East Technical University, 06800 Çankaya, Ankara, Turkey
e-mail: figeno@metu.edu.tr

H.M. Ozaktas
Department of Electrical Engineering, Bilkent University, 06800 Bilkent, Ankara, Turkey
e-mail: haldun@ee.bilkent.edu.tr

M.A. Kutay
The Scientific and Technological Research Council of Turkey, 06100 Kavaklıdere, Ankara, Turkey
e-mail: alper.kutay@tubitak.gov.tr

J.J. Healy et al. (eds.), *Linear Canonical Transforms*, Springer Series in Optical Sciences 198, DOI 10.1007/978-1-4939-3028-9_10

by names such as generalized Huygens integrals [2], generalized Fresnel transforms [3, 4], special affine Fourier transforms [5, 6], extended fractional Fourier transforms (FRTs) [7], and Moshinsky–Quesne transforms [8], among other names. The so-called *ABCD* systems widely used in optics [9] are also represented by LCTs.

One-dimensional (1D) LCTs [8, 10] constitute a three-parameter class of linear integral transforms [1, 11, 12] which include among its special cases, the one-parameter subclasses of FRTs, scaling operations, and chirp multiplication (CM) and chirp convolution (CC) operations, the latter also known as Fresnel transforms.

LCTs appear widely in optics [2, 10, 11], electromagnetics, classical and quantum mechanics [8, 13, 14], as well as in computational and applied mathematics [15]. The application areas of LCTs include, among others, the study of scattering from periodic potentials [16–18], laser cavities [2, 19, 20], and multilayered structures in optics and electromagnetics [21]. They can also be used for fast and efficient realization of filtering in LCT domains [22].

Generalizations to two-dimensional (2D) transforms and complex-parametered transforms are also present in the literature. Classification of first-order optical systems and their representation through LCTs are studied in [23–27] for 1D and 2D cases, respectively. Bilateral Laplace transforms, Bargmann transforms, Gauss–Weierstrass transforms [8, 28, 29], fractional Laplace transforms [30, 31], and complex-ordered fractional Fourier transformations (CFRTs) [32–35] are all special cases of complex linear canonical transforms (CLCTs).

The LCTs are of great importance in electromagnetic, acoustic, and other wave propagation problems since they represent the solution of the wave equation under a variety of circumstances. At optical frequencies, LCTs can model a broad class of optical systems including thin lenses, sections of free space in the Fresnel approximation, sections of quadratic graded-index media, and arbitrary concatenations of any number of these, sometimes referred to as first-order optical systems [1, 5, 6, 10, 12].

Given its ubiquitous nature and numerous applications, the discretization, sampling, and efficient digital computation of LCTs are of considerable interest.

The 1D LCT of $f(u)$ with parameter matrix $\mathbf{M}$ is denoted as $f_{\mathbf{M}}(u) = (\mathcal{C}_{\mathbf{M}}f)(u)$:

$$(\mathcal{C}_{\mathbf{M}}f)(u) = \sqrt{\beta}e^{-i\pi/4} \int_{-\infty}^{\infty} \exp\left[i\pi(\alpha u^2 - 2\beta uu' + \gamma u'^2)\right] f(u')\, du', \qquad (10.1)$$

where α, β, γ are real parameters independent of u and u' and where $\mathcal{C}_{\mathbf{M}}$ is the LCT operator. The transform is unitary. The 2×2 matrix $\mathbf{M}$ whose elements are A, B, C, D represents the same information as the three parameters α, β, γ which uniquely define the LCT:

$$\mathbf{M} = \begin{bmatrix} A & B \\ C & D \end{bmatrix} = \begin{bmatrix} \gamma/\beta & 1/\beta \\ -\beta + \alpha\gamma/\beta & \alpha/\beta \end{bmatrix} = \begin{bmatrix} \alpha/\beta & -1/\beta \\ \beta - \alpha\gamma/\beta & \gamma/\beta \end{bmatrix}^{-1}. \qquad (10.2)$$

The unit-determinant matrix $\mathbf{M}$ belongs to the class of unimodular matrices. More on the group-theoretical structure of LCTs may be found in [8, 10].

The result of repeated application (concatenation) of LCTs can be handled easily with the above-defined matrix. When two or more LCTs are cascaded, the resulting transform is again an LCT whose matrix is given by the product of the matrices of the cascaded LCTs. For instance, if two LCTs with matrices $\mathbf{M}_1$ and $\mathbf{M}_2$ operate successively, then the equivalent transform is an LCT with matrix $\mathbf{M}_3 = \mathbf{M}_2\mathbf{M}_1$. LCTs are not commutative. The matrix of the inverse of an LCT is simply the inverse of the matrix of the original LCT [8, 10].

There has been considerable work on defining discrete/finite FRTs and, to a lesser degree, discrete/finite LCTs [36–55]. Definitions of the discrete FRT (DFRT) [45, 53, 55] are more established and recognized than definitions of the discrete LCT (DLCT). In this chapter the primary emphasis is not on sampling and the definition of discrete transforms. We concentrate on fast and accurate algorithms for digitally computing continuous LCTs, with careful attention to sampling issues, so as to produce results that are nearly as accurate and fast as is theoretically possible. Some approaches do involve the definition of a discrete transform, others do not.

Historically, computation of the Fresnel diffraction integral, which is a special case of LCTs, has received the greatest attention since it describes the propagation of light in free space (see [56, 57] and the references therein). Since the Fresnel integral is space-invariant and takes the form of a convolution, it can be computed in $O(N \log N)$ time. It is important to note that despite the fact that general LCTs are not space-invariant (not in convolution form), so that a standard Fourier domain approach cannot be used to obtain an $O(N \log N)$-time algorithm, the algorithms presented in this chapter are $O(N \log N)$-time algorithms.

The Fourier transform (FT) is the most prominent special case of LCTs. Most often, the continuous FT is approximated by the discrete Fourier transform (DFT) and the DFT is computed with the fast Fourier transform (FFT) algorithm [58] in $O(N \log N)$ time. The FRT, another important special case of LCTs, is a generalization of the FT. A fast algorithm for digital computation of the continuous FRT was first developed in [59]. This fast FRT algorithm paved the way for the development of fast algorithms for more general transforms, leading to the algorithms for arbitrary LCTs that are discussed in Sects. 10.4, 10.7, and 10.8. The algorithm in [59] serves as a basic building block within these fast linear canonical transform (FLCT) algorithms.

In the next section, we present some preliminary material. In Sect. 10.3, we discuss fast computation of the FRT. In Sect. 10.4, we turn our attention to decomposition-based approaches to LCT computation. Next, in Sect. 10.5, we present DLCT based methods. This is followed by Sect. 10.6, where we discuss hybrid algorithms that involve the DLCT but employ a decomposition-based method to compute it. Finally, we discuss extensions to two-dimensional (2D) and complex transforms.

10.2 Preliminaries

We begin by reviewing the concepts of compactness and the relationship of LCTs to the Wigner distribution.

10.2.1 *Compactness in Space, Frequency, and Phase Space*

A function will be referred to as compact if its support is so. The support of a function is the subset of the real axis in which the function is not equal to zero. In other words, a function is compact if and only if its nonzero values are confined to a finite interval. It is well known that a function and its Fourier transform cannot both be compact (unless they are identically zero). In practice however, it seems that we are always working with a finite space interval and a finite bandwidth. This discrepancy between our mathematical idealizations and the real world is usually not a problem when we work with signals of large space-bandwidth product. The space-bandwidth product can be crudely defined as the product of the spatial extent of the signal and its (double-sided) bandwidth. It is equal to the number of degrees of freedom, the number of complex numbers required to uniquely characterize the signal among others of the same space-bandwidth product.

We will assume that the space-domain representation of our signal is approximately confined to the interval $[-\Delta x/2, \Delta x/2]$ and that its frequency-domain representation is confined to the interval $[-\Delta\sigma/2, \Delta\sigma/2]$. With this statement we mean that a sufficiently large percentage of the signal energy is confined to these intervals. For a given class of functions, this can be ensured by choosing Δx and $\Delta\sigma$ sufficiently large. We then define the space-bandwidth product $N \equiv \Delta x \Delta\sigma$, which is always greater than unity, because of the uncertainty relation.

Let us now introduce the scaling parameter s with the dimension of space and introduce scaled coordinates $u = x/s$ and $\mu = \sigma s$. With these new coordinates, the space and frequency domain representations will be confined to intervals of length $\Delta x/s$ and $\Delta\sigma s$. Let us choose $s = \sqrt{\Delta x/\Delta\sigma}$ so that the lengths of both intervals are now equal to the dimensionless quantity $\sqrt{\Delta\sigma\Delta x}$ which we will denote by Δu. In the newly defined coordinates, our signal can be represented in both domains with $N = \Delta u^2$ samples spaced $\Delta u^{-1} = 1/\sqrt{N}$ apart.

From now on we will assume that this dimensional normalization has been performed and that the coordinates appearing in the definition of the FRT, Wigner distribution, etc. are all dimensionless quantities.

For a signal with rectangular space-frequency support, the space-bandwidth product is equal to the number of degrees of freedom. This is not true for signals with other support shapes. While LCTs do not change the number of degrees of freedom of a signal, they may change its space-bandwidth product.

10.2.2 Relationship of LCTs to the Wigner Distribution

The relationship between first-order optical systems (quadratic-phase systems or LCTs) and the Wigner distribution has been extensively studied [1, 11, 12, 60, 61]. The Wigner distribution $W_f(u, \mu)$ of a signal $f(u)$ can be defined as [62, 63]

$$W_f(u, \mu) = \int_{-\infty}^{\infty} f(u + u'/2) f^*(u - u'/2) e^{-2\pi i \mu u'} \, du'. \quad (10.3)$$

Roughly speaking, $W_f(u, \mu)$ is a function which gives the distribution of signal energy over space and frequency. Its infinite integral over space and frequency, expressed as $\int_{-\infty}^{\infty} \int_{-\infty}^{\infty} W_f(u, \mu) \, du \, d\mu$, gives the signal energy.

Let f denote a signal and $f_{\mathbf{M}}$ be its LCT with parameter matrix $\mathbf{M}$. Then, the Wigner distribution (WD) of $f_{\mathbf{M}}$ can be expressed in terms of the WD of f as [10]

$$W_{f_{\mathbf{M}}}(u, \mu) = W_f(Du - B\mu, -Cu + A\mu). \quad (10.4)$$

This means that the WD of the transformed signal is a linearly distorted version of the original distribution. The Jacobian of this coordinate transformation is equal to the determinant of the matrix $\mathbf{M}$, which is unity. Therefore this coordinate transformation does not change the support area of the Wigner distribution. (A precise definition of the support area is not necessary for the purpose of this paper; it may be defined as the area of the region where the values of the Wigner distribution are non-negligible, or the area of a region containing a certain high percentage of the total energy.) The invariance of support area means that LCTs do not concentrate or deconcentrate energy; that is, they do not carry energy in or out of the defined support area, keeping the total energy within the support area constant. The support area of the Wigner distribution can also be approximately interpreted as the number of degrees of freedom of the signal. Therefore, the number of samples needed to represent the signal does not change after an LCT operation.

10.3 Fast Computation of Fractional Fourier Transforms

Here we review the fast algorithm for computing the continuous FRT presented in [59], both for historical reasons and since it constitutes an important building block of the LCT algorithms we will later present in Sects. 10.4 and 10.7.

Let $\{\mathcal{F}f\}(u)$ denote the Fourier transform of $f(u)$. Integral powers $\mathcal{F}^j$ of the operator $\mathcal{F} \equiv \mathcal{F}^1$ may be defined as its successive applications. Then we have $\{\mathcal{F}^2 f\}(u) = f(-u)$ and $\{\mathcal{F}^4 f\}(u) = f(u)$. The ath order FRT $\{\mathcal{F}^a f\}(u)$ of the function $f(u)$ may be defined for $0 < |a| < 2$ as

$$\mathcal{F}^a[f(u)] \equiv \{\mathcal{F}^a f\}(u) \equiv \int_{-\infty}^{\infty} K_a(u,u')f(u')\,du',$$
$$K_a(u,u') \equiv A_\phi \exp\left[i\pi(u^2\cot\phi - 2uu'\csc\phi + u'^2\cot\phi)\right],$$
$$A_\phi \equiv \frac{\exp(-i\pi\,\mathrm{sgn}(\sin\phi)/4 + i\phi/2)}{|\sin\phi|^{1/2}}, \tag{10.5}$$

where $\phi \equiv a\pi/2$ and i is the imaginary unit. The kernel approaches $K_0(u,u') \equiv \delta(u-u')$ and $K_{\pm 2}(u,u') \equiv \delta(u+u')$ for $a = 0$ and $a = \pm 2$, respectively. The definition is easily extended outside the interval $[-2,2]$ by remembering that $\mathcal{F}^{4j}$ is the identity operator for any integer j and that the FRT operator is additive in index, that is, $\mathcal{F}^{a_1}\mathcal{F}^{a_2} = \mathcal{F}^{a_1+a_2}$.

The FRT, like all LCTs, can be broken down into a succession of simpler operations, such as chirp multiplication, chirp convolution, scaling, and ordinary Fourier transformation. Here we will concentrate on two particular decompositions which lead to two distinct algorithms. By ensuring that the sampling interval satisfies the Nyquist criterion at each stage of the decomposition, it becomes possible to use the output samples to reconstruct good approximations of the continuous FRT.

First, we consider decomposing the FRT into a chirp multiplication followed by a chirp convolution followed by another chirp multiplication [59]. We assume $a \in [-1,1]$. Manipulating Eq. (10.5), we can write

$$f_a(u) = \exp[-i\pi u^2\tan(\phi/2)]g'(u), \tag{10.6}$$
$$g'(u) = A_\phi \int_{-\infty}^{\infty} \exp[i\pi\beta(u-u')^2]g(u')\,du', \tag{10.7}$$
$$g(u) = \exp[-i\pi u^2\tan(\phi/2)]f(u), \tag{10.8}$$

where $g(u)$ and $g'(u)$ represent intermediate results and $\beta = \csc\phi$. In the first step [Eq. (10.8)] we multiply the function $f(u)$ by a chirp function. As shown in [59], the bandwidth and space-bandwidth product of $g(u)$ can be as large as twice that of $f(u)$. Thus, we require samples of $g(u)$ at intervals of $1/2\Delta u$. If the samples of $f(u)$ spaced at $1/\Delta u$ are given to begin with, we can interpolate these by a factor of two and then multiply by the samples of the chirp function to obtain the desired samples of $g(u)$. The next step is to convolve $g(u)$ with a chirp function, as given in Eq. (10.7). To perform this convolution, we note that since $g(u)$ is bandlimited, the chirp function can also be replaced with its bandlimited version without any effect. That is,

$$g'(u) = A_\phi \int_{-\infty}^{\infty} \exp[i\pi\beta(u-u')^2]g(u')\,du' = A_\phi \int_{-\infty}^{\infty} h(u-u')g(u')\,du', \tag{10.9}$$

where

$$h(u) = \int_{-\Delta u}^{\Delta u} H(\mu)\exp(i2\pi\mu u)\,d\mu, \tag{10.10}$$

where

$$H(\mu) = \frac{1}{\sqrt{\beta}} e^{i\pi/4} \exp(-i\pi\mu^2/\beta), \tag{10.11}$$

is the Fourier transform of $\exp[i\pi\beta u^2]$. It is possible to express $h(u)$ explicitly in terms of the Fresnel integral defined as $F(z) = \int_0^z \exp(\pi z^2/2)\, dz$. Now, Eq. (10.7) can be sampled, giving

$$g'\left(\frac{m}{2\Delta u}\right) = \sum_{n=-N}^{N} h\left(\frac{m-n}{2\Delta u}\right) g\left(\frac{n}{2\Delta u}\right). \tag{10.12}$$

This convolution can be evaluated using a FFT. Then, after performing the last step [Eq. (10.6)], we obtain the samples of $f_a(u)$ spaced at $1/2\Delta u$. Since we assumed that all transforms of $f(u)$ are bandlimited to the interval $[-\Delta u/2, \Delta u/2]$, we finally decimate these samples by a factor of 2 to obtain samples of $f_a(u)$ spaced at $1/\Delta u$. Then the continuous function $f_a(u)$ can be reconstructed from these samples.

The second method does not require Fresnel integrals [59]. Equation (10.5) can be alternatively put in the form:

$$\{\mathcal{F}^a f\}(u) = A_\phi e^{i\pi\alpha u^2} \int_{-\infty}^{\infty} e^{-i2\pi\beta u u'} \left[e^{i\pi\alpha u'^2} f(u')\right] du', \tag{10.13}$$

where $\alpha = \cot\phi$ and $\beta = \csc\phi$. We again assume that the Wigner distribution of $f(\cdot)$ is zero outside a circle of diameter Δu centered around the origin. Under this assumption, and by limiting the order a to the interval $0.5 \leq |a| \leq 1.5$, the amount of vertical shear in Wigner space resulting from the chirp modulation is bounded by $\Delta u/2$. Then the modulated function $e^{i\pi\alpha u'^2} f(u')$ is band-limited to Δu in the frequency domain. Thus $e^{i\pi\alpha u'^2} f(u')$ can be represented by Shannon's interpolation formula:

$$e^{i\pi\alpha u'^2} f(u') = \sum_{n=-N}^{N} e^{i\pi\alpha(\frac{n}{2\Delta u})^2} f\left(\frac{n}{2\Delta u}\right) \operatorname{sinc}\left(2\Delta u \left(u' - \frac{n}{2\Delta u}\right)\right), \tag{10.14}$$

where $N = (\Delta u)^2$. The summation goes from $-N$ to N since $f(u')$ is assumed to be zero outside $[-\Delta u/2, \Delta u/2]$. By using Eqs. (10.14) and (10.13), and changing the order of integration and summation we obtain

$$\{\mathcal{F}^a f\}(u) = A_\phi e^{i\pi\alpha u^2} \sum_{n=-N}^{N} e^{i\pi\alpha(\frac{n}{2\Delta u})^2} f\left(\frac{n}{2\Delta u}\right) \times \int_{-\infty}^{\infty} e^{-i2\pi\beta u u'} \operatorname{sinc}\left(2\Delta u \left(u' - \frac{n}{2\Delta u}\right)\right) du'. \tag{10.15}$$

The integral is equal to $e^{-i2\pi\beta u \frac{n}{2\Delta u}}(1/2\Delta u)\mathrm{rect}(\beta u/2\Delta u)$. For the range of $0.5 \leq |a| \leq 1.5$, $\mathrm{rect}(\beta u/2\Delta u)$ will always be equal to unity on the support $|u| \leq \Delta u/2$ of the transformed function. Hence we can write

$$\{\mathcal{F}^a f\}(u) = \frac{A_\phi}{2\Delta u} \sum_{n=-N}^{N} e^{i\pi\alpha u^2} e^{-i2\pi\beta u \frac{n}{2\Delta u}} e^{i\pi\alpha(\frac{n}{2\Delta u})^2} f\left(\frac{n}{2\Delta u}\right). \tag{10.16}$$

Then, the samples of the transformed function are obtained as

$$\{\mathcal{F}^a f\}\left(\frac{m}{2\Delta u}\right) = \frac{A_\phi}{2\Delta u} \sum_{n=-N}^{N} e^{i\pi\left(\alpha(\frac{m}{2\Delta u})^2 - 2\beta\frac{mn}{(2\Delta u)^2} + \alpha(\frac{n}{2\Delta u})^2\right)} f\left(\frac{n}{2\Delta u}\right) \tag{10.17}$$

which is a finite summation allowing us to obtain the samples of the fractional transform in terms of the samples of the original function. Direct computation of this form would require $O(N^2)$ multiplications. An $O(N \log N)$ algorithm can be obtained as follows. We put Eq. (10.17) into the following form after some algebraic manipulations:

$$\{\mathcal{F}^a f\}\left(\frac{m}{2\Delta u}\right) = \frac{A_\phi}{2\Delta u} e^{i\pi(\alpha-\beta)(\frac{m}{2\Delta u})^2} \sum_{n=-N}^{N} e^{i\pi\beta(\frac{m-n}{2\Delta u})^2} e^{i\pi(\alpha-\beta)(\frac{n}{2\Delta u})^2} f\left(\frac{n}{2\Delta u}\right). \tag{10.18}$$

It can be recognized that the summation is the convolution of $e^{i\pi\beta(n/2\Delta u)^2}$ and the chirp modulated function $f(\cdot)$. The convolution can be computed in $O(N \log N)$ time by using the FFT. The output samples are then obtained by a final chirp modulation. Hence the overall complexity is $O(N \log N)$.

10.4 Decomposition-Based LCT Algorithms

This section discusses decomposition-based approaches to fast and accurate digital computation of continuous LCTs. These approaches begin with samples of the continuous input signal and compute samples of the continuous LCT output signal such that the continuous output can be interpolated from the computed output samples. This is accomplished by decomposing the LCT operation into basic building blocks that already have fast algorithms. The main approach we discuss is based on the following decomposition involving the FRT, scaling, and chirp multiplication [64, 65]:

$$\mathbf{M} = \begin{bmatrix} A & B \\ C & D \end{bmatrix} = \begin{bmatrix} 1 & 0 \\ -q & 1 \end{bmatrix} \begin{bmatrix} M & 0 \\ 0 & 1/M \end{bmatrix} \begin{bmatrix} \cos\theta & \sin\theta \\ -\sin\theta & \cos\theta \end{bmatrix}. \tag{10.19}$$

Here $\theta = a\pi/2$ where a is the order of the FRT, q is the chirp multiplication parameter, and M is the scaling factor. As we will see, these three parameters are

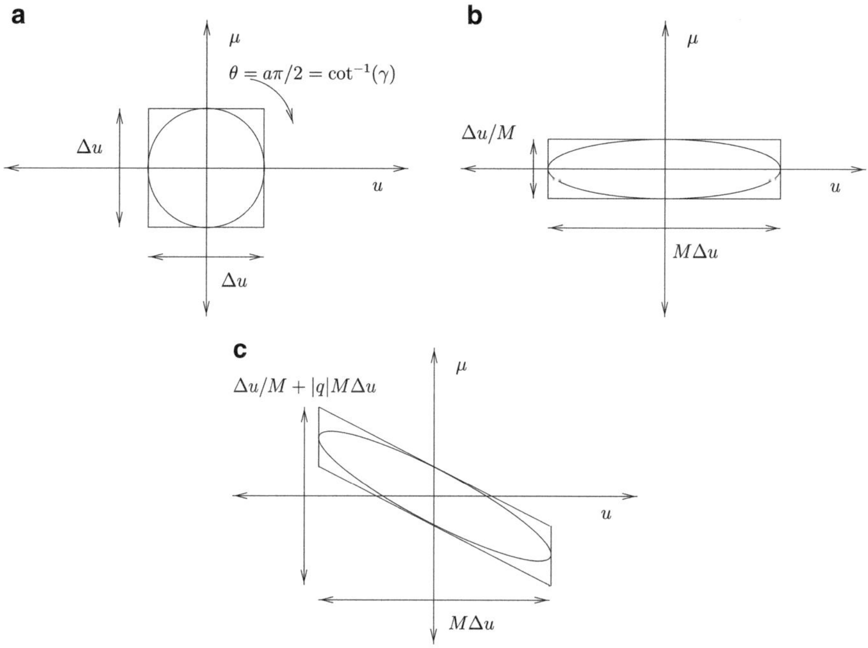

Fig. 10.1 Sequence of geometrical distortions for the decomposition in Eq. (10.19) [65]. (**a**) After the first stage: FRT; (**b**) after the second stage: scaling; (**c**) after the third stage: CM

sufficient to satisfy the above equality for arbitrary ABCD matrices, so that this decomposition is capable of representing arbitrary LCTs. Since the fast method proposed in [59] and reviewed in the previous section can be used for fast computation of the FRT, this decomposition directly leads to a fast algorithm for arbitrary LCTs. This decomposition was inspired by the optical interpretation in [66] and is also a special case of the widely known Iwasawa decomposition [26, 67, 68]. It was also proposed later in [64, 69]. Figure 10.1 illustrates the sequence of geometrical distortions in phase space corresponding to this decomposition, which is rotation, scaling, and shearing, respectively. The initial space-frequency support is a circle of diameter Δu.

To obtain the decomposition parameters in terms of the LCT parameters, we multiply out the right-hand side of Eq. (10.19) and replace the matrix entries A, B, C, D with α, β, γ, we obtain:

$$\begin{bmatrix} \gamma/\beta & 1/\beta \\ -\beta + \alpha\gamma/\beta & \alpha/\beta \end{bmatrix} = \begin{bmatrix} M\cos\theta & M\sin\theta \\ -qM\cos\theta - \sin\theta/M & -qM\sin\theta + \cos\theta/M \end{bmatrix}, \tag{10.20}$$

which is equivalent to four equations which we can solve for a, q, M:

$$a = (2/\pi)\cot^{-1}\gamma, \tag{10.21}$$

$$M = \begin{cases} \sqrt{1+\gamma^2}/\beta, \ \gamma \geq 0, \\ -\sqrt{1+\gamma^2}/\beta, \ \gamma < 0, \end{cases} \tag{10.22}$$

$$q = \gamma\beta^2/(1+\gamma^2) - \alpha. \tag{10.23}$$

The ranges of the square root and the $\cot^{-1}$ both lie in $(-\pi/2, \pi/2]$. In operator notation this algorithm can be expressed as

$$\mathcal{C}_{\mathbf{M}} = \mathcal{Q}_q \, J_k \, \mathcal{M}_M \, \mathcal{F}^a_{\text{lc}}. \tag{10.24}$$

In this method, the first operation is an FRT, whose fast computation in $O(N \log N)$ time is presented in [59, 70]. Other works dealing with fast computation of the FRT include [71, 72]. The first algorithm presented in [59] and reviewed above was based on decomposing the FRT into a CM followed by a CC followed by a final CM, and computed the samples of the continuous FRT in terms of the samples of the original signal. Care was taken to ensure that the output samples uniquely represented the continuous FRT in the Nyquist–Shannon sense. The presently discussed LCT algorithm employs that algorithm as a subroutine. The only approximation in this subroutine comes from the step involving chirp convolution in which a DFT/FFT is used to approximate the samples of the continuous FT. No other approximation is made, either in this subroutine or in any of the other operations that we employ. Thus the only source of approximation can be traced to the evaluation of a continuous FT by use of a DFT (implemented with an FFT), which is a consequence of the fundamental fact that the signal energy cannot be confined to finite intervals in both domains. The second operation in this method is scaling, which only involves a reinterpretation of the same samples with a scaled sampling interval. The final operation is CM which takes $O(N)$ time, leading to an overall complexity of $O(N \log N)$. As in the first method, it is again necessary to ensure that the final output samples are sufficient to represent the transformed signal in the Nyquist–Shannon sense. Since LCTs distort the original space-frequency support, both the space and frequency extent of the signal, as well as its space-bandwidth product may increase, despite the fact that the area of the support remains the same. Therefore, a greater number of samples than Δu^2 may be needed to represent the transformed signal in the Nyquist–Shannon sense (unless we use some specialized basis to represent the signals) [64].

Delaying confrontation with the necessity to deal with this greater number of samples until the very last step is a significant advantage of the present method. Since the FRT corresponds to rotation, and scaling only to reinterpretation of the samples, these steps do not require us to increase the number of samples. At the last CM step, if we multiply the samples of the intermediate result with the samples of the chirp, the samples obtained will be good approximations of the true samples of the transformed signal at that sampling interval. If these samples are sufficient for our purposes, nothing further need be done. However, in general these samples will be below the Nyquist rate for the transformed signal and will not be sufficient for full recovery of the continuous function. To obtain a sufficient number of samples that will allow full recovery, we must interpolate the intermediate result before chirp multiplication at least by a factor k corresponding to the increase in space-bandwidth product [65]:

$$k \geq 1 + \left|\gamma - \alpha(1+\gamma^2)/\beta^2\right|. \tag{10.25}$$

For convenience we choose k to be the smallest integer satisfying this inequality.

We have considered several examples to illustrate and compare the presented methods. We consider the chirped pulse function $\exp(-\pi u^2 - i\pi u^2)$, denoted F1, and the trapezoidal function $1.5\text{tri}(u/3) - 0.5\text{tri}(u)$, denoted F2 ($\text{tri}(u) = \text{rect}(u) * \text{rect}(u)$). Since these two functions are well confined to a circle with diameter $\Delta u = 8$ we take $N = 8^2$. We also consider the binary sequence 01101010 occupying $[-8, 8]$ with each bit 2 units in length, so that $N = 16^2$. This binary sequence is denoted by F3 and the function shown in Fig. 10.2 is denoted by F4, again with $N = 16^2$. These choices for Δu result in ~0 %, 0.0002 %, 0.47 %, 0.03 % of the energies of F1, F2, F3, F4, respectively, to fall outside the chosen frequency extents. The chosen space extents include all of the energies of F2, F3, F4 and virtually all of the energy of F1. We consider two transforms, the first (T1) with parameters $(\alpha, \beta, \gamma) = (-3, -2, -1)$, and the second (T2) with parameters $(-4/5, 1, 2)$. The LCTs T1 and T2 of the functions F1, F2, F3, F4 have been computed by the presented fast method (referred to as A2), another fast method which will be presented shortly (referred to as A1), and by a highly inefficient brute force numerical approach based on composite Simpson's rule, which is here taken as a reference.

The results are tabulated in Table 10.1 for both transforms (T1, T2). Also shown are the errors that arise when using the DFT in approximating the FT of the same functions, which serves as a reference. (The error is defined as the energy of the difference normalized by the energy of the reference, expressed as a percentage.) Results are shown for two algorithms denoted as A1 and A2, both of which appear in [65] as Method I and Method II, respectively. The algorithm outlined above is A2. The algorithm A1 will be summarized below after we discuss the results.

The key observations that can be made from this table are as follows. The errors obtained depend on the function, since different functions have different amounts of energy contained in their tails which fall outside the assumed space and frequency extents (or assumed space-frequency region). For those cases in which the error is large, such as F3, this means that we have determined the space-bandwidth product less conservatively than the other examples, and the error can be reduced by increasing N. Generally speaking, the errors obtained depend very little on the transform parameters or which method we use, and are comparable to the error arising when we use the DFT to approximate the FT. Since a DFT lies at the heart of both methods, this is the smallest error one could hope for to begin with.

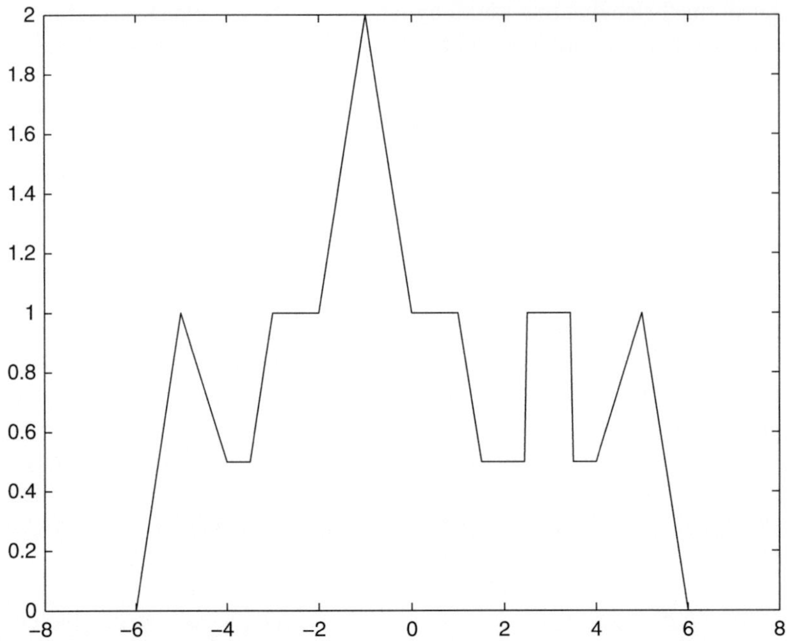

Fig. 10.2 Example function F4 [65]

Table 10.1 Percentage errors for different functions F, transforms T, and algorithms A [65]

	A1 T1	A1 T2	A2 T1	A2 T2	DFT
F1	**3.2×10^{-22}**	**9.5×10^{-22}**	**2.7×10^{-17}**	**6.6×10^{-17}**	**2.0×10^{-21}**
F2	**7.8×10^{-4}**	**8.1×10^{-4}**	**11×10^{-4}**	**9.9×10^{-4}**	**6.2×10^{-4}**
F3	**1.5**	**1.6**	**1.4**	**1.5**	**1.2**
F4	**9.7×10^{-2}**	**11×10^{-2}**	**8.9×10^{-2}**	**9.9×10^{-2}**	**8.3×10^{-2}**

Figure 10.3 shows the error versus number of sample points N for selected functions and transforms. We observe that the error decreases steeply at first with increasing N as expected, but saturates when we approach and exceed the space-bandwidth product of the signals (here 64). This demonstrates that the number of samples N can be chosen comparable to the space-bandwidth product, which is the smallest number we can expect to work with, and need not be chosen larger. Algorithm A2 was used to obtain this plot for illustration purposes but similar results can also be obtained when we use Algorithm A1.

We now briefly summarize Algorithm A1, appearing in [65], where the use of matrix factorizations to decompose LCTs into cascade combinations of elementary LCT blocks has been studied exhaustively. Since each stage in such a decomposition

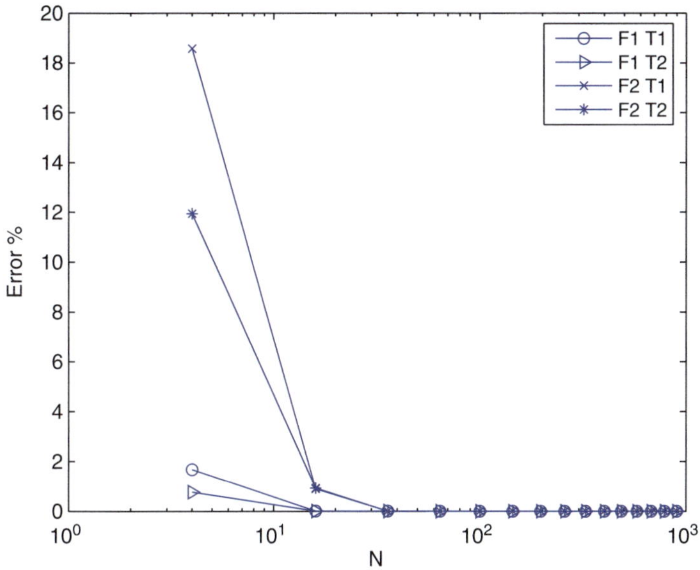

Fig. 10.3 Percentage errors versus N for selected functions and transforms [65]

can be computed in at most $O(N \log N)$ time, the overall LCT can also be. Numerous such decompositions are possible [10, 73], but they are not equally suited for numerical purposes. For instance, direct naive application of the decomposition of chirp multiplication, Fourier transformation, scaling (magnification), and again chirp multiplication, which suggests itself upon inspection of Eq. (10.1), will in general lead to very high sampling rates if conventional Shannon–Nyquist sampling is employed. We have carried out a systematic exhaustive analysis of all possible decompositions of arbitrary LCTs into the three basic operations of scaling, chirp multiplication (CM), and Fourier transformation (FT). All possible decompositions with three, four, and five cascade blocks have been considered and every permutation has been checked to see if that decomposition is capable of expressing an LCT with arbitrary parameters. The resulting algorithm can be summarized as follows:

- If $|\gamma| \leq 1$, use the decomposition:

$$\mathbf{M} = \begin{bmatrix} 1 & 0 \\ \alpha & 1 \end{bmatrix} \begin{bmatrix} 0 & 1 \\ -1 & 0 \end{bmatrix} \begin{bmatrix} 1 & 0 \\ \gamma/\beta^2 & 1 \end{bmatrix} \begin{bmatrix} \beta & 0 \\ 0 & 1/\beta \end{bmatrix},$$

$$\mathcal{C}_{\mathbf{M}} = \mathcal{Q}_{-\alpha}\, J_{k/2}\, \mathcal{F}_{\mathrm{lc}}\, \mathcal{Q}_{-\gamma/\beta^2}\, J_2\, \mathcal{M}_{\beta}, \tag{10.26}$$

where J_z represents the $\times z$ oversampling operation. The minimum value of k is $k \geq 1 + |\gamma| + |\alpha|(1 + |\gamma|)^2/\beta^2$.

- If $|\gamma| > 1$, use the decomposition:

$$\mathbf{M} = \begin{bmatrix} 1 & 0 \\ \alpha - \beta^2/\gamma & 1 \end{bmatrix} \begin{bmatrix} 0 & 1 \\ -1 & 0 \end{bmatrix} \begin{bmatrix} 1 & 0 \\ -\gamma/\beta^2 & 1 \end{bmatrix} \begin{bmatrix} 0 & 1 \\ -1 & 0 \end{bmatrix} \begin{bmatrix} -\gamma/\beta & 0 \\ 0 & -\beta/\gamma \end{bmatrix},$$

$$\mathcal{C}_{\mathbf{M}} = \mathcal{Q}_{-\alpha+\beta^2/\gamma}\, J_{k/2}\, \mathcal{F}_{\mathrm{lc}}\, \mathcal{Q}_{\gamma/\beta^2}\, J_2\, \mathcal{F}_{\mathrm{lc}}\, \mathcal{M}_{-\gamma/\beta}. \tag{10.27}$$

The minimum value of k is $k \geq 1 + 1/|\gamma| + (1 + |\gamma|)^2|\alpha - \beta^2/\gamma|/\beta^2$.

The above algorithm, A1, has been compared with the earlier presented FRT-based algorithm A2, both in terms of computational complexity and accuracy, but no significant difference has been found [65].

Another work dealing with decomposition-based methods to digitally compute the fractional Fourier, Fresnel, and general LCTs is [69]. This work is of significance because of two reasons. First, it reviews previous fast algorithms presented in the literature for computing the several important special cases of LCTs such as FRT and Fresnel transformation (FST), and distills them into a single decomposition-based approach covering the general case. Second, it emphasizes the importance of tracking the space-bandwidth product through successive stages of the decomposition as also stressed in [65]. It also sets forth a systematic, elegant, uniform, and general approach to determining the overall increase in space-bandwidth product of the final transformed signal and hence the number of samples needed for Nyquist–Shannon interpolation. It is assumed that the input signal energy is initially contained within some arbitrary four-sided shape in phase space that is defined by the coordinates of the four corners, $u_1, \mu_1,\ u_2, \mu_2,\ u_3, \mu_3,\ u_4, \mu_4$. The spatial and frequency extents of the signal are denoted by W_0 and B_0, respectively. The resulting number of regularly placed samples required to represent the signal in the Nyquist–Shannon sense is given by $N_0 = W_0 B_0$. Given the four corners of the initial support as

$$\mathbf{S} = \begin{bmatrix} u_1 & u_2 & u_3 & u_4 \\ \mu_1 & \mu_2 & \mu_3 & \mu_4 \end{bmatrix}, \tag{10.28}$$

the new coordinates of the four corners of the region after the LCT is given by

$$\mathbf{S}' = \begin{bmatrix} Au_1 + B\mu_1 & Au_2 + B\mu_2 & Au_3 + B\mu_3 & Au_4 + B\mu_4 \\ Cu_1 + D\mu_1 & Cu_2 + D\mu_2 & Cu_3 + D\mu_3 & Cu_4 + D\mu_4 \end{bmatrix}, \tag{10.29}$$

where A,B,C,D are the LCT parameters.

Then, the spatial and frequency extents are the maximum distance between any two of the u and any two of the μ coordinates, respectively. To represent this, the $\mathrm{Max}(\cdots)$ notation is defined to denote the maximum element on each row of the matrix it operates on Hennelly and Sheridan [69]. Finally, the resultant spatial, W_{LCT}, and frequency, B_{LCT}, extents after the LCT are given by

$$\begin{bmatrix} W_{\text{LCT}} \\ B_{\text{LCT}} \end{bmatrix} = \text{Max} \begin{bmatrix} |A(u_1-u_2)+B(\mu_1-\mu_2)| & |A(u_1-u_3)+B(\mu_1-\mu_3)| \\ |C(u_1-u_2)+D(\mu_1-\mu_2)| & |C(u_1-u_3)+D(\mu_1-\mu_3)| \end{bmatrix}$$
$$\begin{matrix} |A(u_1-u_4)+B(\mu_1-\mu_4)| & |A(u_2-u_3)+B(\mu_2-\mu_3)| \\ |C(u_1-u_4)+D(\mu_1-\mu_4)| & |C(u_2-u_3)+D(\mu_2-\mu_3)| \end{matrix}$$
$$\left. \begin{matrix} |A(u_2-u_4)+B(\mu_2-\mu_4)| & |A(u_3-u_4)+B(\mu_3-\mu_4)| \\ |C(u_2-u_4)+D(\mu_2-\mu_4)| & |C(u_3-u_4)+D(\mu_3-\mu_4)| \end{matrix} \right]. \tag{10.30}$$

By using the above procedure, the space-bandwidth of the signal can be tracked through the intermediate stages of the LCT decomposition and the number of samples required to represent the signal in the Nyquist–Shannon sense can be determined.

The Fresnel transform has received a lot of attention, since it models free-space propagation of waves under the Fresnel approximation. Hennelly and Sheridan [69] also studies the problem of fast computation of the Fresnel transform and reviews restrictions of several earlier approaches. Based on these observations, two decompositions are proposed that work for arbitrary LCTs. One of them is based on the Iwasawa decomposition that we have presented above [see Eq. (10.19)]. The second is given by the following matrix equation:

$$\begin{bmatrix} A & B \\ C & D \end{bmatrix} = \begin{bmatrix} 1 & 0 \\ C/A & 1 \end{bmatrix} \begin{bmatrix} A & 0 \\ 0 & 1/A \end{bmatrix} \begin{bmatrix} 1 & B/A \\ 0 & 1 \end{bmatrix}. \tag{10.31}$$

This is a chirp convolution, scaling, chirp multiplication decomposition. Although it can indeed realize arbitrary LCTs, it uses both a chirp convolution and a chirp multiplication that requires an unnecessary increase in the sampling rate. This is partly avoidable by using more suitable decompositions [65].

10.5 DLCT Based Algorithms

In the previous section, we focused on algorithms for fast computation of continuous LCTs that are based on decomposition of the LCT operation. In this section, we discuss the approach involving the definition of a DLCT, but in which computation is not based on decomposition into basic operations as in the previous section. In this approach, fast digital algorithms for LCTs, which are often referred to as FLCT algorithms, are derived by first defining a DLCT, which is to the continuous LCT what the DFT is to the continuous Fourier transform, and then developing a fast algorithm to compute the DLCT similar to the FFT algorithm.

The key points of the derivation of [74] are given here. We first review the derivation of the DLCT, which was first given in [75], and discuss the shifting properties of this DLCT, which is the key component in the development of the FLCT [74]. We start with the sampled version of a continuous signal $f(u)$ as

$$f^{\delta u}(u) = f(u)\delta_{\delta u}(u) = \sum_{n=-\infty}^{\infty} f(n\delta u)\delta(u - n\delta u) \tag{10.32}$$

or in Fourier series form

$$f^{\delta u}(u) = \frac{1}{\delta u} f(u) \sum_{n=-\infty}^{\infty} \exp(i2\pi k f_s u), \tag{10.33}$$

where $\delta(u)$ is Dirac's impulse, δu is the sampling interval, and f_s is the sampling frequency. Let us recall the definition of the LCT with parameters α, β, γ, of a function $f(u)$ as

$$(\mathcal{C}_{\mathbf{M}} f)(u) = \int_{-\infty}^{\infty} A \exp[i\pi(\alpha u^2 - 2\beta u u' + \gamma u'^2] f(u')\, du'. \tag{10.34}$$

$\mathcal{C}_{\mathbf{M}}$ is the LCT operator. We note that in [74] and later in [76], $\alpha \to -\gamma$, $\gamma \to -\alpha$, $\beta \to -\beta$, since the authors work with the inverse of the LCT matrix as defined here. We omit the complex constant A in what follows. The key properties to be utilized are the shifting theorems [74]:

$$\begin{aligned} \mathcal{C}_{\mathbf{M}}[\exp(i2\pi\xi u') f(u')](u) &= \exp(-i\pi\alpha\xi^2/\beta^2)\exp(i2\pi u\xi\alpha/\beta) \\ &\quad \times \mathcal{C}_{\mathbf{M}}[f(u')](u - \xi/\beta), \end{aligned} \tag{10.35}$$

$$\begin{aligned} \mathcal{C}_{\mathbf{M}}[f(u' - \xi)](u) &= \exp[i\pi\xi^2(\gamma - \alpha\gamma^2/\beta^2)]\exp[i2\pi u\xi(\alpha\gamma/\beta - \beta)] \\ &\quad \times \mathcal{C}_{\mathbf{M}}[f(u')](u - \xi\gamma/\beta). \end{aligned} \tag{10.36}$$

When the LCT operator operates on the sampled signal in Eq. (10.32) we get:

$$\begin{aligned} \mathcal{C}_{\mathbf{M}}[f^{\delta u}(u')](u) &= \int_{-\infty}^{\infty} \left[\sum_{n=-\infty}^{\infty} f(n)\delta(u' - n\delta u) \right] \exp(-i2\pi\beta u u') \\ &\quad \times \exp[-i\pi(\gamma u'^2 + \alpha u^2)] du' \\ &= \exp(i\pi\alpha u^2) \sum_{n=-\infty}^{\infty} f(n\delta u)\exp[i\pi\gamma(n\delta u)^2]\exp(-i2\pi\beta u n\delta u). \end{aligned} \tag{10.37}$$

Equation (10.37) is the discrete time (or space) LCT (DTLCT) which is analogous to the discrete time (or space) Fourier transform (DTFT). We can also apply the LCT operator to the alternative expression of a sampled function as given in Eq. (10.33), to obtain:

$$\mathcal{C}_{\mathbf{M}}[f^{\delta u}(u')](u) = \frac{1}{\delta u} \sum_{n=-\infty}^{\infty} \exp[-i\pi(k\delta u)^2\alpha/\beta^2] \times \exp[i2\pi(k/\delta u)\alpha u/\beta]\mathcal{C}_{\mathbf{M}}[f(u')](u - k/\delta u\beta). \quad (10.38)$$

The above expression is important since it reveals the periodicity of the LCT of the sampled signal, which is a crucial property in deriving the FLCT. The magnitude of $\mathcal{C}_{\mathbf{M}}[f^{\delta u}(u')](u)$ is equal to the magnitude of $\mathcal{C}_{\mathbf{M}}[f(u')](u)$ repeated periodically with the period $1/\delta u\beta$ and the phase of $\mathcal{C}_{\mathbf{M}}[f^{\delta u}(u')](u)$ is also equal to that of $\mathcal{C}_{\mathbf{M}}[f(u')](u)$ repeated periodically with the same period $1/\delta u\beta$.

Denoting the DTLCT operator by $\mathcal{DTLCT}$, analysis yields the following shift theorem for the DTLCT [74]:

$$\begin{aligned}&\mathcal{DTLCT}_{\mathbf{M}}[f((n-l)\delta u)](u)\\ &= \exp[i\pi l^2\delta u^2(\gamma - \alpha\gamma^2/\beta^2)]\\ &\quad \times \exp[i2\pi u l\delta u(\alpha\gamma/\beta - \beta)]\mathcal{DTLCT}_{\mathbf{M}}[f(n\delta u)](u - l\delta u\gamma/\beta), \quad (10.39)\end{aligned}$$

which implies that when the input function is shifted and then a DTLCT is applied to the result, the output will be the same as the DTLCT being directly applied to the original input function, except that the result is shifted by an amount proportional to the shift in the input function and has a linear phase factor and a constant phase factor dependent on the shift amount.

For the DTLCT in Eq.(10.37), the summation is still infinite and the output variable u is continuous. In order to achieve a true discrete transform definition which maps discrete signals to discrete signals, we replace the infinite summation with a finite summation over N samples, but we assume that N is chosen sufficiently large so that the summation covers the interval over which the signal is non-negligible:

$$\exp(i\pi\alpha u^2) \sum_{n=-N/2}^{N/2-1} f(n\delta u) \exp[i\pi\gamma(n\delta u)^2] \exp(-i2\pi\beta u n\delta u). \quad (10.40)$$

Now, we discretize the u variable by taking N samples in the following range:

$$-\frac{1}{2\delta u\beta} \leq u \leq \frac{1}{2\delta u\beta} - \frac{1}{N\delta u\beta} \quad (10.41)$$

in steps of $\delta u_{\mathbf{M}} = 1/N\delta u\beta$ [74]. The discrete transform thus obtained repeats itself outside of this range. Thus we finally obtain the DLCT as a function of the discrete output variable m as:

$$
\begin{aligned}
&\mathcal{DLCT}_{\mathbf{M}}[f(n\delta u)](m\delta u_{\mathbf{M}}) \\
&\quad = \exp\left[i\pi\alpha(m\delta u_{\mathbf{M}})^2\right] \sum_{n=-N/2}^{N/2-1} f(n\delta u)\exp[i\pi\gamma(n\delta u)^2]\exp[-i2\pi\beta(n\delta u)(m\delta u_{\mathbf{M}})] \\
&\quad = \exp\left[i\pi\alpha\left(\frac{m}{N\delta u\beta}\right)^2\right] \sum_{n=-N/2}^{N/2-1} f(n\delta u)\exp[i\pi\gamma(n\delta u)^2]\exp\left(-\frac{i2\pi nm}{N}\right),
\end{aligned}
\tag{10.42}
$$

where m covers the range $-N/2 \leq m \leq N/2-1$. A few remarks are in order at this point. First, note that this DLCT definition [75] contains a finite sum which arises from the sampling of the continuous input function and the continuous transform kernel. Second, the only string attached to the definition is $\delta u_{\mathbf{M}} = 1/(N\delta u|\beta|)$; hence, there are many ways to choose the parameters N, δu, and $\delta u_{\mathbf{M}}$, which correspond to the number of samples, and the sampling intervals in the input and output domains. In order to use this DLCT definition in practice to approximately compute the continuous LCT, it is necessary to know how to choose the number of samples N, and the sampling intervals δu and $\delta u_{\mathbf{M}}$ based on some prior information about the signal. The relationship between the DLCT and the continuous LCT is also needed to provide a foundation for how to use the DLCT to approximately compute the samples of the LCT of a continuous signal. The answers to these issues will be provided in the next section based on [77, 78]. The resulting LCT computation method allows us to work with the same number of samples in both the input and output domains without requiring any oversampling. On the other hand, the use of the Shannon–Nyquist sampling theorem instead of the LCT sampling theorem leads to problems such as the need to use a greater number of samples, requiring different sampling rates at intermediate stages of the computation, or different numbers of samples at the input and output domains [79, 80].

Now we go back to our discussion of the derivation of the FLCT algorithm. In [74], two important shifting properties of (10.42) have been derived. These two properties are essential in deriving the FLCT from the DLCT, much like the derivation of the FFT algorithm. These are:

$$
\begin{aligned}
&\exp\left[i\pi\alpha\left(\frac{m}{N\delta u\beta}\right)^2\right] \sum_{n=-\infty}^{\infty} [f(n\delta u)\exp(i2\pi\xi n\delta u)] \\
&\quad \exp[i\pi\gamma(n\delta u)^2]\exp\left(-\frac{i2\pi nm}{N}\right)
\end{aligned}
$$

$$= \exp\left[-i\pi\alpha\left(\frac{\xi}{\beta}\right)^2\right]\exp\left(\frac{i2\pi\alpha m\xi}{N\delta u\beta^2}\right)\exp\left[i\pi\alpha\left(\frac{m-\xi N\delta u}{N\delta u\beta}\right)^2\right]$$
$$\times\sum_{n=-\infty}^{\infty} f(n\delta u)\exp[i\pi\gamma(n\delta u)^2]\exp\left[-\frac{i2\pi n(m-\xi N\delta u)}{N}\right],\tag{10.43}$$

$$\exp\left(\frac{i2\pi\xi m}{N\delta u\beta}\right)\exp\left[i\pi\alpha\left(\frac{m}{N\delta u\beta}\right)^2\right]$$
$$\sum_{n=-\infty}^{\infty} f(n\delta u)\exp[i\pi\gamma(n\delta u)^2]\exp\left(-\frac{i2\pi nm}{N}\right)$$
$$= \exp\left[i\pi\alpha\left(\frac{m}{N\delta u\beta}\right)^2\right]\sum_{n=-\infty}^{\infty} f(n\delta u)\exp\left(\frac{i2\pi\gamma n\xi\delta u}{\beta}\right)\exp\left(-\frac{i\pi\gamma\xi^2}{\beta^2}\right)$$
$$\exp\left\{i\pi\gamma\left[\left(n-\frac{\xi}{\delta u\beta}\right)\delta u\right]^2\right\}\exp\left[-\frac{i2\pi m(n-\xi/\delta u\beta)}{N}\right].\tag{10.44}$$

The properties given in Eqs. (10.43) and (10.44), in conjunction with the chirp periodicity of the DLCT, enable us to use the time (or space) decomposition and frequency (or spatial frequency) decomposition to derive the FLCT. The rest of the derivation is quite long and may be found in [74]. The resulting algorithm converts an N-point DLCT into four $N/4$-point DLCTs. Recursively applying this and by properly choosing $N = 2^n$ for some integer n, we finally obtain 2×2 DLCTs to calculate. This gives an $O(N \log N)$ method to compute the DLCT. The procedure is similar to that for the conventional FFT.

Another significant work on fast computation of LCTs is [76]. In this paper, the DLCT and its fast algorithm, the FLCT, have been further studied by deriving an advanced FLCT that can work with an input sample vector of almost any length. The FLCT given in [74] is a radix-2 algorithm, so there is a restriction on the length of the input vector such that $N = 2^n$. For other lengths of the input vector, [76] proposes the following DLCT that can be implemented in any radix:

$$\mathcal{DLCT}_{\mathbf{M}}[f(n\delta u)](m\delta u_{\mathbf{M}}) = \sqrt{\frac{\beta}{2\pi}}\exp\left(\frac{-i\pi}{4}\right)\sum_{n=-N/2}^{N/2-1} f(n\delta u)W_N^{n,m},\tag{10.45}$$

where

$$W_N^{n,m} = \exp\{i\pi[\alpha(m/N\delta u\beta)^2 - 2nm/N + \gamma(n\delta u)^2]\}\tag{10.46}$$

and $\mathcal{DLCT}_{\mathbf{M}}$ stands for DLCT operator with LCT parameters (α, β, γ), δu is the sampling interval at the input, $\delta u_{\mathbf{M}}$ is the sampling interval at the output, and N is the length of the input vector. The details of the derivation can be found in [76]. The important point about this FLCT is that it can be implemented to work with any radix, so any input vector of any length can be DLCT-transformed with any desired radix in a way that the computation cost is minimized. An arbitrary vector length can be decomposed into its prime dividers and the corresponding radixes can be chosen to optimize the computation. Note, however, that this does not work with vectors of prime length. No LCT algorithm exists, as yet, that can work on prime-length input vectors [76].

10.6 Hybrid LCT Algorithms

In Sect. 10.4, fast algorithms for computing continuous LCTs are discussed, which are obtained by decomposing continuous LCTs. These algorithms produce output vectors which are good approximations to the samples of the continuous transform, limited only by the fundamental fact that a signal cannot have finite extent in more than one domain. Since the sampling interval is ensured to satisfy the Nyquist criterion at each stage of the decomposition, the output samples can be used to reconstruct good approximations of the continuous LCT. While this approach is nearly optimal in terms of speed and accuracy, it has two disadvantages. Although it implicitly defines a discrete mapping from the vector of input samples to the vector of transform samples, this discrete mapping does not constitute an analytically elegant DLCT. More importantly, this type of approach sometimes requires an increase in the sampling rate, either during computation at intermediate stages, or in representation of the output. This is contrary to the fact that the area of the time- or space-frequency support, and hence the total number of samples required to represent a signal, remains unchanged when a signal undergoes linear canonical transformation.

In Sect. 10.5, we reviewed works that first propose a definition of the DLCT and then develop a fast algorithm for computing it. However, the issue of how to relate this DLCT to the continuous LCT we are trying to approximate with it has not been addressed in these works. In order to use these fast computation approaches in practice to approximately compute the continuous LCT, it is necessary to know how to choose the number of samples N, and the sampling intervals δu and $\delta u_{\mathbf{M}}$. Note that the DLCT algorithm described in the last section has been developed with the following condition only: $\delta u_{\mathbf{M}} = 1/(N\delta u|\beta|)$, leaving many ways to choose the parameters N, δu, and $\delta u_{\mathbf{M}}$. Works that use the Shannon–Nyquist sampling theorem instead of the LCT sampling theorem lead to problems such as the need to use a greater number of samples or different numbers of samples at the input and output domains [79, 80].

In this section, we will present a FLCT computation method [77, 78] based on the DLCT defined in [75], which overcomes many of the limitations of previous algorithms. This approach works with the minimum number of samples as determined by the LCT sampling theorem without requiring any sampling rate change at the intermediate stages. Moreover, it provides a good approximation to the continuous LCT as ensured by an exact relation between the DLCT and the continuous LCT [78].

We first review the exact relation between the DLCT and the continuous LCT given in [78], which provides a solid basis for how to use the DLCT to approximately compute the samples of the LCT of a continuous signal. This exact relation helps us use the DLCT to obtain a good and efficient approximation of the continuous LCT by properly choosing the number of samples. We want to obtain as good an approximation as possible, limited only by the fundamental fact that a signal cannot have strictly finite extent in more than one domain. The answer to this problem first appeared in [77], but became better established when it was formulated as an exact relation between the DLCT and the continuous LCT [78] in a manner similar to the classical theorem relating the DFT to the continuous Fourier transform [81].

The DLCT of $f(k\,\delta u)$ is defined as follows for $m = -N/2, \ldots, N/2-1$ [75]:

$$\mathcal{DLCT}_{\mathbf{M}}[f(k\delta u)](m\delta u_{\mathbf{M}}) \equiv \delta u \sum_{k=-N/2}^{N/2-1} f(k\,\delta u)\, K_{\mathbf{M}}(m\,\delta u_{\mathbf{M}}, k\,\delta u),$$

$$K_{\mathbf{M}}(m\,\delta u_{\mathbf{M}}, k\,\delta u) = \sqrt{\beta}\, e^{-\frac{i\pi}{4}}\, e^{\frac{i\pi}{N|\beta|}\left(\alpha \frac{\delta u_{\mathbf{M}}}{\delta u} m^2 - 2\beta k m + \gamma \frac{\delta u}{\delta u_{\mathbf{M}}} k^2\right)}, \tag{10.47}$$

where $\delta u_{\mathbf{M}} = (|\beta| N \delta u)^{-1}$. Here δu and $\delta u_{\mathbf{M}}$ are the sampling intervals in the time (or space) and LCT domains, respectively. N is the number of samples. This definition of DLCT can be made unitary by including an additional factor $\sqrt{\delta u_{\mathbf{M}}/\delta u}$ [78]. Let $f(u)$ and $f_{\mathbf{M}}(u)$ be a continuous-time signal and its LCT with parameters α, β, γ. Define the following periodically replicated functions where each period has been modulated with varying phase terms:

$$\bar{f}(u)_{(\mathbf{M}^{-1},\Delta u)} \equiv \sum_{n=-\infty}^{\infty} f(u - n\Delta u) e^{-i\pi\gamma n \Delta u(2u - n\Delta u)}, \tag{10.48}$$

$$\bar{f}_{\mathbf{M}}(u)_{(\mathbf{M},\Delta u_{\mathbf{M}})} \equiv \sum_{n=-\infty}^{\infty} f_{\mathbf{M}}(u - n\Delta u_{\mathbf{M}}) e^{i\pi\alpha n \Delta u_{\mathbf{M}}(2u - n\Delta u_{\mathbf{M}})}, \tag{10.49}$$

where Δu and $\Delta u_{\mathbf{M}}$ are arbitrary. These functions are chirp-periodic in the sense of [42]. Then, the exact relation between continuous and discrete LCTs can be stated as follows: *The samples of the chirp periodic functions defined in Eqs.* (10.48) *and* (10.49) *are exactly related to each other through the samples of the continuous kernel [the DLCT matrix in Eq.* (10.47)*]*:

$$\bar{f}_{\mathbf{M}}(m\,\delta u_{\mathbf{M}})_{(\mathbf{M},\Delta u_{\mathbf{M}})} = \delta u \sum_{k \in \langle N \rangle} \bar{f}(k\,\delta u)_{(\mathbf{M}^{-1},\Delta u)} K_{\mathbf{M}}(m\,\delta u_{\mathbf{M}}, k\,\delta u), \tag{10.50}$$

for any m and any interval of length N denoted by $\langle N \rangle$*, and where the sampling intervals and the number of samples depends on the periods* Δu *and* $\Delta u_{\mathbf{M}}$ *as follows:*

$$\delta u = \frac{1}{|\beta|\Delta u_{\mathbf{M}}}, \quad \delta u_{\mathbf{M}} = \frac{1}{|\beta|\Delta u}, \quad N = \Delta u \Delta u_{\mathbf{M}} |\beta|. \tag{10.51}$$

This exact relation provides the underlying foundation for approximately computing the samples of the LCT of a continuous signal by replacing the transform integral with a finite sum. Because this exact relation generalizes the corresponding relation for Fourier transforms, which has been regarded as a fundamental theorem by Papoulis [81], this DLCT approximates the continuous LCT in the same sense that the DFT approximates the continuous FT. We can use this exact relation to see how this DLCT provides a good approximation of the continuous LCT. Consider an arbitrary signal $f(u)$ that is *not* chirp-periodic. Let us assume that a large percentage of the total energy of the signal is concentrated in the intervals $[-\Delta u/2, \Delta u/2]$ and $[-\Delta u_{\mathbf{M}}/2, \Delta u_{\mathbf{M}}/2]$, in the time (or space) and LCT domains, respectively. Then, $\bar{f}(u)_{(\mathbf{M}^{-1},\Delta u)} \approx f(u)$ and $\bar{f}_{\mathbf{M}}(u)_{(\mathbf{M},\Delta u_{\mathbf{M}})} \approx f_{\mathbf{M}}(u)$ in the respective intervals, and from (10.50) the DLCT of the samples of the function are the approximate samples of the continuous LCT of that function:

$$f_{\mathbf{M}}(m\,\delta u_{\mathbf{M}}) \approx \delta u \sum_{k=-N/2}^{N/2-1} f(k\,\delta u)\, K_{\mathbf{M}}(m\,\delta u_{\mathbf{M}}, k\,\delta u), \tag{10.52}$$

where δu, $\delta u_{\mathbf{M}}$, and N are as in (10.51). If both the functions $f(u)$ and $f_{\mathbf{M}}(u)$ could be identically zero outside of the given intervals, the mapping between the samples of these functions would be exact. But, since the extent of a function and its LCT cannot both be finite for $\beta \neq \infty$ [61, 82], there will be overlaps between the periodically replicated and phase modulated functions, and the DLCT will be an approximation between the samples of the continuous signals. This approximation for the LCT and FRT is similar to that for the FT, where in the FT case the limitation is that the extent of the signal and its Fourier transform cannot both be finite. The functions (10.48) and (10.49) reveal the precise nature of overlap and aliasing that occurs, which is somewhat different than the Fourier case which is a pure periodic replication. As with the DFT, the approximation improves with increasing N since this decreases the overlap between the replicas.

Note that we need chirp-periodic functions in order to have an exact equivalence between continuous and discrete LCTs, just as we need periodic functions to have an exact equivalence between continuous and discrete FTs. However, our interest in using this DLCT is mostly for *non*-chirp-periodic functions (just as we use the DFT for non-periodic functions), in which case the DLCT provides a good approximation to the continuous LCT, limited only by the fact that the extent of a function and its LCT cannot both be finite for $\beta \neq \infty$ [61, 82]. Moreover, the proper choice of the sampling intervals and the number of samples for a good approximation is given by the relations in (10.51).

This elegant and natural, fast and accurate LCT computation method provides a very different approach than that given in Sect. 10.4 [78]. In Sect. 10.4, we assumed that the time (or space) and frequency extents of the signals are specified in the conventional manner, defining an initial rectangular time- or space-frequency support. The number of samples were determined from the standard Nyquist–Shannon sampling theorem [59, 64, 65, 69, 79, 80, 83]. In this section, we assume the extents are specified in the input and output LCT domains; that is, in the original time or space domain, and the target LCT domain. This leads to an initial parallelogram-shaped time- or space-frequency support [84, 85]. The number of samples is determined from the LCT sampling theorem [78]. The minimum number of samples required for computation is given by the so-called bicanonical width product [78], which is also equal to the area of the parallelogram support. The DLCT defined in [75] works with this minimum number of samples without requiring any interpolation or oversampling at the intermediate stages of the computation, in contrast to previously given approaches [79, 80] for the same DLCT. Also recall that, sampling the input and output by using the Nyquist–Shannon sampling theorem, as in these works and Sect. 10.4, usually leads to a greater number of samples and sometimes requires different numbers of samples for the input and output signals.

So far, we discussed how this DLCT can be used to accurately compute the samples of a continuous DLCT. Now, let us discuss how this DLCT can be implemented in a fast way. As discussed previously, the DLCT in (10.47) can be evaluated with either of two fast methods:

1. A direct approach by successively performing a chirp multiplication, a FFT and a second chirp multiplication (by taking advantage of the simple form of the DLCT) [42, 75, 78].
2. A radix-type approach that generalizes the FFT for the Fourier transform to the LCT (by taking advantage of the shifting properties of the DLCT) [74, 76].

Both approaches yield efficient computation in $O(N \log N)$ time. The first of these is simpler, since by employing the FFT as a building block, it does not require us to get into the complications of a radix-type approach [78]. This also has the advantage of relying on widely available, highly optimized FFT implementations, whereas such optimized implementations do not exist for the radix-type approach to fast LCT computation.

The DLCT definition given in Eq. (10.47) has many desirable properties when used with the parameters satisfying Eq. (10.51) [78]. It has a simple analytic expression and is unitary. It can be efficiently computed in $O(N \log N)$ time by successively performing a chirp multiplication, a FFT, and a second chirp multiplication. It has a well-defined relationship to the continuous LCT [77, 78]. Its accuracy is only limited by the fundamental fact that a signal cannot have strictly finite extent in more than one domain. Therefore it is an important candidate for being a widely accepted definition of the discrete version of the LCT.

We refer to this type of approach to LCT computation as a hybrid algorithm. It involves both an analytically desirable definition of the DLCT and fast decomposition-based computation by successively performing a chirp multiplication, a FFT, and a second chirp multiplication. Thus, it combines the best features of DLCT based algorithms and decomposition based algorithms. It conserves computational resources by eliminating the interpolation/decimation steps, as well as keeping the number of samples at the minimum possible [78].

This approach has been revisited in [85], where an interpretation of the method has been given through phase-space diagrams, revealing the decomposition-based nature of the algorithm (similar to the decompositions of the continuous LCT discussed in Sect. 10.4). The decomposition consists of a chirp multiplication, magnified FT, and a second chirp multiplication [63]. It first shears the parallelogram-shaped initial support into a rectangle, then rotates it by 90°, and then again shears it back to a parallelogram with the same area as the initial parallelogram. During these stages, no overlap is introduced between the replicated supports (arising from sampling). Hence this phase-space picture allows us to see from yet another perspective how this elegant and accurate LCT computation method [77, 78] works with the minimum number of samples, without requiring interpolation.

Finally, before ending this section, we also mention another fast algorithm for LCT computation which is based on an alternative form for the unitary DLCT [86]. In this approach, a convergent quadrature definition for the continuous Fourier transform is used to decompose arbitrary LCTs as a scaling, chirp, DFT, chirp, scaling decomposition. The decomposition and the DFT definition developed in this paper results in the following algorithm for fast digital computation of LCTs: Let the vector **g**, of length N, stand for the samples of the LCT of $f(u)$ with parameters A, B, C, D, evaluated at the points $v_j = 4Bu_j/\pi$, where $u_j = (\frac{2j-N-1}{\sqrt{2N}})\pi/2$. Now,

1. Set up the vector **p** with $p_k = e^{-i\pi\frac{(k-1)(N-1)}{N}} e^{iAu_k^2/2B} f(u_k)$, where $k = 1, 2, \ldots, N$. This step corresponds to first scaling the input samples and then multiplying with the chirp in the decomposition.
2. Take the conventional DFT of the vector **p** using the FFT algorithm and denote the result as **q**.
3. Construct the diagonal matrix **S** as $\mathbf{S}_{jk} = \frac{\pi e^{i\frac{\pi}{2}\frac{(N-1)^2}{N}}}{\sqrt{2N}} \frac{e^{\frac{iDv_j^2}{2B}} e^{-i\pi\frac{N-1}{N}(j-1)}}{\sqrt{2\pi iB}} \delta_{jk}$ where $j, k = 1, 2, \ldots, N$. This matrix serves to implement the second chirp multiplication. δ_{jk} stands for the Kronecker delta.
4. Lastly, obtain the result vector **g** by multiplying the vector **q** with the diagonal matrix **S**. The elements of the vector **g** correspond to samples of the continuous LCT of $f(u)$ at the points $v_j = 4Bu_j/\pi$.

None of the above steps takes more than $O(N \log N)$ time, so that the overall complexity is $O(N \log N)$.

10.7 Computation of Two-Dimensional LCTs

Two-dimensional *separable* LCTs are addressed in [8, 10, 87–91]. The most special case is the isotropic 2D LCT in which the system is fully symmetric, orthogonal, and the parameters for both dimensions are identical. This case can be represented by only three parameters as in a 1D LCT [24]. When the system is still orthogonal but the parameters for the orthogonal dimensions differ, the system becomes a 2D separable LCT, which is represented by six parameters [24]. Separable 2D transforms do not pose much difficulty because the separable transform is essentially two independent 1D transforms along the two dimensions and the dimensions can be treated independently. However, the *non-separable* transform (2D NS LCT) is significantly more general. The two dimensions are coupled to each other by four additional cross-parameters, increasing the total number of parameters to ten. This general case is non-separable, non-axially symmetric, non-orthogonal, and anamorphic/astigmatic [2, 24, 27, 68, 73]. 2D NS LCTs are able to represent not only systems involving anamorphic/astigmatic components and reference surfaces, but also other interesting systems such as optical mode convertors and resonators since they can represent the coupling between the dimensions [24, 92–94]. Another prominent feature of 2D NS LCTs is their ability to represent systems with rotations between any arbitrary planes in phase-space, like rotations and gyrations [24, 27]. These systems are collected under the general name of gyrators and are useful in two-dimensional image processing, signal processing, mode transformation, etc. [27, 95–98].

Given an algorithm for efficiently computing 1D LCTs [64, 65, 74], the efficient computation of separable 2D transforms is straightforward because the kernel can be separated and the 2D transform can be reduced to two successive 1D LCTs. On the other hand, in the non-separable case, the two dimensions are coupled and handling this case requires special attention. An alternative representation of LCTs is presented and studied in [67]. This decomposition is based on the well-known Iwasawa decomposition [99]. In [67], the authors further decompose the first matrix of the Iwasawa decomposition into a two-dimensional *separable* fractional Fourier transform (2D S FRT) that is sandwiched between two coordinate rotators. Earlier in this chapter, we had mentioned the deployment of the 1D version of the Iwasawa decomposition to develop a fast and efficient algorithm for 1D LCTs [64, 65]. By using the 2D Iwasawa-type decomposition of [67], it becomes likewise possible to derive an efficient algorithm for the computation of 2D NS LCTs [83].

The 2D NS LCT with parameter matrix $\mathbf{M}$, of an input function $f(\mathbf{u})$, can be expressed as [67, 100]

$$\begin{aligned} f_{\mathbf{M}}(\mathbf{u}) &= (\mathcal{C}_{\mathbf{M}}f)(\mathbf{u}) \\ &= \frac{1}{\sqrt{\det i\mathbf{B}}} \int_{-\infty}^{\infty}\int_{-\infty}^{\infty} \exp[i\pi(\mathbf{u}'^{\mathrm{T}}\mathbf{B}^{-1}\mathbf{A}\mathbf{u}' \\ &\quad -2\mathbf{u}'^{\mathrm{T}}\mathbf{B}^{-1}\mathbf{u} + \mathbf{u}^{\mathrm{T}}\mathbf{D}\mathbf{B}^{-1}\mathbf{u})]f(\mathbf{u}')\,d\mathbf{u}', \end{aligned} \tag{10.53}$$

where $\mathbf{u} = [u_x \ u_y]^{\mathrm{T}}$, $\mathbf{u}' = [u'_x \ u'_y]^{\mathrm{T}}$ with T denoting the transpose operation. $\mathbf{A}, \mathbf{B}, \mathbf{C}, \mathbf{D}$ are 2×2 submatrices defining the transformation matrix $\mathbf{M}$ of the system that represents the 2D-LCT, $\mathbf{B}$ being nonsingular. The matrix $\mathbf{M} = [\mathbf{A}\ \mathbf{B};\ \mathbf{C}\ \mathbf{D}]$ is real and symplectic. From a group-theoretical point of view, 2D NS LCTs form the ten-parameter symplectic group $Sp(4, \mathbb{R})$. ($\mathbf{M}$ has 16 parameters with six constraints leaving ten independent parameters.) More on group-theoretical properties of LCTs can be found in [8].

The Iwasawa decomposition is the core of our algorithm. After the dimensional normalization explained in Sect. 10.2.1, any transformation matrix $\mathbf{M}$ can be written in the following Iwasawa form [67, 99]:

$$\mathbf{M} = \begin{bmatrix} \mathbf{A} & \mathbf{B} \\ \mathbf{C} & \mathbf{D} \end{bmatrix} = \begin{bmatrix} \mathbf{I} & \mathbf{0} \\ -\mathbf{G} & \mathbf{I} \end{bmatrix} \begin{bmatrix} \mathbf{S} & \mathbf{0} \\ \mathbf{0} & \mathbf{S}^{-1} \end{bmatrix} \begin{bmatrix} \mathbf{X} & \mathbf{Y} \\ -\mathbf{Y} & \mathbf{X} \end{bmatrix}, \tag{10.54}$$

where

$$\mathbf{G} = -(\mathbf{C}\mathbf{A}^{\mathrm{T}} + \mathbf{D}\mathbf{B}^{\mathrm{T}})(\mathbf{A}\mathbf{A}^{\mathrm{T}} + \mathbf{B}\mathbf{B}^{\mathrm{T}})^{-1}, \tag{10.55}$$

$$\mathbf{S} = (\mathbf{A}\mathbf{A}^{\mathrm{T}} + \mathbf{B}\mathbf{B}^{\mathrm{T}})^{1/2}, \tag{10.56}$$

$$\mathbf{X} = (\mathbf{A}\mathbf{A}^{\mathrm{T}} + \mathbf{B}\mathbf{B}^{\mathrm{T}})^{-1/2}\mathbf{A}, \tag{10.57}$$

$$\mathbf{Y} = (\mathbf{A}\mathbf{A}^{\mathrm{T}} + \mathbf{B}\mathbf{B}^{\mathrm{T}})^{-1/2}\mathbf{B}. \tag{10.58}$$

Given the 4×4 matrix $\mathbf{M}$, we can determine 2×2 matrices $\mathbf{G}, \mathbf{S}, \mathbf{X}, \mathbf{Y}$ by using Eqs. (10.55)–(10.58). If we are able to develop a fast algorithm to compute the three stages in $O(\tilde{N} \log \tilde{N})$ time, the overall transform can also be calculated in $O(\tilde{N} \log \tilde{N})$ time where $\tilde{N}$ stands for the total number of samples in a 2D signal. In this decomposition, the first operation is an orthosymplectic system, followed by a scaling (magnification) system, finally followed by a two-dimensional chirp multiplication (2D CM). (Note that each of the stages of the algorithm are special cases of 2D NS LCTs.)

The first and the most sophisticated stage of the decomposition is the orthosymplectic system. This stage of the decomposition can be further decomposed into a 2D S FRT that is sandwiched between two coordinate rotators [67]:

$$\begin{bmatrix} \mathbf{X} & \mathbf{Y} \\ -\mathbf{Y} & \mathbf{X} \end{bmatrix} = \mathbf{R}_{\mathbf{r}_2} \mathbf{F}_{\mathbf{a}_x, \mathbf{a}_y} \mathbf{R}_{\mathbf{r}_1}, \tag{10.59}$$

where the 4×4 matrices $\mathbf{R}_{\mathbf{r}_1}$, $\mathbf{F}_{\mathbf{a}_x, \mathbf{a}_y}$, $\mathbf{R}_{\mathbf{r}_1}$ are defined as:

$$\mathbf{R}_{\mathbf{r}_1} = \begin{bmatrix} \cos(r_1) & \sin(r_1) & 0 & 0 \\ -\sin(r_1) & \cos(r_1) & 0 & 0 \\ 0 & 0 & \cos(r_1) & \sin(r_1) \\ 0 & 0 & -\sin(r_1) & \cos(r_1) \end{bmatrix}, \tag{10.60}$$

$$\mathbf{R_{r_2}} = \begin{bmatrix} \cos(r_2) & \sin(r_2) & 0 & 0 \\ -\sin(r_2) & \cos(r_2) & 0 & 0 \\ 0 & 0 & \cos(r_2) & \sin(r_2) \\ 0 & 0 & -\sin(r_2) & \cos(r_2) \end{bmatrix}, \tag{10.61}$$

$$\mathbf{F_{a_x,a_y}} = \begin{bmatrix} \cos(a_x\pi/2) & 0 & \sin(a_x\pi/2) & 0 \\ 0 & \cos(a_y\pi/2) & 0 & \sin(a_y\pi/2) \\ -\sin(a_x\pi/2) & 0 & \cos(a_x\pi/2) & 0 \\ 0 & -\sin(a_y\pi/2) & 0 & \cos(a_y\pi/2) \end{bmatrix}. \tag{10.62}$$

$\mathbf{R_{r_1}}$ and $\mathbf{R_{r_2}}$ are rotation matrices that impose rotations of angles r_1 and r_2, respectively, through the spatial variables (u_x, u_y) and through their frequency variables (μ_x, μ_y). Unlike these traditional rotators which rotate within space and spatial frequency separately, the FRT rotates within the space-frequency planes of each dimension. $\mathbf{F_{a_x,a_y}}$ stands for the 2D S FRT that makes separable rotations of angle $a_x\pi/2$ in the (u_x, μ_x) plane and of angle $a_y\pi/2$ in the (u_y, μ_y) plane. Since this 2D FRT operation is separable, it corresponds to two 1D FRT operations performed over each of the dimensions. This amounts to first performing 1D FRTs with the fractional order a_x for each of the rows (or columns) and then performing 1D FRTs with the fractional order a_y for each of the columns (or rows) of the sampling grid. It is this observation that enables us to implement this stage of the decomposition efficiently in $O(\tilde{N}\log\tilde{N})$ time.

The interpretation of the coordinate rotators requires care. When we are working with sampled functions, we know the value and coordinates (the location where the particular sample is taken) of all the samples we have. A coordinate rotation can be interpreted in this situation as a rotation of the locations of the samples resulting in a new sampling grid, rather than a change in the sample values. If we assume we start with a regular rectangular grid, after the coordinate rotation, the grid would no longer coincide with the original grid unless the rotation is an integer multiple of $\pi/2$. Unfortunately, in order to perform FRT operations along the horizontal and vertical directions, we need the samples to be on a regular rectangular grid in order to employ available fast algorithms. Therefore, we must carry out an interpolation operation to determine the values of the function on a regular rectangular grid. In summary, the first stage of our algorithm involves determining the angle parameters for the first coordinate rotation, followed by two 1D FRTs over each of the dimensions, and then followed by the second coordinate rotation. These angles can be computed by equating the LCT matrix with the decomposition matrix and solving for the angles. All these steps can be calculated in $O(\tilde{N}\log\tilde{N})$ time.

The second stage is the scaling operation and it seems to be the simplest of the three stages. It is not, however, as trivial as in the 1D case [65]. In 1D, it corresponds to only a reinterpretation of the spacing between the samples. The sampling interval scales with the scaling parameter. Intuitively, it squeezes in or stretches out the total number of samples as the word scaling implies. This means there is no change in the total number of samples and thus no need to oversample the input samples. The analogue of the 1D scalar scaling parameter in the 2D case is the matrix **S**. When **S** is diagonal, which means there is no coupling between the two dimensions of the function for scaling purposes, the scaling is separable. Due to this separability, this situation does not impose an increase in the space-bandwidth products and thus does not require oversampling, just as in the 1D case. But when the off-diagonal elements of **S** are non-zero, the scaling operation is no longer so trivial. Computationally, such a scaling operation amounts to modifying the information that tells us which coordinates the samples belong to. Nevertheless, since it requires only the reinterpretation of the coordinates of the samples plus a possible oversampling, it does not impose much computational load. The matrix **S** can be easily used to determine the output samples by using the input–output relation of the scaling operation:

$$f_{sc}(\mathbf{u}) = \frac{1}{\sqrt{\det \mathbf{S}}} f(\mathbf{S}^{-1}\mathbf{u}), \tag{10.63}$$

where f is the function to be scaled and f_{sc} is the scaled function, and $\mathbf{u} = [u_x \ \ u_y]^{\mathrm{T}}$.

The last stage of the main Iwasawa decomposition is the 2D CM operation whose parameters are given by the matrix **G** as defined in Eq. (10.55). The input–output relation of this 2D-CM is given as:

$$f_{ch}(\mathbf{u}) = e^{-i\pi(G_{11}u_x^2+(G_{12}+G_{21})u_xu_y+G_{22}u_y^2)} f(\mathbf{u}), \tag{10.64}$$

where f_{ch} stands for the chirp-multiplied function. The 2D CM operation is the stage that is mainly burdened with any shears inherent in the 2D NS LCT to be computed. Such shears may considerably increase the space-bandwidth products of the function. Thus, before the 2D CM operation, the space-bandwidth products of the function should be calculated carefully and any necessary oversampling should be performed. This CM operation may turn out to be non-separable or separable for particular 2D NS LCTs but regardless, it requires only one multiplication for each sample, resulting in $O(\tilde{N})$ time computation.

As in the 1D case, this algorithm also has the ability to track the space-bandwidth product of the function through each step in order to control the sampling rate of the function with the goal of having enough samples to be able to reconstruct the continuous function without information loss, and at the same time without needlessly increasing the number of samples to maintain efficiency. However, the sampling rate control mechanism is quite involved so we refer the reader to [83] for details.

Above, we presented the 2D-NS-LCT fast computation algorithm based on the Iwasawa decomposition, which is a generalization of the algorithm denoted as A2 in Sect. 10.4. In the same section, we also mentioned another algorithm, denoted A1, which is also based on basic decompositions. In [101], generalization of the A1 algorithm to 2D-NS-LCT computation has been considered. The results of both approaches are comparable to each other both in 1D and 2D.

10.8 Computation of Complex-Parametered Linear Canonical Transforms

Extension of real-parametered LCTs to complex-parametered LCTs (CLCTs) is rather involved. Bilateral Laplace transforms, Bargmann transforms, Gauss–Weierstrass transforms, [8, 28, 29], fractional Laplace transforms [30, 31], and CFRTs [32–35] are all special cases of complex LCTs. More on the mathematical foundations and theory of CLCTs can be found in [8, 28, 102, 103].

Complex-parametered LCTs allow several kinds of optical systems to be represented, including lossy as well as lossless ones. Magnification (scaling), Fourier transformation (FT), real fractional Fourier transformation (RFRT), real chirp multiplication (CM), complex chirp multiplication (CCM), Gauss–Weierstrass Transform, CFRT are all special cases of CLCTs that have optical realizations. The CFRT is the generalization of the FRT where the order of the transformation is allowed to be a complex number, and consequently the *ABCD* matrix elements are in general complex. The optical interpretation of the CFRT, its properties and optical realizations can be found in [32–35, 104].

The CLCT of $f(u)$ with complex parameter matrix $\mathbf{M}_{\mathrm{C}}$ is denoted as $f_{\mathbf{M}_{\mathrm{C}}}(u) = (\mathcal{C}_{\mathbf{M}_{\mathrm{C}}} f)(u)$:

$$(\mathcal{C}_{\mathbf{M}_{\mathrm{C}}} f)(u) = \int_{-\infty}^{\infty} K_{\mathrm{C}}(u,u') f(u')\, du',$$

$$K_{\mathrm{C}}(u,u') = e^{-i\pi/4}\sqrt{\overline{\beta}}\, \exp\left[i\pi(\overline{\alpha}u^2 - 2\overline{\beta}uu' + \overline{\gamma}u'^2)\right], \tag{10.65}$$

where $\overline{\alpha}, \overline{\beta}, \overline{\gamma}$ are complex parameters independent of u and u' and where $\mathcal{C}_{\mathbf{M}_{\mathrm{C}}}$ is the CLCT operator. $\mathbf{M}_{\mathrm{C}}$ again has unit-determinant and is given by

$$\mathbf{M}_{\mathrm{C}} = \begin{bmatrix} A & B \\ C & D \end{bmatrix} = \begin{bmatrix} A_r + iA_c & B_r + iB_c \\ C_r + iC_c & D_r + iD_c \end{bmatrix} = \begin{bmatrix} \overline{\gamma}/\overline{\beta} & 1/\overline{\beta} \\ -\overline{\beta} + \overline{\alpha}\overline{\gamma}/\overline{\beta} & \overline{\alpha}/\overline{\beta} \end{bmatrix}, \tag{10.66}$$

where $A_r, A_c, B_r, B_c, C_r, C_c, D_r, D_c$ are real numbers.

Fast digital computation algorithms based on decomposition approaches have been developed. Here we will only show how given *ABCD* matrices can be decomposed in a manner that leads to a fast algorithm for computation of CLCTs. For further details, we refer to [105]. In the most general case, the matrix $\mathbf{M}_{\mathrm{C}}$ is composed of the four complex parameters *A*,*B*,*C*,*D*, whose real and imaginary parts add up to a total of eight parameters. These eight parameters are restricted by the unimodularity condition on $\mathbf{M}_{\mathrm{C}}$, which requires the real part of the determinant to be 1 and the imaginary part to be 0. Because of these two equations, the total number of independent parameters of a general CLCT is 6. The following decomposition will be the basis of the fast algorithm for CLCTs:

$$\mathbf{M}_{\mathrm{C}} = \begin{bmatrix} 1 & 0 \\ -q_{3r} & 1 \end{bmatrix} \begin{bmatrix} 1 & 0 \\ -iq_{3c} & 1 \end{bmatrix} \begin{bmatrix} 0 & -1 \\ 1 & 0 \end{bmatrix} \begin{bmatrix} 1 & 0 \\ -q_{2r} & 1 \end{bmatrix} \begin{bmatrix} 1 & 0 \\ -iq_{2c} & 1 \end{bmatrix}$$
$$\times \begin{bmatrix} 0 & 1 \\ -1 & 0 \end{bmatrix} \begin{bmatrix} 1 & 0 \\ -q_{1r} & 1 \end{bmatrix} \begin{bmatrix} 1 & 0 \\ -iq_{1c} & 1 \end{bmatrix}. \tag{10.67}$$

The above algorithm can efficiently compute any arbitrary CLCT with complex parameters but the output plane should be real, i.e. the variable at the transform output is not complex. This is indeed the case for most of the real-world cases, where one is interested in the output field on a particular plane. In [106], the above algorithm has been further generalized to cover any complex variable at the output. The sampling issues of CLCTs have also been studied in detail in [106].

Before ending this section, we mention that LCTs are a special case of the more general family of oscillatory integrals of the form $F(w) = \int f(u)e^{iwg(u)}du, \quad w \gg 1$. Generally speaking, computation of such more general integrals is time consuming. In [107], a method to convert any general oscillatory integral to a canonical form $F(w) = \int_{-1}^{1} f(u)e^{iwu}du$ and then compute it in $O(N^2)$ time is presented.

10.9 Conclusion

In this chapter, we discussed algorithms for computation of LCTs from the N samples of the input signal in $O(N \log N)$ time. Our approach is based on concepts from signal analysis and processing rather than conventional numerical analysis. With careful consideration of sampling issues, N can be chosen very close to the theoretical minimum required to represent the signals. The transform output may have a higher time- or space-bandwidth product due to the nature of the transform family, though it has the same bicanonical width product with the input.

We considered three groups of algorithms. In the first, the LCT operation is decomposed into more elementary operations. Most elegant among these is based on the Iwasawa decomposition, which involves the FRT. In the second, one first defines a DLCT, and then obtains a FLCT algorithm for this DLCT, much like the FFT algorithm for the DFT. In the third group, a DLCT is defined, but the computation procedure is based on decompositions.

The algorithms can relate the samples of the input function to the samples of the continuous LCT of this function in the same sense that the FFT implementation of the DFT computes the samples of the continuous FT of a function. Since the sampling rates are carefully controlled, the output samples obtained are accurate approximations to the true ones and the continuous LCT can be recovered via interpolation of these samples. The only inevitable source of deviation from exactness arises from the fundamental fact that a signal and its transform cannot both be of finite extent. This is the same source of deviation encountered when using the DFT/FFT to compute the continuous FT. Thus the algorithms compute LCTs with a performance similar to the DFT/FFT in computing the Fourier transform, both in terms of speed and accuracy. This limitation affects not only LCT algorithms, but also the computation of Fourier transforms using the DFT. Thus this is a source of error we cannot hope to overcome.

Compared to earlier approaches, these algorithms not only handle a much more general family of integrals, but also effectively address certain difficulties, limitations, or tradeoffs that arise in other approaches to computing the Fresnel integral, which is of importance in the theory of diffraction.

Acknowledgement H.M. Ozaktas acknowledges partial support of the Turkish Academy of Sciences.

References

1. M.J. Bastiaans, Wigner distribution function and its application to first-order optics. J. Opt. Soc. Am. **69**, 1710–1716 (1979)
2. A.E. Siegman, *Lasers* (University Science Books, Mill Valley, 1986)
3. D.F.V. James, G.S. Agarwal, The generalized Fresnel transform and its application to optics. Opt. Commun. **126**(4–6), 207–212 (1996)
4. C. Palma, V. Bagini, Extension of the Fresnel transform to ABCD systems. J. Opt. Soc. Am. A **14**(8), 1774–1779 (1997)
5. S. Abe, J.T. Sheridan, Generalization of the fractional Fourier transformation to an arbitrary linear lossless transformation an operator approach. J. Phys. A Math. Gen. **27**(12), 4179–4187 (1994)
6. S. Abe, J.T. Sheridan, Optical operations on wavefunctions as the Abelian subgroups of the special affine Fourier transformation. Opt. Lett. **19**, 1801–1803 (1994)
7. J. Hua, L. Liu, G. Li, Extended fractional Fourier transforms. J. Opt. Soc. Am. A **14**(12), 3316–3322 (1997)
8. K.B. Wolf, Construction and properties of canonical transforms, Chap. 9, in *Integral Transforms in Science and Engineering* (Plenum Press, New York, 1979)
9. E. Hecht, *Optics*, 4th edn. (Addison Wesley, Reading, 2001)
10. H.M. Ozaktas, Z. Zalevsky, M.A. Kutay, *The Fractional Fourier Transform with Applications in Optics and Signal Processing* (Wiley, New York, 2001)
11. M.J. Bastiaans, The Wigner distribution function applied to optical signals and systems. Opt. Commun. **25**(1), 26–30 (1978)
12. M.J. Bastiaans, Applications of the Wigner distribution function in optics, in *The Wigner Distribution: Theory and Applications in Signal Processing*, ed. by W. Mecklenbräuker, F. Hlawatsch (Elsevier, Amsterdam, 1997), pp. 375–426
13. M. Moshinsky, Canonical transformations and quantum mechanics. SIAM J. Appl. Math. **25**(2), 193–212 (1973)
14. C. Jung, H. Kruger, Representation of quantum mechanical wavefunctions by complex valued extensions of classical canonical transformation generators. J. Phys. A Math. Gen. **15**, 3509–3523 (1982)
15. B. Davies, *Integral Transforms and Their Applications* (Springer, New York, 1978)
16. D.J. Griffiths, C.A. Steinke, Waves in locally periodic media. Am. J. Phys. **69**(2), 137–154 (2001)
17. D.W.L. Sprung, H. Wu, J. Martorell, Scattering by a finite periodic potential. Am. J. Phys. **61**(12), 1118–1124 (1993)
18. L.L. Sanchez-Soto, J.F. Carinena, A.G. Barriuso, J.J. Monzon, Vector-like representation of one-dimensional scattering. Eur. J. Phys. **26**(3), 469–480 (2005)

19. S. Baskal, Y.S. Kim, Lens optics as an optical computer for group contractions. Phys. Rev. E **67**(5), 056601 (2003)
20. S. Baskal, Y.S. Kim, ABCD matrices as similarity transformations of Wigner matrices and periodic systems in optics. J. Opt. Soc. Am. A **26**(9), 2049–2054 (2009)
21. E. Georgieva, Y.S. Kim, Slide-rule-like property of Wigner's little groups and cyclic S matrices for multilayer optics. Phys. Rev. E **68**(2), 026606 (2003)
22. B. Barshan, M.A. Kutay, H.M. Ozaktas, Optimal filtering with linear canonical transformations. Opt. Commun. **135**(1–3), 32–36 (1997)
23. S.C. Pei, J.J. Ding, Eigenfunction of linear canonical transform. IEEE Trans. Signal Process. **50**, 11–26 (2002)
24. T. Alieva, M.J. Bastiaans, Properties of the canonical integral transformation. J. Opt. Soc. Am. A **24**, 3658–3665 (2007)
25. M.J. Bastiaans, T. Alieva, Classification of lossless first-order optical systems and the linear canonical transformation. J. Opt. Soc. Am. A **24**, 1053–1062 (2007)
26. J. Rodrigo, T. Alieva, M. Luisa Calvo, Optical system design for orthosymplectic transformations in phase space. J. Opt. Soc. Am. A **23**, 2494–2500 (2006)
27. R. Simon, K.B. Wolf, Structure of the set of paraxial optical systems. J. Opt. Soc. Am. A **17**(2), 342–355 (2000)
28. K.B. Wolf, Canonical transformations I. Complex linear transforms. J. Math. Phys. **15**(8), 1295–1301 (1974)
29. K.B. Wolf, On self-reciprocal functions under a class of integral transforms. J. Math. Phys. **18**(5), 1046–1051 (1977)
30. A. Torre, Linear and radial canonical transforms of fractional order. J. Comput. Appl. Math. **153**, 477–486 (2003)
31. K.K. Sharma, Fractional Laplace Transform. Signal Image Video Process. **4**(3), 377–379 (2009)
32. C.C. Shih, Optical interpretation of a complex-order Fourier transform. Opt. Lett. **20**(10), 1178–1180 (1995)
33. L.M. Bernardo, O.D.D. Soares, Optical fractional Fourier transforms with complex orders. Appl. Opt. **35**(17), 3163–3166 (1996)
34. C. Wang, B. Lu, Implementation of complex-order Fourier transforms in complex ABCD optical systems. Opt. Commun. **203**(1–2), 61–66 (2002)
35. L.M. Bernardo, Talbot self-imaging in fractional Fourier planes of real and complex orders. Opt. Commun. **140**, 195–198 (1997)
36. N.M. Atakishiyev, K.B. Wolf, Fractional Fourier–Kravchuk transform. J. Opt. Soc. Am. A **14**(7), 1467–1477 (1997)
37. S.C. Pei, M.H. Yeh, Improved discrete fractional Fourier transform. Opt. Lett. **22**(14), 1047–1049 (1997)
38. N.M. Atakishiyev, S.M. Chumakov, K.B. Wolf, Wigner distribution function for finite systems. J. Math. Phys. **39**(12), 6247–6261 (1998)
39. N.M. Atakishiyev, L.E. Vicent, K.B. Wolf, Continuous vs. discrete fractional Fourier transforms. J. Comput. Appl. Math. **107**(1), 73–95 (1999)
40. S.C. Pei, M.H. Yeh, C.C. Tseng, Discrete fractional Fourier transform based on orthogonal projections. IEEE Trans. Signal Process. **47**(5), 1335–1348 (1999)
41. S.C. Pei, M.H. Yeh, T.L. Luo, Fractional Fourier series expansion for finite signals and dual extension to discrete-time fractional Fourier transform. IEEE Trans. Signal Process. **47**(10), 2883–2888 (1999)
42. T. Erseghe, P. Kraniauskas, G. Carioraro, Unified fractional Fourier transform and sampling theorem. IEEE Trans. Signal Process. **47**(12), 3419–3423 (1999)
43. M.A. Kutay, H. Ozaktas, H.M. Ozaktas, O. Arikan, The fractional Fourier domain decomposition. Signal Process. **77**(1), 105–109 (1999)
44. A.I. Zayed, A.G. Garcķa, New sampling formulae for the fractional Fourier transform. Signal Process. **77**(1), 111–114 (1999)

45. C. Candan, M.A. Kutay, H.M. Ozaktas, The discrete fractional Fourier transform. IEEE Trans. Signal Process. **48**(5), 1329–1337 (2000)
46. I.S. Yetik, M.A. Kutay, H. Ozaktas, H.M. Ozaktas, Continuous and discrete fractional Fourier domain decomposition, in *Proceedings of the IEEE International Conference on Acoustics, Speech, and Signal Processing (ICASSP'00)*, vol. 1 (2000), pp. 93–96
47. S.C. Pei, M.H. Yeh, The discrete fractional cosine and sine transforms. IEEE Trans. Signal Process. **49**(6), 1198–1207 (2001)
48. G. Cariolaro, T. Erseghe, P. Kraniauskas, The fractional discrete cosine transform. IEEE Trans. Signal Process. **50**(4), 902–911 (2002)
49. L. Barker, Continuum quantum systems as limits of discrete quantum systems, IV. Affine canonical transforms. J. Math. Phys. **44**(4), 1535–1553 (2003)
50. C. Candan, H.M. Ozaktas, Sampling and series expansion theorems for fractional Fourier and other transforms. Signal Process. **83**(11), 2455–2457 (2003)
51. J.G. Vargas-Rubio, B. Santhanam, On the multiangle centered discrete fractional Fourier transform. IEEE Signal Process. Lett. **12**(4), 273–276 (2005)
52. M.H. Yeh, Angular decompositions for the discrete fractional signal transforms. Signal Process. **85**(3), 537–547 (2005)
53. K.B. Wolf, Finite systems, fractional Fourier transforms and their finite phase spaces. Czech. J. Phys. **55**, 1527–1534 (2005)
54. K.B. Wolf, Finite systems on phase space. J. Mod. Phys. B **20**(11), 1956–1967 (2006)
55. K.B. Wolf, G. Krötzsch, Geometry and dynamics in the fractional discrete Fourier transform. J. Opt. Soc. Am. A **24**(3), 651–658 (2007)
56. D. Mendlovic, Z. Zalevsky, N. Konforti, Computation considerations and fast algorithms for calculating the diffraction integral. J. Mod. Opt. **44**(2), 407–414 (1997)
57. D. Mas, J. Garcia, C. Ferreira, L.M. Bernardo, F. Marinho, Fast algorithms for free-space diffraction patterns calculation. Opt. Commun. **164**(4–6), 233–245 (1999)
58. J.W. Cooley, J.W. Tukey, An algorithm for the machine calculation of complex Fourier series. Math. Comput. **19**, 297–301 (1965)
59. H.M. Ozaktas, O. Arıkan, M.A. Kutay, G. Bozdağı, Digital computation of the fractional Fourier transform. IEEE Trans. Signal Process. **44**, 2141–2150 (1996)
60. M.J. Bastiaans, The Wigner distribution function and Hamilton's characteristics of a geometric-optical system. Opt. Commun. **30**(3), 321–326 (1979)
61. F.S. Oktem, H.M. Ozaktas, Equivalence of linear canonical transform domains to fractional Fourier domains and the bicanonical width product: a generalization of the space-bandwidth product. J. Opt. Soc. Am. A **27**(8), 1885–1895 (2010)
62. F. Hlawatsch, G.F. Boudreaux-Bartels, Linear and quadratic time-frequency signal representations. IEEE Signal Process. Mag. **9**(2), 21–67 (1992)
63. L. Cohen, *Time-Frequency Analysis* (Prentice Hall, Englewood Cliffs, 1995)
64. H.M. Ozaktas, A. Koç, I. Sari, M.A. Kutay, Efficient computation of quadratic-phase integrals in optics. Opt. Lett. **31**, 35–37 (2006)
65. A. Koç, H.M. Ozaktas, C. Candan, M.A. Kutay, Digital computation of linear canonical transforms. IEEE Trans. Signal Process. **56**(6), 2383–2394 (2008)
66. H.M. Ozaktas, M.F. Erden, Relationships among ray optical, Gaussian beam, and fractional Fourier transform descriptions of first-order optical systems. Opt. Commun. **143**, 75–86 (1997)
67. T. Alieva, M.J. Bastiaans, Alternative representation of the linear canonical integral transform. Opt. Lett. **30**(24), 3302–3304 (2005)
68. M.J. Bastiaans, T. Alieva, Synthesis of an arbitrary ABCD systemwith fixed lens positions. Opt. Lett. **31**, 2414–2416 (2006)
69. B.M. Hennelly, J.T. Sheridan, Generalizing, optimizing, and inventing numerical algorithms for the fractional Fourier, Fresnel, and linear canonical transforms. J. Opt. Soc. Am. A **22**, 917–927 (2005)

70. X. Yang, Q. Tan, X. Wei, Y. Xiang, Y. Yan, G. Jin, Improved fast fractional-Fourier-transform algorithm. J. Opt. Soc. Am. A **21**(9), 1677–1681 (2004)
71. J. García, D. Mas, R.G. Dorsch, Fractional-Fourier-transform calculation through the fast-Fourier-transform algorithm. Appl. Opt. **35**(35), 7013–7018 (1996)
72. F.J. Marinho, L.M. Bernardo, Numerical calculation of fractional Fourier transforms with a single fast-Fourier-transform algorithm. J. Opt. Soc. Am. A **15**(8), 2111–2116 (1998)
73. X. Liu, K.H. Brenner, Minimal optical decomposition of ray transfer matrices. Appl. Opt. **47**(22), E88–E98 (2008)
74. B.M. Hennelly, J.T. Sheridan, Fast numerical algorithm for the linear canonical transform. J. Opt. Soc. Am. A **22**, 928–937 (2005)
75. S.C. Pei, J.J. Ding, Closed-form discrete fractional and affine Fourier transforms. IEEE Trans. Signal Process. **48**, 1338–1353 (2000)
76. J.J. Healy, J.T. Sheridan, Fast linear canonical transforms. J. Opt. Soc. Am. A **27**(1), 21–30 (2010)
77. A. Stern, Why is the linear canonical transform so little known?, in *AIP Conference Proceedings*, 2006, pp. 225–234
78. F.S. Oktem, H.M. Ozaktas, Exact relation between continuous and discrete linear canonical transforms. IEEE Signal Process. Lett. **16**(8), 727–730 (2009)
79. J.J. Healy, B.M. Hennelly, J.T. Sheridan, Additional sampling criterion for the linear canonical transform. Opt. Lett. **33**(22), 2599–2601 (2008)
80. J.J. Healy, J.T. Sheridan, Sampling and discretization of the linear canonical transform. Signal Process. **89**(4), 641–648 (2009)
81. A. Papoulis, *Signal Analysis* (McGraw-Hill, New York, 1977)
82. J.J. Healy, J.T. Sheridan, Cases where the linear canonical transform of a signal has compact support or is band-limited. Opt. Lett. **33**(3), 228–230 (2008)
83. A. Koç, H.M. Ozaktas, L. Hesselink, Fast and accurate computation of two-dimensional non-separable quadratic-phase integrals. J. Opt. Soc. Am. A **27**(6), 1288–1302 (2010)
84. F.S. Oktem, Signal representation and recovery under partial information, redundancy, and generalized finite extent constraints. M.S. thesis, Bilkent University, Turkey, 2009
85. J.J. Healy, J.T. Sheridan, Reevaluation of the direct method of calculating Fresnel and other linear canonical transforms. Opt. Lett. **35**(7), 947–949 (2010)
86. R.G. Campos, J. Figueroa, A fast algorithm for the linear canonical transform. Signal Process. **91**(6), 1444–1447 (2011)
87. A. Sahin, H.M. Ozaktas, D. Mendlovic, Optical implementations of two-dimensional fractional Fourier transforms and linear canonical transforms with arbitrary parameters. Appl. Opt. **37**, 2130–2141 (1998)
88. M. Moshinsky, C. Quesne, Linear canonical transformations and their unitary representations. J. Math. Phys. **12**(8), 1772–1780 (1971)
89. M. Nazarathy, J. Shamir, First-order optics—a canonical operator representation: lossless systems. J. Opt. Soc. Am. **72**(3), 356–364 (1982)
90. A. Sahin, H.M. Ozaktas, D. Mendlovic, Optical implementation of the two-dimensional fractional Fourier transform with different orders in the two dimensions. Opt. Commun. **120**, 134–138 (1995)
91. M.F. Erden, H.M. Ozaktas, A. Sahin, D. Mendlovic, Design of dynamically adjustable anamorphic fractional Fourier transformer. Opt. Commun. **136**(1–2), 52–60 (1997)
92. E.G. Abramochkin, V.G. Volostnikov, Generalized Gaussian beams. J. Opt. A Pure Appl. Opt. **6**, S157–S161 (2004)
93. L. Allen, M.W. Beijersbergen, R.K.C. Spreeuw, J.P. Woerdman, Orbital angular momentum of light and the transformation of Laguerre-Gaussian laser modes. Phys. Rev. A **45**(11), 8185–8189 (1992)
94. R. Pratesi, L. Ronchi, Generalized Gaussian beams in free space. J. Opt. Soc. Am. **67**(9), 1274–1276 (1977)
95. J.A. Rodrigo, T. Alieva, M.L. Calvo, Experimental implementation of the gyrator transform. J. Opt. Soc. Am. A **24**(10), 3135–3139 (2007)

96. J.A. Rodrigo, T. Alieva, M.L. Calvo, Gyrator transform: properties and applications. Opt. Express **15**(5), 2190–2203 (2007)
97. J.A. Rodrigo, T. Alieva, M.L. Calvo, Applications of gyrator transform for image processing. Opt. Commun. **278**(2), 279–284 (2007)
98. K.B. Wolf, T. Alieva, Rotation and gyration of finite two-dimensional modes. J. Opt. Soc. Am. A **25**(2), 365–370 (2008)
99. K.B. Wolf, *Geometric Optics on Phase Space* (Springer, Berlin, 2004)
100. G.B. Folland, *Harmonic Analysis in Phase Space* (Princeton University Press, Princeton, 1989)
101. J. Ding, S. Pei, C. Liu, Improved implementation algorithms of the two-dimensional nonseparable linear canonical transform. J. Opt. Soc. Am. A **29**(8), 1615–1624 (2012)
102. K.B. Wolf, Canonical transformations II. Complex radial transforms. J. Math. Phys. **15**(12), 2102–2111 (1974)
103. P. Kramer, M. Moshinsky, T.H. Seligman, Complex extensions of canonical transformations and quantum mechanics, in *Group Theory and Its Applications*, vol. 3, ed. by E.M. Loebl (Academic, New York, 1975), pp. 249–332
104. A.A. Malyutin, Complex-order fractional Fourier transforms in optical schemes with Gaussian apertures. Quantum Electron. **34**(10), 960–964 (2004)
105. A. Koç, H.M. Ozaktas, L. Hesselink, Fast and accurate algorithm for the computation of complex linear canonical transforms. J. Opt. Soc. Am. A **27**(9), 1896–1908 (2010)
106. C. Liu, D. Wang, J.J. Healy, B.M. Hennelly, J.T. Sheridan, M.K. Kim, Digital computation of the complex linear canonical transform. J. Opt. Soc. Am. A **28**(7), 1379–1386 (2011)
107. Y. Liu, Fast evaluation of canonical oscillatory integrals. Appl. Math. Inf. Sci. **6**(2), 245–251 (2012)

Part III
Applications

Chapter 11
Deterministic Phase Retrieval Using the LCT

Unnikrishnan Gopinathan, John J. Healy, Damien P. Kelly, and John T. Sheridan

Abstract Phase retrieval is one of the most interesting and widely researched inverse problems in optical sciences. Since all the forward problems based on real-world physical phenomenon involve some loss of information, the corresponding inverse problem is ill-posed unlike its forward counterpart. This makes the solution of inverse problems quite challenging. A majority of the approaches to solve the phase retrieval problem broadly fall into two categories. In the first category, one uses a priori information of the signal to find one of the many possible approximate solutions to the problem within an acceptable error criteria. One of the first and the most widely used method in this category is the Gerchberg–Saxon (GS) algorithm that uses a priori information of the signal at the input and output plane of an optical system that performs a Fourier Transformation. The second category consists of a direct approach based on a deterministic algorithm. The present chapter is concerned with these approaches. Typically, these algorithms are based on linear transformations on a set of intensity measurements of the signal. These intensity measurements are performed at different planes as the signal propagates through an optical system. In this chapter, we discuss the complex signal determination from multiple intensity measurements. We derive a generic algorithm that can retrieve a complex valued signal from two intensity measurements, one at the input plane and the second at the output plane, of an arbitrary optical system that consists of thin lenses (refractive or GRIN elements) separated by sections of free space.

U. Gopinathan
Department of Applied Physics, Defence Institute of Advanced Technology, Girinagar, Pune-411025, India
e-mail: unni.gopinathan@gmail.com

J. Healy • D.P. Kelly • J.T. Sheridan (✉)
School of Electrical and Electronic Engineering, University College Dublin, Belfield, Dublin 4, Ireland
e-mail: john.sheridan@ucd.ie

J.J. Healy et al. (eds.), *Linear Canonical Transforms*, Springer Series in Optical Sciences 198, DOI 10.1007/978-1-4939-3028-9_11

11.1 Introduction

Phase retrieval is one of the most interesting and widely researched inverse problems in optical sciences [1]. Since all the forward problems based on real-world physical phenomenon involve some loss of information, the corresponding inverse problem is ill-posed unlike its forward counterpart. This makes the solution of inverse problems quite challenging. In simple terms, the phase retrieval problem can be stated as follows. Consider a signal propagating through an optical system. The optical system may consist of a stretch of free space or many optical elements separated by stretches of free space. From the signal intensity at the output of the optical system as sensed by an irradiance detector, can the complete signal information (characterized by amplitude and phase) be retrieved? It is not difficult to see that this inverse problem is ill-posed since the phase information is lost during the detection process, whereas the corresponding forward problem is well posed if the transformation that characterizes the optical system is known.

A majority of the approaches to solve the phase retrieval problem broadly fall into two categories. In the first category, one uses a priori information of the signal to find one of the many possible approximate solutions to the problem within an acceptable error criteria. One of the first and the most widely used method is the Gerchberg–Saxon (GS) algorithm [2]. This method uses a priori information of the signal at the input and output plane of an optical system that performs a Fourier Transformation. The optical system may be a single refractive lens with two stretches of free space corresponding to the focal length of the lens before and after the lens. It could also be a stretch of free space with the distance large enough to satisfy the Fraunhofer approximation. A commonly used attribute of the signal in the input plane of the optical system is that the signal is real valued. At the output Fourier plane, the intensity of the signal is known. The method finds an approximate solution within acceptable error that satisfies the constraints in the input and Fourier domain by iterating back and forth between these domains and applying the a priori information of the signal as constraints. The starting guess for the unknown phase of the signal that has to be determined is chosen as random in most cases. Many variants of the GS algorithm have been proposed but all these algorithms have some features in common. Firstly, these algorithms solve the following nonlinear equation:

$$|T\,[f\,(x, y, z_\mathrm{i})]|^2 = I\,(x, y, z_\mathrm{o})\,, \tag{11.1}$$

where T is the transformation that relates the input and output domain, $f(x, y, z_\mathrm{i})$ is the complex valued signal to be estimated at the input plane $z = z_\mathrm{i}$, and $I(x, y, z_\mathrm{o})$ is the intensity measured at the output plane $z = z_\mathrm{o}$. Secondly, these algorithms are iterative, non-deterministic, and seek approximate solution to the nonlinear equation (11.1). The final solution arrived at depend on the initial starting point which is a random guess in most cases.

The second set of approaches uses a direct approach based on a deterministic algorithm [3–25]. The present chapter is concerned with these approaches. Typically, these algorithms are based on linear transformations on a set of intensity measurements of the signal. These intensity measurements are performed at different planes as the signal propagates through an optical system. Equivalently, the measurements could be done at the outputs of different optical systems for the same signal propagating through these systems. These methods estimate the complex valued signal based on a linear transformation as given by:

$$T\left[I_0, I_1, \ldots, I_n\right] = f\left(x, y\right), \tag{11.2}$$

where T is a linear transformation and I_0, $I_1, \ldots, I_n$ are the multiple intensity measurements. These methods have been broadly referred to as non-interferometric deterministic phase retrieval methods. The first of these methods were proposed by Teague [3, 4] and Streibl [5] in separate works around the same time. They proposed a method to determine the complex optical signal at the input plane from two intensity measurements in sufficiently close planes at distances z and $z + \Delta z$. Since then non-interferometric deterministic phase retrieval methods have received considerable attention. These methods have been used for a broad range of applications from microscopy [6, 7] to astronomy [8], both with fully [9, 10] and partially coherent sources [11, 12], and wavelengths ranging from the visible to X-rays [13]. In a broad sense, all these methods extract the phase and thereby the complete signal information from single or multiple intensity measurements using a deterministic algorithm based on the underlying physical model. Most of the deterministic phase retrieval methods are based on the Transport of Intensity model including initial methods of Teague and Streibl.

As discussed in this chapter, non-interferometric deterministic phase retrieval methods that reconstruct phase from multiple intensity measurements can also be interpreted in terms of sampling phase space distribution functions [25] like the Ambiguity Function (AF) [14–23, 26]. Hence these approaches are also generically referred to as Phase Space Tomography. Since the phase space distributions functions like AF contain the entire signal information, it is possible to reconstruct the signal information by sampling the entire phase space distribution function in a grid. Semichaevsky and Testorf [18] and Nugent [13] have presented a description of the various deterministic phase retrieval techniques using phase space distribution functions. Methods have been proposed to extract signal information by intensity measurements in fractional Fourier transform (FRT) domains. Alieva et al. [20–22] proposed a method to reconstruct the signal from intensity measurements in two close FRT domains (i.e. different FRT order outputs from the same input). In the context of metrology, intensity measurements in FRT domains have been used to extract information regarding object tilt and translation [27–29].

It would appear that the signal extraction methods based on output intensity measurements at two close FRT domains lead to a lesser number of necessary samples. Unnikrishnan et al. [23] showed that the choice of the necessary fractional order separation needs some a priori knowledge of the input signal. This is because: (1) the bandwidth of the signal results in an upper bound on the fractional order separation and, (2) more importantly, being a differential method the effect of noise in the system is deleterious and leads to a lower bound on the fractional order separation.

In this chapter, we discuss the complex signal determination from multiple intensity measurements. We derive a generic method that can retrieve a complex valued signal from two intensity measurements one at the input plane and the second at the output plane of a Quadratic Phase System (QPS) that consists of thin lenses (refractive or GRIN elements) separated by sections of free space. A QPS performs a Linear Canonical Transform (LCT) on the signal [30]. It is shown that the Fourier transform of the input signal and output of the QPS (Linear Canonical Transformed signal) corresponds to two slices of the Ambiguity Function map separated by an angle related to the QPS parameters. The experimental result using an optical system that performs a scale invariant fractional Fourier transform with a quadratic phase signal generated by a lens is discussed.

11.2 Signal Description Using Ambiguity Function

An optical signal is most commonly described in the space domain or spatial frequency domain. A space domain description gives the distribution of signal energy in two or three dimensional space, whereas the spatial frequency domain descriptions gives signal energy distribution in terms of spatial frequency content of the signal. Space description of a signal is a function of spatial coordinates, whereas spatial frequency description is a function of spatial frequency. In some situations need arises for representing the signal in space as well as spatial frequency domains simultaneously. This is especially true for the case of non-stationary signals where the spatial frequency of the signal varies with spatial coordinates. In such cases, one is interested in local distribution of energy as a function of frequency, or in other words, local frequency spectrum. Of the many joint space–frequency signal representations that are available in literature, we will be using the Ambiguity Function as it is related to the Mutual Intensity and intensity of a signal. The Ambiguity Function (AF) representation, which is referred to as AF map henceforth, is a 2-D map for a 1-D signal and a 4-D map for a 2-D signal. Henceforth we will use 1-D signal for our discussion for simplicity but extensions to 2-D is straightforward. Consider a quasi-monochromatic signal $f(x)$ which may be fully or partially coherent. The signal is characterized by Mutual Intensity which is a second order statistical function:

$$\Gamma(x_1, x_2) = \langle f(x_1), f^*(x_2) \rangle, \tag{11.3}$$

where brackets indicate an ensemble average over a set of realizations of the function. The Mutual Intensity function can be expressed in terms of sum and difference coordinates as follows:

$$J\left(x,\overline{x}\right)=\Gamma\left(x_1,x_2\right), \tag{11.4}$$

where $x=\frac{x_1+x_2}{2};\overline{x}=x_1-x_2$.

The Mutual Intensity and the AF are related by a Fourier Transform:

$$A\left(x,\overline{\nu}\right)=\int J\left(x,\overline{x}\right)\exp\left(-i2\pi\overline{\nu}\overline{x}\right)d\overline{x}, \tag{11.5a}$$

$$J\left(x,\overline{x}\right)=\int A\left(\overline{x},\overline{\nu}\right)\exp\left(i2\pi\overline{\nu}x\right)d\overline{\nu}. \tag{11.5b}$$

The intensity of the signal can be obtained from the Mutual Intensity Function using the relation:

$$I(x)=J\left(x,0\right). \tag{11.6}$$

From Eqs. (11.5b) and (11.6), the intensity of the signal is related to the AF as:

$$I(x)=\int A\left(0,\overline{\nu}\right)\exp\left(i2\pi\overline{\nu}x\right)d\overline{\nu}. \tag{11.7}$$

In the AF plane, $A\left(0,\overline{\nu}\right)$ corresponds to a slice along the $\overline{\nu}$ axis, and the intensity of the signal, $I(x)$, is the inverse Fourier transform of this slice.

11.3 Propagation of the Signal Through an Optical System and Its AF Representation

Let us consider the propagation of a signal through an optical system that consists of thin lenses (refractive or GRIN) separated by sections of free space. Such a system is referred to as a QPS. In the paraxial approximation, a QPS can be shown to perform a LCT on the signal. The transformed signal is given by:

$$\begin{aligned}f_{\{\alpha,\beta,\gamma\}}\left(x'\right)&=L_{\{\alpha,\beta,\gamma\}}\left\{f(x)\right\}\left(x'\right)\\&=\exp\left(-\tfrac{j\pi}{4}\right)\sqrt{\beta}\int f(x)\exp\left[j\pi\left(\alpha x^2-2\beta xx'+\gamma x'^2\right)\right]dx.\end{aligned} \tag{11.8}$$

It is to be noted that the propagation of a signal through a QPS does not change the total energy of the signal as the LCT that characterizes the QPS is a unitary transformation. Now, we consider the following question. How does the AF representation of a signal change after propagation through a QPS? Interestingly, the signal propagation through a QPS does not change the individual values of the AF. The $\left(\overline{x},\overline{\nu}\right)$ coordinates of the AF map undergoes an affine transformation as given by:

$$\begin{bmatrix} \bar{\bar{x}} \\ \bar{\bar{\nu}} \end{bmatrix} = \begin{bmatrix} a\ b \\ c\ d \end{bmatrix} \begin{bmatrix} \bar{x} \\ \bar{\nu} \end{bmatrix} = \begin{bmatrix} \frac{\gamma}{\beta} & \frac{1}{\beta} \\ -\beta + \frac{\alpha\gamma}{\beta} & \frac{\alpha}{\beta} \end{bmatrix} \begin{bmatrix} \bar{x} \\ \bar{\nu} \end{bmatrix}. \tag{11.9}$$

It may be noted that the transformation matrix is the ABCD matrix that characterizes the QPS whose determinant has unity value. The AF map of the signal after QPS propagation is given by:

$$A\left(\bar{\bar{x}}, \bar{\bar{\nu}}\right) = A\left(a\bar{x} + b\bar{\nu}, \overline{cx} + \overline{d\nu}\right). \tag{11.10}$$

The Mutual Intensity of the signal at the output of the QPS is given by:

$$J_1\left(x, \bar{\bar{x}}\right) = \int A\left(\bar{\bar{x}}, \bar{\bar{\nu}}\right) \exp\left(j2\pi\bar{\bar{\nu}}x\right) d\bar{\bar{\nu}}. \tag{11.11}$$

The intensity of the signal at the output of the QPS is given by:

$$I_1(x) = J_1(x, 0) = \int A\left(0, \bar{\bar{\nu}}\right) \exp\left(j2\pi\bar{\bar{\nu}}x\right) d\bar{\bar{\nu}}\Big). \tag{11.12}$$

The intensity of the signal at the output of the QPS is the inverse Fourier Transform of the slice of the AF map along the $\bar{\bar{\nu}}$ axis in the transformed $(\bar{\bar{x}}, \bar{\bar{\nu}})$ coordinate system. Equivalently, this corresponds to a slice of the AF along the line $a\bar{x} + b\bar{\nu} = 0$ in the $(\bar{x}, \bar{\nu})$ coordinate system. Hence, we have seen that propagation through QPS is equivalent to taking a slice along the line $a\bar{x} + b\bar{\nu} = 0$ in the AF map. In terms of QPS parameters, this corresponds to the line, $\gamma\bar{x} + \bar{\nu} = 0$. Along this line the $\bar{x}$ and $\bar{\nu}$ are related as:

$$\bar{x} = -k\bar{\nu}, \tag{11.13}$$

where $k = \frac{b}{a} = \frac{1}{\gamma}$.

Thus the Fourier transform of the two intensity measurements at the input and output of a QPS system corresponds to two slices in the AF map separated by an angle $\Delta\phi = \tan^{-1}\left(\frac{b}{a}\right) = \tan^{-1}\left(\frac{1}{\gamma}\right)$. This situation is illustrated in Fig. 11.1. Let $\tilde{I}_0(.)$ and $\tilde{I}_1(.)$ represent the Fourier transform of the signal intensities at the input and output of a QPS. Then we have:

$$\tilde{I}_0\left(\bar{\nu}\right) = A\left(\bar{x}, \bar{\nu}\right)\big|_{\bar{x}=0}, \tag{11.14a}$$

$$\tilde{I}_1\left(\bar{\nu}\right) = A\left(\bar{x}, \bar{\nu}\right)\Big|_{\bar{x}=-k\bar{\nu}}. \tag{11.14b}$$

11.4 Signal Recovery from Intensity Measurements

We have seen that the Fourier Transform of the intensity measurements at input and output plane of a QPS gives two slices of the AF separated by an angle $\Delta\phi = \tan^{-1}\left(\frac{1}{\gamma}\right)$. Multiple intensity measurements performed at different planes in a QPS are equivalent to different slices in the AF plane and all of them passing through the origin. The angle between any two slices is determined by the section of QPS between the two planes where the corresponding intensity measurements were performed. In this manner, by doing sufficient intensity measurements in planes suitably separated, one can reconstruct the entire AF which contains complete signal information. But, for signal recovery, it is not necessary to reconstruct the entire AF as the AF representation contains redundancy. In this section, we show that it is possible to reconstruct the entire signal by doing two intensity measurements at the input and output plane of a QPS that corresponds to two slices in the AF map provided that the angle between the two slices is sufficiently small.

From Eqs. (11.14a) and (11.14b) we have:

$$\frac{\tilde{I}_1(\overline{\nu}) - \tilde{I}_0(\overline{\nu})}{k\overline{\nu}} = \frac{A(\overline{x}, \overline{\nu})\,|_{\overline{x}=-k\overline{\nu}} - A(\overline{x}, \overline{\nu})|_{\overline{x}=0}}{k\overline{\nu}}. \tag{11.15}$$

In the above equation, $k = \tan\ \Delta\phi$ where $\Delta\phi$ is the angle between the two slices. When the angle between the two slices $\Delta\phi$ is sufficiently small, $\Delta\phi,\ k \to 0$. Then Eq. (11.15) can be written as:

$$\frac{\tilde{I}_1(\overline{\nu}) - \tilde{I}_0(\overline{\nu})}{k\overline{\nu}} = \left.\frac{\partial A(\overline{x}, \overline{\nu})}{\partial \overline{x}}\right|_{\overline{x}=0}. \tag{11.16}$$

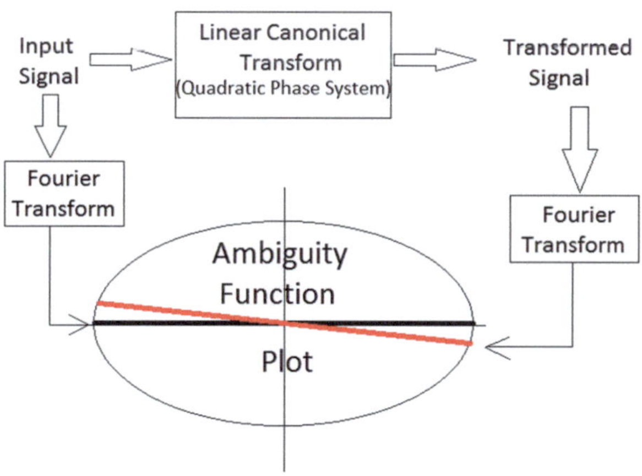

Fig. 11.1 A schematic diagram showing the relationship of the two slices of the Ambiguity Function to the signal intensities at the input and output of a QPS

In the Appendix, it is shown that:

$$\left.\frac{\partial A\left(\overline{x},\overline{\nu}\right)}{\partial\overline{x}}\right|_{\overline{x}=0} = i\int I_0(x)\frac{d\varphi(x)}{dx}\exp\left(-j2\pi\overline{\nu}x\right)dx. \tag{11.17}$$

From Eqs. (11.16) and (11.17), we get:

$$\frac{d\varphi(x)}{dx} = \frac{\pi}{I_0(x)}\left\{\left[\frac{I_1(x)-I_0(x)}{k}\right]\otimes sgn\left(-x\right)\right\}, \tag{11.18}$$

where sgn denotes signum function and ⊗ denotes convolution operation. The signum function is defined as:

$$sgn(x) = \begin{cases} -1 & for\ x < 0 \\ 0 & for\ x = 0 \\ 1 & for\ x > 0 \end{cases}. \tag{11.19}$$

To arrive at Eq. (11.18), we have used the following relation:

$$\frac{1}{\pi i}\int\frac{1}{\overline{\nu}}\exp\left(-i2\pi\overline{\nu}x\right)d\overline{x},\overline{\nu} = sgn(x). \tag{11.20}$$

From Eq. (11.18), we can obtain $\varphi(x)$ to within a constant C as:

$$\varphi(x) = \int\left\{\frac{\pi}{I_0(x)}\left\{\left[\frac{I_1(x)-I_0(x)}{k}\right]\otimes sgn\left(-x\right)\right\}\right\}dx + C. \tag{11.21}$$

Having obtained $\varphi(x)$ from Eq. (11.21), the complete signal information may be obtained as $\sqrt{I_0(x)}\,\varphi(x)$.

11.5 Free Space Propagation and Fraction Fourier Transformation

When the QPS corresponds to a stretch of free space given by distance Z, $\Delta\phi = \tan^{-1}(\lambda Z)$. When the QPS corresponds to an optical system that performs a scale invariant fractional Fourier transform θ, $\Delta\phi = \theta$.

11.6 How Close Should the Two Slices Be?

To retrieve complete signal information from two intensity measurements that corresponds to two slices in AF plane, the angle between the two slices should be small. We proceed to quantify this requirement.

Let $\delta\overline{x}$ be the sampling period along the $\overline{x}$ axis of the AF map and $\overline{\nu}_B$ be the full bandwidth of the signal. $\overline{\nu}_B$ may be obtained from the AF map by first projecting the AF distribution appropriately:

$$\frac{\overline{\nu}_B}{2}\tan\,\Delta\phi \le \delta\overline{x}. \tag{11.22}$$

To satisfy Nyquist criteria:

$$\delta\overline{x} \le \frac{1}{\overline{\nu}_B}. \tag{11.23}$$

Combining Eqs. (11.22) and (11.23), we get an expression for the upper bound of $\Delta\phi$:

$$\tan\,\Delta\phi \le \frac{2}{\overline{\nu}_B{}^2}. \tag{11.24}$$

The lower bound of $\Delta\phi$ is dictated by the noise in the system. Let us assume that the noise is a stationary random process. Let N_0 be the noise in the system. The angle between the two slices should satisfy the inequality:

$$\int_{-\frac{\overline{\nu}_B}{2}}^{\frac{\overline{\nu}_B}{2}} \left|I_1\left(\overline{\nu}\right) - I_0\left(\overline{\nu}\right)\right|^2 d\overline{\nu} > N_0\overline{\nu}_B. \tag{11.25}$$

The above equation suggests that the choice of $\Delta\phi$ should be such that the energy difference signal $|I_1 - I_0|\,(.)$ should be greater than the noise power $N_0\overline{\nu}_B$.

11.7 Estimating the Phase Profile of a Lens

Let us consider the above-mentioned method to estimate the phase profile of a thin lens. A thin lens can be characterized by a quadratic phase signal $f(x) = \exp\left[2\pi i\left(2b_2x^2 + b_1x + b_0\right)\right]$. For this signal, the Ambiguity Function is given by:

$$A\left(\overline{x},\overline{\nu}\right) = \int \exp\left[2\pi i\left(2b_2x\overline{x} + b_1\overline{x}\right)\right]\exp\left(-2\pi i\overline{\nu}x\right)dx. \tag{11.26}$$

From Eq. (11.14a), we have $I_0(x) = 1$.

Substituting Eq. (11.26) into Eq. (11.17) and after some algebraic manipulations we get:

$$\nu(x) = \frac{d\varphi(x)}{dx} = 2b_2x + b_1. \tag{11.27}$$

For this system, let us find out the upper bound and lower bound of $\Delta\phi$, the angle between the two slices in the Ambiguity Function space. If W is the width of the detector array used for intensity measurements, the full bandwidth of the signal $\overline{\nu}_B$ is restricted to $\overline{\nu}_B = 2Wb_2$. From Eq. (11.24) we get the upper bound on $\Delta\phi$ as:

$$\tan\ \Delta\phi \le \frac{1}{2W^2b_2^2}. \tag{11.28}$$

To determine the lower bound on $\Delta\phi$, let us first determine $\tilde{I}_0(.)$and $\tilde{I}_1(.)$ using Eqs. (11.14a) and (11.14b) and (11.26).

$$\tilde{I}_0\left(\overline{\nu}\right) = -\delta\left(\overline{\nu}\right), \tag{11.29a}$$

$$\tilde{I}_1\left(\overline{\nu}\right) = \frac{-1}{(2b_2k+1)}\exp\left[\frac{-2\pi b_1 k\overline{\nu}i}{(2b_2k+1)}\right]\delta\left(\overline{\nu}\right), \tag{11.29b}$$

Substituting the values of $\tilde{I}_0(.)$ and $\tilde{I}_1(.)$ in Eq. (11.25) and after some algebraic manipulations, we get the lower bound for the angle $\Delta\phi$:

$$\tan\ \Delta\phi > \frac{1}{2b_2}\left[\left(\sqrt{2N_0Wb_2}\right)^{-1} - 1\right]^{-1}. \tag{11.30}$$

The above equation merits a close look at the three parameters that decide the lower bound on the angle $\Delta\phi$, namely N_0, the system noise power; W, the width of the detector array; and b_2, the quadratic phase signal parameter which in turn depends on the focal length of the lens used to generate the signal. As the system noise is higher, N_0 is higher resulting in a higher lower bound. The same is the case with the detector width W. But a higher b_2 results in a lower value for the lower bound. Since b_2 is inversely proportional to the focal length of the lens used to generate the quadratic phase signal, a higher focal length lens results in a higher value for the lower bound. Hence, a QPS system designed to measure the profile of a lower focal length lens may not work for a higher focal length lens is the condition given in Eq. (11.30) is not satisfied. This is actually demonstrated through an experiment described below.

The optical system that performs a scale invariant fractional Fourier transform on the input signal [31] (shown in Fig. 11.2) was used to estimate the curvature of a field created by a thin spherical lens (focal length 8 mm, clear aperture 3 mm). For a plane-wave input, under the paraxial approximation the field immediately after the

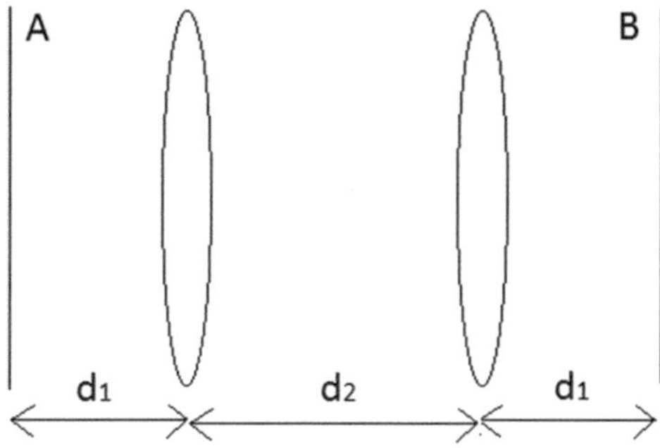

Fig. 11.2 Schematic of a two lens optical system that can perform a two dimensional scale invariant FRT. The two lenses have the same focal lengths. Planes A and B represent input and output planes of the system

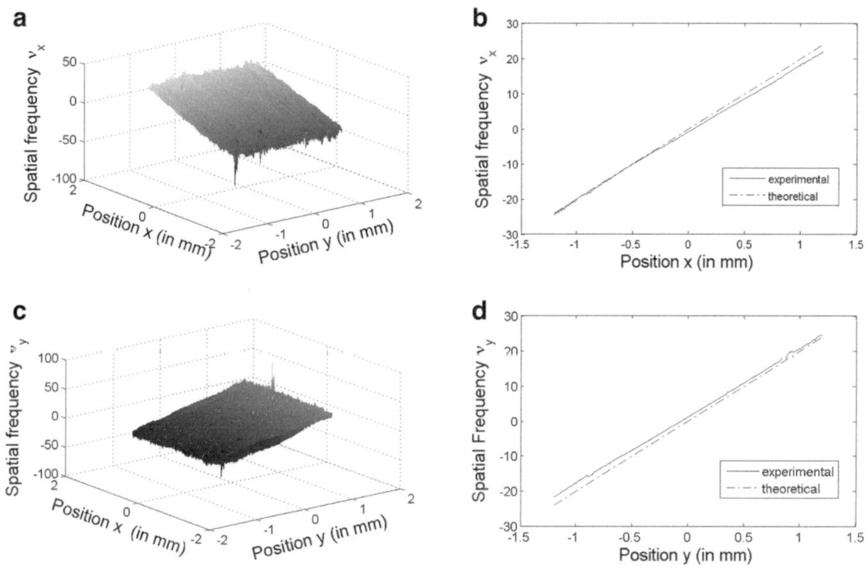

Fig. 11.3 Experimental results for 8 mm focal length lens (diameter 3 mm) (**a**) spatial frequency ν_x (**b**) spatial frequency ν_x with y direction averaged out; (**c**) spatial frequency ν_y (**d**) spatial frequency ν_y with x direction averaged out; *broken line* in (**b**) and (**d**) indicate the predicted spatial frequency corresponding to an 8 mm focal length lens

spherical lens is given by

$$f(x,y) = \exp\left[i\pi\left(\frac{x^2+y^2}{\lambda f_i}\right)\right],$$

where f_i is the focal length of the lens in the two orthogonal x and y directions. If the lens is tilted with respect to the optical axis of the system, the value of b_1 is non-zero.

The optical system shown in Fig. 11.2 performs a scale invariant optical FRT with the input and output coordinates scaled by a factor $s = \sqrt{\lambda f}$, where f is the focal length of the lens used in the FRT system. Accounting for this scaling, $b_2 = f/2f_i$ and the spatial frequency as measured by the system without any tilt is f/f_i. With the 8 mm lens placed at the input, the optical system is first configured to obtain a FRT order $\alpha = \pi$ ($f = 160$ mm, $d_1 = 160$ mm and $d_2 = 320$ mm) and the first intensity measurement is performed. This is preferred, for practical reasons, to $\alpha = 0$ system configuration, where the input and output planes coincide with the plane of the lenses. The second intensity measurement is done by perturbing the fractional order by a small value $\Delta\alpha = 0.006$ corresponding to a perturbation in d_1 by 0.5 mm and d_2 by 1 mm. The intensity measurements were done using a CCD (Imperx model 1024×1024 pixels, 7.6 μm square pixel). The width of the sensor array is $W = 8$ mm in both x and y directions. Figure 11.3a shows the value of ν_x estimated using Eq. (11.21). Figure 11.3b shows ν_x averaged along the y direction. Plotted in broken line is the theoretically expected curve with the value $b_2 = 20$ and $b_1 = 0$ for $f = 160$ mm and $f_i = 8$ mm. The difference in the experimentally determined curve and the theoretically expected curve may be attributed to various experimental errors. The main source of experimental errors is the positioning errors of the lenses leading to errors in the distances d_1 and d_2, errors in collimation of the beam and tilt of components with respect to the optical axis of the system. The estimated value of focal length from the experimentally determined curve was $f_y = 8.36$ mm and $f_x = 8.3$ mm.

Figure 11.4 shows the phase estimated from the spatial frequency shown in Fig. 11.3. Figure 11.5 shows the plots corresponding to a lens of focal length 20 cm. The measurements were carried out for the same FRT order separation $\Delta\alpha = 0.006$, used in the previous case. As can be seen from the plots this fractional order separation leads to an erroneous detection of spatial frequency. For this case, the value of $b_2 = 0.8$ leads to a higher value of the lower bound as predicted by Eq. (11.30).

For completeness, in Fig. 11.6 we show the results for the case when there is no input signal. The measurements were carried out for the same FRT order separation $\Delta\alpha = 0.006$, used in the previous case and collimated illumination. The result therefore corresponds to noise and drifts in the system and places a lower threshold on performance.

11.8 Conclusion

In this chapter, we have described a method to reconstruct complete signal information—amplitude and phase—from two intensity measurements. The Fourier transform of the two intensity measurements at the input and output of a QPS may be interpreted as two slices in the AF plane. The angular separation between the two slices is determined by the QPS parameters. It is to be noted that some a priori information of the signal regarding the bandwidth is required to chose an appropriate QPS system that would ensure that the two intensity measurements does not lead to an erroneous signal estimation. We discussed an experiment performed using an

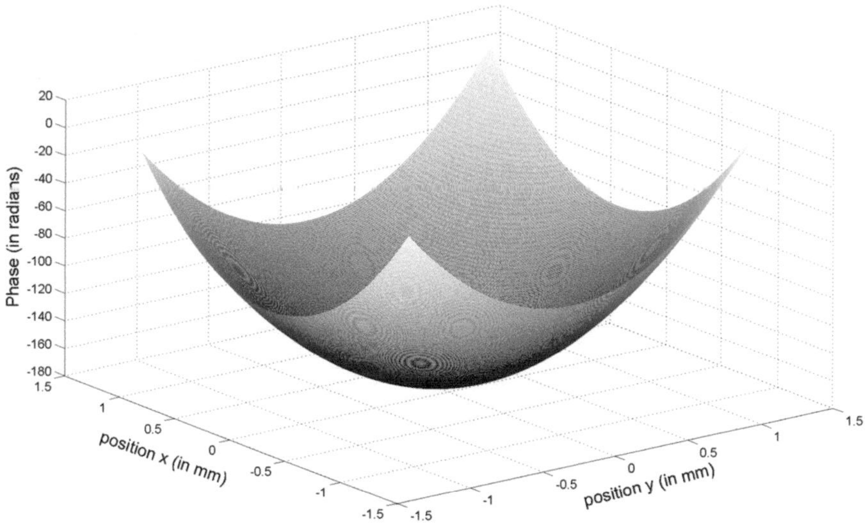

Fig. 11.4 Phase calculated from the estimated spatial frequency for 8 mm focal length lens

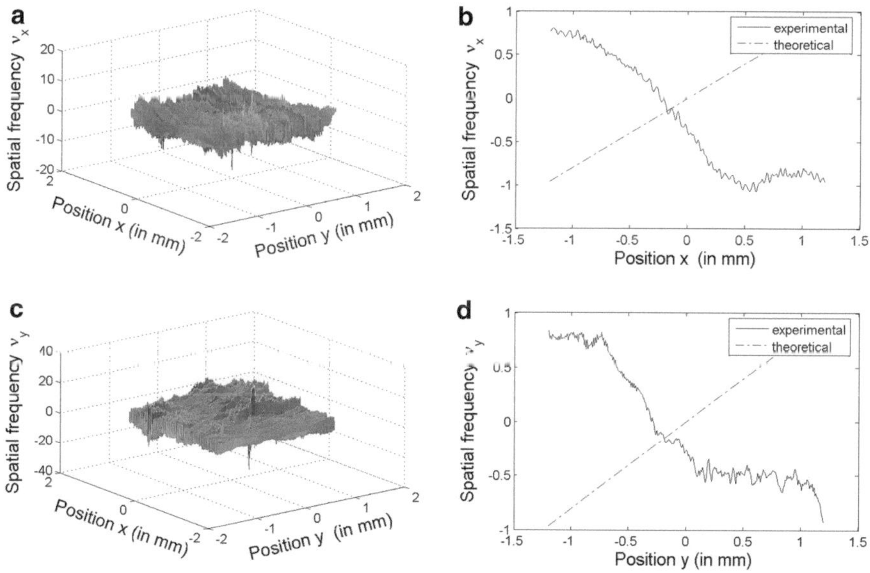

Fig. 11.5 Experimental results for 20 cm focal length lens: (**a**) spatial frequency ν_x (**b**) spatial frequency ν_x with y direction averaged out; (**c**) spatial frequency ν_y (**d**) spatial frequency ν_y with x direction averaged out. The *broken line* in (**b**) and (**d**) indicate the predicted spatial frequency corresponding to an 20 cm focal length lens

optical system that performs scale invariant FRT to retrieve the phase information of a quadratic phase signal generated using a lens.

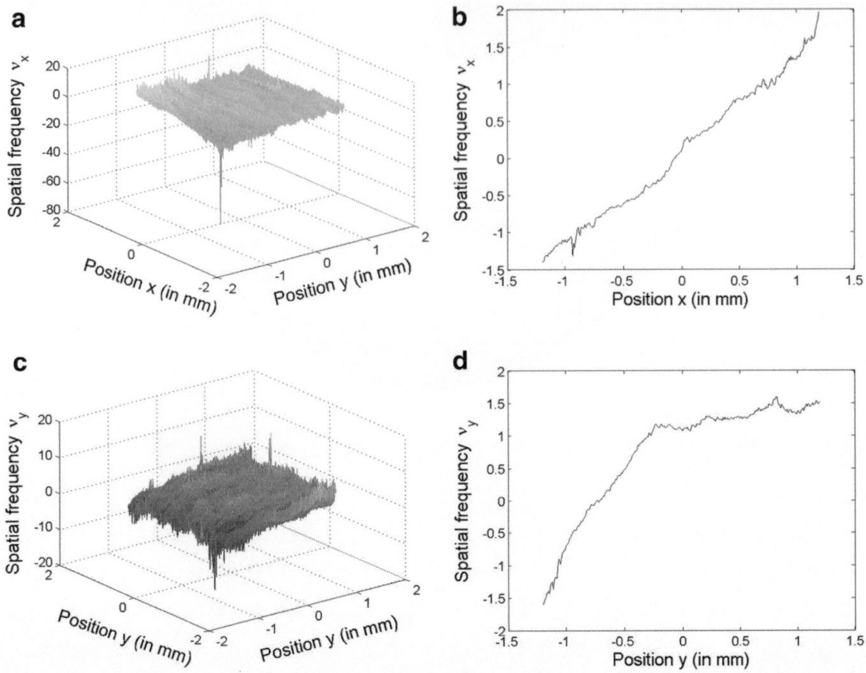

Fig. 11.6 Collimate beam input (**a**) spatial frequency ν_x (**b**) spatial frequency ν_x with y direction averaged out; (**c**) spatial frequency ν_y (**d**) spatial frequency ν_y with x direction averaged out

Appendix

The AF can be written as:

$$A(\overline{x},\overline{\nu}) = \int f\left(x+\frac{\overline{x}}{2}\right) f^*\left(x-\frac{\overline{x}}{2}\right)\exp\left(-j2\pi\overline{\nu}x\right)dx. \tag{11.A1}$$

$$\left.\frac{dA(\overline{x},\overline{\nu})}{d\overline{x}}\right|_{\overline{x}=0} = \int \left.\frac{d\left[f\left(x+\frac{\overline{x}}{2}\right)f^*\left(x-\frac{\overline{x}}{2}\right)\right]}{d\overline{x}}\right|_{\overline{x}=0}\exp\left(-j2\pi\overline{\nu}x\right)dx. \tag{11.A2}$$

Let $f(x) = A(x)\exp\left[i\varphi(x)\right]$, then

$$f\left(x+\frac{\overline{x}}{2}\right)f^*\left(x-\frac{\overline{x}}{2}\right) = A\left(x+\frac{\overline{x}}{2}\right)A\left(x-\frac{\overline{x}}{2}\right)\exp\left\{i\left[\varphi\left(x+\frac{\overline{x}}{2}\right)-\varphi\left(x-\frac{\overline{x}}{2}\right)\right]\right\}. \tag{11.A3}$$

Differentiating both sides with respect to $\overline{x}$and substituting $\overline{x} = 0$, we get:

$$\left.\frac{d\left[f\left(x+\frac{\overline{x}}{2}\right)f^*\left(x-\frac{\overline{x}}{2}\right)\right]}{d\overline{x}}\right|_{\overline{x}=0} = iI_0(x)\frac{d\varphi(x)}{dx}, \qquad (11.A4)$$

where $I_0(x) = |A(x)|^2$

$$\left.\frac{dA\left(\overline{x},\overline{\nu}\right)}{d\overline{x}}\right|_{\overline{x}=0} = i\int I_0(x)\frac{d\varphi(x)}{dx}\exp\left(-i2\pi\overline{\nu}x\right)dx. \qquad (11.A5)$$

References

1. M. Bertero, P. Boccacci, *Introduction to Inverse Problems in Imaging* (Institute of Physics Publishing, London, 1998)
2. R.W. Gerchberg, W.O. Saxton, A practical algorithm for the determination of the phase from image and diffraction plane pictures. Optik **35**, 237–246 (1972)
3. M.R. Teague, Deterministic phase retrieval: a Green's function solution. J. Opt. Soc. Am. **73**(11), 1434–1441 (1983)
4. M.R. Teague, Image formation in terms of transport equation. J. Opt. Soc. Am. A **2**(11), 2019–2026 (1985)
5. N. Streibl, Phase imaging by the transport equation of intensity. Opt. Commun. **49**(1), 6–10 (1984)
6. A. Barty, K.A. Nugent, D. Paganin, A. Roberts, Quantitative optical phase microscopy. Opt. Lett. **23**(1), 817–819 (1998)
7. M. Jayashree, G.K. Dutta, R.M. Vasu, Optical tomographic microscope for quantitative imaging of phase objects. Appl. Opt. **39**, 277–283 (2000)
8. F. Roddier, Wavefront sensing and the irradiance transport equation. Appl. Opt. **29**, 1402–1403 (1990)
9. K. Ichikawa, A.W. Lohmann, M. Takeda, Phase retrieval based on irradiance transport equation and the Fourier transport method. Appl. Opt. **27**(16), 3433–3436 (1988)
10. G.K. Datta, R.M. Vasu, Non-interferometric methods of phase estimation for application in optical tomography. J. Mod. Opt. **46**(9), 1377–1388 (1999)
11. D. Paganin, K.A. Nugent, Non-interferometric phase imaging with partially coherent light. Phys. Rev. Lett. **80**(12), 2586–2589 (1998)
12. T.E. Gureyev, A. Roberts, K.A. Nugent, Partially coherent fields, the transport-of-intensity equation, and the phase uniqueness. J. Opt. Soc. Am. A **12**(9), 1942–1946 (1995)
13. K.A. Nugent, X-ray non-interferometric phase imaging. J. Opt. Soc. Am. A **24**(2), 536–547 (2007)
14. J. Tu, S. Tomura, Wave field determination using tomography of the Ambiguity function. Phys. Rev. E **55**(2), 1946–1949 (1997)
15. J. Tu, S. Tomura, Analytical relation for recovering the mutual intensity by means of intensity information. J. Opt. Soc. Am. A **15**(1), 202–206 (1998)
16. D.F. McAlister, M. Beck, L. Clarke, A. Mayer, M.G. Raymer, Optical phase retrieval by phase-space tomography and fractional-order Fourier transforms. Opt. Lett. **20**(10), 1181–1183 (1995)

17. M.G. Raymer, M. Beck, D.F. McAlister, Complex-wavefield reconstruction using phase-space tomography. Phys. Rev. Lett. **72**(8), 1137–1140 (1994)
18. A. Semichaevsky, M. Testorf, Phase-space interpretation of deterministic phase retrieval. J. Opt. Soc. Am. A **21**(11), 2173–2179 (2004)
19. D. Dragoman, Redundancy of phase-space distribution functions in complex field recovery problems. Appl. Opt. **42**(11), 1932–1937 (2003)
20. T. Alieva, M.J. Bastiaans, Phase-space distributions in quasi-polar coordinates and the fractional Fourier transform. J. Opt. Soc. Am. A **17**(12), 2324–2329 (2000)
21. T. Alieva, M.J. Bastiaans, On fractional Fourier moments. IEEE Signal Process. Lett. **7**(11), 320–323 (2000)
22. T. Alieva, M.J. Bastiaans, L. Stankovic, Signal reconstruction from two close fractional Fourier power spectra. IEEE Trans. Signal Process. **51**(1), 112–123 (2003)
23. U. Gopinathan, G. Situ, T.J. Naughton, J.T. Sheridan, Non-interferometric phase retrieval using a fractional Fourier system. J. Opt. Soc. Am. A **25**(1), 108–115 (2008)
24. J. Frank, S. Altmeyer, G. Wernicke, Non-interferometric, non-iterative phase retrieval by Green's function. J. Opt. Soc. Am. A **27**, 2244–2251 (2010)
25. L. Preda, Fractional derivative applied to the retrieval of phase information from an interferogram. Appl. Phys. B **108**, 533–538 (2012)
26. M.J. Bastiaans, Application of Wigner distribution function in optics, in *Wigner Distribution—Theory and Applications in Signal Processing*, ed. by W. Mecklenbraüker, F. Hlawatsch (Elsevier Science, Amsterdam, 1997), pp. 375–426
27. J.T. Sheridan, R. Patten, Holographic interferometry and the fractional Fourier transformation. Opt. Lett. **25**, 448–450 (2000)
28. J.T. Sheridan, B.M. Hennelly, D.P. Kelly, Motion detection, the Wigner Distribution Function and the optical fractional Fourier transform. Opt. Lett. **28**, 884–886 (2003)
29. D.P. Kelly, J.E. Ward, U. Gopinathan, B.M. Hennelly, F.T. O'Neill, J.T. Sheridan, Paraxial speckle based metrology system with an aperture. J. Opt. Soc. Am. A **23**, 2861–2870 (2006)
30. B.M. Hennelly, J.T. Sheridan, Generalizing, optimizing, and inventing numerical algorithms for the fractional Fourier, Fresnel and linear canonical transforms. J. Opt. Soc. Am. A **22**(5), 917–927 (2005)
31. L.Z. Cai, Y.Q. Yang, Optical implementation of scale invariant fractional Fourier transform of continuously variable orders with a two-lens system. Opt. Laser Technol. **34**, 249–252 (2002)

Chapter 12
Analyzing Digital Holographic Systems with the LCT

Damien P. Kelly and John T. Sheridan

Abstract Digital holography is an electronic means of measuring the complex amplitude of an optical wavefield using CCD or CMOS arrays. With an appropriate reconstruction algorithm the intensity values registered by the pixels produce images of a particular (object) scene. These holographic systems are very sensitive to even small vibrations or deformations of an object, mainly due to the phase information that is also recovered by the measurement. Hence these systems are useful in a wide array of different metrology problems. It is important that we somehow quantify the information that can be recovered with such a detection scheme; better if we can provide a theoretical framework to optimize an optical system design for a given metrology problem. In this manuscript we show how the Linear Canonical Transform can fulfill this role and can optimally match the space-bandwidth product (SBP) of a signal to be measured with the SBP of a CCD/CMOS detector array. We provide formulae that determine the performance of generalized holographic optical systems (containing lenses and sections of free space), taking into account the finite extent of the CCD array, the size of the pixels, and the spacing between them. Some illustrative examples are presented, with associated numerical simulations.

12.1 Introduction

Recording holograms, using electronic detectors, has been of interest to the optical community, since the first publications on this topic appeared, nearly 45 years ago [1–3]. From the early nineties improvement in digital cameras revived interest in modern digital holography, see for example [4–17]. Holography provides a means of recovering the amplitude and phase of an optical wavefield by mixing the object wave of interest with a known reference wave and recording the resulting intensity

D.P. Kelly (✉)
Technische Universität Ilmenau, Ilmenau, Germany
e-mail: damien-peter.kelly@tu-ilmenau.de

J.T. Sheridan
School of Electrical and Electronic Engineering, University College Dublin, Dublin, Ireland
e-mail: john.sheridan@ucd.ie

J.J. Healy et al. (eds.), *Linear Canonical Transforms*, Springer Series in Optical Sciences 198, DOI 10.1007/978-1-4939-3028-9_12

distribution. When the reference and object wave interfere with each other the resulting intensity field at the camera plane can be written as [12–14]:

$$I_c(x) = |u(x) + u_r(x)|^2,$$
$$I_c(x) = I(x) + I_r(x) + c(x) + c^*(x), \quad (12.1)$$

where $I(x)$ and $I_r(x)$ are the object and reference field intensities, respectively, while $c(x) = u(x)u_r^*(x)$ and represents the real image term that we are interested in recovering. The "*" indicates a complex conjugate operation. In digital holography a CCD or CMOS camera is used to record this intensity distribution replacing the traditional holographic plate. A significant advantage of the digital approach is that the resulting electronic hologram can be processed in real time to give reconstructed images of the amplitude and phase of the object of interest. Many different optical setups for recording digital holograms exist such as the Fresnel or lens-less Fourier systems [16, 18, 19]. Importantly, holograms recorded with these various optical setups, exhibit different properties, and so it is useful to provide a general theoretical framework for the analysis, and comparison, of different optical systems [16, 20]. In this chapter the use of the Linear Canonical Transform (LCT) to provide such a general framework is reviewed and explored. Using this transform the imaging performance of different setups can be compared, and indeed this approach also provides a means of optimizing a given system to perform a particular imaging task.

The chapter is organized as follows: In Sect. 12.2.1, we begin with a description of the optical systems examined and how an LCT-based analysis can include the sphericity of the reference wave, and take into account the effect of illuminating the object with either a converging or diverging spherical wave. We show how several different optical systems including the Fresnel, Fourier, and inline Gabor holographic systems can all be described in terms of the LCT. In Sect. 12.2.2 we discuss the effect of the camera on recovering the third term in Eq. (12.1), in particular for the case when the Phase Shifting Interferometric (PSI) techniques have been used to remove the DC and twin image terms [21–24]. It is assumed that at the camera plane a series of lossy operations are performed: (1) the intensity distribution recorded is limited by the finite extent of the camera which has significant implications for the spatial frequencies that can be recovered; (2) the averaging of the light intensity incident on the camera, by the finite size of the camera pixels, tends to attenuate by differing amounts spatial frequencies present in the object signal generally resulting in poorer imaging performance; and finally (3) the effect of sampling the captured intensity produces an infinite series of replicas in the reconstruction plane. If PSI techniques have been employed, then the sole effect of the sampling operation is to define the distance between neighboring replicas in the reconstruction plane. Provided that this distance is less than the actual extent of the object field, then the recoverable detail in the numerically reconstructed field is not affected by the sampling operation. Hence it is possible to recover frequencies far higher than the Nyquist limit [13, 14, 16, 25–28]. In Sect. 12.2.3, we examine the role of the DC and twin image terms, see Eq. (12.1), and how they impact on the performance of the optical system. It is demonstrated that if an off-axis reference is used for single-shot capture then the sampling rate of the camera plays a different role to that described in Sect. 12.2.2, and tends to be the dominant factor in determining the spatial frequencies of the signal that can be recovered (see also Section 3, in Ref. [14]). In Sect. 12.3, we present the result of some numerical simulations that highlight several of the theoretical issues addressed in the previous section. We finish with a brief conclusion.

12.2 Analysis

In this section we examine how the LCT may be used to describe the behavior of several different types of digital holographic imaging systems.

12.2.1 A Description of the Problem

In Fig. 12.1, we present a schematic depicting a typical optical setup used to record a digital hologram. Laser light is spatially filtered to improve the spatial coherence of the light source (and hence the contrast of the recorded holograms), before being collimated to form an approximately flat plane wave segment. This plane wave segment passes through the first beam splitter (BS1) where it is separated into what becomes the reference and object beams. A lens can be inserted in the reference arm,

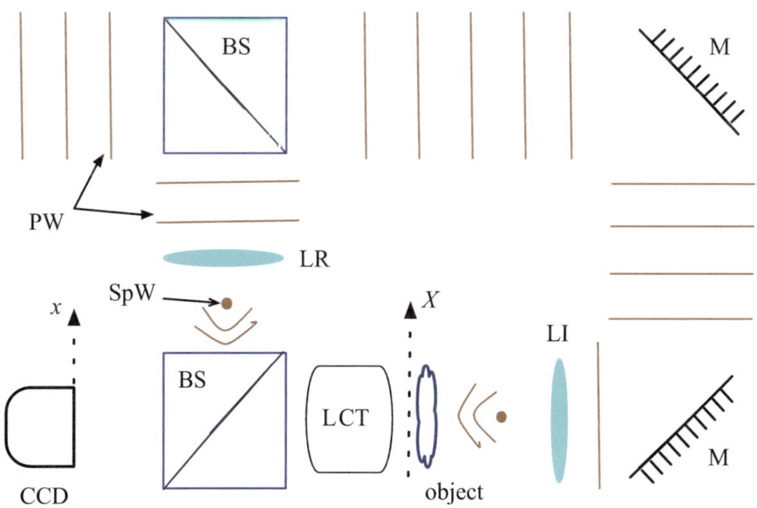

Fig. 12.1 A schematic depicting an LCT based digital holographic setup: *BS* beam splitter, *M* mirror, *LR* reference field lens, *LI* illumination field lens

see Fig. 12.1, to vary the curvature of the reference wave. In the object arm the plane wave segment illuminates and interacts with the sample that we wish to examine. Here we assume that we are operating in the paraxial regime and hence that the Fresnel transform can be used to describe the process of free-space diffraction. We also assume that the "thin lens" approximation holds and that a lens can be described in the manner outlined by Goodman in Chapter 5 of Ref. [29]. The illuminating plane wave segment interacts with the sample of interest in a complicated manner to produce a scattered electromagnetic field in the 3D volume after the object. In the analysis presented here we assume that this 3D field in (x, y, z) is generated by a 2D complex amplitude distribution in a plane $(x, y, z=0)$ that lies immediately behind the sample.

In what follows, for the sake of brevity we provide a 1D analysis (in x), and note that the results can be extended in a straightforward manner to the 2D case. Hence the field in a plane just after the sample can be described mathematically as:

$$U(X) = U_i(X)T(X), \tag{12.2}$$

where $U_i(X)$ is the wavefield that illuminates the sample which is modeled by the transmittance function, $T(X)$. We now wish to find a relationship between the field $U(X)$ and the diffracted object wavefield, $u(x)$, that is incident on the camera. Our analysis allows for the possibility that a series of lenses (of infinite extent) and sections of free space can be inserted between plane X and the CCD plane, see Fig. 12.1. This freedom means that the signal $u(x)$ can be processed (by the optical system) prior to capture. In this way the spatial and spatial frequency extents of the signal can be controlled, so that the impact of the CCD array (e.g. any filtering operations) can be minimized and the maximum amount of information about $U(X)$ preserved.

In geometrical optics, ray-tracing procedures are used to design optical systems. This ray-tracing procedure can be described using a combination of different matrices, each describing the effect of a specific sequential part of the optical system. In the next section this formulism is used to describe: (1) the properties of the optical system as they relate the object (input) plane X and the camera (output) plane, and (2) the effect of the illumination and reference beams.

12.2.1.1 Bulk Optical System

Using the geometrical optics matrix formalism, the effect of a section of free space propagation of length (z), i.e. the Fresnel transformation (FST), is described using the following matrix:

$$M_{\mathrm{FST}}^{z} = \begin{pmatrix} 1 & z \\ 0 & 1 \end{pmatrix}. \tag{12.3}$$

The matrix describing a converging lens of focal length, f, is given by:

$$M_{\mathrm{LENS}}^{f} = \begin{pmatrix} 1 & 0 \\ -1/f & 1 \end{pmatrix}. \tag{12.4}$$

A diverging spherical beam can be described using Eq. (12.4) by simply changing the sign of the focal length, f. These matrices can be combined, to describe the effect of a series of lenses and sections of free space, giving an overall optical system matrix. For example, a Fresnel transform, z_1, followed by a lens, f_1, and another Fresnel transform, z_2, produces the following total matrix:

$$\begin{pmatrix} A_s & B_s \\ C_s & D_s \end{pmatrix} = M_{\mathrm{FST}}^{z1} M_{\mathrm{LENS}}^{f1} M_{\mathrm{FST}}^{z1}. \tag{12.5}$$

For the special case when $z_1 = z_2 = f_1 = f$, Eq. (12.5) describes an optical Fourier transform (OFT) system:

$$M_{\mathrm{OFT}}^{f} = \begin{pmatrix} 0 & f \\ -1/f & 0 \end{pmatrix}. \tag{12.6}$$

Many other types of optical systems can be described in a similar manner, including the optical fractional Fourier transform (OFRT) [15]. Following the formalism of Collins in Ref. [30], these ray-tracing matrices with unit determinants are related to corresponding diffraction-type integrals. In the most general linear lossless case the integral corresponds to the LCT. The resulting relationship between the fields in the object, $U(X)$, and camera planes, $u(x)$, is:

$$u(x) = \mathrm{LCT}_{A_s B_s D_s}\{U(X)\}(x) = \frac{1}{\sqrt{j\lambda B_s}}$$

$$\int_{-\infty}^{\infty} U(X) \exp\left[\frac{j\pi}{\lambda B_s}\left(D_s x^2 - 2xX + A_s X^2\right)\right] dX, \tag{12.6}$$

where the parameters A_s, B_s, and D_s are those appearing in the matrix derived using the geometrical analysis of the optical system. For example, assuming that we have both plane wave illumination and a plane wave reference beam, a Fresnel relationship between the input plane and the camera plane can be described by setting the values of *ABCD* to those given in Eq. (12.3). In the next Sect. 12.2.1.2 we show how variations in the curvature of the illumination and reference beams can be included in the overall system *ABCD* parameters.

12.2.1.2 The Illumination and Reference Beams

Examining Fig. 12.1, we note that we can change the form of the reference wave, $u_r(x)$, by inserting a lens into the reference arm. In this case:

$$u_r(x) = \exp\left(\frac{j\pi}{\lambda z_r}x^2\right), \tag{12.7}$$

where z_r is the distance to the focus of the lens LR, see Fig. 12.1, i.e. the spatial location where a plane wave segment incident on lens LR would be focused. Similarly we can place another lens, LI, see Fig. 12.1, in the illumination beam to change the curvature of the wavefield incident on the sample, giving:

$$U_i(X) = \exp\left(\frac{-j\pi}{\lambda z_i}X^2\right), \tag{12.8}$$

where z_i is the distance to the focus of the lens LI. Inserting Eqs. (12.2) and (12.8) into Eq. (12.6) and using the conjugate of Eq. (12.7), we can directly write an expression for the $c(x)$ term in Eq. (12.1), as:

$$c(x) = \exp\left(-\frac{j\pi}{\lambda z_r}x^2\right) \mathrm{LCT}_{A_sB_sD_s}\left\{\exp\left(\frac{j\pi}{\lambda z_i}X^2\right)T(X)\right\}(x) = \mathrm{LCT}_{ABD}\left\{T(X)\right\}(x). \tag{12.9}$$

From Eq. (12.9), we note that the function $T(X)$ is multiplied by a lens function which can be included using Eq. (12.4). Similarly we can also account for the conjugate of the reference beam using Eq. (12.4). Hence the overall system *ABCD* parameters are given by:

$$\begin{pmatrix} A & B \\ C & D \end{pmatrix} = \begin{pmatrix} 1 & 0 \\ -1/z_r & 1 \end{pmatrix}\begin{pmatrix} A_s & B_s \\ C_s & D_s \end{pmatrix}\begin{pmatrix} 1 & 0 \\ -1/z_i & 1 \end{pmatrix}. \tag{12.10}$$

The LCT-based model for digital holography systems presented in Eq. (12.10) is suitable for a wide range of different types of optical arrangements. In the following Sect. 12.2.1.3 we will examine three special cases to illustrate this feature.

12.2.1.3 Three Special Systems

In this subsection we will examine the *ABCD* parameters for 3 different optical systems: System (1), Fresnel; System (2), Fourier; and System (3), inline Gabor. In the case of the Fresnel setup, (1), plane waves are used to illuminate the object under investigation and for the reference field. In this instance both z_i and z_r are set to infinity, and a Fresnel transform relates the fields at the sample and camera planes. Therefore the matrix describing the system, MS1, is given by:

Fig. 12.2 A schematic depicting a typical inline Gabor holographic setup LCT

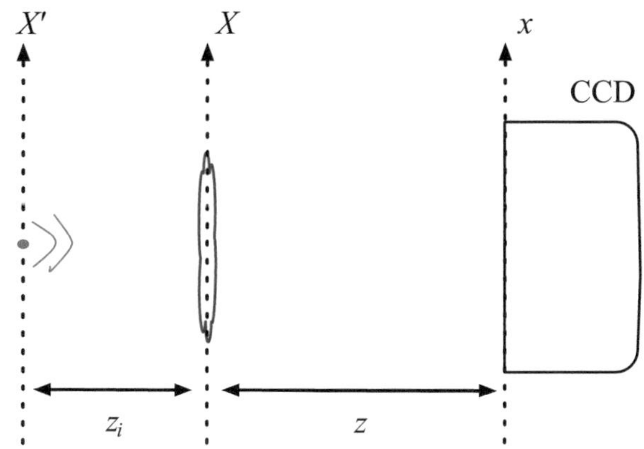

$$\begin{pmatrix} A_{\text{MS1}} & B_{\text{MS1}} \\ C_{\text{MS1}} & D_{\text{MS1}} \end{pmatrix} = \begin{pmatrix} 1 & z \\ 0 & 1 \end{pmatrix}. \tag{12.11}$$

For the case of Fourier system, (2), a spherical reference wave is used. A thin lens is introduced which produces a diverging spherical wave, whose focal plane is located at the same distance from the CCD, as from the object plane. We note that this technique is sometimes referred to as a *lens-less Fourier system* since the diverging spherical reference wave can also be formed by light emerging from a point-like source, e.g. the tip of an optic fiber. In both instances the reference wave is described using Eq. (12.7) and the overall system parameters change to become MS2, i.e.:

$$\begin{pmatrix} A_{\text{MS2}} & B_{\text{MS2}} \\ C_{\text{MS1}} & D_{\text{MS1}} \end{pmatrix} = \begin{pmatrix} 1 & 0 \\ -1/z_r & 1 \end{pmatrix} \begin{pmatrix} 1 & z \\ 0 & 1 \end{pmatrix} = \begin{pmatrix} 1 & z \\ -1/z_r & 1 - z/z_r \end{pmatrix}. \tag{12.12}$$

Let us now consider our third case, (3), the inline Gabor holographic system [11]. Such a system is depicted in Fig. 12.2.

Here a spherical wave illuminates the sample that is assumed to be a weakly scattering transparent object, e.g. a small biological sample. Making such a weak phase approximation allows us to write $T(X)$ as:

$$T(X) = \exp\left[j\varphi(X)\right] \approx 1 + \tilde{T}(X), \tag{12.13}$$

where $\tilde{T}(X) = j\varphi(X)$. From Eq. (12.13) we see that two important things happen: First, the illuminating beam passes through the sample negligibly changed, and is then incident on the camera where it serves as a reference wave. Second the $\tilde{T}(X)$ term acts to scatter the incident illumination field creating a weak secondary field that contains the phase information about the sample. This is the object field, which is illuminated with a diverging spherical wave. The situation where the illuminating field produces both the reference field and the very weak scattered field follows from the important approximation in Eq. (12.13). We can describe a Gabor setup with the following mathematical expression:

$$\begin{pmatrix} A_{\text{MS3}} & B_{\text{MS3}} \\ C_{\text{MS3}} & D_{\text{MS3}} \end{pmatrix} = \begin{pmatrix} 1 & 0 \\ -\frac{1}{z_i+z} & 1 \end{pmatrix} \begin{pmatrix} 1 & z \\ 0 & 1 \end{pmatrix} \begin{pmatrix} 1 & 0 \\ \frac{1}{z_i} & 1 \end{pmatrix}. \tag{12.14}$$

Above we have shown how to use the LCT to systematically derive the system parameters for three different types of commonly used holographic systems. Using this approach many other types of holographic setups can be designed having specific system parameters. These parameters play a crucial role in defining the performance of a given holographic setup. As noted at the CCD plane a series of filtering and sampling operations are performed that remove information from the signal we wish to recover. By varying the system parameters prior to holographic capture, we can change how the power of the object signal is distributed in phase space, and control this process to recover as much information about that signal as possible. To understand how to do this we more closely examine the role of the digital CCD array in holographic imaging systems, in the next section.

12.2.2 The Role of the Digital Camera

The reference and object wavefields overlap and interfere with each other at the camera plane, and the resulting intensity pattern is recorded, see Eq. (12.1). Several filtering operations occur during this process. First the finite extent of the camera means that only a portion of the intensity field is recorded. Second the camera pixels themselves are of finite extent and record the average light intensity incident upon their active area. Third and finally, the intensity field is sampled returning a 2D array (x and y) of real values to the user. We express these mathematically as follows:

$$W^n = \beta p_L(x)\delta_\Delta(x)\left[I_c(x) * p_\gamma(x)\right], \tag{12.15}$$

where β is a constant of proportionality which is neglect from now on. The functions $p_L(x)$ and $p_\gamma(x)$ represent the effects of the finite extents of the camera and pixels, respectively. We define them as:

$$p_Q(x) = \begin{cases} 1 & \text{when } |x| < Q \\ 0 & \text{otherwise} \end{cases}. \tag{12.16}$$

The finite extents of the camera and each individual pixel are $2L$ and 2γ, respectively. The comb function $\delta_\Delta(x)$ is defined as:

$$\delta_\Delta(x) = \sum_{n=-\infty}^{\infty} \delta\left(x - n\Delta\right), \tag{12.17}$$

where Δ is the spacing between the centers of adjacent pixels and where $\delta(x)$ is a Dirac delta functional. We now wish to gain insights into how these filtering and sampling operations act together to modify the information captured. To do this we follow the approach outlined in Ref. [16], whereby a specially selected function is chosen to be the input to the system. This signal has finite extent and well-defined spatial frequency content. Therefore by examining the effects of the capture process at the digital camera on we can see how the field of view in the reconstruction plane varies as a function of the sampling rate at the camera plane. We can also determine the range of spatial frequencies that can be recovered by the imaging system. We choose our input signal to be:

$$U(X) = \frac{1}{\sqrt{\alpha_i}} \exp\left(\frac{-X^2}{\alpha_i{}^2}\right) \cos\left(2\pi f_x X\right). \tag{12.18}$$

From Eq. (12.18) we see that most of the input signal energy is contained within the range $-2\alpha_i < X < 2\alpha_i$ and is centered at f_x, spatial frequency component. Substituting Eq. (12.18) into Eq. (12.6) and integrating produces the following result [16]:

$$u(x) = u_{f_x}(x) + u_{-f_x}(x), \tag{12.19}$$

where

$$u_{f_x}(x) = K_c \exp\left[-\frac{(x-\lambda B f_x)^2}{\alpha_c^2}\right] \exp\left[j\left(\phi_c^l x + \phi_c^q x^2\right)\right], \tag{12.20}$$

and where K_c is a complex constant. $\phi_c^l x$ and $\phi_c^q x^2$ are linear and quadratic phase terms, respectively. From Eq. (12.20), we see that the power of the input signal has been equally split between two Gaussian limited functions centered at $x = \lambda B f_x$ and $x = -\lambda B f_x$, each of extent, α_c, where

$$\alpha_c = \left(\frac{\pi\alpha_i}{\sqrt{A^2\pi^2\alpha_i^4 + B^2\lambda^2}}\right)^{-1}. \tag{12.21}$$

Whether a particular spatial frequency can be recovered depends on both the filtering effect of the finite size pixels and on whether or not the power associated with a spatial frequency has "walked off" the optical axis to such an extent that it is no longer incident on the CCD array. Let us first examine the role of the finite camera aperture. Since we know that the power associated with a particular spatial frequency is mapped to a definite spatial location in the camera plane, we must ensure that this spatial location lies within the CCD aperture. This condition is satisfied when:

$$L \geq \lambda B f_x + 2\alpha_c. \tag{12.22}$$

We now examine the role that averaging over the finite size pixel area has on the light incident upon them. This is modeled as a convolution operation in the spatial domain, see Eq. (12.15). The impact of this operation is most easily analyzed in the Fourier domain. Our analysis allows us to illustrate the flexibility of the LCT approach used here, since we can conveniently describe the effect of a general LCT operation (that describes the physical LCT system), followed by a Fourier transform as:

$$\begin{pmatrix} C/\lambda & D/\lambda \\ -A\lambda & -B\lambda \end{pmatrix} = \begin{pmatrix} 1/\lambda & 0 \\ 0 & -\lambda \end{pmatrix} \begin{pmatrix} 0 & 1 \\ -1 & 0 \end{pmatrix} \begin{pmatrix} A & B \\ C & D \end{pmatrix}. \tag{12.23}$$

The extent of the signal in the Fourier domain is given by $2BW$ where:

$$BW = Df_x + 2\tilde{\alpha}_c, \tag{12.24}$$

and

$$\tilde{\alpha}_c = \left(\frac{\pi\alpha_i}{\sqrt{(C/\lambda)^2 \pi^2 \alpha_i^4 + D^2}} \right)^{-1}. \tag{12.25}$$

Now that we understand how the signal's power is distributed in the spatial frequency domain we need to examine how the pixel acts to attenuate and eliminate spatial frequencies. The Fourier transform of $p_\gamma(x)$ is:

$$\tilde{p}_\gamma(v) = 2\gamma \ \sin\ c\,(2\pi\gamma v)\,, \tag{12.26}$$

where $\sin c(x) = \sin(x)/x$. In the text we use a "tilde" over a function and the use of the spatial frequency variable, v, to indicate that we are in the Fourier domain. Assuming a fill factor of 100 %, i.e. all the pixel area is active, then $\Delta = 2\gamma$. Examining Eq. (12.26), we see that $\tilde{p}_\gamma\,[1/\,(2\gamma)] = 0$, which was used to define the maximum spatial frequency that could be recovered. Here we assume that spatial frequencies higher than this limit are so severely attenuated that recovery is not possible. Hence we have another constraint that can be summarized as the following rule of thumb,

$$BW \leq 1/\,(2\gamma)\,. \tag{12.27}$$

In relation to Eq. (12.27) we note that the bandwidth, BW, can be controlled, by varying the parameters C and D. Significantly such variations can be quite easily achieved by simply varying the curvature of the reference and illumination beams.

The important role of sampling in general LCT systems has been discussed by several authors, see Refs. [31–35]. In Ref. [14] experimental results were presenting showing that it was possible to recover spatial frequencies higher than the Nyquist limit in a Fresnel inline holographic setup. Based on our results we can now restate the results derived elsewhere, see Refs. [8, 13]. When the term $c(x)$ is sampled at the camera plane at a rate of Δ, then a series of replicas are generated in the reconstruction domain that are separated from each other by $\lambda B_N/\Delta$. Provided that the input extent of the signal is less than this value, replicas in the reconstruction domain will not overlap with each other. This condition can be expressed with the following equation:

$$4\alpha_i \leq \lambda B_N/\Delta, \tag{12.28}$$

where B_N is as defined in Eq. (12.29). To reconstruct the real image term requires an inverse discrete LCT implementation. The parameters for this numerical LCT operation are given by:

$$\begin{pmatrix} A_N & B_N \\ C_N & D_N \end{pmatrix} = \begin{pmatrix} D & -B \\ -C & A \end{pmatrix}. \tag{12.29}$$

In this section we have now discussed the limits on the performance of general optical systems imposed by the use of a digital camera to capture the output field. Some of the implications of these theoretical results will be examined in Sect. 12.3 where numerical simulations are presented to compare and contrast the performance of the three different holographic systems discussed in Sect. 12.2.1.3.

12.2.3 The Twin Image and DC Terms

In the context of our analysis we now examine how the twin image and DC terms affect the imaging performance of holographic systems.

12.2.3.1 Twin Image Term

We first note that the model used to describe the role of the camera is linear and hence we can consider how the filtering and sampling operations act on each term in Eq. (12.1) individually. From this statement we conclude that spatial frequencies present in the twin image term, the fourth term in Eq. (12.1), will be filtered in the same manner as the real image term. From Eq. (12.9) we find that $c^*(x)$ can be expressed as:

$$c^*(x) = \mathrm{LCT}_{\overline{ABD}}\{T^*(X)\}(x), \tag{12.30}$$

where

$$\begin{pmatrix} \overline{A} & \overline{B} \\ \overline{C} & \overline{D} \end{pmatrix} = \begin{pmatrix} A & -B \\ -C & D \end{pmatrix}. \tag{12.31}$$

The flat line over the parameters indicates that we are referring to the twin image term. If this term is present during the numerical reconstruction then the final reconstructed form of the twin image term, $\bar{U}_s(X)$ is given by:

$$\overline{U}_s(X) = \text{LCT}_{\overline{A_N B_N D_N}} \{c^*(x)\} (X), \tag{12.32a}$$

where

$$\begin{pmatrix} \overline{A}_N & \overline{B}_N \\ \overline{C}_N & \overline{D}_N \end{pmatrix} = \begin{pmatrix} D & -B \\ -C & A \end{pmatrix} \begin{pmatrix} A & -B \\ -C & D \end{pmatrix} = \begin{pmatrix} AD + BC & -2BD \\ -2AC & BC + AD \end{pmatrix}. \tag{12.32b}$$

Eqs. (12.32a) and (12.32b) imply that the reconstructed twin image will be defocused and smeared out over the reconstructed plane, where it acts as a coherent noise source. Exceptions to this include "lens-less" Fourier and object arm Fourier transform systems, see Refs. [15, 20]. In both these systems the real and twin images appear as focused and structured distributions.

We would like to make one final comment about the numerical reconstruction of the twin image term. As was noted in the previous section, the sampling operation that is performed in the CCD plane will define the spatial separation between the resulting replicas that lie in the (numerical) reconstruction plane. This replica distance or reconstruction window is given by $\lambda B_N/\Delta$. In some cases, for example, the inline Gabor setup, the extent of the twin image term will in fact be greater than this replica separation distance. For our example we may estimate the extent of the twin image, TE, from the following formula:

$$TE = \lambda \overline{B}_N f_x + 2\overline{\alpha}_c, \tag{12.33}$$

where

$$\overline{\alpha}_c = \left(\frac{\pi \alpha_i}{\sqrt{\overline{A}_N^2 \pi^2 \alpha_i^4 + B_N^2 \lambda^2}} \right)^{-1}. \tag{12.34}$$

The effect of this is that some of the power associated with the reconstructed twin image will move outside this reconstruction window to cause aliasing in higher order replicas. We note however that the numerical algorithms used to calculate the LCT distributions conserve power over the output reconstruction window, i.e. over $\lambda B_N/\Delta$. This means that when power associated with a reconstructed twin image moves outside the reconstruction window, an equal amount of power from higher order replicas will move into the reconstruction window causing significant aliasing but also ensuring that the total power over the reconstruction window remains constant. A similar effect is discussed in Section 10.2.3 of Ref. [17]. In many practical implementations of the Gabor inline setup this does not lead to significant distortion of the reconstructed hologram as the power of the twin is spread over the whole reconstruction window acting as a low constant background noise.

12.2.3.2 DC Image Terms

The DC image terms also act to degrade the quality of any reconstructed holograms if steps are not taken to remove them. In some instances it is straightforward to remove the $I_r(x)$ term. For example in a Frensel system this can be achieved by using a narrow filter in the Fourier domain that zeroes the values of a small number of pixels at and around the DC spatial frequency location. Due to the nature of the $I_r(x)$ most of its power will be concentrated in a small region and hence this simple filtering operation is quite effective. In the case of a Fourier-type system this removal can be performed in the reconstruction domain using a digital notch filter (a similar approach was taken in Ref. [24]). Other cases can be more problematic since the power of the reference wave will not necessarily be localized in a Fourier domain; however, in this case different phase space filtering strategies may be adopted.

Removing the object intensity, $I(x)$, tends to be a more complicated problem. This is because $I(x)$ tends to have a structured distribution whose power tends not be localized in any phase space plane. This makes notch filtering quite difficult to implement effectively. One strategy that can be employed is to reduce the power of the object arm relative to the reference arm. This will lead to a reduction in the power in $I(x)$, and when analyzing inline Gabor setups this term is often neglected entirely for this reason.

This approach will also work in a standard Fresnel-type system. The relatively strong reference wave acts to preserve the information contained in the interference terms, $c(x)$ and $c^*(x)$. However as the strength of object wave is decreased relative to the reference wave, it has been observed experimentally that the quality of the reconstructed hologram (particularly the phase distribution) will reduce [36].

Another approach to removing these two terms is to record $I(x)$ and $I_r(x)$ separately and to numerically subtract these intensity distributions from Eq. (12.1) leaving the interference terms as the remainder. While this approach does work if done carefully it can lead to an increase in the noise level in the holographic reconstructions [36].

12.2.3.3 Off-Axis Holography

The technique proposed by Leith and Upatnieks to overcome the difficulties encountered when using the inline geometry is to employ an off-axis reference wave to spatially separate the object, twin, and DC terms in the Fourier plane [37]. This approach allows for single-shot holographic capture. An off-axis reference wave contains a linear phase term and can be expressed as:

$$u_{or}(x) = u_r(x)\exp(j2\pi\kappa x), \tag{12.35}$$

where κ is the carrier frequency. Replacing $u_r(x)$ in Eq. (12.1) with $u_{or}(x)$ from Eq. (12.35), and making use of the Fourier shift theorem, the Fourier transform of $I_c(x)$ is given by:

$$\tilde{I}_c(v) = \tilde{I}_r(v) + \tilde{I}(v) + \tilde{c}\,(v-\kappa) + \tilde{c}^*\,(v-\kappa). \tag{12.36}$$

From Eq. (12.36) we see that the twin and real image terms are now separated from each other in the Fourier domain. By digitally filtering out the unwanted terms we can recover the desired real image information. We must however make an assumption about the nature of the $c(x)$ term in Eq. (12.1), namely that its spatial frequency extent, $2BW$, is bounded. The $c^*(v)$ term also has a spatial frequency extent of $2BW$. Since $\tilde{I}(v) = c(v) * c^*(v)$, the spatial frequency extent of $\tilde{I}(v)$ is $4BW$.

Because we use a single-shot capture to spatially separate the individual terms in the Fourier domain the sampling rate at the camera plane limits the total spatial frequency extent to $1/\Delta$. Thus for an off-axis arrangement the maximum spatial frequency extent that can be recovered is lower and bounded by:

$$f_{\max}^{\text{offaxis}} \leq \frac{1}{4T}. \tag{12.37}$$

12.3 Numerical Simulations

In this section we present some numerical simulations that illustrate several different features that arise when imaging the same object using the three holographic systems that were discussed in Sect. 12.2.1.3: (1) Fresnel system, (2) "lens-less" Fourier system, and (3) an inline Gabor setup.

In order to compare the relative imaging performance of the three systems we must choose a realistic set of physical system parameter values. We proceed by examining their ability to image a point source located a distance $z_{\text{ps}} = 14$ mm from a digital CCD array, at a wavelength of $\lambda = 633$ nm. We also assume that $c(x)$ has been perfectly recovered using PSI. For System (1), the reference wave is a flat plane wave normally incident on the camera. In System (2) the reference wave is a spherical wave emerging from a point located at $z_{r2} = 14$ mm. Finally in System (3) the point source used to illuminate the object is located at a distance of $z_i = 40$ mm behind the object, while the reference wave is located a distance $z_{r3} = z_i + z_{\text{ps}}$ from the camera plane. Using these values we can calculate the $ABCD$ matrix for each system. The results are presented in Table 12.1.

Since we expect the image of a point source to be relatively well localized in space in the reconstruction domain, it will in general be well separated from any neighboring replicas. Given a camera with 1200 pixels, separated from each other by $\Delta = 10$ μm with a 100 % fill factor, the camera extent is $2L = 12$ mm. This indicates that replicas arising due to the numerical reconstruction process are located a distance of $\lambda z_{ps}/\Delta$ 886 mm from each other in the reconstruction domain. To estimate what might limit the imaging performance for each system we return to Eqs. (12.22), (12.24) and (12.27), and use them to determine the maximum spatial frequency limits imposed by each filtering effect, i.e. the camera's finite extent and the finite pixel size.

Neglecting the α_c term from Eq. (12.22), the maximum spatial frequency that can be imaged is:

$$f_{\max}^{L} = \frac{L}{\lambda\, z_{ps}}. \tag{12.38}$$

Once again neglecting the $\tilde{\alpha}_c$ term in Eq. (12.24), and noting that $Df_{\max}^{\gamma} = BW$ and $\Delta = 2\gamma$, Eq. (12.27) can be rearranged to give:

$$f_{\max}^{\gamma} = \frac{1}{2\gamma D}. \tag{12.39}$$

In Table 12.1 values for Eqs. (12.38) and (12.39), for each system are presented. For the system parameter values examined the averaging of the light intensity by the pixels is the dominant factor limiting the system performance for both Systems 1 and 3. System 2 performs best, since $D = 0$ for this special case and therefore pixel averaging does not limit the system performance at all. Instead it acts to attenuate the complex amplitude of the signal in the reconstruction plane [16, 17, 20].

In Figs. 12.3, 12.4, and 12.5 we plot the intensity of the point spread functions, $h(X, z_{nf})$, for each of these systems. When $z_{nf} = 0$, we are in the reconstructed object plane and by varying z_{nf} we can numerically propagate the signal field to either side of this plane.

From Fig. 12.3, we can see that the reconstructed point spread function intensity distribution looks unusual in that the maximum intensity lies a small distance from the object plane. This occurs because of the complex interaction of the pixel

Table 12.1 System parameters for the numerical results presented in Figs. 12.3, 12.4 and 12.5

System	1	2	3
$\begin{pmatrix} A & B \\ C & D \end{pmatrix}$	$\begin{pmatrix} 1 & 7/500 \\ 0 & 1 \end{pmatrix}$	$\begin{pmatrix} 1 & 7/500 \\ -500/7 & 0 \end{pmatrix}$	$\begin{pmatrix} 13/20 & 7/500 \\ -1000/27 & 20/27 \end{pmatrix}$
$f_{\max}^{L}$	667 lines/mm	667 lines/mm	667 lines/mm
$f_{\max}^{\gamma}$	100 lines/mm	∞	135 lines/mm

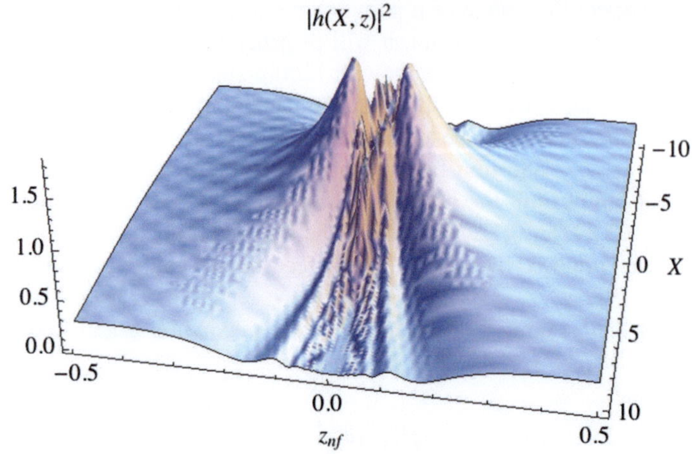

Fig. 12.3 The magnitude distribution of the PSF for System 1: a Fresnel system

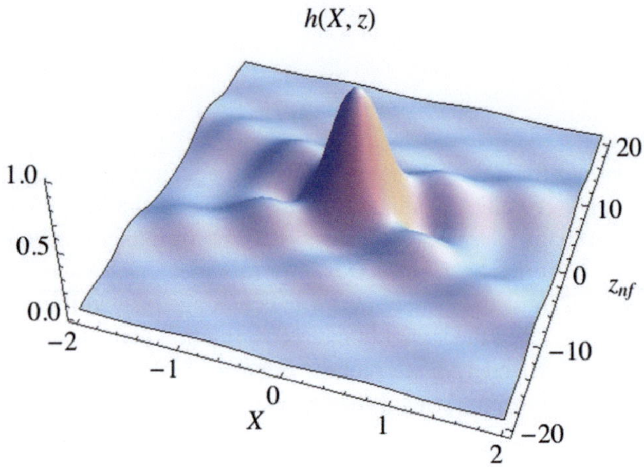

Fig. 12.4 The magnitude distribution of the PSF for System 2: a "lens-less" Fourier system

averaging and the finite camera extent which both act simultaneously to filter the spatial frequencies of the input signal, see Section 10.3 of Ref. [15].

The superior performance of System 2 can be clearly seen in Fig. 12.4. The point spread function appears well behaved with a clear maximum intensity peak located at the actual point source. In Fig. 12.5, the intensity distribution of the point spread function looks similar to that in Fig. 12.3. Once again two planes exist, located symmetrically about $z_{nf} = 0$, where two different but equal maximum peak intensity peaks are located. We note that the positions of these peaks in Figs. 12.3 and 12.5 differ. In Fig. 12.5, the peaks are located in planes that are closer to the correct in-focus object plane. This perhaps reflects the fact that we expect a marginally

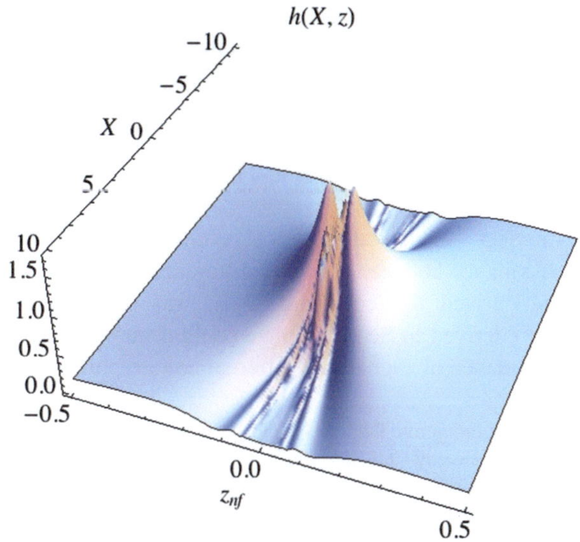

Fig. 12.5 The magnitude distribution of the PSF for System 3: an inline Gabor system

improved performance by System 3 compared to System 1, since System 3 has a D value given by $D = 22/27$ and therefore can allows a higher spatial frequencies to pass through, see Table 12.1.

12.4 Conclusion

In this manuscript we examined how the LCT can be used to analyze and design modern digital holographic systems. Using a matrix formulism it is possible to determine the imaging performance of a wide range of different types of optical systems in a very convenient manner. The limiting factors determining the performance of these systems arise due to the nature of digital cameras. The capture of intensity values by these cameras acts to remove information from the signal that has passed through the holographic system. This happens in three ways: First, the finite extent of the camera aperture means that light not incident on the camera cannot contribute to the final reconstructed image; Second, the finite extent of the camera pixels acts to average the light intensity incident upon them again limiting the maximum spatial frequencies that can be imaged, and finally; Third, the period sampling by the camera pixels produces an infinite number of replicas in the reconstruction domain separated from each other by a distance $\lambda B_N/\Delta$. In the case of PSI holography this sampling operation will not effect the spatial frequencies that can be imaged by the holographic system.

The role of the DC and twin image terms has also been examined. It has been shown that in the case of off-axis holography, the maximum spatial frequency allowed through the system will in most cases be limited by the sampling rate. In contrast to PSI techniques, all of the camera bandwidth, $1/\Delta$, is used to spatially separate the DC, twin, and real terms from each other. The main theoretical results presented in Sect. 12.2 have been summarized in a set of 4 simple rules of thumb [camera aperture: Eq. (12.22), pixel averaging: Eq. (12.27), sampling: Eq. (12.28), and finally an off axis reference wave system: Eq. (12.37)] that can be used to quantify the performance of any LCT holographic system, including PSI or off-axis setups.

In Sect. 12.3, some numerical results were presented contrasting the difference in performance between the three different holographic systems whose characteristics were derived in Sect. 12.2.1.3. It was shown that using the simple rules of thumb derived in Sect. 12.2, we could accurately predict the expected and relative performance of each of the different holographic systems.

Acknowledgements DPK is a Junior-Stiftungsprofessor of "Optic design, modeling and simulation" supported by Carl-Zeiss-Stiftung (FKZ: 21-0563-2.8/121/1). JTS acknowledges the support of the Science Foundation Ireland and Enterprise Ireland under the National Development Plan.

References

1. J.W. Goodman, R.W. Lawrence, Digital image formation from electronically detected holograms. Appl. Phys. Lett. **11**, 77–79 (1967)
2. M.A. Kronrod, N.S. Merzlyakov, L.P. Yaroslavskii, Reconstruction of a hologram with a computer. Sov. Phys. Tech. Phys. **17**, 333–334 (1972)
3. L. Onural, P.D. Scott, Digital decoding of in-line holograms. Opt. Eng. **26**, 261124 (1987)
4. U. Schnars, W. Jueptner, Direct recording of holograms by a CCD target and numerical reconstruction. Appl. Opt. **33**, 179–181 (1994)
5. T. Kreis, W. Juptner, Suppression of the dc term in digital holography. Opt. Eng. **36**, 2357–2360 (1997)
6. T.M. Kreis, Frequency analysis of digital holography. Opt. Eng. **41**, 771–778 (2002)
7. T.M. Kreis, Frequency analysis of digital holography with reconstruction by convolution. Opt. Eng. **41**, 1829–1839 (2002)
8. A. Stern, B. Javidi, Analysis of practical sampling and reconstruction from Fresnel fields. Opt. Eng. **43**, 239–250 (2004)
9. P. Picart, J. Leval, D. Mounier, S. Gougeon, Some opportunities for vibration analysis with time averaging in digital Fresnel holography. Appl. Opt. **44**, 337–343 (2005)
10. L. Xu, X. Peng, Z. Guo, J. Miao, A. Asundi, Imaging analysis of digital holography. Opt. Express **13**, 2444–2452 (2005)
11. J. Garcia-Sucerquia, W. Xu, S.K. Jericho, P. Klages, M.H. Jericho, H.J. Kreuzer, Digital in-line holographic microscopy. Appl. Opt. **45**, 836–850 (2006)
12. P. Picart, J. Leval, General theoretical formulation of image formation in digital Fresnel holography. J. Opt. Soc. Am. A **25**, 1744–1761 (2008)
13. D.P. Kelly, B.M. Hennelly, C. McElhinney, T.J. Naughton, A practical guide to digital holography and generalized sampling. Proc. SPIE **7072**, 707215 (2008)
14. D.P. Kelly, B.M. Hennelly, N. Pandey, T.J. Naughton, W.T. Rhodes, Resolution limits in practical digital holographic systems. Opt. Eng. **48**, 095801 (2009)
15. Y. Hao, A. Asundi, Resolution analysis of a digital holography system. Appl. Opt. **50**, 183–193 (2011)

16. D.P. Kelly, B.M.H.J.J. Healy, J.T. Sheridan, Quantifying the 2.5D imaging performance of digital holographic systems. J. Eur. Opt. Soc. Rap. Public. **6**(11034) (2011)
17. M. Leclercq, P. Picart, Digital Fresnel holography beyond the Shannon limits. Opt. Express **20**, 18303–18312 (2012)
18. D. Claus, D. Iliescu, P. Bryanston-Cross, Quantitative space-bandwidth product analysis in digital holography. Appl. Opt. **50**, H116–H127 (2011)
19. D. Claus, D. Iliescu, J. Watson, J. Rodenburg, Comparison of different digital holographic setup configurations, in *Digital Holography and Three-Dimensional Imaging* (Optical Society of America, 2012), p. DM4C.3
20. H. Jin, H. Wan, Y. Zhang, Y. Li, P. Qiu, The influence of structural parameters of CCD on the reconstruction image of digital holograms. J. Mod. Opt. **55**, 2989–3000 (2008)
21. I. Yamaguchi, T. Zhang, Phase-shifting digital holography. Opt. Lett. **22**, 1268–1270 (1997)
22. G.-S. Han, S.-W. Kim, Numerical correction of reference phases in phase-shifting interferometry by iterative least-squares fitting. Appl. Opt. **33**, 7321–7325 (1994)
23. Z. Wang, B. Han, Advanced iterative algorithm for phase extraction of randomly phase-shifted interferograms. Opt. Lett. **29**, 1671–1673 (2004)
24. D.S. Monaghan, D.P. Kelly, N. Pandey, B.M. Hennelly, Twin removal in digital holography using diffuse illumination. Opt. Lett. **34**, 3610–3612 (2009)
25. F. Gori, Fresnel transform and sampling theorem. Opt. Commun. **39**, 293–297 (1981)
26. L. Onural, Sampling of the diffraction field. Appl. Opt. **39**, 5929–5935 (2000)
27. A. Stern, B. Javidi, Sampling in the light of Wigner distribution. J. Opt. Soc. Am. A **21**, 360–366 (2004)
28. A. Stern, B. Javidi, Sampling in the light of Wigner distribution: errata. J. Opt. Soc. Am. A **21**, 2038 (2004)
29. J. Goodman, *Introduction to Fourier Optics*, 2nd edn. (McGraw-Hill, New York, 1966)
30. S.A. Collins Jr., Lens-system diffraction integral written in terms of matrix optics. J. Opt. Soc. Am. **60**, 1168–1177 (1970)
31. J.J. Healy, B.M. Hennelly, J.T. Sheridan, Additional sampling criterion for the linear canonical transform. Opt. Lett. **33**, 2599–2601 (2008)
32. J.J. Healy, J.T. Sheridan, Uniform sampling for signals with finite instantaneous bandwidth, in *Proceedings of the Irish Systems and Signals Conference*, 2009
33. J.J. Healy, J.T. Sheridan, Fast linear canonical transforms. J. Opt. Soc. Am. A **27**, 21–30 (2010)
34. F. Oktem, H.M. Ozaktas, Exact relation between continuous and discrete linear canonical transforms. IEEE Signal Process. Lett. **16**, 727–730 (2009)
35. A. Koc, H.M. Ozaktas, L. Hesselink, Fast and accurate algorithm for the computation of complex linear canonical transforms. J. Opt. Soc. Am. A **27**, 1896–1908 (2010)
36. Personal communication with Lysann Megel, Masters thesis, Technical University of Ilmenau, 2012
37. E.N. Leith, J. Upatnieks, Reconstructed wavefronts and communication theory. J. Opt. Soc. Am. **52**, 1123–1128 (1962)

Chapter 13
Double Random Phase Encoding Based Optical Encryption Systems Using Some Linear Canonical Transforms: Weaknesses and Countermeasures

Pramod Kumar, Joby Joseph, and Kehar Singh

Abstract During the last two decades or so, a large number of optical information security systems have been proposed by researchers, by exploiting various inherent advantages of optics, with a view to gaining superiority over the existing digital security systems. Among them, double random phase encoding (DRPE) scheme is the oldest and most exhaustively explored optical scheme. However, symmetric nature and linearity prove to be bane of the DRPE scheme, and give an open invitation to unauthorized users to crack the system by mounting different type of attacks. Due to this fact, security of the DRPE scheme has been compromised, making the system vulnerable to attacks. A number of schemes have also been introduced based on the use of fractional Fourier-, Fresnel-, gyrator-, Hartley-, and other transforms. However, vulnerabilities are not limited only to the Fourier domain based systems. Fresnel- and fractional Fourier domain based systems have also been found weak against the chosen-, and the known-plaintext attacks. Resistance of many of the security enhanced DRPE schemes in some other linear canonical transform domains has also been found to be weak against these attacks or their modified forms. When linearity is claimed to be broken with introduction of an amplitude mask at the Fourier plane, the DPRE scheme is able to nullify the high level known-plaintext attack but susceptibility is found against a simple impulse function attack. Recently, some strategies have been adopted to improve the resistance of the DRPE based schemes against the impulse function attacks. Some aspects of the problem have been discussed in the present chapter.

P. Kumar (✉)
Physics Department, Ben-Gurion University of the Negev, Beer-Sheva 84105, Israel
e-mail: pramodkumar.24@gmail.com

J. Joseph
Physics Department, IIT Delhi, New Delhi 110016, India
e-mail: joby@physics.iitd.ac.in

K. Singh
The Northcap University (Formerly ITM University), Gurgaon, Haryana, India
e-mail: keharsiitd@gmail.com

J.J. Healy et al. (eds.), *Linear Canonical Transforms*, Springer Series in Optical Sciences 198, DOI 10.1007/978-1-4939-3028-9_13

13.1 Introduction

History tells us that each and every human civilization has always tried to achieve "perfect" secrecy with their own methods like Greek with "Skytale," Roman with "Caeser cipher" and Indian with "Nirabhasha," etc. But impossibility to attain it in practice has consistently challenged human beings to develop newer methods with enhanced security. Moreover from earlier days to the modern day society, the necessity of information security has been never ending.

Cryptography, the science of securing information plays an important role in information security, and is one of the tools that ensures security, integrity, and authentication of data. In the present era, cryptography has been of immense use in many application areas such as communication, defense, multimedia, e-commerce, computer networks, software and hardware products, medical records, document authentication, and intellectual property and copyright protection to name a few. The evolution of secure information systems to curb the forgery and counterfeiting, etc. rests on the state of the art of the relevant period. Therefore, newer methods/schemes are continuously devised to overcome the existing vulnerabilities in the older methods.

In the modern times, frauds have become increasingly common due to the rapid advancements of technology in computers, charge-coupled devices, printers, scanners, copiers, and image-processing software. As a result of this, it has become possible to produce authentic-looking counterfeits portraits, logos, symbols, money bills, and other complex patterns. Due to the increased threat and incidents of terrorism worldwide, inspection requirements in public places, ports of entry, and transportation hubs around the globe have increased. Therefore it is necessary that the secret information sent from one place to another should be in a coded form.

13.1.1 Necessity of Optical Security Systems

In the present age, often termed the "information age," flow of information is playing an increasingly prominent role in our daily lives. Information security has therefore become of utmost concern following a sharp rise in the rate at which the information is being generated and disseminated. Storage and transmission of information runs the risk of making sensitive and valuable information vulnerable to unauthorized access. Nowadays digital information security systems, with the advent of specialized electronic data processing machines and superfast computers, have become more capable of protecting information from theft, damage, and noise, etc. It would be interesting to know that most of the information protection ideas have originated from the mathematics and computer science, but physics has also played a major role in this area.

Although a large number of commercial digital encryption systems are playing a major role in our daily lives, the required computing speed when large amount of data (e.g., high resolution images) is involved remains the bottleneck of these systems. The quest for the optical information security techniques was therefore undertaken to overcome the drawbacks of the digital security systems. Massively parallel processing in optics, of two-dimensional image data, provides an edge over the electronic processors which are limited to serial processing. Optics provides a large number of degrees of freedom, as data can be encoded in various attributes of light such as phase, amplitude, wavelength, spatial frequencies, polarization, and orbital angular momentum. Speed of digital encryption systems becomes poor as an increase in key length is required to make the system safer from attacks, whereas the optical systems are inherently faster having multi-dimensional capability. Further to break an optical security system, a hacker needs to employ sophisticated optical techniques, which are less accessible and flexible compared to the digital techniques. As a result, optical cryptography has become a burgeoning area of research. Advantages of both digital and optical techniques can be combined in hybrid optical-digital systems. Further, the use of photo refractive and photopolymer materials has given ample space to record immense amount of data three-dimensionally within these holographic media which afford high storage densities.

13.1.2 Encryption Schemes Based on Various Optical Transforms

A large number of optical security systems/schemes [1–8] have evolved over the years, which are based on a variety of principles/methods. However, in the present chapter it is impossible to include all of these. As a result, it has been possible to mention only a limited number of techniques making use of some linear canonical transforms (LCTs) which are a three parameter class of linear integral transforms. LCTs can be used to describe the effect of propagation of a wave field through any general quadratic phase system. Fortunately, optical implementation of most of the LCTs with arbitrary parameters is now possible.

With the introduction of a number of transforms, security enhanced optical encryption systems have been further explored to improve resistance and add new encryption keys. Fractionalization of transforms gives us a new degree of freedom for encoding information. Therefore a number of schemes have also been introduced based on the use of fractional Fourier-, Fresnel-, gyrator-, Hartley-, Hilbert-, Mellin-, Hadamard-, Gabor-, Hankel-, random sine and cosine, wavelet, and other transforms. Of these, the security enhanced schemes based on Fourier-, [9–45] Fresnel-, [21, 46–58], and fractional Fourier transform (FrFT) [59–113] have been very widely reported in the literature. Since the optical security systems are, to some extent, capable of providing resistance against frauds and counterfeiting to ensure integrity, a tremendous intensification in the development of growth of optics based identification systems useful in detecting fake passports, drivers' licenses, passes, credit cards, and other documents has been witnessed.

The techniques using various canonical transforms, particularly Fourier-, Fresnel-, and FrFT, and their variants, have also been used for information hiding and watermarking. We had to leave out the studies on the techniques using holography, polarization, and joint transform correlators (JTCs) in which various transforms have been extensively used. Due to the constraint (framework) of this chapter, a number of techniques developed in the area of optical information security could not be listed. Already a large number of book chapters, special issues of journals, and review papers in various scientific journals and conferences have reported about different approaches and applications of the optical security schemes.

A number of publications have illustrated the exhaustive research carried out with successful experimental demonstration of the double random phase encoding (DRPE) scheme in the Fourier domain. This technique has indeed provided a foundation for many of the subsequent developments. Research in the field of DPRE is confined not only to optics, but ideas from other areas have also been incorporated to affect further improvements. Therefore, in the following sections, we will try to draw an outline of the DPRE scheme in some LCT domains, namely Fourier-, Fresnel-, and FrFT domains. Most recent concern is about security aspects of the DRPE scheme which is not limited only to the Fourier domain, but has penetrated the Fresnel-, fractional Fourier-, and other transform domains also. The attacks and their countermeasures adopted to improve the security of the DRPE scheme will also be discussed here. Although we have made sincere efforts to include all milestones in the progress of the DRPE scheme, we express sincere regret for not being able to mention/cite many of the advances which have been reported in the literature.

During the last two decades or so, a large number of optical encoding techniques have been formulated; among them, the DRPE scheme [9] is the oldest and most exhaustively investigated scheme. In the DRPE (Fig. 13.1), two statistically independent random phase masks (RPMs) employed at the input-, and the Fourier planes in a 4-f optical spatial filtering setup, encode a two-dimensional input image into a stationary white noise. The input plane RPM $R_{\mathrm{in}}(x, y)$ is required to convert an input image $I(x, y)$ into non-stationary white noise after the first Fourier transformation, while the Fourier plane RPM $R_{\mathrm{f}}(u, v)$ encodes the information into stationary white noise. The encryption process is mathematically written as

$$E(x, y) = \mathrm{FT}^{-1}\left[\left(\mathrm{FT}\left[I(x, y)\, R_{\mathrm{in}}(x, y)\right]\right) R_{\mathrm{f}}(u, v)\right] \tag{13.1}$$

where FT[] and FT^{-1} [] represent, respectively, the Fourier-, and inverse Fourier transform operations.

On the basis of form of the input image, the DRPE technique has two modes. In the amplitude-based DRPE, the input image is of amplitude nature, due to which only the Fourier plane RPM serves as the encryption key during decryption because quantity of interest at the decryption plane is only the intensity of the input image. In the phase-based DRPE, both the RPMs are required to retrieve the original phase image. Fully phase-based DRPE provides more resistance against the additive noise

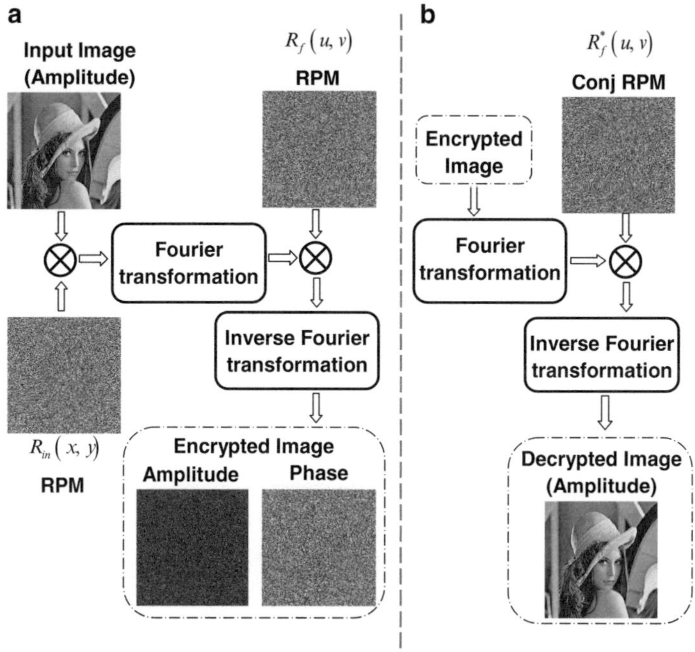

Fig. 13.1 Schematic diagram for the DRPE scheme: (**a**) encryption and (**b**) decryption

in comparison with the amplitude-based DRPE encryption [16]. The phase nature of the input information makes the phase-based DRPE resistant against copying, as naked eye and intensity sensitive device like CCD which cannot record it. Detailed description of the DRPE scheme can be found in a number of research papers and review articles in the literature cited in the list of references. It is remarkable to note that in spite of existence of a large body of literature, a formal treatment of the optical technique was published [41] only in 2012. In this paper, the authors proposed a new discrete model of the DRPE system that retains the properties of its continuous counterpart and ensures correct sampling so as to avoid aliasing at every stage of the simulation. The two diffusers are created so as to approximate the transmission function of physical narrow band diffusers.

In recent years, double or multiple images have been encoded with the DRPE technique into single encrypted image [4, 21, 26, 29, 36]. Encryption is also performed in various domains like Fresnel-, and Fractional Fourier domain, etc., with different multiplexing techniques like wavelength- and position multiplexing. But the quality of decrypted images with some of these multiplexing methods has been found to deteriorate due to cross-talk among various decrypted images. Initially, the DRPE scheme was designed for encoding binary-, and gray-scale images into stationary white noise, but later color images were also encrypted [17]. For encryption of a color image, each decomposed constituent primary color is encoded individually with DPRE using wavelength multiplexing. In comparison with gray-scale or monochromatic images, color images are able to offer more information, therefore research related to color information has escalated in recent years. Drawback of using three color channels is removed in an alternative and more simplified method [77], where the indexed image formats of color image is encrypted instead of each primary color image.

It may be pointed out that experimental optical implementation of many of the proposals described in the literature on the DRPE or its variants has not yet been demonstrated. However, the schemes have been validated by presenting results of computer simulations.

13.1.3 Architecture for Optical Implementation

For the optical realization of the DRPE scheme, two classical optical pattern recognition correlator architectures, i.e. 4-f vanderLugt- (VL) and JTC architectures are adopted [2]. In both of these architectures, the input image is displayed either in the form of a transparency or on a spatial light modulator (SLM). Due to continuous technological improvements in both amplitude- and phase modulating electrically addressed liquid crystal SLMs, these devices have been widely used for the purpose. The random phase variation is obtained with the introduction of ground glass or from the non-uniform random coating of gelatin on glass plates. Display of the RPMs on the SLMs provides ease of use during the encryption and decryption processes. Output of the DRPE encryption process is a stationary white noise, which has complex nature. And this encrypted information obtained at the output plane has random amplitude and phase, due to which holographic process is required for recording.

Although both of the architectures (VLC and JTC) for the DRPE scheme require two RPMs to convert amplitude or phase image into stationary white noise, the complex nature of the encrypted image, stringent alignment for decryption, and requirement of conjugate RPMs for decryption are the main drawbacks of the DRPE in the VLC architecture. Problem arising due to complex nature of the encrypted image is alleviated in the JTC architecture, as the encrypted information is real and non-negative. Due to the shift invariance property of the JTC, a shift of the key RPM only changes the position of the decrypted image, and therefore the problem of stringent alignment of the RPM is alleviated. Employment of the same key RPM for encryption and decryption further removes the necessity for the design of conjugate RPM during decryption. Therefore, the DRPE performed in the JTC architecture shows superiority over the DRPE in the VLC architecture.

Decryption process can be performed with two methods. In one of the methods, original image is decrypted simply by reversing the optical setup, and employing conjugate RPMs in place of the original RPMs. In this method, the encrypted image is Fourier transformed, and the complex conjugate RPM at the Fourier plane nullifies the effect of the encryption mask $R_f(u, v)$. The subsequent inverse

Fourier transform gives the decrypted image multiplied with the mask $R_{in}(x, y)$ at the decryption plane. Depending on the type of amplitude or phase-based DRPE, either decrypted image is captured on CCD or multiplied with conjugate of input plane RPM.

In the second method [14], need of conjugate RPMs during the decryption process is eliminated, as the same set of encryption RPM keys are employed to retrieve the decrypted image. This method for the decryption process is based on the optical phase conjugation technique in which the conjugation of the encrypted image is obtained with the use of optical phase conjugation in a photorefractive crystal through four-wave mixing. The counter propagating phase conjugate beam is able to nullify the effect of the RPMs in the decryption process.

13.2 Extension from the Fourier- to Other Transform Domains

Security of the DRPE scheme is restricted either by a single or both RPMs depending on the type of input image considered for encryption. Therefore new encryption parameters in different members of the LCT family have been explored to enlarge the key space by adding new encryption keys. Substitution of various transforms in place of the Fourier transform results in improved strength of the DRPE scheme.

13.2.1 Fractional Fourier Domain

FrFT has been widely used for encryption after its use was first demonstrated [59, 60] in the DRPE scheme. It is well known that the zero-order FrFT is equal to an identity operation, whereas complete Fourier transform is of first order. From an optical point of view, the FrFT is executed either by using quadratic graded index media or by utilizing optical setup consisting of lenses. On the basis of using lenses Lohmann's type I and Lohmann's type II systems are designed. In type I setup, a lens is sandwiched between two equal free-space propagation distances, while in type II system, between two lenses of equal focal length is a free-space propagation distance.

Encryption process in the fractional domain is written as

$$E\left(x_o, y_o\right) = \mathrm{FrFT}^{\beta}\left[\left(\mathrm{FrFT}^{\alpha}\left[I\left(x, y\right) R_{\mathrm{in}}\left(x, y\right)\right]\right) R_{\mathrm{fr}}\left(u_1, v_1\right)\right] \tag{13.2}$$

where $R_{fr}(u_1, v_1)$: fractional plane RPM.

DRPE in the fractional domain (Fig. 13.2) is viewed as a generalization of the DRPE in the Fourier plane method; here, the fractional-orders act as new encryption keys in addition to the RPMs. Perfect decryption is possible only with correct

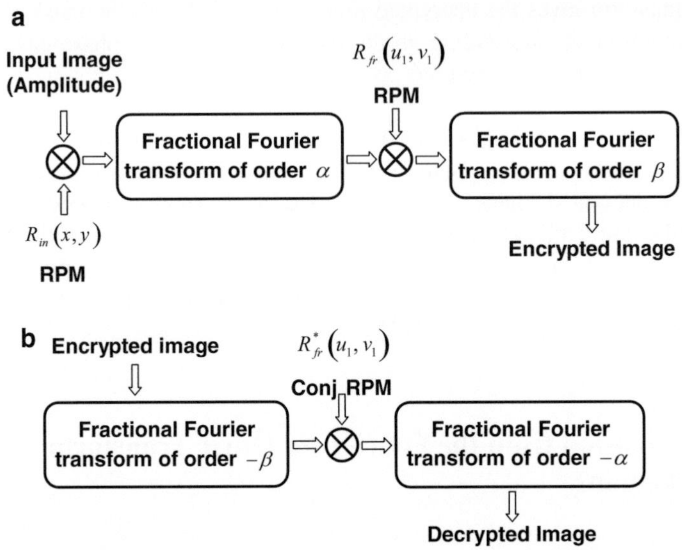

Fig. 13.2 DRPE in the fractional Fourier domain: (**a**) encryption process and (**b**) decryption process

fractional-orders in addition to the RPMs which are specified during the encryption process. A slight variation of 0.05 in the fractional-order [59] would preserve the information security, even if an attacker is able to acquire knowledge about the RPMs.

13.2.2 Anamorphic FrFT Domain

Substitution of spherical lenses with a pair of cylindrical lenses having different focal length in the Lohmann's system transforms a function into an anamorphic fractional Fourier domain. Therefore, the DRPE in the anamorphic domain encodes a two-dimensional image with two different fractional-orders along the *x*- and *y*-axes.

Introduction of anamorphic FrFT [61] further enlarges the encryption key space of the DRPE scheme, as in this domain more freedom is available to choose different fractional-orders along two orthogonal directions of an optical system. Therefore, it becomes more difficult for an unauthorized user to crack the security of the DRPE system.

13.2.3 Anamorphic FrFT at an Angle

Usually, the cylindrical lenses are aligned either along x- or y-axis to perform the FT or FrFT with these lenses in an optical systems. An arbitrary in-plane rotation of the cylindrical lenses allows one to perform FT or FrFT with the modified Lohmann's second-type system (MLST) at that angle [80]. The FrFT at an angle obtained from the MLST system is given by

$$G^{(\alpha_1,\alpha_2;\phi)}(u_1,v_1) = C_P \exp\left\{i\pi\left(\frac{u_1^2}{\tan[\alpha_1(\pi/2)]} + \frac{v_1^2}{\tan[\alpha_2(\pi/2)]}\right)\right\} \int_{-\infty}^{\infty}\int_{-\infty}^{\infty} g_1(x_1,y_1)\exp\left\{-i\pi\left(\frac{x_1^2}{\tan[\alpha_1(\pi/2)]} + \frac{y_1^2}{\tan[\alpha_2(\pi/2)]}\right)\right\} \exp\left\{-i2\pi\left(\frac{x_1u_1}{\sin[\alpha_1(\pi/2)]} + \frac{y_1v_1}{\sin[\alpha_2(\pi/2)]}\right)\right\} dx_1dy_1 \tag{13.3}$$

where α_1 and α_2 are the fractional-orders in the range $[0, 1]$

$$g(x,y) = g_1(x_1,y_1) \tag{13.4}$$

$g_1(x_1,y_1)$ is new representation of image $g(x,y)$ in rotated (x_1,y_1) plane.

The constant factor is

$$C_P = \frac{\exp(-i\pi\ \mathrm{sgn}\{\sin[\alpha_1(\pi/2)]\}/4 + i\alpha_1(\pi/4))}{|\sin[\alpha_1(\pi/2)]|^{1/2}} \times \frac{\exp(-i\pi\ \mathrm{sgn}\{\sin[\alpha_2(\pi/2)]\}/4 + i\alpha_2(\pi/4))}{|\sin[\alpha_2(\pi/2)]|^{1/2}} \tag{13.5}$$

The relationship between the rotated and original axes is $x_1 = x\sin\phi + y\cos\phi$, and $y_1 = -x\cos\phi + y\sin\phi$.

The anamorphic FrFT by the MLST system in operator form is denoted as

$$G(u_1,v_1) = \mathrm{FrFT}^{(\alpha_1,\alpha_2;\phi)}[g_1(x_1,y_1)] \tag{13.6}$$

where α_1 and α_2 are fractional-orders along x_1- and y_1-axes at an angle ϕ, with respect to the x-, and y-axes.

The anamorphic FrFTs of a zero padded input image (Lena; size 256 × 256 pixels Fig. 13.3), when the rotation angle is kept constant at 30° for fractional-orders (0.0, 0.25), (0.0, 0.50), (0.0, 0.75), and (0.0, 1.0) are illustrated, respectively, in Fig. 13.3.

The encryption process in the anamorphic FrFT domain employing the MLST systems S_1 and S_2 is written as

$$E(x_o,y_o) = \mathrm{FrFT}^{(\alpha_3,\alpha_4;\phi_2)}\left[\left(\mathrm{FrFT}^{(\alpha_1,\alpha_2;\phi_1)}[I(x,y)R_{\mathrm{in}}(x,y)]\right)R_{\mathrm{fr}}(u_1,v_1)\right] \tag{13.7}$$

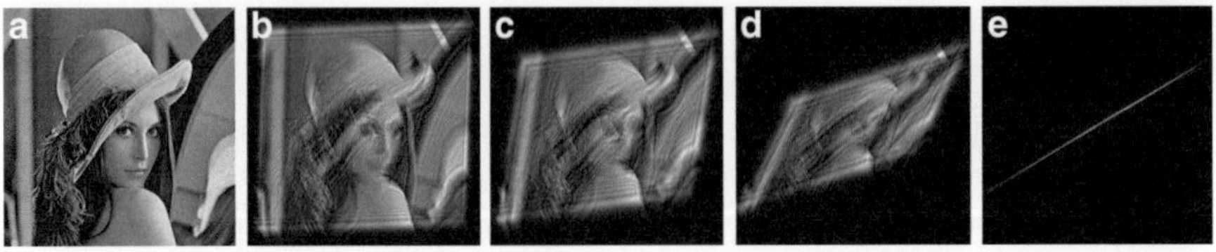

Fig. 13.3 (**a**) Input image; FrFT at same angle of zero padded image at (**b**) (0.0, 0.25) (**c**) (0.0, 0.50) (**d**) (0.0, 0.75) and (**e**) (0.0, 1.0)

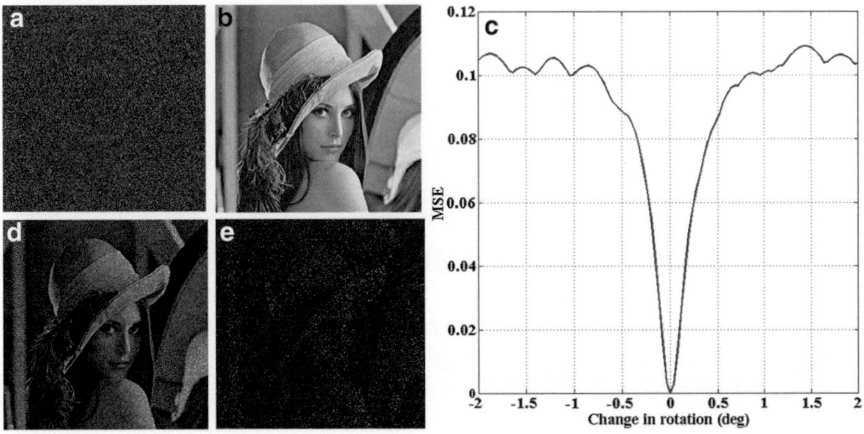

Fig. 13.4 Numerical simulation results: (**a**) encrypted image, (**b**) decrypted image, (**c**) mean-square-error (MSE) between the decrypted and the original image, with change in the rotation angle of the MLST system S_2 (**d**) decryption with change in the rotation of the MLST system S_2: by a 0.08°, and (**e**) 0.2° from the actual value

The anamorphic fractional-orders with in-plane rotation of the MLST systems S_1 and S_2 are represented, respectively, as $(\alpha_1, \alpha_2; \phi_1) = (0.5, 0.9; 45°)$ and $(\alpha_3, \alpha_4; \phi_2) = (0.7, 0.3; 60°)$. It can be seen that the original image shows no resemblance to the encrypted image as illustrated in Fig. 13.4. Nearly perfect decrypted image (Fig. 13.4) is obtained by using the conjugate of R_{fr}, with the inverse anamorphic fractional-orders and keeping the in-plane rotation angle same as during the encryption process. Decryption sensitivity with respect to the change in the in-plane rotation of the MLST system S_2 is plotted in Fig. 13.4.

It can be seen that a slight deviation of 0.08° and 0.2° from the original angle for S_2 results in partial decryption of the image (Fig. 13.4).

Qualitative experimental results for encryption and decryption process have been performed with an arrangement shown in Fig. 13.5. Results are illustrated in Fig. 13.6. A rotation of the MLST system S_2 by a very small angle = 0.1° results in complete noise, having no information about the original image.

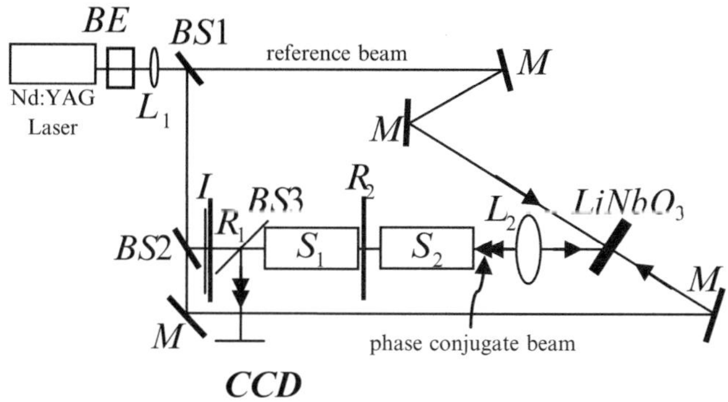

Fig. 13.5 Experimental setup, BE: beam expander; BS1, BS2, BS3: beam splitter; M: mirror; S_1, S_2: MLST system; L_2: imaging lens; $R_1 = R_{in}$ and $R_2 = R_{fr}$: RPM; and $LiNbO_3$: photorefractive crystal

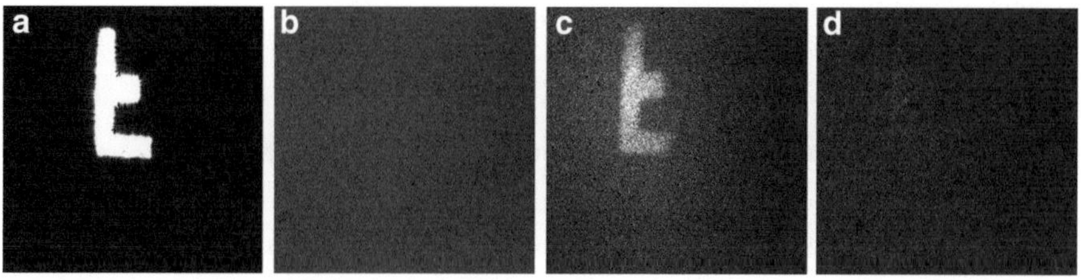

Fig. 13.6 Experimental result image: (**a**) original, (**b**) encrypted, (**c**) decrypted, and (**d**) decrypted with rotation of 0.1° of S_2 from the original

13.3 Strength Estimation of the DRPE in the Fractional Domains

The DRPE scheme operates on real-valued amplitude images or phase-only input images to encode them into stationary white noise. As only the Fourier plane RPM is necessary for perfect decryption in the amplitude-based DRPE, the strength can be estimated numerically in terms of the total number of attempts required to breach the security of the system. Usually out of 256 phase levels any level can be chosen for a pixel, therefore an RPM of size 100×100 pixels will have $256^{(100\times100)}$ number of attempts for decryption, while in phase-based DRPE, both the RPMs are of utmost importance. Therefore, for an authorized user, the total number of attempts becomes $256^{2\times(100\times100)}$.

Substitution of the FrFT in place of the FT enhances security of the DRPE system with the addition of fractional-orders as new encryption keys. During the decryption, the FrFT domain DRPE is found sensitive to a deviation of 0.05 from the actual order [59]. An unauthorized user could plan to vary the order in step of 0.01 as the fractional-order are in range [0, 1], and this leads to only 100 possible values in this range [80]. Inclusion of the second fractional-order with the same sensitivity will lead to $(100)^2 \times 256^{2\times(100\times100)}$ total number of orders. The anamorphic FrFT further increases the task of an attacker as $(100)^4 \times 256^{2\times(100\times100)}$ orders are required to know the exact order. In-plane rotation angle of the MLST system [80] adds rotation angle as new encryption key, whose sensitivity is found to be 0.2°. When varied in steps of 0.1°, the numbers of possible attempts added due to rotation angle are 10×180, as the MLST system could regain its original position only after 180° rotation. Therefore, the total number of attempts required for an anamorphic DRPE system employing the MLST system [80] becomes $(10 \times 180) \times (100)^4 \times 256^{2\times(100\times100)}$.

Introduction of more number of keys results in enhancing the security of the system, as the total number of attempts necessary to break the system becomes extremely large. With the increase in computational load, for an attacker to try all those combinations becomes intractable, but actually there is no need to attempt all those combinations. The main conclusion drawn from this strength estimation [80] is that there are only limited or fixed numbers of fractional-orders to break the system, which was not earlier calculated.

13.4 Vulnerability of Various Schemes to Attacks

The security of an optical encryption scheme can only be trusted if these systems are able to endure various attacks [28, 114–138] under specified conditions. Although the DRPE scheme has shown great strength against different attacks like brute force-, occlusion-, noise-, and misalignment, etc., but recent studies have exposed the weaknesses against some of the attacks. Evaluation of the DRPE against blind decryption is also not sufficient to judge its strength, as only a limited number of keys are considered from a large key space. Further introduction of additional keys in the form of the fractional-orders, and free-space propagation distance modify the DRPE scheme into a computationally more complex cryptosystems, but search within the limited key space is inadequate to evaluate the resistance of the DPRE-based systems against attacks [119, 122].

13.4.1 Brute Force Attack

One of the simplest and obvious attacks is the brute force attack, where an exhaustive search for the right key is carried out with every possible key combination in the entire search space [119]. Due to the discrete nature of these cryptosystems, an attacker with adequate knowledge about size of the RPMs, phase levels, wavelength, and transform domain could make all possible set of probable combinations to find one workable combination. But the number of combinations to search the required key is very large, which further shoots up with the addition of more keys or an increase in size of keys. Although these number of combinations seems to be intractable and huge search is required for the correct key, in practice all those combinations are not required, as partially recognizable image can be sufficient for many purposes.

13.4.2 Chosen-Cipher Text Attack

Realization of the DRPE employing the 4-f setup gives rise to the possibility of impulse function (low level) attack, which is due to the transformation of an impulse function into a unity amplitude function at the Fourier plane. Subsequent transformation reveals the complete information about the Fourier plane RPM in form of encrypted image [114]. When the attacker is denied the encryption system, and accessible decryption system can only record the intensity, in this case the repeated probing with a set of cipher texts could be designed to crack the system [119]. Linearity of the DRPE system also gives rise to a variant form of the delta function attack, where two similar looking input images differ only by the center pixel, i.e. difference of these two images is a centered delta function [119]. This less suspicious form of the delta function attack is also applicable in the phase-based DRPE system where one-time impulse attack is not possible.

Applicability of the impulse function attack is not limited to only the Fourier domain but it also works well in the Fresnel domain DRPE. If the attacker possesses details about the Fresnel distance and the wavelength, then both the input- and the Fresnel plane keys, i.e. RPMs are cracked with the attack, choosing a series of impulse functions as specific plaintexts.

13.4.3 Chosen-Plaintext Attack

In comparison with the chosen-cipher text attack, more opportunities are available to the attacker in choosing the input images during the chosen-plaintext attack. Therefore with this attack, an unauthorized user is able to choose or design a series of plaintexts necessary to encrypt, and corresponding encoded images are obtained. The Fourier plane RPM, which is the most important key for the amplitude-based DRPE scheme is easily divulged with simple and effective selection of plaintexts with this attack. Security enhanced DRPE scheme in the Fresnel domain is also found prone to this attack [118], as an unauthorized user cracks both input- and transform plane keys, when one possesses information about the wavelength and two free-space propagation distances.

Success of this attack is limited not only to the DRPE employed in optical 4-f architecture, but security of the DRPE scheme in the joint transform architecture is also broken with a single predefined chosen-plaintext and corresponding ciphertexts [128]. In another method, acquisition of three encrypted images corresponding to their respective predesigned chosen plaintexts with accessibility of the encryption machine completely determines the RPM employed during encryption process. Information about the separation between the input image and encryption phase mask is also not required in this attack.

13.4.4 Known-Plaintext Attack

Symmetric and linear natures are the inherent features of the DRPE scheme, and due to these properties, an unauthorized user is able to try different attacks. A more advanced attack, which requires very few resources in comparison with the earlier mentioned attacks is the known-plaintext attack [117, 119, 121] in which only a pair of plaintext-ciphertext is sufficient to retrieve the encryption keys. Recovery of these input- and Fourier plane RPMs is performed with an ease by employing the phase retrieval techniques, which is mainly carried out with Gerchberg–Saxton-, Fienup's hybrid input–output-, or other modified algorithms. With the advancement in the computational algorithms, much faster iterative algorithms are coming up, and attackers are able to breach the scheme in lesser time. This attack [119, 121] provides more relaxation as no predefined input-ciphertext is necessary. The DRPE in JTC architecture is also found vulnerable to the known-plaintext attack.

In the known-plaintext heuristic attack [116], an estimation of the Fourier plain key is obtained with the application of simulated annealing heuristic algorithm. With the convergence of simulated annealing algorithm, an unauthorized user is able to obtain the cracked key, and information about the real-valued input image is obtained with very low threshold error. The combination of hill climbing algorithm with simulated annealing known as hybrid heuristic algorithm [124] significantly reduces the searching time and decryption error. These heuristic approach based attacks do not search for exact solution, but cracked key providing decryption with slight error is acceptable.

Although security of the DRPE scheme in the FrFT domain is enhanced with the introduction of fractional-orders as additional encryption keys, the scheme still lacks the strength to endure the known-plaintext attack [125]. Here, a properly designed strategy to determine the exact fractional-orders is needed to exhaustively search in subdivided fractional domain space step-by-step with a very small fractional-order. This search process is time-consuming, but powerful computer makes it possible in reasonable time. Thereafter phase retrieval algorithm retrieves the two phase keys from the known-plaintext pair.

Security flaws of the DRPE scheme employing JTC architecture also remain unsettled, as the known-plaintext attack is able to shatter the resistivity with ease [127]. This attack employs heuristic hybrid algorithm in combination with the Gerchberg–Saxton algorithm, due to the intensity nature of the encrypted image in the JTC architecture. Resistance of the DRPE scheme in JTC architecture is also found feeble against this attack.

13.4.5 Alternative Search Algorithm for Keys

As explained earlier, a number of attacks designed for cryptanalysis of the DRPE reveal the weaknesses. Introducing additional encryption parameters using various techniques further enlarges the key space, but such security enhanced systems are also found vulnerable to attacks. Usually with these strategies an attacker tries to extract the original key. Therefore, either the attacker must have proper knowledge about the geometric keys like wavelength, free-space propagation distance, and fractional-orders, etc. or an exhaustive search should be made to obtain these geometric keys. However these exhaustive searches are time consuming, and it becomes impractical to search the exact key in expanded large key space. Some numerical technique could significantly lessen the computational complexity, but increase in the image size exponentially enlarges complexity. Success of these attacks is only possible, when all other geometric keys are known in advance.

Recently, a new method [129] has been described, where all the key parameters are assumed to be wrong, so that the mismatch due to one key can be compensated with other keys. The method is based on a strategy to derive an alternative region of the possible keys. The keys retrieved from this method are completely different from the original set of keys for the Fresnel domain DRPE. No constraint is imposed on the choice of these parameters. The phase mask keys are retrieved with the phase retrieval algorithms, when the attacker has information about a large number of ciphertexts and their corresponding plaintexts. After several round of iterations, the decrypted image is obtained with recognizable features.

13.5 Attack-Resistant DRPE Scheme

In order to improve the security of the conventional DRPE scheme against various existing attacks, there is an urgent need to deeply investigate and understand the behavior of these attacks in the scenario of this scheme. And with these understandings, new methods could be developed to nullify these attacks. Recent cryptanalysis relates the origin of these attacks due to the optical linearity of the DRPE scheme. But, in recent years, with modification in the conventional DRPE scheme, various methods have been formulated to overcome these attacks.

13.5.1 Impulse Attack-Free DRPE Scheme

As is well known, the Fourier transformation of a delta function yields a unity function. As the conventional DRPE scheme employs two RPMs in 4-f optical setup at input and transform plane, the consequence of this universal Fourier transformation property and the 4-f geometry for encryption make the DRPE scheme vulnerable to a one time simple impulse function attack. Therefore to negate such an attack, transformation of impulse function into a random function is necessary. This could be achieved [31, 32] in the following ways.

13.5.1.1 Randomized Lens-Phase Function

Immunity against the impulse function attack is achieved with lens, L_{Ra} having randomized surface, when the lens-phase function is modified by multiplying it with a random function, which is similar to modification of the lens-phase function in the presence of aberrations. Here the random lens phase-based FT is denoted as $\mathrm{MFT}^{L_{\mathrm{R}}}$ [31]. Due to the randomized lens-phase function, the $\mathrm{MFT}^{L_{\mathrm{R}}}$ of a delta function transforms into a random function as depicted in Fig. 13.7.

In the impulse attack-free DRPE scheme (Fig. 13.8) the usual FT lenses are replaced with lenses $L_{\mathrm{Ra}}, L_{\mathrm{Rb}}$ having randomized lens-phase functions in the conventional DRPE scheme. The encryption scheme then becomes,

$$E(x, y) = \mathrm{MFT}^{L_{\mathrm{Rb}}}\left[\left(\mathrm{MFT}^{L_{\mathrm{Ra}}}\left[I(x, y)\, R_{\mathrm{in}}(x, y)\right]\right) R_{\mathrm{f}}(u, v)\right] \tag{13.8}$$

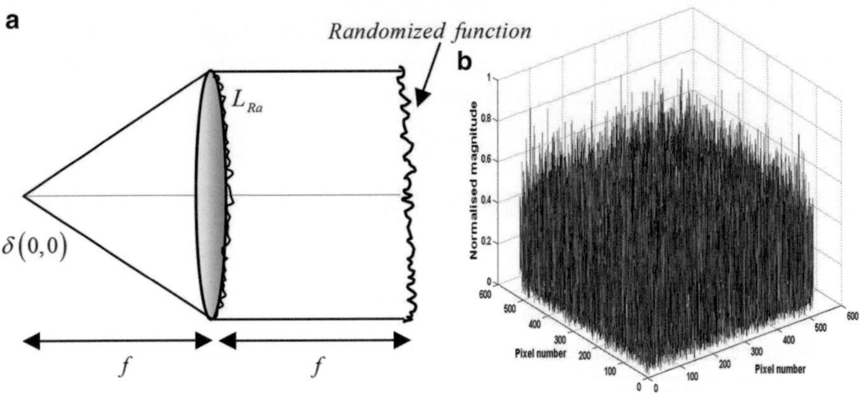

Fig. 13.7 (**a**) $\mathrm{MFT}^{L_{\mathrm{R}}}$ of a delta function into randomized function, (**b**) normalized magnitude of the randomized function

Fig. 13.8 Schematic diagram for the impulse attack-free DRPE scheme

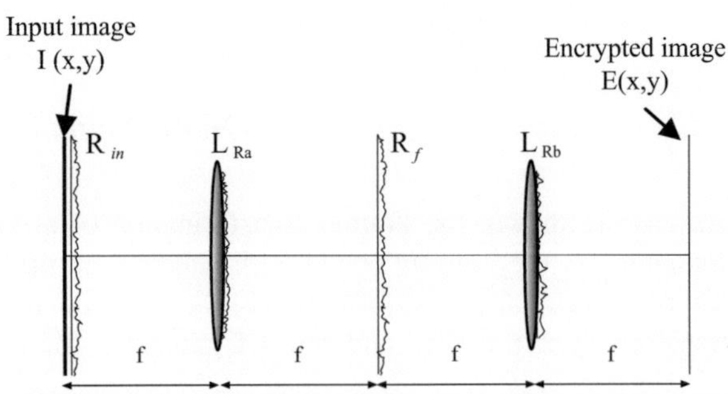

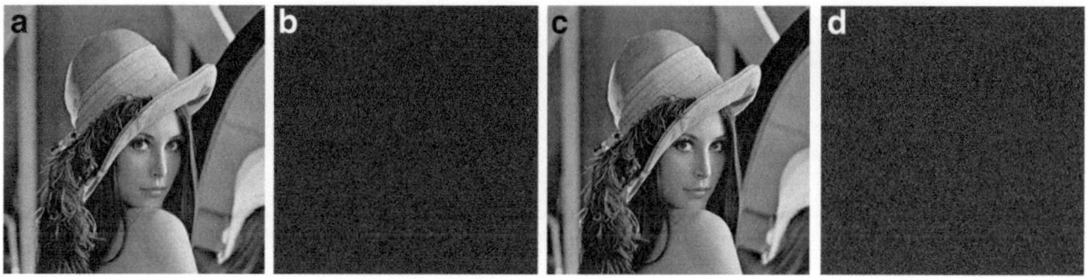

Fig. 13.9 Numerical simulation results: (**a**) input image, (**b**) encrypted image, (**c**) decrypted image, (**d**) decrypted with the Fourier plane mask obtained from the FT of the impulse encrypted image

In a case of the impulse attack on this scheme, the encrypted image is not the inverse Fourier transform of the Fourier plane mask, but a complex function having random amplitude as well as random-phase distribution, which is due to the two $\mathrm{MFT}^{L_\mathrm{R}}$ lenses.

$$E(x,y) = \mathrm{MFT}^{L_\mathrm{Rb}}\left[\left(\mathrm{MFT}^{L_\mathrm{Ra}}\left[\delta(0,0)\,R_\mathrm{in}(x,y)\right]\right)R_\mathrm{f}(u,v)\right] \tag{13.9}$$

The Fourier plane mask is not cracked even with the variant form of the impulse function employing two similar looking images.

$$E(x,y) = \mathrm{MFT}^{L_\mathrm{Rb}}\left[\left(\mathrm{MFT}^{L_\mathrm{Ra}}\left[\left(I_1(x,y) - I_2(x,y)\right)R_\mathrm{in}(x,y)\right]\right)R_\mathrm{f}(u,v)\right] \tag{13.10}$$

For an input gray-scale image (Lena, 512 × 512 pixels, Fig. 13.9), the encrypted- and decrypted images are shown, respectively, in Fig. 13.9. When an unauthorized user is able to encrypt the impulse function as an input image, the noise pattern (Fig. 13.9) is obtained in the form of decrypted image from the FT operations to retrieve the key mask. As the randomization due to $\mathrm{MFT}^{L_\mathrm{R}}$ lenses cannot be nullified from the FT, the security is enhanced in addition to the immunity from the impulse function attack and its variant form.

Schematic diagram for experimental demonstration of this impulse attack-free scheme is illustrated in Fig. 13.10. Random phase variation is obtained from the FT lenses of focal length 18 and 20 cm, when these lenses are coated with gelatin to have random surface variations. Complex demagnified encrypted image is recorded holographically in the $\mathrm{Fe:LiNbO_3}$ crystal, and decryption is performed with the phase conjugation technique. Experimental results are illustrated in Fig. 13.11.

13.5.1.2 Introduction of an Additional RPM in Front of the Lenses

As the cause for the viability of an impulse function attack has become known, another method to avoid the formation of a unity function at the Fourier plane is also adopted, in which the RPMs are placed in front of both the FT lenses in a 4-f optical setup. These RPMs in front of the FT lenses protect the DRPE scheme from

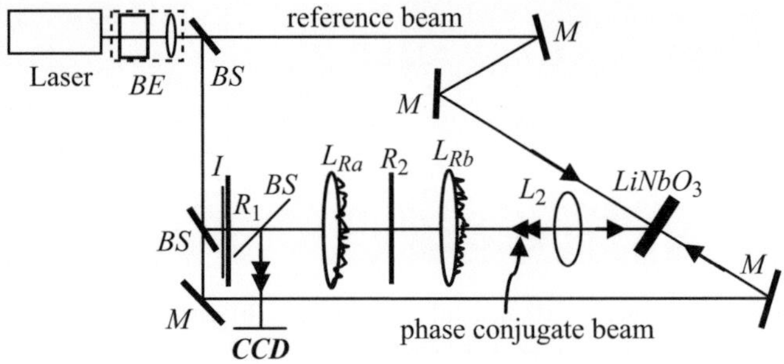

Fig. 13.10 Experimental setup for the impulse attack-free scheme, BE: beam expander; BS: beam splitter; M: mirror; $R_1 = R_{in}$ and $R_2 = R_f$: RPM; L_{Ra} and L_{Rb}: lenses with randomized phase functions; L_2: imaging lens; and $LiNbO_3$: photorefractive crystal

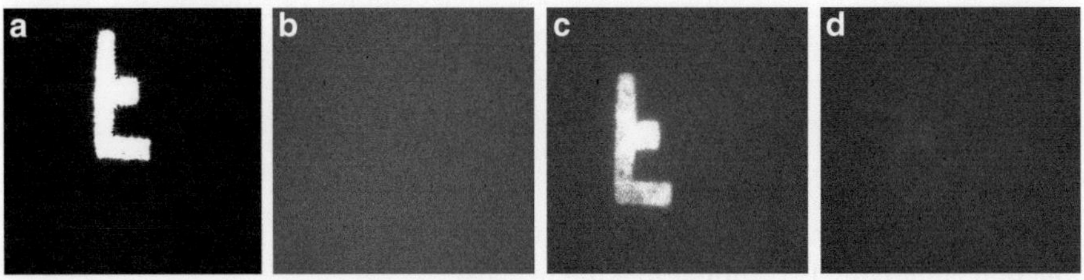

Fig. 13.11 Experimental results: (**a**) image to be encrypted; (**b**) encrypted image; (**c**) correctly decrypted image; (**d**) decryption with lateral shift of 0.01 mm in lens L_{Rb}

the impulse function attack. Schematic diagram for the impulse attack-free DRPE [32] is illustrated in Fig. 13.12. The encryption process is written as

$$E(x,y) = \frac{1}{i\lambda f}\exp\left\{\frac{ik}{2f}\left(1-\frac{d_4}{f}\right)\left(x^2+y^2\right)\right\} \times \mathrm{FT}\left[\left(\mathrm{FRT}_{\lambda,d_3}\left[U_f(u,v)\,R_f(u,v)\right]\right)\times R_4(u_1,v_1)\right] \tag{13.11}$$

where

$$U_f(u,v) = \frac{1}{i\lambda f}\exp\left\{\frac{ik}{2f}\left(1-\frac{d_2}{f}\right)\left(u^2+v^2\right)\right\} \times \mathrm{FT}\left[\left(\mathrm{FRT}_{\lambda,d_1}\left[I(x,y)\,R_{in}(x,y)\right]\right)\times R_2(x_1,y_1)\right] \tag{13.12}$$

Suppose an unauthorized user is able to encrypt the impulse function into an encrypted image, it is represented as,

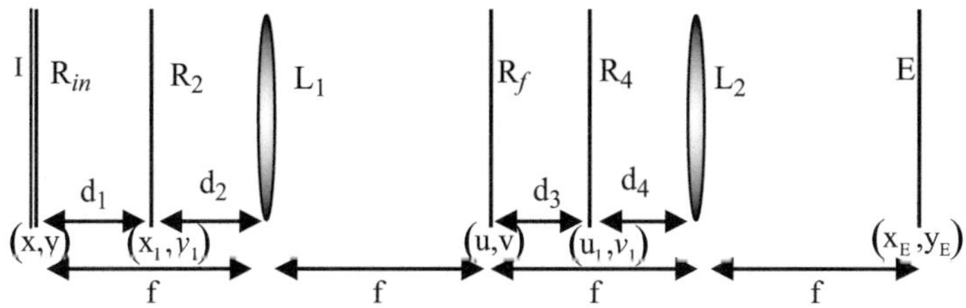

Fig. 13.12 Schematic diagram, $R_{\text{in}}, R_2, R_{\text{f}}$, and R_4: RPMs; I: original image to be encrypted; E: encrypted image; L_1 and L_2: lenses of focal length f each; d_1, d_2, d_3 and d_4: free-space propagation distances

$$U_{\text{f}}^{\delta}(u,v) = \frac{1}{i\lambda f}\exp\left\{\frac{ik}{2f}\left(1-\frac{d_2}{f}\right)\left(u^2+v^2\right)\right\} \times \text{FT}\left[\left(\text{FRT}_{\lambda,d_1}\left[\delta(0,0)R_{\text{in}}(x,y)\right]\right)\times R_2(x_1,y_1)\right] \tag{13.13}$$

$$E^{\delta}(x,y) = \frac{1}{i\lambda f}\exp\left\{\frac{ik}{2f}\left(1-\frac{d_4}{f}\right)\left(x^2+y^2\right)\right\} \times \text{FT}\left[\left(\text{FRT}_{\lambda,d_3}\left[U_{\text{f}}^{\delta}(u,v)R_{\text{f}}(u,v)\right]\right)\times R_4(u_1,v_1)\right] \tag{13.14}$$

While in the conventional DRPE, the encrypted image is the inverse FT of the Fourier plane, this modified DRPE scheme provides a random function due to additional RPMs in front of the lenses in the 4-f optical setup.

An image of Lena (Fig. 13.13) (256 × 256 pixels) embedded in a window of size (512 × 512 pixels) is assumed to be an input image. The size of the image, RPMs, and the encrypted image are 1 cm × 1 cm each. Lenses are of focal length 20 cm each, while free-space propagation distances d_1, d_2, d_3 and d_4 taken to be 10 cm each. The encrypted image does have resemblance to the original image, and perfect decryption results in the retrieval of decrypted image (Fig. 13.13). Suppose an attacker is able to encrypt an impulse function, and assume this scheme to be the conventional DRPE scheme. In that case without nullifying the effect of RPMs in front of the lenses, the decryption performed with the cracked Fourier mask in the conventional DRPE appears as noise shown in Fig. 13.14. In the absence of any of the RPMs, only noise will be obtained as depicted respectively in Fig. 13.14.

13.5.2 Known-Plaintext Resistant Scheme

As pointed out earlier, in comparison with the various attacks on the DRPE scheme, the known-plaintext attack is considered to be the most effective and destructive. The success of this attack is not restricted to only the Fourier domain, as Fractional

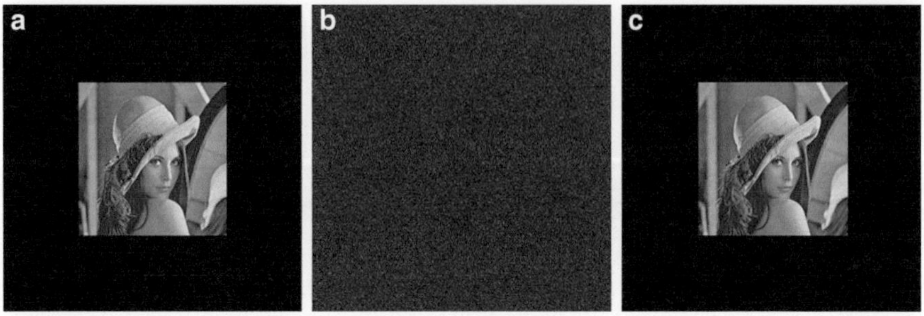

Fig. 13.13 Numerical simulation results: (**a**) input image; (**b**) encrypted image; and (**c**) perfectly decrypted image

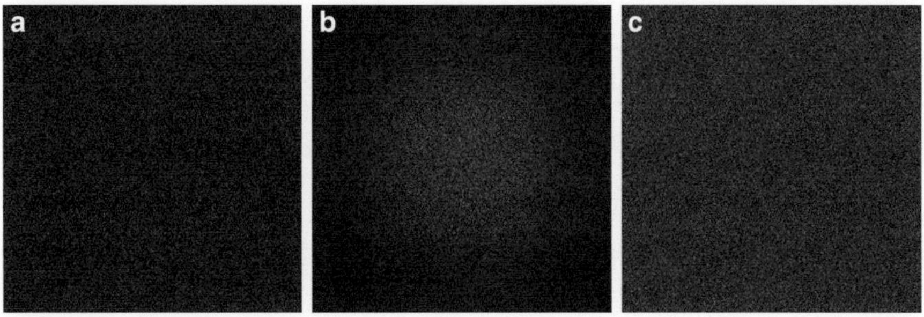

Fig. 13.14 Decryption performed under different conditions when the Fourier plane mask is obtained with an impulse attack, (**a**) using classical DRPE setup; (**b**) when only R_4 is known, and (**c**) when only R_2 is known

domain is also unable to endure this attack. To negate such an attack, amplitude modulation has been performed at the Fourier plane in the DRPE scheme. This could be achieved in the following ways [25, 130]:

13.5.2.1 Amplitude Modulation

Lapses in the security of the DRPE scheme originate due to existence of the linearity. To make the scheme immune against the known-plaintext attack, linearity is claimed to be broken with the introduction of undercover amplitude-modulation operation, and the modified scheme is termed "double random phase-amplitude encoding" (DRPAE) [25]. In addition to both the statistically independent RPMs $R_{\text{in}}(x, y)$ and $R_{\text{f}}(u, v)$, a binary spatially random amplitude mask $A(u, v)$ is bonded with the Fourier plane mask in the known-plaintext attack-free DRPAE scheme (Fig. 13.15). Encoding of the amplitude input image is mathematically represented as

$$E^{i}_{\text{drpae}}(x, y) = \text{FT}^{-1}\left[\left(\text{FT}\left[I(x, y)\, R_{\text{in}}(x, y)\right]\right)\left\{R_{\text{f}}(u, v)\, A(u, v)\right\}\right] \tag{13.15}$$

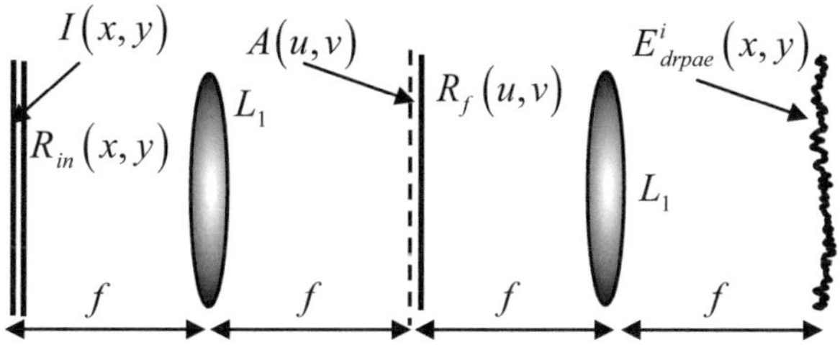

Fig. 13.15 Schematic diagram for the DRPAE scheme

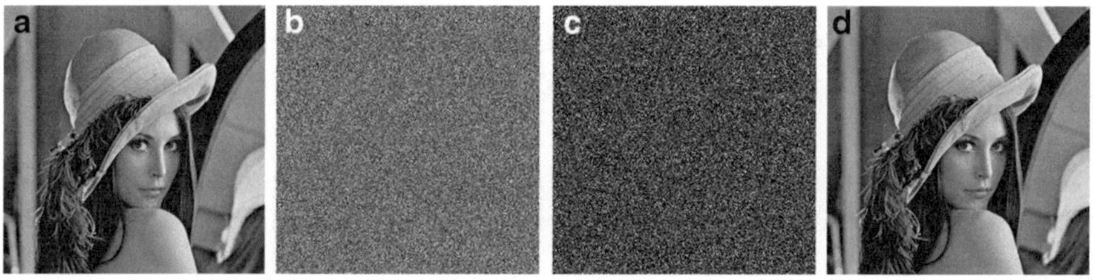

Fig. 13.16 Numerical simulation results for DRPAE: (**a**) input image, (**b**) Fourier plane amplitude mask, (**c**) Fourier plane RPM, (**d**) decrypted image with all correct parameters

where the amplitude mask is represented as

$$A(u,v) = \begin{cases} 1 & \text{if } (u,v) \in \gamma \\ k & \text{if } (u,v) \notin \gamma \end{cases} \tag{13.16}$$

A predefined constant p denotes the ratio between the number of pixels having value 1, and k for the amplitude γ is the chosen point set in the mask.

Enhancement in the security of the DRPAE scheme is claimed due to fact that the existence of the undercover amplitude mask remains hidden to the unauthorized user, and skipping the correct form of the AM results in a false decryption. Even if the attacker is able to possess the correct information about the values of k and p, the decrypted image will result in noise only, due to inaccurate distribution of the amplitude transmittance. The main characteristic feature of the DRPAE scheme is the ability to thwart the powerful known-plaintext attack. The undercover amplitude mask plays a significant role in enhancing the security as a key with removing the linearity of the conventional DRPE scheme. Thus the original image is protected, when the attacker is devoid of exact information about the amplitude mask. For numerical simulation, a gray-scale image of Lena (Fig. 13.16) having 512×512 pixels is assumed to be an input image. The $R_f(u,v)$ and $A(u,v)$ are illustrated in Fig. 13.16 respectively. The amplitude mask has random distribution of 1 and $k = 0.01$, while the factor p is chosen to be 30. Decrypted image (Fig. 13.16) is retrieved only when all the correct parameters are considered.

13.5.2.2 Uniform Amplitude at the Fourier Plane

The known-plaintext is able to crack the DPRE scheme, due to the fact that two necessary inputs for the phase retrieval algorithm are amplitude components of the input- and the encrypted image. In order to avoid this attack, amplitude of the encrypted image could be made constant [130]. Due to this, an attacker fails to breach the security, and encryption key could not be retrieved with the phase retrieval algorithm.

13.6 Resistance of the Known-Plaintext Attack-Free DRPAE Scheme against Impulse Attack

The DRPAE scheme [25] achieves immunity against the most powerful known-plaintext attack, when linearity of the conventional DRPE scheme is broken with the introduction of an undercover amplitude mask at the Fourier plane. Security of the DRPE scheme is improved only due to the fact that without any prior knowledge of the amplitude mask at the Fourier plane, an attacker would retrieve noise as the cracked decrypted image. Although resistance of the non-linear DRPAE scheme is assessed against the advanced and powerful known-plaintext attack, the vulnerability to the impulse function attack [134] is found to persist, and mathematically demonstrated in the following steps,

$$E_{\text{drpae}}^{\delta}(x, y) = \text{FT}^{-1}\left[\left(\text{FT}\left[\delta(0,0)\, R_{\text{in}}(x, y)\right]\right)\{R_{\text{f}}(u, v)\, A(u, v)\}\right] \tag{13.17}$$

The Fourier transformation of the above Eq. (13.15) becomes

$$\{R_{\text{f}}(u, v)\, A(u, v)\} = \text{FT}\left[E_{\text{drpae}}^{\delta}(x, y)\right] \tag{13.18}$$

Complete information with correct values of k, and p about the undercover amplitude mask is cracked from the absolute of cracked mask. The Fourier plane RPM is also obtained

$$R_{\text{f}}(u, v) = \frac{\{R_{\text{f}}(u, v)\, A(u, v)\}}{\text{abs}\,\{R_{\text{f}}(u, v)\, A(u, v)\}} \tag{13.19}$$

In the DRPAE scheme, the delta function is also Fourier transformed into a unity function at the Fourier plane, and the subsequent inverse Fourier transformation divulges the complete information about encryption keys in the form of the encrypted image, similarly to that in the DRPE scheme. Therefore, an unauthorized user is able to obtain the set of encryption keys, i.e. the amplitude mask (Fig. 13.17) and Fourier plane RPM (Fig. 13.17). Therefore, with these keys the original image is cracked (Fig. 13.17)

Fig. 13.17 Impulse attack: (**a**) cracked RPM, (**b**) cracked amplitude mask, and (**c**) cracked original image

$$\delta(0,0) = I_1(x,y) - I_2(x,y) \tag{13.20}$$

The encrypted images $E_1(x,y)$ and $E_2(x,y)$ obtained, respectively, from the input images $I_1(x,y)$ and $I_2(x,y)$ are denoted as follows:

$$E_1(x,y) = \mathrm{FT}^{-1}\left[\left(\mathrm{FT}\left[I_1(x,y)\, R_{\mathrm{in}}(x,y)\right]\right)\{R_{\mathrm{f}}(u,v)\, A(u,v)\}\right] \tag{13.21}$$

$$E_2(x,y) = \mathrm{FT}^{-1}\left[\left(\mathrm{FT}\left[I_2(x,y)\, R_{\mathrm{in}}(x,y)\right]\right)\{R_{\mathrm{f}}(u,v)\, A(u,v)\}\right] \tag{13.22}$$

$$E_1(x,y) - E_2(x,y) = \mathrm{FT}^{-1}\left[\left(\mathrm{FT}\left[\{I_1(x,y) - I_2(x,y)\}\, R_{\mathrm{in}}(x,y)\right]\right)\{R_{\mathrm{f}}(u,v)\, A(u,v)\}\right] \tag{13.23}$$

By replacement of the $\{I_1(x,y) - I_2(x,y)\}$ with the delta function, the above-mentioned equation becomes

$$E_1(x,y) - E_1(x,y) = \mathrm{FT}^{-1}\left[\left(\mathrm{FT}\left[\{\delta(0,0)\}\, R_{\mathrm{in}}(x,y)\right]\right)\{R_{\mathrm{f}}(u,v)\, A(u,v)\}\right] \tag{13.24}$$

Taking FT on both sides of Eq. (13.24) and according to Eq. (13.18), it becomes

$$\{R_{\mathrm{f}}(u,v)\, A(u,v)\} = \mathrm{FT}\left[E_1(x,y) - E_2(x,y)\right] \tag{13.25}$$

This shows that the linearity persists in the DRPAE, similar to that in the conventional DRPE scheme. And the DRPAE scheme is unable to nullify the impulse function attack, it remains vulnerable this attack as the conventional DRPE scheme.

13.7 Conclusions

In conclusion, we have re-emphasized the importance of cryptography to protect information for certain applications in the present era. The progress made in the area of optical cryptographic techniques, particularly using three special classes of the LCTs, namely the Fourier-, fractional Fourier-, and Fresnel transforms, has then been outlined. Many degrees of freedom to manipulate the physical parameters of optical waves can be used for optical encryption, resulting in higher degree of security.

A brief outline is also given of various attacks such as the brute force-, chosen-ciphertext, chosen-plaintext-, known-plaintext-, and some heuristic attacks. Some techniques developed recently by us, especially from the point of view of making the optical cryptographic techniques immune to certain type of attacks, have been discussed briefly.

It is now well recognized that the cryptosystems are indispensable to ensure privacy and authentication of data. The field of optical cryptography is still growing with newer techniques being reported. Hybrid systems based on digital and optical systems make use of the advantages offered by the two techniques. However at the same time, it is also well known from the Kerckhoff's principle that the security of a cryptosystem should depend only on the secret key(s) and not on the secrecy of any part of the system. It has been found that some of the schemes published in the literature do not meet this criterion. Only future investigations would reveal as to how many of the digital, optical, and hybrid techniques will pass this test. Also some methods are more suitable than others for transmission of secure information via the Internet.

Acknowledgements The authors are thankful to SPIE, OSA, and IOP for granting permission to reproduce Figs. (13.3-13.6), (13.7-13.14), and (13.15-13.17) respectively.

References

1. K. Singh, G. Unnikrishnan, N.K. Nishchal, Photorefractive optical processing for optical security. Proc. SPIE **4803**, 205–219 (2002)
2. B. Javidi (ed.), *Optical and Digital Techniques for Information Security* (Springer, NY, 2005)
3. B. Javidi (ed.), *Optical Imaging Sensors and Systems for Homeland Security Applications* (Springer, NY, 2006)
4. A. Alfalou, C. Brosseau, Optical image compression and encryption methods. Adv. Opt. Photon. **1**, 589–636 (2009)
5. O. Matoba, T. Nomura, E. Perez-Cabre, M.S. Millan, B. Javidi, Optical techniques for information security. Proc. IEEE **97**, 1128–1148 (2009)
6. J. Gluckstad, J. Palima, Optical encryption and decryption, in *Generalized Phase Contrast: Applications in Optics and Photonics* (Springer, New York, 2009), Chap. 11, pp. 273–298

7. A. Kumar, M. Singh, K. Singh, Speckle coding for optical and digital data security applications, in: *Advances in Speckle Metrology and Related Techniques*, ed. by G.H. Kaufmann (Wiley-VCH, Weinheim, 2011), Chap. 6, pp. 239–299
8. M.S.M. Garcia-Varela, E. Perez-Cabre, Optical data encryption, in *Optical and Digital Image Processing: Fundamentals and Applications*, ed. by G. Cristobal, P. Schelkens, H. Thienpont (Wiley-VCH Verlag, Weinheim, 2011), Chap. 33, pp. 739–767
9. P. Refregier, B. Javidi, Optical image encryption based on input plane and Fourier plane random encoding. Opt. Lett. **20**, 767–769 (1995)
10. B. Javidi, G. Zhang, J. Li, Encrypted optical memory using double-random phase encoding. Appl. Opt. **36**, 1054–1058 (1997)
11. B. Javidi, A. Sergent, G. Zhang, L. Guibert, Fault tolerance properties of a double phase encoding encryption technique. Opt. Eng. **36**, 992–998 (1997)
12. B. Javidi, A. Sergent, E. Ahouzi, Performance of double phase encoding encryption technique using binarized encrypted images. Opt. Eng. **37**, 565–569 (1998)
13. F. Goudail, F. Bollaro, B. Javidi, P. Refregier, Influence of a perturbation in a double phase-encoding system. J. Opt. Soc. Am. A **15**, 2629–2638 (1998)
14. G. UnniKrishnan, J. Joseph, K. Singh, Optical encryption system that uses phase conjugation in photorefractive crystal. Appl. Opt. **37**, 8181–8186 (1998)
15. B. Javidi, L. Bernard, N. Towghi, Noise performance of double-phase encryption compared to XOR encryption. Opt. Eng. **38**, 9–19 (1999)
16. N. Towghi, B. Javidi, Z. Luo, Fully phase encrypted image processor. J. Opt. Soc. Am. A **16**, 1915–1927 (1999)
17. S. Zhang, M.A. Karim, Color image encryption using double random phase encoding. Microwave Opt. Technol. Lett. **21**, 318–323 (1999)
18. B. Javidi, N. Towghi, N. Maghzi, S.C. Verrall, Error-reduction techniques and error analysis for fully phase- and amplitude based encryption. Appl. Opt. **39**, 4117–4130 (2000)
19. B. Wang, C.-C. Sun, W.-C. Su, A.E.T. Chou, Shift-tolerance property of an optical double-random phase-encoding encryption system. Appl. Opt. **39**, 4788–4793 (2000)
20. H.T. Chang, W.C. Lu, C.J. Kuo, Multiple-phase retrieval for optical security systems by use of random-phase encoding. Appl. Opt. **41**, 4825–4834 (2002)
21. J.F. Barrera, R. Henao, M. Tebaldi, R. Torroba, N. Bolognini, Multiplexing encryption–decryption via lateral shifting of a random phase mask. Opt. Commun. **259**, 532–536 (2006)
22. T. Nomura, E. Nitanai, B. Javidi, Design of input phase mask for the space bandwidth of the optical encryption system. Opt. Eng. **45**, 017006-1/5 (2006)
23. H. Suzuki, M. Yamaguchi, M. Yachida, N. Ohyama, Experimental evaluation of fingerprint verification system based on double random phase encoding. Opt. Express **14**, 1755–1766 (2006)
24. C. Cuadrado-Laborde, Time-variant signal encryption by lensless dual random phase encoding applied to fiber optic links. Opt. Lett. **32**, 2867–2869 (2007)
25. X.C. Cheng, L.Z. Cai, Y.R. Wang, X.F. Meng, H. Zhang, X.F. Xu, X. Shen, G.Y. Dong, Security enhancement of double-random phase encryption by amplitude modulation. Opt. Lett. **33**, 1575–1577 (2008)
26. M. Singh, A. Kumar, K. Singh, Multiplexing in optical encryption by using an aperture system and a rotating sandwich random phase diffuser in the Fourier plane. Opt. Lasers Eng. **46**, 243–251 (2008)
27. M. Singh, A. Kumar, K. Singh, Secure optical system that uses fully phase-based encryption and lithium niobate crystal as phase contrast filter for decryption. Opt. Laser Technol. **40**, 619–624 (2008)
28. G.-h. Situ, D.S. Monaghan, T.J. Naughton, J.T. Sheridan, G. Pedrini, W. Osten, Collision in double random phase encoding. Opt. Commun. **281**, 5122–5125 (2008)
29. A. Alfalou, A. Mansour, Double random phase encryption scheme to multiplex and simultaneous encode multiple images. Appl. Opt. **48**, 5933–5947 (2009)
30. J.F. Barrera, R. Torroba, Efficient encrypting procedure using amplitude and phase as independent channels to display decoy objects. Appl. Opt. **48**, 3120–3129 (2009)

31. P. Kumar, A. Kumar, J. Joseph, K. Singh, Impulse attack free double-random phase encryption scheme with randomized lens-phase functions. Opt. Lett. **34**, 331–333 (2009)
32. P. Kumar, J. Joseph, K. Singh, Impulse attack-free four random phase mask encryption based on a 4-*f* optical system. Appl. Opt. **48**, 2356–2363 (2009)
33. D.S. Monaghan, U. Gopinathan, G. Situ, T.J. Naughton, J.T. Sheridan, Statistical investigation of the double random phase encoding technique. J. Opt. Soc. Am. A **26**, 2033–2042 (2009)
34. D.S. Monaghan, G.-h. Situ, U. Gopinathan, T.J. Naughton, J.T. Sheridan, Analysis of phase encoding for optical encryption. Opt. Commun. **282**, 482–492 (2009)
35. D.S. Monaghan, U. Gopinathan, D.P. Kelly, T.J. Naughton, J.T. Sheridan, Systematic errors of an optical encryption system due to the discrete values of a spatial light modulator. Opt. Eng. **48**, 027001-1/7 (2009)
36. J.F. Barrera, R. Torroba, One step multiplexing optical encryption. Opt. Commun. **283**, 1268–1272 (2010)
37. B.-C. Chen, H.-Z. Wang, Optically-induced-potential-based image encryption. Opt. Express **19**, 22619–22627 (2011)
38. P. Kumar, J. Joseph, K. Singh, Optical image encryption using a jigsaw transform for silhouette removal in interference-based methods and decryption with a single spatial light modulator. Appl. Opt. **50**, 1805–1811 (2011)
39. H.-j. Li, Y.-r. Wang, Double-image encryption based on discrete fractional random transform and chaotic maps. Opt. Lasers Eng. **49**, 753–757 (2011)
40. C. Lingel, J.T. Sheridan, Optical cryptanalysis: metrics of robustness and cost functions. Opt. Lasers Eng. **49**, 1131–1138 (2011)
41. S. Liu, B.M. Hennelly, J.T. Sheridan, Numerical simulation of double random phase encoding. Opt. Eng. **51**, 128201-1/10 (2012)
42. E. Perez-Cabre, H.C. Abril, M.S. Millan, B. Javidi, Photon-counting double-random-phase encoding for secure image verification and retrieval. J. Opt. (IOP) **14**, 094001-1/11 (2012)
43. T. Sarkadi, P. Koppa, Optical encryption using pseudorandom complex spatial modulation. Appl. Opt. **51**, 8068–8073 (2012)
44. X.-g. Wang, D. Zhao, Double images encryption method with resistance against the specific attack based on an asymmetric algorithm. Opt. Express **20**, 11994–12003 (2012)
45. C. Lin, X. Shen, Z.-y. Li, Cryptographic analysis on the key space of optical phase encryption algorithm based on the design of discrete random phase mask. Opt. Laser Technol. **49**, 108–117 (2013)
46. B.M. Hennelly, J.T. Sheridan, Random phase and jigsaw encryption in the Fresnel domain. Opt. Eng. **43**, 2239–2249 (2004)
47. G.-h. Situ, J.-j. Zhang, Double random-phase encoding in the Fresnel domain. Opt. Lett. **29**, 1584–1586 (2004)
48. G.-h. Situ, J.-j. Zhang, Multiple image encryption by wavelength multiplexing. Opt. Lett. **30**, 1306–1308 (2005)
49. G.-h. Situ, J.-j. Zhang, Position multiplexing for multiple-image encryption. J. Opt. A: Pure Appl. Opt. **8**, 391–397 (2006)
50. A. Nelleri, J. Joseph, K. Singh, Digital Fresnel field encryption for three-dimensional information security. Opt. Eng. **46**, 045801-1/8 (2007)
51. H.E. Hwang, H.T. Chang, W.N. Lie, Fast double-phase retrieval in Fresnel domain using modified Gerchberg–Saxton algorithm for lensless optical security systems. Opt. Express **17**, 13700–13710 (2009)
52. P. Kumar, J. Joseph, K. Singh, Optical image encryption based on interference under convergent random illumination. J. Opt. (IOP) **12**, 095402-1/11(2010)
53. P. Kumar, J. Joseph, K. Singh, Holographic encryption system in the Fresnel domain with convergent random illumination. Opt. Eng. **49**, 095803-1/6 (2010)
54. H.T. Chang, H.-E. Hwang, C.-L. Lee, M.-T. Lee, Wavelength multiplexing multiple image encryption using cascaded phase-only masks in the Fresnel transform domain. Appl. Opt. **50**, 710–716 (2011)

55. H.T. Chang, H.-E. Hwang, C.L. Lee, Position multiplexing multiple-image encryption using cascaded phase-only masks in Fresnel transform domain. Opt. Commun. **284**, 4146–4151 (2011)
56. J.-J. Huang, H.-E. Hwang, C.-Y. Chen, C.-M. Chen, Optical multiple-image encryption based on phase encoding algorithm in the Fresnel transform domain. Opt. Laser Technol. **44**, 2238–2344 (2012)
57. S. Yuan, Y.-h. Xin, M. t. Liu, X.-x. Yao, X.-j. Sun, An improved method to enhance the security of double random-phase encoding in the Fresnel domain. Opt. Laser Technol. **44**, 51–56 (2012)
58. S.K. Rajput, N.K. Nishchal, Known-plaintext attack-based optical cryptosystem using phase-truncated Fresnel transform. Appl. Opt. **52**, 871–878 (2013)
59. G. Unnikrishnan, K. Singh, Double random fractional Fourier-domain encoding for optical security. Opt. Eng. **39**, 2853–2859 (2000)
60. G. Unnikrishnan, J. Joseph, K. Singh, Optical encryption by double random phase encoding in the fractional Fourier domain. Opt. Lett. **25**, 887–889 (2000)
61. G. Unnikrishnan, J. Joseph, K. Singh, Fractional Fourier domain encrypted holographic memory by use of an anamorphic optical system. Appl. Opt. **40**, 299–306 (2001)
62. S. Liu, L. Yu, B. Zhu, Optical image encryption by cascaded fractional Fourier transforms with random phase filtering. Opt. Commun. **187**, 57–63 (2001)
63. S. Liu, Q. Mi, B. Zhu, Optical image encryption with multistage and multichannel fractional Fourier-domain filtering. Opt. Lett. **26**, 1242–1244 (2001)
64. Y. Zhang, C.-H. Zhang, N. Tanno, Optical encryption based on iterative fractional Fourier transform. Opt. Commun. **202**, 277–285 (2002)
65. B.M. Hennelly, J.T. Sheridan, Image encryption and the fractional Fourier transform. Optik **114**, 251–265 (2003)
66. B.M. Hennelly, J.T. Sheridan, Optical image encryption by random shifting in fractional Fourier domains. Opt. Lett. **28**, 269–271 (2003)
67. N.K. Nishchal, G. Unnikrishnan, J. Joseph, K. Singh, Optical encryption using a localized fractional Fourier transform. Opt. Eng. **42**, 3566–3571 (2003)
68. N.K. Nishchal, J. Joseph, K. Singh, Fully phase encryption using fractional Fourier transform. Opt. Eng. **42**, 1583–1588 (2003)
69. N.K. Nishchal, J. Joseph, K. Singh, Fully phase-based encryption using fractional order Fourier domain random phase encoding: error analysis. Opt. Eng. **43**, 2266–2273 (2004)
70. N.K. Nishchal, J. Joseph, K. Singh, Fully phase-encrypted memory using cascaded extended fractional Fourier transform. Opt. Laser Eng. **42**, 141–151 (2004)
71. B.M. Hennelley, J.T. Sheridan, Optical encryption and the space bandwidth product. Opt. Commun. **247**, 291–305 (2005)
72. T. Alieva, M.J. Bastiaans, M.L. Calvo, Fractional transforms in optical information processing. EURASIP J. Appl. Signal Process. **10**, 1498–1519 (2005)
73. J.-l. Zhao, H. Lu, X. Song, J. Li, Y. Ma, Optical image encryption based on multistage fractional Fourier transforms and pixel scrambling techniques. Opt. Commun. **249**, 493–499 (2005)
74. M. Joshi, C. Shakher, K. Singh, Color image encryption and decryption using fractional Fourier transform. Opt. Commun. **279**, 35–42 (2007)
75. Z.-j. Liu, S. Liu, Random fractional Fourier transform. Opt. Lett. **32**, 2088–2090 (2007)
76. R. Tao, Y. Xin, Y. Wang, Double image encryption based on random phase encoding in the fractional Fourier domain. Opt. Express **15**, 16067–16079 (2007)
77. M. Joshi, C. Shakher, K. Singh, Color image encryption and decryption for twin images in fractional Fourier domain. Opt. Commun. **281**, 5713–5720 (2008)
78. M. Joshi, C. Shakher, K. Singh, Image encryption and decryption using fractional Fourier transform and radial Hilbert transform. Opt. Lasers Eng. **46**, 522–526 (2008)
79. H.-j. Li, Y.-r. Wang, Double-image encryption by iterative phase retrieval algorithm in fractional Fourier domain. J. Mod. Opt. **55**, 3601–3609 (2008)

80. P. Kumar, J. Joseph, K. Singh, Double random phase encryption with in-plane rotation of a modified Lohmann's second type system in the anamorphic fractional Fourier domain. Opt. Eng. **47**, 117001-1/7 (2008)
81. R. Tao, J. Lang, Y. Wang, Optical image encryption based on the multiple-parameter fractional Fourier transform. Opt. Lett. **33**, 581–583 (2008)
82. L.-f. Chen, D. Zhao, Color image encoding in dual fractional Fourier-wavelet domain with random phases. Opt. Commun. **282**, 3433–3438 (2009)
83. M. Joshi, C. Shakher, K. Singh, Logarithms-based RGB image encryption in the fractional Fourier domain: a nonlinear approach. Opt. Lasers Eng. **47**, 721–727 (2009)
84. Z.-j. Liu, J.-m. Dai, X.-g. Sun, S. Liu, Triple image encryption scheme in fractional Fourier transform domains. Opt. Commun. **282**, 518–522 (2009)
85. L.-f. Chen, D.-m. Zhao, F. Ge, Gray images embedded in a color image and encrypted with FRFT and region shift encoding methods. Opt. Commun. **283**, 2043–2049 (2010)
86. Q. Guo, Z.-j. Liu, S. Liu, Color image encryption by using Arnold and discrete fractional random transforms in HIS space. Opt. Lasers Eng. **48**, 1174–1181 (2010)
87. M. Joshi, C. Shakher, K. Singh, Image encryption using radial Hilbert transform filter bank as an additional key in the modified double random fractional Fourier encoding architecture. Opt. Lasers Eng. **48**, 605–615 (2010)
88. M. Joshi, C. Shakher, K. Singh, Fractional Fourier transform based image multiplexing and encryption technique for four-color images using input images as keys. Opt. Commun. **283**, 2496–2505 (2010)
89. J. Lang, R. Tao, Y. Wang, Image encryption based on the multiple-parameter discrete fractional Fourier transform and chaos function. Opt. Commun. **283**, 2092–2096 (2010)
90. Z.-j. Liu, J.-m. Dai, X.-g. Sun, S. Liu, Single phase encoding method based on the fractional Fourier transform. Optik **121**, 1748–1751 (2010)
91. N. Saini, A. Sinha, Key management of the double random-phase encoding method using public-key encryption. Opt. Lasers Eng. **48**, 329–334 (2010)
92. N. Singh, A. Sinha, Chaos based multiple image encryption using multiple canonical transforms. Opt. Laser Technol. **42**, 724–731 (2010)
93. N. Zhou, T. Dong, J. Wu, Novel image encryption algorithm based on multiple-parameter discrete fractional random transform. Opt. Commun. **283**, 3037–3042 (2010)
94. W. Chen, X.-d. Chen, Optical cryptography topology based on a three-dimensional particle-like distribution and diffractive imaging. Opt. Express **19**, 9008–9019 (2011)
95. D. Liu, W. Jin, Color image encryption based on joint fractional Fourier transform correlator. Opt. Eng. **50**, 068201-1/6 (2011)
96. X.-g. Wang, D. Zhao, Security enhancement of a phase-truncation based image encryption algorithm. Appl. Opt. **50**, 6645–6651 (2011)
97. N.-r. Zhou, Y.-x. Wang, L.-h. Gong, H. He, J.-h. Wu, Novel single channel color image encryption algorithm based on chaos and fractional Fourier transform. Opt. Commun. **284**, 2789–2796 (2011)
98. W. Chen, X.-d. Chen, C.J.R. Sheppard, Optical image encryption based on phase retrieval combined with three-dimensional particle-like distribution. J. Opt. (IOP) **14**, 075402-1/9 (2012)
99. J. Lang, Image encryption based on the reality-preserving multiple parameter fractional Fourier transform. Opt. Commun. **285**, 2584–2590 (2012)
100. J. Lang, A no-key exchange secure image sharing scheme based on Shamir's three-pass cryptography protocol and the multiple-parameter fractional Fourier transform. Opt. Express **20**, 2386–2398 (2012)
101. Z.-j. Liu, L. Xu, J.-m. Dai, S. Liu, Image encryption by using local random phase encoding in fractional Fourier transform domains. Optik **123**, 428–432 (2012)
102. Z.-j. Liu, S. Li, M. Yang, W. Liu, S. Liu, Image encryption based on the random rotation operation in the fractional Fourier domains. Opt. Laser Eng. **50**, 1352–1358 (2012)
103. S.K. Rajput, N.K. Nishchal, Image encryption based on interference that uses fractional Fourier domain asymmetric keys. Appl. Opt. **51**, 1446–1452 (2012)

104. S.K. Rajput, N.K. Nishchal, Asymmetric color cryptosystem using polarization selective diffractive optical element and structured phase mask. Appl. Opt. **51**, 5377–5386 (2012)
105. M.-g. Shan, J. Chang, Z. Zhong, B. Hao, Double image encryption based on discrete multiple-parameter fractional Fourier transform and chaotic maps. Opt. Commun. **285**, 4227–4234 (2012)
106. Q. Wang, Q. Guo, J.-y. Zhou, Double image encryption based on linear blend operation and random phase encoding in fractional Fourier transform domain. Opt. Commun. **285**, 4317–4323 (2012)
107. Z. Zhong, J. Chang, M.-g. Shan, B.-g. Hao, Fractional Fourier-domain random encoding and pixel scrambling technique for double image encryption. Opt. Commun. **285**, 18–23 (2012)
108. Z. Zhong, J. Chang, M.-g. Shan, B.-g. Hao, Double image encryption using double pixel scrambling and random phase encoding. Opt. Commun. **285**, 584–588 (2012)
109. W. Chen, X.-d. Chen, Security enhanced interference-based optical image encryption. Opt. Commun. **286**, 123–129 (2013)
110. Z.-j. Liu, S. Li, W. Liu, Y.-h. Wang, S. Liu, Image encryption algorithm by using fractional Fourier transform and pixel scrambling operation based on double random phase encoding. Opt. Laser Eng. **51**, 8–14 (2013)
111. L.-s. Sui, B. Gao, Single-channel color image encryption based on iterative fractional Fourier transform and chaos. Opt. Laser Technol. **48**, 117–127 (2013)
112. J.-h. Wu, X.-z. Luo, N.-r. Zhou, Four-image encryption method based on spectrum truncation, chaos and the MODFrFT. Opt. Laser Technol. **45**, 571–577 (2013)
113. N.-r. Zhou, X.-b. Liu, Y. Zhang, Y.-x. Yang, Image encryption scheme based on fractional Mellin transform and phase retrieval technique in fractional Fourier domain. Opt. Laser Technol. **47**, 341–346 (2013)
114. A. Carnicer, M. Montes-Usategui, S. Arcos, I. Juvells, Vulnerability to chosen-ciphertext attacks of optical encryption schemes based on double random phase keys. Opt. Lett. **30**, 1644–1646 (2005)
115. Y. Frauel, A. Castro, T.J. Naughton, B. Javidi, Security analysis of optical encryption. Proc. SPIE **5986**, 589603-1/10 (2005)
116. U. Gopinathan, D.S. Monaghan, T.J. Naughton, J.T. Sheridan, A known-plaintext heuristic attack on the Fourier plane encryption algorithm. Opt. Express **14**, 3181–3186 (2006)
117. X. Peng, P. Zhang, H. Wei, B. Yu, Known-plaintext attack on optical encryption based on double random phase keys. Opt. Lett. **31**, 1044–1046 (2006)
118. X. Peng, H. Wei, P. Zhang, Chosen-plaintext attack on lensless double-random phase encoding in the Fresnel domain. Opt. Lett. **31**, 3261–3263 (2006)
119. Y. Frauel, A. Castro, T.J. Naughton, B. Javidi, Resistance of the double random phase encryption against various attacks. Opt. Express **15**, 10253–10265 (2007)
120. G. Situ, U. Gopinathan, D.S. Monaghan, J.T. Sheridan, Cryptanalysis of optical security systems with significant output images. Appl. Opt. **46**, 5257–5262 (2007)
121. H. Wei, X. Peng, H. Liu, S. Peng, B.Z. Gao, Known-plaintext attack on the double phase encoding and its implementation with parallel hardware. Proc. SPIE **6837**, 683703-1/7 (2007)
122. A.M. Youssef, On the security of a cryptosystem based on multiple-parameters discrete fractional Fourier transform. IEEE Signal Process. Lett. **15**, 77–78 (2008)
123. M.-z. He, Q.-f. Tan, L.-c. Cao, Q.-h. He, G.-f. Jin, Security enhanced optical encryption system by random phase key and permutation key. Opt. Express **17**, 22462–22473 (2009)
124. W.-s. Liu, G.-l. Yang, H.-y. Xie, A hybrid heuristic algorithm to improve known-plain text attack on Fourier plane encryption. Opt. Express **17**, 13928–13938 (2009)
125. W. Qin, X. Peng, Vulnerability to known-plaintext attack of optical encryption schemes based on two fractional Fourier transform order keys and double random phase keys. J. Opt. A: Pure Appl. Opt. **11**, 075402-1/8 (2009)
126. Q.W. Ran, H.Y. Zhang, J. Zhang, L.Y. Tan, J. Ma, Deficiencies of the cryptography based on multiple-parameter fractional Fourier transform. Opt. Lett. **34**, 1729–1731 (2009)
127. J.F. Barrera, C. Vargas, M. Tebaldi, R. Torroba, N. Bolognini, Known-plaintext attack on a joint transform correlator encrypting system. Opt. Lett. **35**, 3553–3555 (2010)

128. J.F. Barrera, C. Vargas, M. Tebaldi, R. Torroba, Chosen-plaintext attack on a joint transform correlator encrypting system. Opt. Commun. **283**, 3917–3921 (2010)
129. G. Situ, G. Pedrini, W. Osten, Strategy for cryptanalysis of optical encryption in the Fresnel domain. Appl. Opt. **49**, 457–462 (2010)
130. H. Tashima, M. Takeda, H. Suzuki, T. Obi, M. Yamaguchi, N. Ohyama, Known plaintext attack on double random phase encoding using fingerprint as key and a method for avoiding the attack. Opt. Express **18**, 13722–13781 (2010)
131. W. Qin, X. Peng, X.-f. Meng, Cryptanalysis of optical encryption schemes based on joint transform correlator architecture. Opt. Eng. **50**, 028201-1/4 (2011)
132. W. Qin, X. Peng, X.-f. Meng, B.Z. Gao, Vulnerability to chosen-plaintext attack of opto-electronic information encryption with phase-shifting interferometry. Opt. Eng. **50**, 065601-1/5 (2011)
133. W.-q. He, X. Peng, X.-f. Meng, A hybrid strategy for cryptanalysis of optical encryption based on double-random phase amplitude encoding. Opt. Laser Technol. **44**, 1203–1206 (2012)
134. P. Kumar, J. Joseph, K. Singh, Known-plaintext attack-free double random phase-amplitude optical encryption: vulnerability to impulse function attack. J. Opt. (IOP) **14**, 045401-1/8 (2012)
135. P. Kumar, A. Kumar, J. Joseph, K. Singh, Vulnerability of the security enhanced double random phase-amplitude encryption scheme to point spread function attack. Opt. Laser Eng. **50**, 1196–1201 (2012)
136. X.-g. Wang, D. Zhao, A special attack on the asymmetric cryptosystem based on phase-truncated Fourier transforms. Opt. Commun. **285**, 1078–1081 (2012)
137. S. Yuan, Y.-H. Xin, M.-T. Liu, S.-X. Yao, X.-J. Sun, An improved method to enhance the security of double random-phase encoding in the Fresnel domain. Opt. Laser Technol. **44**, 51–56 (2012)
138. T.J. Naughton, B.M. Hennelly, T. Dowling, Introducing secure modes of operation for optical encryption. J. Opt. Soc. Am. A **25**, 2608–2617 (2008)

Chapter 14
Complex-Valued ABCD Matrices and Speckle Metrology

Steen G. Hanson, Michael L. Jakobsen, and Harold T. Yura

Abstract We demonstrate that within the paraxial ray approximation the propagation of light through a complex optical system can be formulated in terms of a Huygens principle expressed with the complete system's ABCD-matrix elements. As such, propagation through an optical system reduces to that of calculating the relevant matrix elements and substituting these into the expressions derived here. We have introduced complex-valued matrix element to represent apertures, thus having diffraction properties inherent in the description.

We have extended the treatments of Baues and Collins to include partially coherent light sources, optical elements of finite size, and distributed random inhomogeneity along the optical path. In many cases (e.g., laser beam propagation and Gaussian optics) we have been able to derive simple analytical expressions for the optical field quantities at an observation plane.

A series of laser-based optical measurement systems have been analyzed and analytical expressions for their main parameters have been given. Specifically, scattering from rough surfaces not giving rise to a fully developed speckle field, various anemometers and systems for measuring rotational velocity have been treated in order to show the benefits of the complex ABCD matrices.

14.1 Introduction

We will in the following focus on the introduction of soft apertures, i.e. Gaussian apodized, as a new and imaginary element in the ABCD-formalism. The primary use of this will be the analysis of speckle-based optical systems and the related issues, i.e. light penetration through optical turbulence, rough surface scattering, the effects of optical misalignment, etc. The matrix formalism is similar to the concept of

S.G. Hanson (✉) • M.L. Jakobsen
Department of Photonics Engineering, DTU Fotonik, P.O. Box 49, 4000 Roskilde, Denmark
e-mail: vsgh@fotonik.dtu.dk

H.T. Yura
Electronics and Photonics Laboratory, The Aerospace Corporation,
P.O. Box 92957, Los Angeles, CA 90009, USA

J.J. Healy et al. (eds.), *Linear Canonical Transforms*, Springer Series in Optical Sciences 198, DOI 10.1007/978-1-4939-3028-9_14

Canonical Transforms, but the mathematical correspondence will not be discussed here, neither will the theoretical evaluation be presented in details. The following treatise thus is meant for giving examples on how the matrix formalism can be used in practice; therefore, most evaluations will be shown for scalar fields within the paraxial approximation, only, and mostly in one dimension. The extension into 2D will be displayed, as will examples treating a full vectorial scenario. Last but not least, this chapter is intended for giving the reader a feel for the benefits—and the shortcomings—of using the complex ABCD-method. To some extent, it is a simple way quickly to analyze rather complicated optical systems giving a first-order approximation to the relevant outcome for the setup. In short, the method specifies a way to contract the entire optical system into one matrix having all the information for the optical systems. In this way, analytical solutions to various parameters can be obtained, and the optimization becomes utterly simplified, here most importantly including diffraction due to inherent limiting apertures. Having obtained analytical solutions to problems usually paves the way not only for increasing the intelligibility but also for optimization. To some extent, one might therefore allocate this method for the lazy scientist and the productive engineer!

Finally, we sincerely apologize for not presenting a full list of contributions to this highly interesting field, but rather focus on our contribution to the field of matrix-optics. Other contributors to this book have done an excellent job in combining Canonical Transforms with Matrix Optics highlighting the history of the field and thereby providing a deeper insight into the fundamentals of the underlying principles. In this chapter of the book we will take a series of assumptions for granted, without referring to previous authors who have performed the cumbersome work of proving the basic results on which we build. A good overview of the use of real-valued ABCD matrices has been given by Kloos [1].

This chapter of the book will be divided into the following sections: First we will introduce the Gaussian aperture as a "new" matrix element and show how diffraction becomes an integral part of the formalism. Section 14.2 shows how to apply the Complex Matrix formalism to the analysis of scattering off rough surfaces. Section 14.2 will be starting out with scattering off a structure giving rise to fully developed speckles. Analytical expressions for integrated speckles, and sum- and differences of integrated speckle signals are derived. Finally, this chapter will present analytical expressions based on the matrix formalism for the dynamics of speckle patterns due to various kinds of displacements of the scattering structures.

Section 14.3 will focus on deriving analytical expressions for light scattered off structures having surface properties that do not give rise to fully developed speckles, i.e. the field scattered off adjacent positions of the structure can no longer be considered delta-correlated, or the rms. phase change due to the scattering process is on the average lower than 2π. Expressions for spatial correlation between the complex field and the intensities will be derived and presented as a function of the complex parameters for the ABCD matrix applicable for the optical system under consideration. Therefore, the results apply for any optical system within the formalism, being it free space, Fourier transform, or imaging, all including apertures and possible defocusing effects.

The main thrust of our work with complex ray matrices has been for analyzing and establishing new sensor schemes. A survey of the findings with focus on the derivation based on matrix operations is presented in Sect. 14.4. Here we concentrate on the evaluation of analytic expressions relevant for treating speckle-based sensors aimed at probing the dynamics of solid targets. In particular, we will discuss the results for speckle fields arising from linearly moving targets, being it axial or lateral movement, and the results for rotating objects, again for all three axes of rotation.

Section 14.5 highlights the way in which the real-valued ABCD-matrix method can be used to gain not only first-order but also higher order moments for a field that has passed spatially distributed optical turbulence. This evaluation is important not only for beam propagation through optical turbulence but also for propagation through scattering media, being it human tissue or scattering by aerosols in the atmosphere.

14.2 ABCD Matrices with Complex Parameters

The physical-optics model employed in this analysis is based on the Huygens–Fresnel formulation for an arbitrary initial source distribution and a general class of optical elements, which have phase- or amplitude changes up to second order in the transverse direction. Here, we consider the propagation of monochromatic beams and, as such, omit the explicit time-dependence in the formulation presented below. In all situations treated here we employ the well-known paraxial approximation in determining the optical field, in which terms up through second order in the transverse coordinates (with respect to the optical axis) are retained. This approximation is valid when the separation of the optical elements is large compared with the transverse spatial extent of the beam.

In this section we first consider an optical system without limiting apertures and jitter and seek to determine the output field that results from an arbitrary given input field. The physical situation is depicted in Fig. 14.1. Let $U_{\text{in}}(x_0, y_0)$ be the field on the input plane denoted by (x_0, y_0) located at $z = 0$. The field propagates through a train of optical elements $E_1, E_2, \ldots.. E_n$ located at arbitrary separations, to an observation plane at distance z_{n+1} from the final optical element E_N. We assume that the index of refraction of the intervening space between the optical elements is a smoothly varying function of z_{n+1}. We also consider that the surface of the optical element at z_i has the form

$$z_i = z_{0i} + \alpha_i(z)\, x_i^2 + \beta_i(z)\, y_i^2, \tag{14.1}$$

where (x_i, y_i) are transverse coordinates of the ith optical element and $\alpha_i(z)$ and $\beta_i(z)$ are real. For simplicity in presentation, we depict all the optical elements as lying along a straight line, the mutual separations indicated in Fig. 14.1 being the actual distance along the optical axis between the corresponding optical elements in a

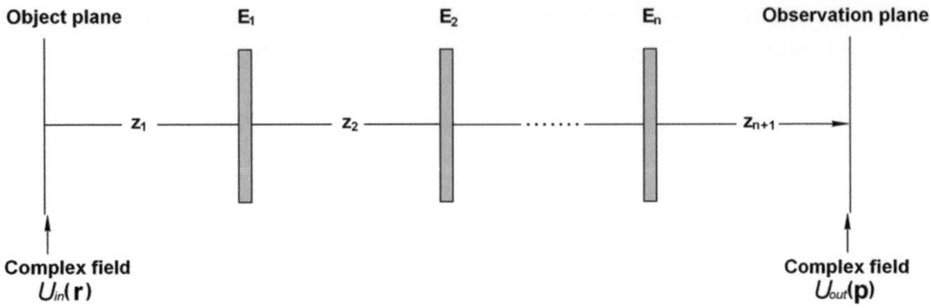

Fig. 14.1 Schematic representation of beam wave propagation through a train of optical elements

situation in which reflection or refraction is present. Under these assumptions it has been shown [2, 3] that the Huygens–Fresnel 2D diffraction integral relating the fields across the input and output planes can be expressed as

$$U_{\text{out}}(\mathbf{p}) = \int G(\mathbf{r}, \mathbf{p})\; U_{\text{in}}(\mathbf{r})\; d\mathbf{r}, \tag{14.2a}$$

where the Green's function is given by:

$$\begin{aligned} G(\mathbf{r}, \mathbf{p}) = \frac{-i\,k}{2\pi\,\sqrt{B_x\,B_y}}\;\exp[-i\,k\,L] \\ \exp\left[-\frac{i\,k}{2B_x}\left(A_x\,x^2 - 2\,xp_x + D_x\,p_x^2\right)\right] \\ \exp\left[-\frac{i\,k}{2B_y}\left(A_y\,y^2 - 2\,yp_y + D_y\,p_y^2\right)\right]. \end{aligned} \tag{14.2b}$$

Here k is the optical wave number $2\pi/\lambda$, λ being the free-space optical wavelength, L is the optical distance along the z-axis, $A_{x,y}$, $B_{x,y}$, and $D_{x,y}$ are the x- and y-axis geometrical ray-matrix elements for the complete optical system between input and output planes. For the case of media with no absorption or gain the ABCD matrix is real-valued, and, when the input and output planes are in free space, its determinant is unity: $AD - BC = 1$. The ray-transfer matrices of some elementary optical structures are given in Fig. 14.2.

Examination of Eq. (14.2) reveals that if $\left|kA_{x,y}\,d_{x,y}^2/B_{x,y}\right|^2 \ll 1$, where $d_{x,y}$ are the characteristic lengths of the optical field in the object plane along the x- and y-axes, respectively (i.e., in the far field), then the far-field optical wave function, apart from a multiplicative phase factor, can be obtained directly from the corresponding wave function for line-of-sight propagation by replacing the line-of-sight propagation distance by $B_{x,y}$. In particular, the far-field irradiance pattern that is obtained after propagation through an optical system that can be characterized by a ray-transfer matrix can be obtained directly from the corresponding pattern for line-of-sight propagation by replacing the propagation distance with B.

Fig. 14.2 Relevant real-valued ray matrices [4]

Table 1. Ray Matrices for Various Simple Optical Elements[a]

Structure	Diagram	Matrix
Straight section (length L)	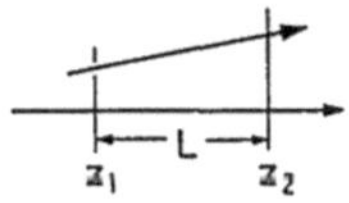	$\begin{bmatrix} 1 & L \\ 0 & 1 \end{bmatrix}$
Thin lens [focal length f: ($f > 0$, converging; $f < 0$, diverging)]	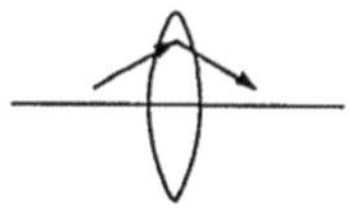	$\begin{bmatrix} 1 & 0 \\ -1/f & 1 \end{bmatrix}$
Flat dielectric interface (refractive indices n_1, n_2)	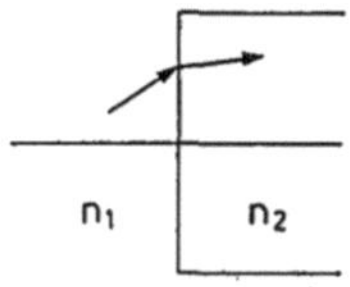	$\begin{bmatrix} 1 & 0 \\ 0 & n_1/n_2 \end{bmatrix}$
Curved dielectric interface (radius R)	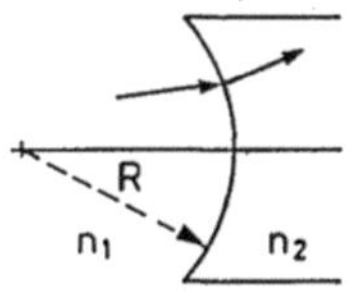	$\begin{bmatrix} 1 & 0 \\ \dfrac{n_2 - n_1}{n_2 R} & \dfrac{n_1}{n_2} \end{bmatrix}$
Curved mirror (radius of curvature R)	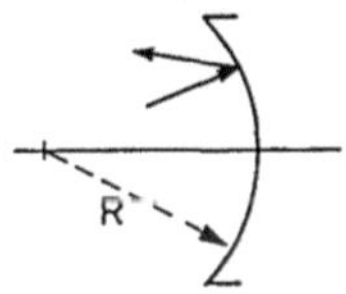	$\begin{bmatrix} 1 & 0 \\ -2/R & 1 \end{bmatrix}$
Thin lens (focal length f followed by a distance L)	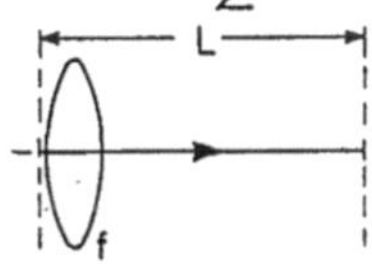	$\begin{bmatrix} 1 - \dfrac{L}{f} & L \\ -\dfrac{1}{f} & 1 \end{bmatrix}$

For a rotationally symmetric optical system, $A_x = A_y \equiv A$ etc., and in this case Eq. (14.2) reduces to

$$U_{\mathrm{out}}(\mathbf{p}) = \int G(\mathbf{r}, \mathbf{p})\ U_{\mathrm{in}}(\mathbf{r})\ \mathrm{d}\mathbf{r}, \tag{14.3a}$$

where

$$G(\mathbf{r}, \mathbf{p}) = \frac{-i\,k}{2\pi\, B} \exp[-i\,k\,L] \exp\left[-\frac{i\,k}{2B}\left(A\,\mathbf{r}^2 - 2\,\mathbf{r}\cdot\mathbf{p} + D\,\mathbf{p}^2\right)\right]. \tag{14.3b}$$

Thus the propagation of light near the optical axis of an astigmatic optical system can be described in the paraxial approximation by means of a Huygens–Fresnel integral with the aid of the geometrical-optics ray matrices (ABCD matrices) for the complete optical system. This formulation of propagation presents a rather simple way of determining refraction effects involving arbitrary optical elements. Needless to say, diffraction effects are not present although phase information is retained. It is necessary merely to multiply the relevant ray matrices together to find the complete system matrix and then to perform the integration indicated in Eq. (14.2) or (14.3).

With the form of the ellipsoidal surfaces given in Eq. (14.1), one obtains two independent 2×2 matrix transformations for the x- and y-directions, respectively. In the more general case in which a term proportional to $x \cdot y$ is added to the right-hand side of Eq. (14.1) one must deal with a symplectic 4×4 matrix transformation [2], and for the case of an initial Gaussian-shaped beam Collins has obtained an analytic form for the resulting complex field amplitude [3]. These results are algebraically complex, and the interested reader is referred to the article for the details. Here, we deal with the ellipsoidal form given by Eq. (14.1) and first summarize some previous results of refs. [1, 2] valid for free-space laser beam propagation.

Without loss of generality we assume that the initial complex field amplitude at the input plane is of the form [4, 5]:

$$U_i(x,y) = A_0 \exp\left[-\frac{x^2}{\omega_{xi}^2}\right] H_m\left[\sqrt{2}\frac{x}{\omega_{xi}}\right] \exp\left[-\frac{y^2}{\omega_{yi}^2}\right] H_n\left[\sqrt{2}\frac{y}{\omega_{yi}}\right], \quad (14.4)$$

where ω_{xi} and ω_{yi} are the initial $1/e$ field spot radii along the x- and y-axes, respectively, H_i are Hermite-polynomials of order i, and A_0 is a (complex) constant. It is to be noted that the radius of curvature of the input beam is here taken to be infinite (i.e., the input plane is assumed to be at the location of the beam waist). This assumption does not limit our analysis, as a thin astigmatic lens of relevant focal length in the x- and y-directions is assumed to be situated immediately to the right of the input plane and is the first optical element that the laser beam passes through. Substituting Eq. (14.4) into (14.2) and performing the integration yields, at the output plane:

$$\begin{aligned} U_i\left(p_x, p_y\right) &= A_0 \left(\tfrac{\omega_{yi}\omega_{yi}}{\omega_y\omega_y}\right) \exp\left[-i\left(kL + \left(\tfrac{1}{2}+m\right)\delta_x + \left(\tfrac{1}{2}+n\right)\delta_y\right)\right] \\ &\exp\left[-i\,\tfrac{\pi p_x^2}{\lambda q_x}\right] H_m\left[\sqrt{2}\tfrac{x}{\omega_x}\right] \exp\left[-i\,\tfrac{\pi p_y^2}{\lambda q_y}\right] H_n\left[\sqrt{2}\tfrac{y}{\omega_y}\right], \end{aligned} \quad (14.5)$$

where:

$$q_{x,y} = \frac{A_{x,y}q_{ix,y} + B_{x,y}}{C_{x,y}q_{ix,y} + D_{x,y}} \text{ and } \frac{1}{q_{ix,y}} = -\frac{i\lambda}{\pi\,\omega_{ix,y}^2}. \quad (14.6)$$

The parameters in the output plane are

$$\begin{aligned} &\text{Spot size}: \ \omega_{x,y} = \omega_{ix,y}\left(A_{x,y}^2 + \left(2B_{x,y}/k\,\omega_{ix,y}^2\right)^2\right)^{1/2}, \\ &\text{Radius of curvature}: \ \tfrac{1}{R_{x,y}} = \left(\tfrac{\omega_{ix,y}}{\omega_{x,y}}\right)^2 \left(A_{x,y}\,C_{x,y} + B_{x,y}\,D_{x,y}\left(\tfrac{2}{k\,\omega^2{}_{ix,y}}\right)^2\right), \\ &\text{and the phase shift } \delta_{x,y} = \text{Arctg}\left[\frac{\lambda B_{x,y}}{\pi\,\omega_{ix,y}^2\,A_{x,y}}\right]. \end{aligned} \quad (14.7)$$

The propagation of a simple TEM00 mode is a simple subset of the above equations. We can now get the important parameters for propagation of a TEM00 Gaussian beam through a train of optical elements. A simple relation for the so-called q-parameter for connecting the output of a Gaussian beam from input (q_{in}) to output (q_{out}) exists. We have $q_{\mathrm{out}} = (Aq_{\mathrm{in}} + B) / (Cq_{\mathrm{in}} + D)$. This relation remains valid even when we as shown in the following introduce complex matrix elements due to the introduction of apertures. It should also be noted that the concept of complex q-parameters lends itself to the analysis of laser resonators. If we assign a matrix for an entire passage in a cavity, the solution to a possible eigenmode is derived by finding the q-parameter q_0 given by $q_0 = (Aq_0 + B) / (Cq_0 + D)$.

14.2.1 Gaussian Apertures

Next, we will introduce the Gaussian apodized limiting aperture as a new element in the matrix formalism. Remembering that the transmission function for a circular symmetric lens with focal length f and the corresponding ray matrix are given by, respectively:

$$\widehat{T}[\mathbf{r}] = \exp\left[-\frac{i\,k\,\mathbf{r}^2}{2f}\right] \text{ and } \widehat{\mathrm{M}}_{\mathrm{lens}} = \left\{\begin{matrix} 1 & 0 \\ -\frac{1}{f} & 1 \end{matrix}\right\}. \tag{14.8}$$

Based on the above, it seems reasonable to associate the transmission function of a Gaussian aperture with $1/e^2$ intensity radius with a ray matrix, as shown [6, 7]:

$$\widehat{T}[\mathbf{r}] = \exp\left[-\frac{|\mathbf{r}|^2}{\sigma^2}\right] \text{ and } \widehat{\mathrm{M}}_{\mathrm{aperture}} = \left\{\begin{matrix} 1 & 0 \\ -\frac{2i}{k\,\sigma^2} & 1 \end{matrix}\right\}. \tag{14.9}$$

It should be emphasized that this matrix and the previously mentioned all have the determinant equal to one in case the refractive indices are identical in the input- and output plane. The determinant of the product of a series of matrices, all having unity determinant, is unity as well. Thus it usually suffices to use three matrix elements in the expressions for the observed fields. A compound complex ray matrix for an optical system with inherent aperture(s) will usually display complex values for all matrix elements. A simple example of the optical system is shown in Fig. 14.3, consisting of a Gaussian input aperture σ_{in}, free space a distance f, a lens with focal length f, an aperture σ_{Four} followed by free space a distance f. This gives the following compound matrix:

$$\widehat{M} = \left\{\begin{matrix} \frac{2f\left(f+i\,k\left(\sigma_{\mathrm{in}}^2+\sigma_{\mathrm{Four}}^2\right)\right)}{\left(k\sigma_{\mathrm{in}}\sigma_{\mathrm{Four}}\right)^2} & f - \frac{2i\,f^2}{k\sigma_{\mathrm{Four}}^2} \\ -\left(\frac{1}{f} + \frac{2i}{k\sigma_{\mathrm{Four}}^2} + \frac{4f}{\left(k\sigma_{\mathrm{in}}\sigma_{\mathrm{Four}}\right)^2}\right) & -\frac{2i\,f}{k\sigma_{\mathrm{Four}}^2} \end{matrix}\right\}. \tag{14.10}$$

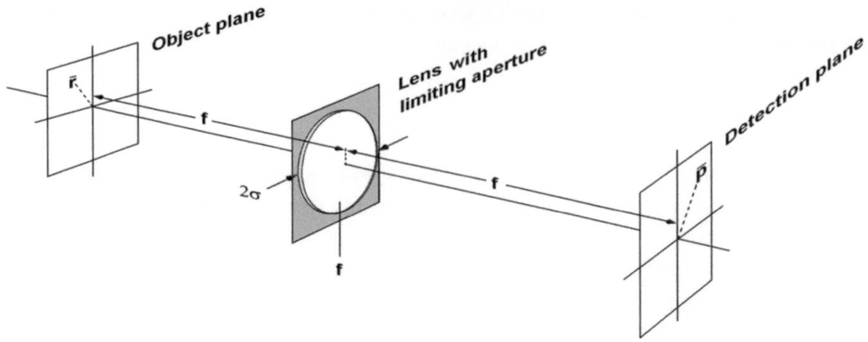

Fig. 14.3 Fourier transforming system

For large apertures, it is noticed that the A-element vanishes, indicating that the output position of a "ray" is independent of its input position. Note that in the complex notation we shall be careful when using the term "ray" if we do not associate diffraction as an inherent property. The same applies for the output slope as a function of the input slope, due to the vanishing of the D-element. On the other hand, the B- and the C-element will survive showing that the output position of a "ray" will depend on its input slope, and vice versa. This indicates that for a system without apertures, we have established an ideal Fourier transforming system. But as apertures are introduced and decreased in diameter, even the real part of the A-element will have a finite value indicating that we partly have "imaging," here pointing towards the onset of a pinhole camera. The leverage point at which the refractive properties of the Fourier-transforming lens are superseded by its diffractive properties is determined by the relation between f and $k\left(\sigma_{\text{Four}}^2 + \sigma_{\text{in}}^2\right)$. A closer examination of this shows that the number of supported modes in the optical system determines the accuracy of the Fourier transformation, i.e. the correctness of this approximation increases with the number of supported modes. This was based on analyzing the A-element. The D-element divided by the B-element controls the field curvature in the output plane, cf. Eq. (14.3). Here, the imaginary part of this ratio defines the radius of the irradiance in the output plane, whereas its real part gives the radius of curvature of the field. Again, for a large aperture of the Fourier-transforming lens, the radius of the curvature will go to infinity, indicating plane wave incidence.

An increased insight in the transformation by the Green's function can be obtained by a simple rewriting of Eq. (14.3):

$$G(\mathbf{r},\mathbf{p}) = \frac{-i\,k}{2\pi\,B}\exp\left[-i\,k\,L\right]\exp\left[-\frac{i\,k}{2AB}\left(\left(A\,\mathbf{r}-\mathbf{p}\right)^2 + \left(AD-1\right)\,\mathbf{p}^2\right)\right], \tag{14.11}$$

dividing the expression into a part responsible for "imaging" including diffraction and a part controlling the curvature and irradiance in the output plane.

14.2.2 Scattering: Fully Developed Speckle Field

Next, we will use the matrix formalism to analyze a simple example, namely the scattering off a structure giving rise to a fully developed speckle field. The scattered field $U_{sc}(\mathbf{r}, t)$ from any surface can be written:

$$U_{sc}(\mathbf{r}, t) = \Psi(\mathbf{r}, t)\ U_{inc}(\mathbf{r}), \tag{14.12}$$

where the incident field is time-independent, and given by $U_{inc}(\mathbf{r})$. The scattering structure is now assumed time-dependent and described by $\Psi(\mathbf{r}, t)$. The model we use for the scattering structure will determine the field in the output plane, as well as its temporal development. The incident field is tacitly assumed time-independent. If the scattering structure is stochastic in nature, we can only express its behavior in terms of its relevant moments. For a structure giving rise to fully developed speckles, we need to know that the field scattered off the structure obeys circular symmetric Gaussian statistics. In other words: if we sample the field immediately after being scattered off the structure, the probability density function (PDF) for the field has to be distributed according to a 2D Gaussian bell, remembering that the field has a real- and imaginary part, defined by its amplitude and phase. To have a field distribution that is circular symmetric—in the complex plane—we need to have a scattered field which includes all phases between 0 and 2π, evenly. In other words, the optical depth of the scattering structure has to exceed the optical wavelength. In case of backscattering from a rough structure, this is equivalent to having a surface RMS roughness exceeding the wavelength. Additionally, in order to invoke the circular symmetric Gaussian PDF, we need to add a fair number of wavelets with random phase in the observation plane. This means that the illuminating spot on the object has to exceed the smallest lateral scale of the surface structure and a certain amount of blurring in the case of perfect imaging. These two constraints being met, we can continue the analysis of the statistics of field parameters propagated through an *ABCD*-system after being scattered off a rough surface, or after having passed a strong diffuser [8].

The correlation between the field at one position and the field at an adjacent position—at a later time—can thus be written:

$$B_{\Psi}[\mathbf{r}_1, \mathbf{r}_2; \tau] \equiv \langle \Psi(\mathbf{r}_1, t)\ \Psi^*(\mathbf{r}_2, t+\tau) \rangle \propto \frac{4\pi}{k^2} \delta(\mathbf{r}_1 - \mathbf{r}_2 + \mathbf{v}\tau). \tag{14.13}$$

Here we have included the time-dependence and assumed the surface structure to be rigid, i.e. it moves as a frozen pattern, usually named the Taylor's hypothesis and assumed stationarity in time and space. Additionally, there is a delta-correlation between scattering from adjacent positions; knowing the scattered field at one position provides no information on the scattered field from a neighboring position. As will be shown later, this assumption may lead to erroneous results in some cases. To avoid this, it would be more correct to assign a smallest scale equal to the wavelength. Finally, the reflection coefficient for the field is unity and the factor $4\pi/k^2$ normalizes the intensity by assuming the power of the scattered light being scattered evenly in a solid angle of 2π. The angular brackets here and in the following stand for "ensemble average."

Having the expression for the field in the input plane of the optical system facilitates calculation of the field in the output plane by inserting Eq. (14.12) into (14.3).

$$U_{\text{out}}(\mathbf{p},t) = \int G(\mathbf{r},\mathbf{p})\ \Psi(\mathbf{r},t)\ U_{\text{inc}}(\mathbf{r})\ d\mathbf{r},$$
$$\text{where} \tag{14.14}$$
$$G(\mathbf{r},\mathbf{p}) = \tfrac{-i\,k}{2\pi\,B}\ \exp\left[-i\,k\,L\right]\exp\left[-\tfrac{i\,k}{2B}\left(A\,\mathbf{r}^2 - 2\,\mathbf{r}\cdot\mathbf{p} + D\,\mathbf{p}^2\right)\right].$$

The ensemble average of the field will vanish due to the ensemble average of the scattering phase function having en ensemble average of zero, i.e. $\langle U_{\text{out}}(\mathbf{p},t)\rangle = \int G(\mathbf{r},\mathbf{p})\ \langle\Psi(\mathbf{r},t)\ \rangle\ U_{inc}(\mathbf{r})\ d\mathbf{r} = 0$. We only know some statistical moments for the scattering structure, usually not the full expression. Therefore, we can express the ensemble average of the intensity:

$$\langle I_{\text{out}}(\mathbf{p},t)\rangle =$$
$$\int_{-\infty}^{\infty}\int_{-\infty}^{\infty} G(\mathbf{r}_1,\mathbf{p})\ G^*(\mathbf{r}_2,\mathbf{p})\ U_{\text{inc}}(\mathbf{r}_1)\ U_{\text{inc}}^*(\mathbf{r}_2)\,\langle\Psi(\mathbf{r}_1,t)\,\Psi^*(\mathbf{r}_2,t)\rangle\,d\mathbf{r}_1 d\mathbf{r}_2\,. \tag{14.15}$$

Inserting the delta-correlation for the scattering structure from Eq. (14.13) immediately gives us an analytic expression for the intensity distribution in the observation plane

$$\langle I_{\text{out}}(\mathbf{p},t)\rangle =$$
$$\tfrac{1}{|B|^2}\ \exp\left[k\ \text{Im}\left[(D/B)\ p^2\right]\right] \tag{14.16}$$
$$\int_{-\infty}^{\infty} d\mathbf{r}\ I_s(\mathbf{r})\ \exp\left[-2k\ \text{Im}\left[1/B\right]\ \mathbf{p}\cdot\mathbf{r}\right]\ \exp\left[k\ \text{Im}\left[A/B\right]r^2\right],$$

where $I_s(\mathbf{r})$ is the source intensity in the input plane. Please note that the above equation is only valid in case we have at least one imaginary element in the matrix. For a matrix with only real-valued elements, the integrated intensity, i.e. the power, in the output plane does not converge. This stems from the fact that we have assumed delta-correlation of the scattering structure. It would have been correct to assume a lateral correlation length not to be below the wavelength [9].

Of special importance when a coherent field is scattered from a rough surface is the so-called speckle pattern that arises due to destructive and constructive interference of light collected in the observation plane, but arising from different parts of the input plane [10, 11]. Here we seek to determine the time-lagged covariance of the photocurrent obtained from two spatially separated point detectors that are located in an arbitrary observation plane, transverse to the optical axis of an ABCD-system. This quantity is given by

$$C\left[\mathbf{p}_1, \mathbf{p}_2; \tau\right] \equiv \langle I_{\text{out}}(\mathbf{p}_1, t)\, I_{\text{out}}(\mathbf{p}_2, t+\tau)\rangle - \langle I_{\text{out}}(\mathbf{p}_1, t)\rangle \langle I_{\text{out}}(\mathbf{p}_2, t+\tau)\rangle\,, \tag{14.17}$$

where τ is the time-lag, the dynamics is stationary in time and $\mathbf{p}_1$ and $\mathbf{p}_2$ are the two-dimensional vector coordinates for the (point) detectors in the observation plane. In case the rigid structure is displaced, being it transversely or axially, linearly translated or rotated about one of the three principal axes, the speckle pattern will undergo some kind of displacement, as well as decorrelation will set in, i.e. the speckle pattern will display boiling properties as it moves, as shown in Fig. 14.4. The time-lagged covariance will include information on the four most important governing parameters for describing the dynamic speckle pattern in the observation plane, viz.

- the intensity distribution
- thc spcckle sizc
- the speckle velocity as a function of the object movement
- the speckle decorrelation (boiling) as a function of the object movement.

We will here increase the complexity and correctness of the scattering structure by assuming its lateral scale being finite but still moving as a solid body, i.e. Taylor's

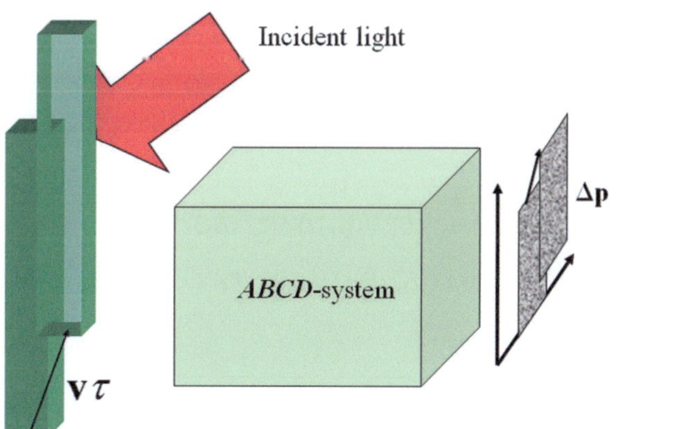

$$\langle I(\mathbf{p},0)\, I(\mathbf{p}+\Delta\mathbf{p},\tau)\rangle = \langle I\rangle^2 + \left|\langle U(\mathbf{p},0)\, U^*(\mathbf{p}+\Delta\mathbf{p},\tau)\rangle\right|^2$$

Fig. 14.4 Space-time correlation of speckle pattern due to displacement of the scattering structure hypothesis. This means that time-lagged scattering phase function reads.

$$B_{\Psi}\left[\mathbf{r}_1, \mathbf{r}_2; \tau\right] = \frac{4\pi}{k^2} \frac{1}{\pi r_{\text{coh}}^2} \exp\left[-\frac{\left(\mathbf{r}_1 - \mathbf{r}_2 - \mathbf{v}_t \tau\right)^2}{r_{\text{coh}}^2}\right], \tag{14.18}$$

where r_{coh} is a measure of the phase correlation length of the objects surface, given by

$$r_{\text{coh}} = r_h / \sigma_{\varphi}. \tag{14.19}$$

Here r_h is the lateral coherence length of the surface-height fluctuations, and σ_{φ} is the standard deviation of the reflected optical phase and $\mathbf{v}_t$ is the transverse velocity of the object. We assume diffuse reflection only. In particular, we take for granted that the specular component of the reflected field is negligible in comparison with the corresponding diffuse component, and hence we can assume that the reflected field obeys circular complex Gaussian statistics. A stringent derivation of the time-lagged intensity covariance would involve following the propagation of four fields (product of two intensities) from the input plane to two positions in the output plane, thus calling for knowledge of the fourth moment of the scattering phase function, which usually is not available unless we assume Gaussian statistics. In Eq. (14.18) we have introduced a model for the second-order statistical moment. In order to proceed, we invoke Gaussian statistics for the field in the output plane, as it involves summation of many statistically independent wavelets. As a result, all higher order moments of the field can be broken into products of second-order correlations [10]. Especially in the case of the covariance, we get:

$$\begin{aligned} C\left[\mathbf{p}_1, \mathbf{p}_2; \tau\right] &\equiv \left\langle I_{\text{out}}\left(\mathbf{p}_1, t\right) I_{\text{out}}\left(\mathbf{p}_2, t+\tau\right)\right\rangle - \left\langle I_{\text{out}}\left(\mathbf{p}_1, t\right)\right\rangle \left\langle I_{\text{out}}\left(\mathbf{p}_2, t+\tau\right)\right\rangle \\ &= \left|\left\langle U_{\text{out}}\left(\mathbf{p}_1, t\right) U_{\text{out}}^*\left(\mathbf{p}_2, t+\tau\right)\right\rangle\right|^2. \end{aligned} \tag{14.20}$$

We can now derive an analytical expression for the time-lagged covariance for any optical system that can be modeled with ABCD matrices. The expression to be solved is:

$$\begin{aligned} C\left[\mathbf{p}_1, \mathbf{p}_2; \tau\right] &= \left|\left\langle U_{\text{out}}\left(\mathbf{p}_1, t\right) U_{\text{out}}^*\left(\mathbf{p}_2, t+\tau\right)\right\rangle\right|^2 = \\ \int_{-\infty}^{\infty}\int_{-\infty}^{\infty} & d\mathbf{r}_1 d\mathbf{r}_2\, U_{\text{inc}}\left(\mathbf{r}_1\right) U_{\text{inc}}^*\left(\mathbf{r}_2\right) B_{\Psi}\left(\mathbf{r}_1, \mathbf{r}_2; \tau\right)\; G\left(\mathbf{r}_1, \mathbf{p}_1\right) G^*\left(\mathbf{r}_2, \mathbf{p}_2\right). \end{aligned} \tag{14.21}$$

Assuming a circular symmetric incident Gaussian beam with intensity radius ω_s and using the expression for the Green's function, Eq. (14.3), we get an analytical expression for the covariance

$$C_{\text{power}}\left[\mathbf{p}_1,\mathbf{p}_2;\tau\right] = C_0\ \exp\left[-2\frac{\mathbf{p}_1^2+\mathbf{p}_2^2}{\omega^2}\right]\ |\gamma\left[\Delta\mathbf{p};\tau\right]|^2, \tag{14.22}$$

$$|\gamma\left(\Delta\mathbf{p};\tau\right)| = \exp\left[-\frac{\left(\mathrm{Im}\left[A\right]-\frac{2\mathrm{Re}\left[B\right]}{k\,\omega_s^2}\right)^2(\mathbf{v}\tau)^2+\left(\Delta\mathbf{p}-\left(\mathrm{Re}\left[A\right]+\frac{2\mathrm{Im}\left[B\right]}{k\,\omega_s^2}\right)(\mathbf{v}\tau)\right)^2}{\rho_0^2}\right], \tag{14.23}$$

were the normalized covariance is given by $|\gamma[\Delta\mathbf{p};\tau]|^2$. This calculation may be derived with the aid of a symbolic computer program, such as *Mathematica* or *MathLab*. The overall factor is given by

$$C_0 = \left(\frac{8\alpha S\,P_0}{\pi\,\omega_s^2k^2\rho_0^2}\right)^2, \tag{14.24}$$

with the speckle size ρ_0 given by

$$\rho_0 = \left(\frac{8|B|^2}{k^2\omega_s^2}+\frac{4}{k}\mathrm{Im}\left[BA^*\right]+r_c^2\left(|A|^2+\frac{4|B|^2}{k^2\omega_s^4}+\frac{4\mathrm{Im}\left[BA^*\right]}{k\omega_s^2}\right)\right)^{1/2}. \tag{14.25}$$

In Eq. (14.24) the detector size is S, and the normalized covariance in Eq. (14.23) is with respect to the signal including the factor α converting incident power to current. A closer look at the speckle size reveals its structure: The first term inversely proportional to the illuminating beam radius stems from diffraction of the incident beam, the second term arises due to diffraction in the optical system, whereas the third term primarily is caused by magnification, $|A|$, of the object's transverse structure. As shown, the three contributions add geometrically. As long as $r_c \ll \omega_s$ the last two terms in the inner parenthesis are negligible. If this was not the case, the assumption of Gaussian statistics would not be fulfilled due to the limited number of independent wavelets being added up.

Due to the movement of the object, here assumed to be a linear translation, the speckle pattern may be translated. The peak position for the time-lagged covariance appears at the position where the two images acquired before and after the translation are most alike. The peak position for a given time-delay τ appears at the position:

$$\Delta\mathbf{p}_{\text{peak}} = \left(\mathrm{Re}\left[A\right]+\frac{2\mathrm{Im}\left[B\right]}{k\omega_s^2}\right)\mathbf{v}\tau. \tag{14.26}$$

Note that the displacement of the speckle pattern is not exactly given by the magnification of the optical system, but somewhat altered. To be more accurate, the peak position of the cross covariance is "squeezed" slightly due to decorrelation. But as the object moves, new scattering structures will be illuminated, and eventually the speckle patterns before and after the movement will lose their similarity. This is named the decorrelation length, and is given by

$$\Delta r_{\text{decorrelation}} = |\mathbf{v}\tau| = \left| \text{Im}\,[A] - \frac{2\text{Re}\,[B]}{k\omega_s^2} \right| \rho_0. \tag{14.27}$$

Finally, the radius if the irradiance in the observation plane becomes:

$$\omega = \rho_0 \frac{|B|^2}{2\left(\text{Im}\,[BD^*]\ \text{Im}\,[BA^*] - \text{Im}[B]^2\right) + 4\text{Im}\,[BD^*]\,|B|^2/\left(k\omega_s^2\right)}, \tag{14.28}$$

The above Eq. (14.28) here simplified for $r_c \to 0$.

We have in the above first assumed that the illuminating field is collimated, its radius of curvature being infinite, and secondly that the object moved laterally. In case the incident beam has a finite curvature, this curvature can merely be included by adding a synthetic lens with the correct focal length as a first optical element. If the field is scattered off a target rotating about a transverse axis of radius R_c, the above formulas apply after having inserted another synthetic lens with a focal length $R_c/2$ and insert a velocity $\Omega\, R_c$ in the above equations, where Ω is the angular velocity of the object. Please note that the transverse axis of radius R_c might not be identical to the radius of curvature of the object, itself. Doing this, a simple set of equations can be established for the analysis of dynamic speckles from a rotating structure.

It is of importance when establishing a measurement system dedicated to measure a certain displacement parameter, to know the relation between speckle velocity—usually instrumental for the accuracy—and speckle decorrelation, which often is detrimental for the measurement.

The expressions related to dynamic speckles apply for the field scattered off a static object, as well, i.e. giving an extension of the Van Cittert–Zernike theorem for optical systems within the ABCD-formalism, and not only for free-space propagation. Therefore we can use Eqs. (14.25) and (14.28) with $\tau = 0$ and get expressions for the intensity distribution in the observation plane and the transverse speckle size, but the axial correlation length of the speckles cannot be derived from these equations. It is well known that the speckles usually are structures that are elongated primarily along the optical axis. To achieve this, we have to derive the covariance between two intensity patterns now derived at two axially displaced distances from the exit of the optical system as shown in Fig. 14.5.

To accomplish this, we again assume that the field in the observation plane obeys circular symmetric Gaussian statistics, so that we can reduce the calculation of intensity correlation (i.e., fourth-order in field) into merely finding the absolute

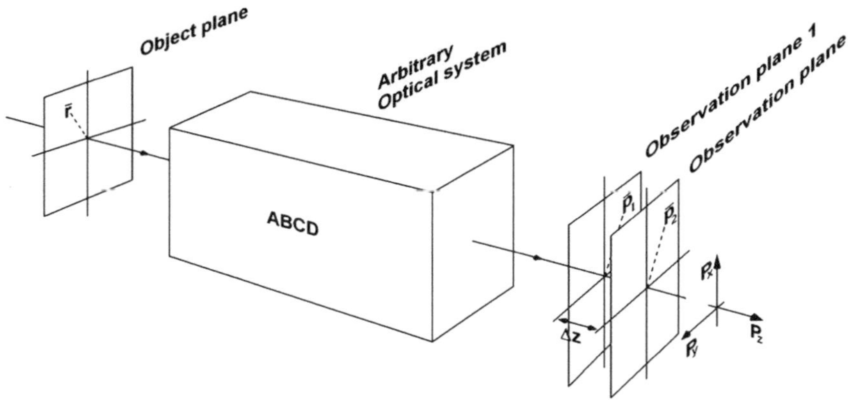

Fig. 14.5 Setup for deriving the axial correlation length of a field

square of the ensemble average of the fields at the two axially displaced distances Δz [12]. The covariance, we have to calculate can be found in the same manner, as was the case for Eq. (14.21):

$$C\left[\mathbf{p}_1, \mathbf{p}_2; \Delta z\right] = \left|\left\langle U_{\text{out}}\left(\mathbf{p}_1, 0\right)\; U^*_{\text{out}}\left(\mathbf{p}_2, \Delta z\right)\right\rangle\right|^2 =$$

$$\int_{-\infty}^{\infty}\int_{-\infty}^{\infty} \mathrm{d}\mathbf{r}_1 \mathrm{d}\mathbf{r}_2\; U_{\text{inc}}\left(\mathbf{r}_1\right)\; U^*_{\text{inc}}\left(\mathbf{r}_2\right)\; B_\Psi\left(\mathbf{r}_1, \mathbf{r}_2\right)\; G\left(\mathbf{r}_1, \mathbf{p}_1; 0\right)\; G^*\left(\mathbf{r}_2, \mathbf{p}_2; \Delta z\right). \tag{14.29}$$

The Green's function for propagation to the exit of the optical system is $G(\mathbf{r}_1, \mathbf{p}_1; 0)$, and the Green's function for the axially displaced observation plane is $G(\mathbf{r}_2, \mathbf{p}_2; \Delta z)$. The first is defined by the complex ABCD matrix as described in Eq. (14.11), whereas the second is found by multiplying the same matrix with the matrix for free-space propagation from the left a distance Δz. If we intended to correlate the intensity in case the object was axially displaced between the two recordings, this would merely be accomplished by multiplying with the free-space matrix from the right. In one dimension, the matrix relevant for $G(\mathbf{r}_2, \mathbf{p}_2; \Delta z)$ becomes:

$$\widehat{M}_{\Delta z} = \begin{Bmatrix} 1 & \Delta z \\ 0 & 1 \end{Bmatrix} \begin{Bmatrix} A & B \\ C & D \end{Bmatrix} = \begin{Bmatrix} A + C\,\Delta z & B + D\,\Delta z \\ C & D \end{Bmatrix}. \tag{14.30}$$

Like for the case of dynamic speckles, we can derive the normalized intensity covariance in analytic form by Eq. (14.29), here derived assuming only a real-valued matrix for the sake of clarity. The normalized covariance becomes:

$$\frac{C\left[\mathbf{p}_1,\mathbf{p}_2;\Delta z\right]}{C\left[\mathbf{p}_1,\mathbf{p}_2;0\right]}=\frac{\exp\left[-\dfrac{\left(D\,k\omega_0\Delta z\right)^2}{B^4\Gamma\left[\Delta z\right]}\mathbf{p}^2\right]}{\Gamma\left[\Delta z\right]} \tag{14.31}$$
$$\text{with }\Gamma\left[\Delta z\right]\equiv\left(\frac{B+D\Delta z}{B}\right)^2+\left(\frac{k\omega_0^2\Delta z}{2B^2}\right)^2.$$

The length of the speckles can be determined by the Δz - value at which the normalized covariance has decreased to 1/*e*. For speckles close to the optical axis, the axial correlation length will become

$$l_{\mathrm{coh}}\cong 2B^2/\left(k\,\omega_0^2\right). \tag{14.32}$$

The speckle length will in loose terms be inversely proportional to the square of the numerical aperture of the optical system, whereas the lateral speckle size is inversely proportional to the numerical aperture. Equation (14.31) further shows that the speckle length will decrease as we move away from the optical axis. The properties for dynamics of speckles in three dimensions arising from a displaced structure have previously been derived and analytic expressions obtained within the complex ABCD-matrix formalism, cf. [12].

When one describes "speckle displacement," the meaning usually is based on observing how the speckles move in a (transverse) plane. This movement could be due to a real transverse movement of a more or less rigid speckle structure, but this could just as well be due to a partly axial movement of a collection of slanted speckle grains. This issue has been analyzed by combining the 3D evaluation and the method for investigating speckle dynamics [12]. It is here shown that in case of linear transverse displacement and out-of-plane rotation of an object, the speckle will suffer a purely transverse displacement, independent of the optical system. But the speckles, themselves, will usually have a tilt with respect to the optical axis.

The derivations shown up until now should be considered as basic examples how to apply the concept of complex ABCD matrices. A suite of issues are open for further investigation, being it for deterministic fields, e.g. singularities, or for fields that can only be described according to their stochastic properties. Of special interest could be the analysis of optical angular momentum, vorticity, energy flow, etc. In addition calculation of the correlation between two "images" taken at different wavelengths, or with different apertures, can easily be calculated, if so desired. Calculation of analytical expressions can be highly simplified by taking advantage of programs for symbolic manipulation, such as *Mathematica* or *MathLab*. Appendix provides a Mathematica program that will give the compound matrix for a given sequence of optical components.

14.3 Scattering from Surfaces Not Giving Rise to Fully Developed Speckles (Field Statistics, Intensity Statistics)

The previously described statistics for scattering off a surface relied on a rather simplistic model for the surface, due to the objective of showing the impact of the optical system for the statistical parameters in the observation plane. Here, we will expand the model to include a more realistic model. After having presented this model and shown how the matrix method can cope with this issue, we will briefly discuss the results. The analysis will here not be based on the assumption of circular symmetric Gaussian statistics for the field.

The first issue that will be treated is field correlations. In the following we have assumed unpolarized, monochromatic (Gaussian-shaped) laser beam illumination of the surface, where the scalar Huygens–Fresnel theory applies. In particular, we obtain image-plane first-order statistics (i.e., the mean and variance of both the real and the imaginary parts of the complex field in the image plane) that are valid for all values of the number of correlation areas of the surface that contribute to the observed intensity. This is in contrast to previous work, which assumed uniform illumination of the surface and was restricted to situations where the corresponding number of correlation areas is much greater than unity (i.e., where the central-limit theorem applies) [13]. As an illustrative example, we present analytic closed-form results, as well as a highly accurate approximation based on elementary functions, for the signal-to-noise ratio in situations where the scattered light is mixed with a strong local oscillator.

Following Goodman [10] we model the surface-height fluctuations as a homogeneous and isotropic Gaussian random process and express the (reflected) optical field correlation function, $\langle \exp\left[i\left(\varphi\left(\mathbf{r}_1\right)-\varphi\left(\mathbf{r}_2\right)\right)\right]\rangle$, in the input plane as (angle brackets denote the ensemble average):

$$B(r) = \langle \exp\left[i\left(\varphi\left(\mathbf{r}_1\right)-\varphi\left(\mathbf{r}_2\right)\right)\right]\rangle = \exp\left[-\sigma_\varphi^2\left(1-\rho_h(r)\right)\right], \tag{14.33}$$

where $\mathbf{r}_{1,2}$ denotes two-dimensional transverse position vectors in the input plane, $r = |\mathbf{r_1} - \mathbf{r}_2|$ and $\varphi\left(\mathbf{r}\right)$ is the reflected optical phase, σ_φ^2 is the corresponding phase variance given by

$$\sigma_\varphi^2 = \left(k\left(1+\cos\left[\beta\right]\right)\right)^2\sigma_h^2, \tag{14.34}$$

where k is the optical wave number, σ_h^2 is the corresponding variance of surface-height fluctuations, $\rho_h(r)$ is the normalized correlation function of the surface-height fluctuations, and β is the angle of incidence of the incident light, cf. Goodman [10].

The form for the correlation function given by Eq. (14.33) is often used in analysis and is reasonably accurate if the surface slopes are small. In obtaining Eq. (14.33) we have assumed that the reflection coefficient is equal to unity over the entire surface. To obtain analytical results, we assume that the normalized correlation function of surface-height fluctuations is given by another Gaussian distribution:

$$\rho_h(r) = \exp\left[-\frac{r^2}{r_h^2}\right]. \tag{14.35}$$

Our intention is to derive expressions for first-order intensity statistics. In order to derive this, we write the scattered field from the scattering surface before propagation through the optical ABCD-system as

$$U(\mathbf{r}) = U_0 \exp\left[-\frac{\mathbf{r}^2}{\omega_s^2}\right] \exp\left[-i\,\varphi(\mathbf{r})\right], \tag{14.36}$$

where U_0 is a constant and ω_s is the $1/e^2$ intensity radius. Assuming a general paraxial ABCD optical system, the corresponding complex field amplitude on the axis in the observation plane $U_{\text{out}}(0)$ can be expressed:

$$U_{\text{out}}(\mathbf{0}) = \int_{-\infty}^{\infty} U(\mathbf{r})\; G(\mathbf{r}, 0)\; d\mathbf{r}, \tag{14.37}$$

where the Green's function is given by $G(\mathbf{r}, 0) = -\frac{ik}{2\pi B} \exp\left[-\frac{ik}{2B} A\, \mathbf{r}^2\right]$.

Here, as in all the derivations, we have tacitly assumed the input and output plane to be in the same medium. As we intend to derive various components of the field statistics, we will derive the deviation from the field U_d that would be present in case there was no phase perturbation imposed by the object, i.e. for $\varphi(\mathbf{r}) = 0$, which corresponds to specular reflection. This perturbation on top of the specular field will be named u and becomes

$$u = -\frac{ik}{2\pi B\, U_d} \int_{-\infty}^{\infty} U_0(\mathbf{r}) \exp\left[-\mathbf{r}^2\left(\frac{1}{\omega_e^2} + \frac{ik}{2z_e}\right)\right] d\mathbf{r}, \tag{14.38}$$

where

$$\omega_e = \sqrt{-2/\left(k\text{Im}\left[A/B\right]\right)} \text{ and } z_e = 1/\text{Re}\left[A/B\right]. \tag{14.39}$$

Examination of Eq. (14.38) reveals that ω_e and z_e can be interpreted as the effective linear dimension of the surface that contributes to the observed intensity under uniform illumination conditions and the effective free-space propagation distance between the input and the output plane, respectively.

Equation (14.38) is the generalization to general ABCD optical systems of the equivalent free-space propagation analog discussed by Jakeman for simple imaging systems [14]. We note that all the stochastic parameters associated with the rough surface are contained in the reduced field quantity u, and hence in the following we determine second-order field statistics of u. The relationship between the second-order field statistics of u and the corresponding statistics of the real intensity is given in Appendix [15].

All the second-order field statistics of u in the output plane can be obtained from the knowledge of

$$u_r \equiv \mathrm{Re}\,[u]\,,\ u_i \equiv \mathrm{Im}\,[u]\,,\ \left\langle |u|^2 \right\rangle \text{ and } \left\langle u^2 \right\rangle. \tag{14.40}$$

Using standard techniques for performing integrals of Gaussian form and noting that $\langle \exp\,[i\varphi] \rangle = \exp\left[-\sigma_\varphi^2/2\right]$, it is straightforward to show that

$$\langle u_r \rangle = \exp\left[-\sigma_\varphi^2/2\right] \text{ and } \langle u_i \rangle = 0. \tag{14.41}$$

To calculate $\langle |u|^2 \rangle$ and $\langle u^2 \rangle$ on the basis of the correlation function given by Eq. (14.1), we follow the procedure outlined in Sect. 2.7.2 of [10] and obtain

$$\begin{aligned} \left\langle |u|^2 \right\rangle &= {}_1F_1\left[1, 1 + \tfrac{1+i\,\delta}{2n}; -\sigma_\varphi^2\right] \text{ and} \\ \left\langle u^2 \right\rangle &= \left(\tfrac{1+i\,\delta}{2n}\right) \frac{\Gamma\left[\frac{1+i\,\delta}{2n}, 0, \sigma_\varphi^2\right]}{\left(\sigma_\varphi^2\right)^{(1+i\delta)/2n}} \exp\left[-\sigma_\varphi^2\right], \end{aligned} \tag{14.42}$$

where ${}_1F_1[a, b; z]$ is the confluent hypergeometric function, $\Gamma[a, b, z]$ is the generalized incomplete gamma function.

$$\begin{aligned} \delta = \tfrac{a}{1+b},\ a = \tfrac{k\omega_e^2}{2\,z_e},\ b = \tfrac{\omega_e}{\omega_s},\ n = (r_\omega/\omega_s)^2 \\ \text{where } \tfrac{1}{r_\omega^2} = \tfrac{1}{\omega_s^2} + \tfrac{1}{\omega_e^2}. \end{aligned} \tag{14.43}$$

From Eqs. (14.37) and (14.38) it follows that r_ω is a measure of the linear dimension of the input plane that contributes to the observed intensity. Hence n can be interpreted as the effective number of independent correlation areas in the input plane that contribute to the observed intensity. Note that the ray-matrix elements of the optical system are contained in the two quantities n and δ [see Eqs. (14.39) and (14.43)].

We get the three most important parameters from the above by using:

$$\begin{aligned} \left\langle u_r^2 \right\rangle &= \tfrac{1}{2}\left(\left\langle |u|^2 \right\rangle + \mathrm{Re}\left\langle u^2 \right\rangle\right) \\ \left\langle u_i^2 \right\rangle &= \tfrac{1}{2}\left(\left\langle |u|^2 \right\rangle - \mathrm{Re}\left\langle u^2 \right\rangle\right) \\ \langle u_r\, u_i \rangle &= \tfrac{1}{2}\mathrm{Im}\left\langle u^2 \right\rangle. \end{aligned} \tag{14.44}$$

The above considerations show that analytical expressions can be derived even for fields that do not obey Gaussian statistics. The direct application of second-order statistics is primarily of importance when considering optical systems based on heterodyning of scattered light with an unperturbed local oscillator, viz. Lased Ranging and Detection (LIDAR), vibrometry, etc.

In cases where scattered light from a surface not giving rise to fully developed speckles, we need information on second-order intensity statistics, which is equal to fourth-order field statistics. If, for instance, the speckled field scattered off a rough surface—giving information on the surface displacement—is to be tracked, the surface giving rise to the most pronounced speckle pattern has to be found. In one extreme, when the surface roughness gives rise to fully developed speckles, the modulation depth will be unity, but the diffusely scattered light will fill out the entire half plane, thus giving a rather highly modulated but weak signal. In the other extreme, when the surface roughness goes to zero, the specularly scattered light will be strong, but having no modulation as the object is translated. The issue therefore arises: What are the characteristics of a surface giving a maximum speckle variation, arising when the diffusely scattered light is coherently mixed with the specularly (i.e., unperturbed) part of the scattered light from the surface. To some extent, the specularly scattered light will act as a coherent local oscillator that enhances the diffusely scattered light. Needless to say, the modulation depth of this speckle pattern is lower than unity, but the modulated part might be stronger than what occurs in case of scattering off a surface giving rise to fully developed speckles. The derivation to follow is based on a previous evaluation [15].

To obtain the expression for the intensity variance in case of scattering off a surface not giving rise to fully developed speckles, we will reuse the previous expressions Eqs. (14.33)–(14.44) obtained for the field statistical parameters. We will here assume Gaussian statistics for the scattered field, limiting the parameter space for which the following results are valid as compared to the previous.

In order to obtain Gaussian statistics for the observed field it is necessary that $n \gg 1$. In addition, it is also necessary that the non-specular reflected (or scattered) light from many independent correlation areas overlap at the observation point. This latter condition will be obtained if $\lambda\ z_e/r_c(\sigma) >> r_c(\sigma)$, where $r_c(\sigma)$ is a measure of the effective lateral scale of the surface-height fluctuations, that is when we are in the far field of the lateral scale of the surface-height fluctuations. For σ much less than or greater than unity we have $r_c \approx r_h$ and r_h/σ, respectively, and a convenient analytic approximation for arbitrary values of σ is given by $r_c = r_h\sqrt{1-\exp[-\sigma^2]}/\sigma$ [16]. Both of these conditions are implicitly assumed here. The corresponding variances of the real, σ_r^2, and imaginary part, σ_i^2, of the field are obtained as

$$\begin{aligned} \sigma_r^2 &= \langle u_r^2\rangle - \langle u_r\rangle^2 = \tfrac{1}{2}\left(\left\langle |u|^2\right\rangle + \mathrm{Re}\left[\langle u^2\rangle\right]\right) - \exp\left[-\sigma^2\right] \\ \sigma_i^2 &= \langle u_i^2\rangle - \langle u_i\rangle^2 = \tfrac{1}{2}\left(\left\langle |u|^2\right\rangle - \mathrm{Re}\left[\langle u^2\rangle\right]\right). \end{aligned} \tag{14.45}$$

Having assumed Gaussian statistics for the scattered field gives the probability function:

$$p\left[u_r, u_i\right] = \frac{1}{2\,\pi\,\sigma_r\sigma_i\sqrt{1-\rho^2}}$$
$$\exp\left[-\frac{1}{2\left(1-\rho^2\right)}\left(\frac{\left(u_r - \langle u_r\rangle\right)^2}{\sigma_r^2} + \frac{u_i^2}{\sigma_i^2} - 2\rho\frac{\left(u_r - \langle u_r\rangle\right)u_i}{\sigma_r\sigma_i}\right)\right] \tag{14.46}$$

where the correlation coefficient is given by

$$\rho = \frac{\langle u_r\,u_i\rangle - \langle u_r\rangle\,\langle u_i\rangle}{\sqrt{\langle u_r^2\rangle - \langle u_r\rangle^2}\sqrt{\langle u_i^2\rangle - \langle u_i\rangle^2}} = \frac{\mathrm{Im}\left[\langle u^2\rangle\right]}{2\,\sigma_r\sigma_i}.$$

Inserting the relevant complex parameters for the optical system through Eqs. (14.39)–(14.45) in the above expression gives the probability distribution function for the real and imaginary part of the field. Having this, we can find the higher order statistical parameters for, e.g., the phase, the intensity and the cross products, if so desired. Examples for these statistical parameters can be found in [15]. Here we will show a single example, viz. the variance of the speckle pattern as it moves across a point detector. As discussed, we expect an increase in the modulation—not the modulation depth—as the specularly reflected field mixes and enhances the diffusely scattered field. The on-axis intensity covariance becomes

$$\sigma_I^2 = \langle I^2\rangle - \langle I\rangle^2 - 4\,I_s\,\sigma_r^2 + 2\left(\sigma_r^4 + \sigma_i^4\right) + 4\rho^2\sigma_r^2\sigma_i^2. \tag{14.47}$$

Figure 14.6 shows the variance of the on-axis intensity for a surface giving rise to partially developed speckle as a function of the surface roughness for three different values of the lateral scale of the height fluctuations, r_h. The example used in this

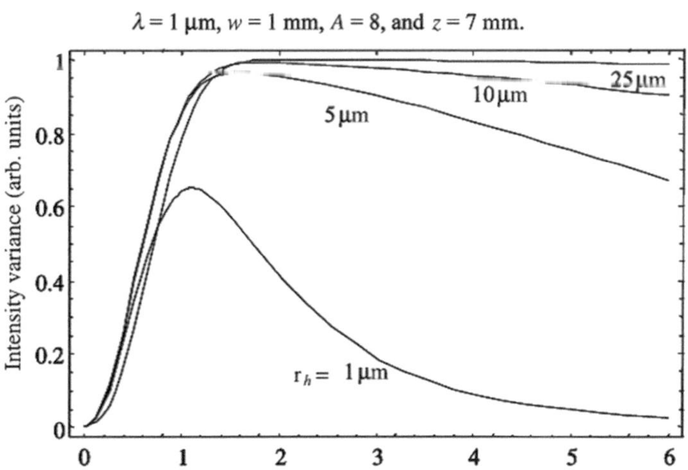

Fig. 14.6 The intensity variance given as a function the standard deviation of phase for various values of r_h

calculation is based on having an illuminating beam radius of 1 mm, a wavelength of 1 μm and a spherical scattering surface of diameter 4 mm placed at a distance of 7 mm from the point detector. Note that although we have free space propagation and expect an A-value of unity, the spherical surface of the object in combination with the distance provides for an A-value of 8.

This chapter has been dedicated to getting analytical expressions for the scattered field statistics up to fourth order for scattering off solid surfaces giving rise to partially developed speckles. The results are valid for optical systems having apertures resulting in complex-valued matrix elements. Likewise, the results are applicable when the partial developed speckle field is due to a low rms. roughness of the target, or is due to the surface having low frequency lateral structures and no high frequency undulations. The result when the surface under investigation has two independent lateral scales has been treated elsewhere, and will not be displayed here [16]. The issue of integrated speckle statistics and the associated differences between displaced detectors has been displayed elsewhere, but will not be repeated here, as the considerations do not highlight the use of complex matrices, but merely is a straightforward use of the method [17]. For the same reason, we will not touch upon the subject of optical misalignment and optical contamination [18] in optical systems treated with the matrix method [19]. Here, as well, the treatment follows directly from the fundamental properties of the formalism, but providing a closed form description of the propagation of the field through the cascaded and perturbed system. Needless to say, the basic formalism does not include optical systems with aberrations, which would call for matrices of higher order.

In case of polarized light, the propagation of the spectral coherence matrix [20, 21] in optical ABCD-systems without apertures has been treated in two dimensions [22, 23].

14.4 Sensor Systems Treated with the ABCD-Formalism

The major applications of complex-valued ABCD matrices for our group have been within optical sensors, and specifically within optical sensors based on dynamic speckles. We will in this chapter show the outcome for only one sensor treated with this approach, namely the torsional vibrometer. A specific discussion of the experimental results and the applications can be found elsewhere, and will not be discussed. The aim here is to show how the formalism can pave the way for analytical expressions that down the line will facilitate the optimization of the measuring system.

Consider the setup shown in Fig. 14.7 [24]. A collimated laser beam is incident on a cylinder with radius R rotating at an angular speed ω.

We consider the distance between the object and the lens, and between the lens and the observation plane, $f + z_1$ and $f + z_2$ respectively, as variables. The tangential peripheral velocity of the object is $R\omega$. We can now use the formalism to derive the

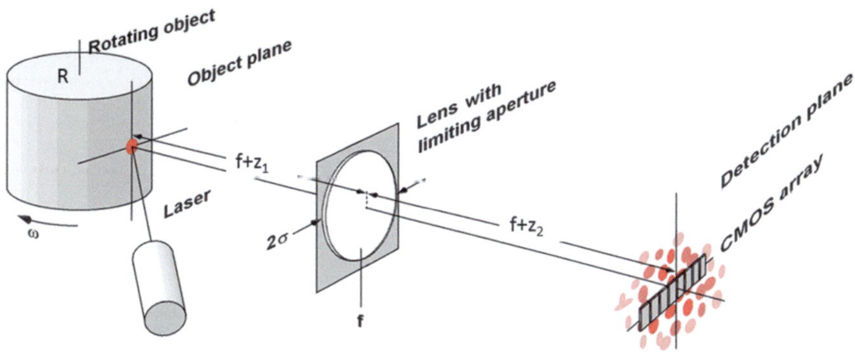

Fig. 14.7 Generic system for measuring rotational and peripheral velocity determined by the distances z_1 and z_2 [25]

pertinent information on the speckle displacement when the object rotates a given angle. These parameters are given directly by Eqs. (14.22)–(14.25) by inserting the pertinent parameters for the optical system. In this way we get the irradiance, the speckle size, the speckle decorrelation, and the speckle displacement as a function of the angular displacement of the object. The compound matrix describing this setup is:

$$\widehat{M}_{\text{total}} = \begin{Bmatrix} 1 & z_2 \\ 0 & 1 \end{Bmatrix} \begin{Bmatrix} 1 & f \\ 0 & 1 \end{Bmatrix} \begin{Bmatrix} 1 & 0 \\ -1/f & 1 \end{Bmatrix} \begin{Bmatrix} 1 & 0 \\ -2i/\left(k\sigma^2\right) & 1 \end{Bmatrix} \cdot \begin{Bmatrix} 1 & f \\ 0 & 1 \end{Bmatrix} \begin{Bmatrix} 1 & z_1 \\ 0 & 1 \end{Bmatrix} \begin{Bmatrix} 1 & 0 \\ -2i/\left(k\omega_0^2\right) & 1 \end{Bmatrix} \begin{Bmatrix} 1 & 0 \\ 2/R & 1 \end{Bmatrix}, \tag{14.48}$$

where ω_0 is the beam size for the incident, collimated Gaussian beam. By incorporating this in the optical system we get a closed expression for the time-lagged spatial covariance but have discarded information on the overall irradiance in the observation plane. By multiplying the matrices, e.g. by applying the program in Appendix, we have the necessary complex matrix elements that can be inserted in Eqs. (14.23) and (14.25) to give the dynamical properties of the speckles and their size, respectively. Note that the first optical element in the optical train is a synthetic lens with focal length $-R/2$ due to the angular rotation of the object. In doing so, the displacement of the target to be inserted in Eq. (14.23) is $R\,\omega$. The compound matrix will not be given here. The peak position Δp_0 for the space time-lagged covariance for $(k\,\sigma\,\omega_0)^2 \gg z_1^2$ for an angular displacement $\Delta\theta$ of the object becomes:

$$\Delta p_0 = A\,R\,\Delta\theta = -2\,\Delta\theta\left(f - \frac{z_2}{f}\left(\frac{R}{2} + z_1\right)\right). \tag{14.49}$$

In the general case, the speckle displacement depends on both the distances and the radius of rotation of the object. A unique measurement of the peripheral or the angular displacement is not obtained. If we adjust the object and observation distances to give imaging and Fourier transformation, i.e. $z_1 \, z_2 = f^2$ and $z_2 = f$, respectively, we get for the peak position of the covariance:

$$\begin{aligned}&\text{Imaging}: \ \Delta p_0 = R \, \Delta\theta \, f / z_1 \text{ for } \left(z_1 \, z_2 = f^2\right) \\ &\text{Fourier transformation}: \ \Delta p_0 = -2 f \, \Delta\theta \text{ for } \left(z_2 = 0\right).\end{aligned} \tag{14.50}$$

Thus placing the CMOS array in Fig. 14.7 in the image plane of the object the speckle displacement becomes proportional to the peripheral displacement; whereas placing the array in the back focal plane ($z_2 = 0$), the speckles move according to the angular displacement. Please note that the measurement of angular displacement is independent of the distance to the object, z_1, the wavelength and as required, independent of the radius of rotation of the object, R. This method can be expanded to derive the distribution of angular twist [26] and the 2D determination of out-of-plane angular displacement [27].

The rotation of target rotating at an angular speed can usually not be followed by a CMOS array, but will require an instantaneous measurement of the speckle velocity. This is also the case for measuring linear velocity based on speckle displacement. Here, spatial filtering velocimetry facilitates a good approach [28, 29]. The speckle pattern is sheared across a spatial filter with period Λ resulting in a temporal modulation of the transmitted irradiance with a frequency v/Λ, where v is the speckle velocity perpendicular to the grating lines. In fact, the spatial filter here acts as a narrow bandpass filter applied to the speckle pattern. Following this concept, the spectral content of the speckle pattern has to have a decent amount of energy in the bandpass region, which in common terms shows that ideally the average speckle size should be $1/\pi$ of the grating period. Generally, the total power in the speckle pattern for a given setup is distributed throughout a range of spatial frequencies, which is significantly larger than the bandwidth of the bandpass filter. Therefore, a more efficient use of the power available in speckle patterns would be beneficial. A method with which to shape a speckle spectrum and analyze this with the matrix method has previously been shown [30].

Measurement of linear displacement is a fundamental problem, and is usually optically performed with either the Laser Doppler Velocimeter or with the Time-of-Flight velocimeter. The first one relies on crossing of two mutually coherent beams in the measuring volume—in the case of flow measurement [31]—or on the object in case of solid structures [32].

The second approach for measuring linear transverse velocity is based on projecting two spots on the target, and recoding the time it takes the object to pass from one spot to the next. Knowing the spot separation and measuring the transit time between the two spots reveals the object velocity component [33]. Theoretical studies of these systems have been conducted based on the complex ABCD-method [16, 34–36]. Specifically [35] includes a discussion on scattering off solid structures having multiple scales.

14.5 Optical Turbulence

In the previous the optical elements have been discretely placed at fixed positions. In the case of field propagating through optical turbulence or being guided through a perturbing channel, the ABCD model might be advantageous. Good overviews of field propagation through optical turbulence have been provided in the past [37].

In this chapter we will recollect the results previously obtained for field and intensity propagation through optical turbulence. The results are based on two previous articles [7, 38]. The formalism presented here has been extended in a number of different scenarios [39] and has formed the basis for the analysis of light penetration through human skin, specifically for analyzing Optical Coherence Tomographical systems (OCT) [40].

We now consider the case in which the light beam propagates through an optical system that contains a random inhomogeneous scattering medium. Here, we treat for simplicity cylindrically symmetrical optical systems that contain no limiting apertures (so that the *ABCD* matrix is real). It can now be shown that the effect of the random distribution of scatters on the propagation of a light beam is to introduce an additional multiplicative factor to the integral of Eq. (14.3). This factor is given by $\exp[\Psi(\mathbf{p},\mathbf{r})]$, where $\mathbf{p} = \left(p_x, p_y\right)$ and $\mathbf{r} = (x_0, y_0)$, and Ψ is the perturbation of the Green's function that is due to the random inhomogeneity along the propagation path. The coordinates $\mathbf{p}$ and $\mathbf{r}$ represent a general observation point and a point in the initial plane, respectively. Because we are dealing with random quantities, it is appropriate to consider the statistical moments of the fields. Here, we shall obtain explicit expressions for the first (i.e., average field) and mean irradiance of the light beam in the observation plane.

The field in the observation plane is thus given by

$$U(\mathbf{p}) = \left(-\frac{ik}{2\pi B}\right) \int \mathrm{d}\mathbf{r}\, U_i(\mathbf{r}) \exp\left[\Psi\left(\mathbf{p},\mathbf{r}\right]\right. \exp\left[-\frac{ik}{2B}\left(A\, r^2 - 2\,\mathbf{p}\cdot\mathbf{r} + D\, p^2\right)\right]. \tag{14.51}$$

Assuming the field Ψ is stationary and taking the ensemble average of the above, we get

$$\langle U(\mathbf{p})\rangle = \langle \exp[\Psi]\rangle\; U_0, \tag{14.52}$$

where U_0 is the field in the absence of the perturbation. Furthermore, it can be shown that, correct to terms of second order in the fluctuations of the index of refraction of the medium,

$$\langle \exp[\Psi]\rangle = \exp\left[\langle\chi\rangle + i\langle\varphi\rangle + \frac{1}{2}\left(\langle\chi^2\rangle + 2i\langle\chi\varphi\rangle - \langle\varphi^2\rangle\right)\right]. \tag{14.53}$$

Here χ and φ are the log-amplitude and phase of the Green's function that is due to the random inhomogeneity. In most cases the mean phase is zero, and additionally $\langle\chi\rangle = -\langle\chi^2\rangle/2$, and $\langle\chi\varphi\rangle$ is much less than $\langle\varphi^2\rangle$. Thus, we obtain

$$\langle \exp[\Psi]\rangle \cong \exp\left[-\frac{1}{2}\langle\varphi^2\rangle\right]. \tag{14.54}$$

The effect of this is not a damping of the power transmitted through the optical channel, but merely a partial destruction of the coherence after passage. In fact, an interference pattern arising from interference between the transmitted wave and an unperturbed part of the beam will suffer a decrease in the modulation depth, as indicated in the above result.

Next, we have to introduce the phase perturbation due to the optical turbulence in the optical path. The three-dimensional spectrum of the index-of-refraction fluctuations is $\Phi_n[\mathbf{Q}, z]$. Here the fluctuations are allowed to vary in strength and spectral contribution along the path, denoted z. The transverse coordinate along the path is $\mathbf{Q}$. Various models for the spectrum $\Phi_n[\mathbf{Q}, z]$ have been presented, based on whether we are dealing with propagation through the atmosphere, plasmas or probing light in human tissue. Here, we will not discuss these spectral models, as they can be found elsewhere for atmospheric turbulence and scattering by aerosols.

The integrated phase disturbance is given by integrating $\Phi_n[\mathbf{Q}, z]$ over all transverse spectral components and along the optical path:

$$\left\langle\sigma_\varphi^2\right\rangle = 2\pi\int_0^L \mathrm{d}z\int_0^\infty \mathrm{d}^2q\ \Phi_n\,[\mathbf{q}; z], \tag{14.55}$$

where L is the total propagation path length through the optical system and $\Phi_n[\mathbf{q}, z]$ is the three-dimensional spectrum of the index-of-refraction fluctuations evaluated at $q_z = 0$. We have included path-length dependence on Φ_n, because in many applications the strength of the fluctuations may vary along the path (e.g., up-propagation through the atmosphere). For example, for propagation through the atmosphere, Φ_n is a sum of the contributions that are due to clear-air turbulence and aerosols. The von karman spectrum describes a well-developed spectrum due to temperature fluctuations. Here, we will assume the spectrum to be given by the simpler spectrum, which is only valid within the inertial sub-range:

$$\Phi_n \cong 0.033\ C_n^2(z)\ Q^{-11/3}. \tag{14.56}$$

For atmospheric propagation, the above spectral approximation for the inertial sub-range is valid for spatial structures scaled from approximately 1 mm to the spectral value corresponding to the largest eddy, which is close to the height above the ground. The strength of the optical turbulence is governed by the parameter "c-sub-n-squared," which attains an increasing value as the atmospheric density and the temperature increase.

An expression for the mean intensity $I[\mathbf{p}] = \left\langle |U(\mathbf{p})|^2 \right\rangle$ involving the Green's function for an arbitrary optical system can now be established by insertion in the expression for the perturbed field, Eq. (14.51). After some simplifications and rearranging we get:

$$\begin{aligned} I[\mathbf{p}] &= \left|\tfrac{k}{2\pi B}\right|^2 \int \mathrm{d}^2 r \, \exp\left[\frac{ik}{B}\mathbf{p}\cdot\mathbf{r}\right] K[\mathbf{r}] \, M[\mathbf{r}], \text{ where} \\ K[\mathbf{r}] &= \int \mathrm{d}^2 R \, u_i[\mathbf{R}+\mathbf{r}/2] \, u_i^*[\mathbf{R}-\mathbf{r}/2] \, \exp\left[-\tfrac{ikA}{B}\mathbf{p}\cdot\mathbf{r}\right] \\ \text{and } M[\mathbf{r}] &= \langle \exp[\Psi(\mathbf{r}_1,\mathbf{p}) + \Psi^*(\mathbf{r}_2,\mathbf{p})] \rangle . \end{aligned} \tag{14.57}$$

Here $M[\mathbf{r}]$ is the mutual coherence function of a point source located at the observation point p and evaluated at position coordinates $\mathbf{r}_1$ and $\mathbf{r}_2$ in the initial plane of the optical system, respectively. In writing Eq. (14.57) we have anticipated the result that for a stationary inhomogeneous media all second-order statistical quantities depend on the difference coordinates only. Hence $M[\mathbf{r}]$ is a function of $\mathbf{r}_1 - \mathbf{r}_2$ only. Furthermore, we have the connection between the mutual coherence function and the structure function $D_w[\mathbf{r}]$ given by $M[\mathbf{r}] = \exp[-D_w[\mathbf{r}]/2]$. Finally, the connection between the three-dimensional spectrum of the index-of-refraction fluctuations and the wave-structure function is given by:

$$D_w[\mathbf{r}] = 4\pi k^2 \int_0^L \mathrm{d}z \int \mathrm{d}^2 Q \, \Phi_n[Q,z] \, (1 - \exp[i\,\mathbf{Q}\cdot\mathbf{r}(z)]) . \tag{14.58}$$

Here $\mathbf{r}(z)$ is the vector separation at a propagation distance z from the point $\mathbf{p}$ between two rays, both of which originate at $\mathbf{p}$ and propagate backwards through the optical system to vector positions $\mathbf{r}_1$ and $\mathbf{r}_2$ in the initial plane of the optical system. This means that $\mathbf{r}(z)$ is given by

$$\mathbf{r}(z) = \left(\frac{B[z]}{B}\right)\mathbf{r}, \tag{14.59}$$

where $\mathbf{r} = \mathbf{r}_1 - \mathbf{r}_2$, B is the ray-transfer-matrix element of the complete system for a ray whose origin is at p, and $B(z)$ is the corresponding matrix element for a ray whose origin is at $\boldsymbol{p}$ and whose endpoint is at a distance z from p; that is, $B(z)$ corresponds to the appropriate matrix element for a ray propagating backward through the system. In short, the ABCD-matrix method is able to track the mutual distance between the interfering rays in the observation plane by keeping track of their mutual distance as they pass the optical system. This distance defines the amount of phase disturbance which the two rays will undergo.

We are now able to give an expression for the irradiance after passage of an optical system in which we have phase disturbances. We assume a Kolmogorov spectrum given by the approximation in Eq. (14.56), and a Gaussian-shaped laser beam with

$$U_i = U_0 \exp\left[-\frac{r^2}{\omega_i^2}\right]. \tag{14.60}$$

Performing the integrations and simplifying yields:

$$I[\mathbf{p}] = \frac{2P_{\mathrm{T}}}{\pi\omega^2} \exp\left[-\frac{2p^2}{\omega^2}\right], \tag{14.61}$$

where P_{T} is the transmitted power,

$$\omega^2 = A^2\omega_i^2 + \frac{4B^2}{k^2\omega_i^2} + \frac{8}{k^2\, q_{\mathrm{T}}^2}, \tag{14.62}$$

and

$$q_{\mathrm{T}} \cong \left(1.46\, k^2 \int_0^L \mathrm{d}z\, C_n^2(z)\, B[z]^{5/3}\right)^{-3/5}. \tag{14.63}$$

In obtaining this expression we have assumed an optical system without apertures, and approximated the 5/3-power dependence in the exponent of M by 2.

14.6 Conclusions

We have demonstrated that within the paraxial ray approximation the propagation of light through a complex optical system can be formulated in terms of a Huygens principle expressed in terms of the complete system ABCD-matrix elements. As such, propagation through an optical system reduces to that of calculating the relevant matrix elements and substituting these into the expressions derived here.

We have extended the treatments of Baues [2] and Collins [3] to include partially coherent light sources, optical elements of finite size, and distributed random inhomogeneity along the optical path. In many cases (e.g., laser beam propagation and Gaussian optics) we have been able to derive simple analytical expressions for the optical field quantities at an observation plane. These expressions are functions of the appropriate ray-transfer-matrix elements of the optical system under consideration and the parameters that describe the initial light source. Hence, for a given initial light source and a given optical system, one can compute the relevant optical parameters at the observation plane (e.g., the irradiance profile). On the other hand, given an initial light source and specified optical beam requirements in the observation plane, the analytical expressions presented here are useful in the synthesis of the appropriate optical system. Specifically in the case of speckle-based systems, the formalism facilitates analytical expressions for various parameters of importance for design and optimization.

A.1 Appendix

MATHEMATICA program for calculation of the compound 1D matrix for a series of optical elements:

```
(* Program: optics.m . Written by Lilybeth *)

(* This program displays the total matrix modeling
the elements in a complex optical system.
It is achieved by accumulating the product of partial
matrices from the predefined functions for a lens,
free space parameter, or a limited aperture parameter.
    Further operations with the total matrix can be
accomplished by assigning it to a variable. *)

(* This program requires the standard package
ReIm.m. Reference:
    Programming in Mathematica, by Maeder. Pag 243 *)

BeginPackage["Optics`", "Algebra`ReIm`"]

Lens::usage = "Lens[f] generates a matrix that
models this element."
Aperture::usage = "Aperture[r] generates a
matrix that models this element."
FreeSpace::usage = "FreeSpace[z] generates
a matrix that models this element."
t = {{1,0}, {0,1}} (* initializing global variable *)
Begin["`Private`"]
Print["This program generates a total matrix according
to the elements of your optical system."];

Print["Please select your choice by entering Lens[f],
Aperture[r,k], or FreeSpace[z]."]
Totalmatrix[m_] := t = m.t;
            Return[t]

        Lens[f_] :=
        Block[{ml},
             f/: Im[f] = 0;
            ml = {{1,0}, {-1/f, 1}};
            Print["Lens matrix is ", ml];
```

```
            m = ml;
            Totalmatrix[m]
            ]
Aperture[r_,k_] :=
        Block[{ma},
            r/: Im[r] = 0; k/: Im[k] = 0;
            ma = {{1,0}, {-(2 I)/(k*r^2), 1}};
            Print["Aperture matrix is", ma];
                  m = ma;
            Totalmatrix[m]
]
FreeSpace[z_] :=
             Block[{mz},
                   z/: Im[z] = 0;
                   mz = {{1,z}, {0,1}};
                   Print["FreeSpace matrix is", mz];
                   m = mz;
Totalmatrix[m]
]
End[]
EndPackage[]
```

References

1. G. Kloos, *Matrix Methods for Optical Layout* (SPIE, Bellingham, 2007)
2. P. Baues, Huygens principle in inhomogeneous isotropic media and a general integral equation applicable to optical resonators. Opto-Electronics **1**, 37–44 (1969)
3. S. Collins, Lens-system diffraction integral written in terms of matrix optics. J. Opt. Soc. Am. **60**, 1168–1177 (1970)
4. H. Kogelnik, T. Li, Laser beams and resonators. Appl. Opt. **5**, 1550 (1966)
5. P. Baues, The connection of geometrical optics with propagation of Gaussian beams and the theory of optical resonators. Opto-Electronics **1**(2), 103–118 (1969)
6. A.E. Siegman, *Lasers* (University Science Books, Mill Valley, 1986)
7. H. Yura, S.G. Hanson, Optical beam wave-propagation through complex optical-systems. J. Opt. Soc. Am. A **4**, 1931–1948 (1987)
8. H. Yura, S.G. Hanson, T. Grum, Speckle-statistics and interferometric decorrelation effects in complex ABCD optical-systems. J. Opt. Soc. Am. A **10**, 316–323 (1993)
9. S.G. Hanson, T.F.Q. Iversen, R.S. Hansen, Dynamical properties of speckled speckles. *Proceedings of SPIE- the International Society for Optical Engineering*, **7387**(1), pp. 738–716 (2010)
10. J.W. Goodman, *Speckle Phenomena in Optics* (Roberts & Company, Englewood, 2007)
11. H. Yura, B. Rose, S.G. Hanson, Dynamic laser speckle in complex ABCD optical systems. J. Opt. Soc. Am. A Opt. Image Sci. Vis. **15**, 1160–1166 (1998)

12. R. Hansen, H. Yura, S.G. Hanson, B. Rose, Three-dimensional speckles: static and dynamic properties, in *Proceedings of the Society of Photo-Optical Instrumentation Engineers (SPIE), Fourth International Conference on Correlation Optics*, 1999, ed. by O. Angelsky
13. R. Hansen, H. Yura, S.G. Hanson, First-order speckle statistics: an analytic analysis using ABCD matrices. J. Opt. Soc. Am. A Opt. Image Sci. Vis. **14**, 3093–3098 (1997)
14. E. Jakeman, W.T. Welford, Speckle statistics in imaging systems. Opt. Commun. **21**, 72–79 (1977)
15. H. Yura, S. Hanson, Variance of intensity for Gaussian statistics and partially developed speckle in complex ABCD optical systems. Opt. Commun. **228**, 263–270 (2003)
16. B. Rose, H. Imam, S. Hanson, H. Yura, Effects of target structure on the performance of laser time-of-flight velocimeter systems. Appl. Opt. **36**, 518–533 (1997)
17. S.G. Hanson, H.T. Yura, Statistics of spatially integrated speckle intensity difference. J. Opt. Soc. Am. A **26**, 371–375 (2009)
18. H. Yura, S. Hanson, Effects of receiver optics contamination on the performance of laser velocimeter systems. J. Opt. Soc. Am. A **13**, 1891–1902 (1996)
19. W. McKinley, H. Yura, S. Hanson, Optical system defect propagation in ABCD systems. Opt. Lett. **13**, 333–335 (1988)
20. T. Shirai, E. Wolf, Coherence and polarization of electromagnetic beams modulated by random phase screens and their changes on propagation in free space. J. Opt. Soc. Am. A Opt. Image Sci. Vis **21**, 1907–1916 (2004)
21. T. Shirai, Some consequences of the van Cittert-Zernike theorem for partially polarized stochastic electromagnetic fields. Opt. Lett. **34**, 3761–3763 (2009)
22. S.G. Hanson, W. Wang, M.L. Jakobsen, M. Takeda, Coherence and polarization of electromagnetic beams modulated by random phase screens and their changes through complex ABCD optical systems. J. Opt. Soc. Am. A **25**, 2338–2346 (2008)
23. M. Takeda, W. Wang, S.G. Hanson, Polarization speckles and generalized Stokes vector wave: a review, in *Speckle 2010: Optical Metrology*, vol. 7387 (2010), p. 73870V
24. B. Rose, H. Imam, S.G. Hanson, H.T. Yura, R.S. Hansen, Laser-speckle angular-displacement sensor: theoretical and experimental study. Appl. Opt. **37**, 2119–2129 (1998)
25. Z. Meng, B. Liu, Research on torsional vibration non-contact measurement of rotary machine, in *Proceedings of SPIE–The International Society for Optical Engineering 6280,* 2006
26. B. Rose, H. Imam, S. Hanson, H. Yura, A laser speckle sensor to measure the distribution of static torsion angles of twisted targets. Meas. Sci. Technol. **9**, 42 (1998)
27. B. Rose, H. Imam, S.G. Hanson, Non-contact laser speckle sensor for measuring one-and two-dimensional angular displacement Capteur non-contact de laser speckle pour mesurer le displacement angulaire: une ou deux dimensions. J. Opt. **29**, 115–120 (1998)
28. Y. Aizu, T. Ushizaka, T. Asakura, Measurements of flow velocity in a microscopic region using a transmission grating: elimination of directional ambiguity. Appl. Opt. **24**, 636–640 (1985)
29. M.L. Jakobsen, H.T. Yura, S.G. Hanson, Spatial filtering velocimetry of objective speckles for measuring out-of-plane motion. Appl. Opt. **51**, 1396–1406 (2012)
30. M.L. Jakobsen, H.T. Yura, S.G. Hanson, Speckles and their dynamics for structured target illumination: optical spatial filtering velocimetry. J. Opt. A Pure Appl. Opt. **11**, 1–9 (2009).
31. J. Bilbro, Atmospheric laser Doppler velocimetry – an overview. Opt. Eng. **19**, 533–542 (1980)
32. L. Drain, Doppler velocimetry. Laser Focus Fiberoptic Technol. **16**, 68–80 (1980)
33. M. Beck, Correlation in instruments – cross-correlation flowmeters. J. Phys. E Sci. Instrum. **14**, 7–19 (1981)
34. B. Rose, H. Imam, L. Lading, S. Hanson, Time-of-flight velocimetry: bias and robustness, in *Technical Digest Series – Optical Society of America*, vol. 14 (1996), pp. 68–70
35. H. Yura, S.G. Hanson, Laser-time-of-flight velocimetry – analytical solution to the optical-system based on ABCD matrices. J. Opt. Soc. Am. A **10**, 1918–1924 (1993)
36. H. Yura, S.G. Hanson, L. Lading, Laser-doppler velocimetry – analytical solution to the optical system including the effects of partial coherence of the target. J. Opt. Soc. Am. A **12**, 2040–2047 (1995)
37. A.D. Wheelon, *Electromagnetic Scintillation* (University Press, Cambridge, 2001)

38. H. Yura, S.G. Hanson, 2nd-order statistics for wave propagation through complex optical systems. J. Opt. Soc. Am. A **6**, 564–575 (1989)
39. L.C. Andrews, R.L. Philips, *Laser Beam Propagation Through Random Media*, 2nd edn. (SPIE, Bellingham, WA 98227–0010 USA, 2005)
40. L. Thrane, H. Yura, S. Hanson, P. Andersen, Optical coherence tomography of heterogeneous tissue: calculation of the heterodyne signal, in *Conference on Anonymous Lasers and Electro-Optics Europe, 1998. 1998 CLEO/Europe.* (IEEE, Glasgow, Scotland, 1998)

Chapter 15
Linear Canonical Transforms on Quantum States of Light

Gabriel F. Calvo and Antonio Picón

Abstract Many quantum information and quantum computation protocols exploit high-dimensional Hilbert spaces. Photons, which constitute the main carrier of information between nodes of quantum networks, can store high-dimensional quantum bits in their spatial degrees of freedom. These degrees of freedom can be tailored by resorting to the symplectic invariant approach based on lossless linear canonical transformations. These transformations enable one to manipulate the transverse structure of a single photon prepared in superpositions of paraxial modes. We present a basic introduction of these transformations acting on photons and discuss some of their applications for elementary quantum information processing.

15.1 Introduction

Most promising approaches for scalable quantum communication (QC) rely on the use of photons as the main carriers of information among remote nodes of quantum networks, where matter-based quantum memories are located [25, 29, 36, 39, 48, 71]. In such nodes the quantum computing schemes are implemented to process the information. Photons, besides being the natural candidate for QC due to their long decoherence time and the relative ease with which they can be manipulated, can actually encode multiple quantum bits of information (qubits) into various degrees of freedom. These include frequency, polarization, and linear momentum.

Beams propagating with a well-confined direction can also store quantum information in their spatial-transverse profile, for example the paraxial modes of the Hermite–Gauss or the Laguerre–Gauss. A photon with a well-defined paraxial mode can be defined analogously to the photon with a well-defined linear momentum [20].

G.F. Calvo
Department of Mathematics and IMACI-Institute of Applied Mathematics in Science and Engineering, UCLM-Universidad de Castilla-La Mancha, Ciudad Real 13071, Spain
e-mail: Gabriel.Fernandez@uclm.es

A. Picón (✉)
Argonne National Laboratory, Argonne, IL 60439, USA
e-mail: apicon@anl.gov

J.J. Healy et al. (eds.), *Linear Canonical Transforms*, Springer Series in Optical Sciences 198, DOI 10.1007/978-1-4939-3028-9_15

In addition to the spin angular momentum related to the polarization of the photon, the paraxial photon also transports an extra angular momentum related to its spatial-transverse profile, the so-called orbital angular momentum (OAM). In particular, the helical wave fronts of the Laguerre–Gaussian modes have a well-defined OAM $\hbar\,\ell$ per photon [5, 32, 64].

The spatial-transverse profile or the OAM has unique features for quantum information tasks as it is possible to encode a qu*d*it, i.e. a quantum bit of information with dimension d, in a single photon. At variance with polarization, in which only qubits ($d = 2$) are allowed, the spatial-transverse profile allows, in principle, an infinity dimension d. The high-dimensional encoding is relevant both for fundamental proof-of-principle experiments [24, 38, 68] and for several quantum information protocols, to cite some of their advantages: creating more complex quantum cryptography protocols [14, 17, 23, 35, 67], optimizing certain computing arquitectures [34], simplifying the quantum logics [41], and performing continuous-variable quantum computation [42, 61].

Here we focus on linear canonical transformations on paraxial modes at the single photon level and their applications on quantum information. The chapter is organized as follows: Sect. 15.2 gives a brief summary of the formalism on symplectic groups and its connection with metaplectic groups. The latter group is related to the linear canonical transformations on the spatial-transverse profile of a paraxial beam. There are different existing approaches to introduce this formalism, we have opted for following the approach described in [57]. Section 15.3 mainly focuses on these transformations at the single photon level. First of all, we introduce the general paraxial expression for a single-photon state and the concept of quantum information bit encoded in its spatial-transverse profile. In the next stage, we introduce the Poincaré Sphere representation for these quantum information bits. Finally, we discuss the construction of single quantum gates for spatial-transverse bit states by means of linear canonical transformations.

15.2 The Symplectic Group in Paraxial Wave Optics

Within a paraxial description of the propagation of a monochromatic wave of frequency ω, two main factors are required: a slowly variant amplitude and a carrier-plane wave, $Ae^{i(kz-\omega t)}$ (without loss of generality, we consider the paraxial wave propagating along the z-direction). The amplitude A varies slowly along z when compared to the wavelength $\lambda = 2\pi/k$, but allows the carrier-plane wave to have a transverse profile with a characteristic beam waist w_0 (where $w_0 \gg \lambda$).

Henceforth we shall refer to $\psi(\mathbf{r}; z)$ as the complex amplitude of a wave (or a complex signal) across a transverse plane at z, and the vector $\mathbf{r} = (x, y)$ to denote the transverse coordinates. It is understood that, unless otherwise explicitly specified, we will be dealing with transverse variables: the vector $\mathbf{r}$ and the two-dimensional spatial frequency vector $\mathbf{q}$. The coordinate z can be conceived as a continuous labeling parameter. Our aim is to translate the symplectic formalism in geometrical optics onto the realm of a wave space, expanded by the Lebesgue space $L^2(\mathbb{R}^2, \mathbb{C})$ of square-integrable complex-valued functions $\psi(\mathbf{r}; z) \in L^2(\mathbb{R}^2, \mathbb{C})$, where

$$L^2(\mathbb{R}^2, \mathbb{C}) = \left\{ \phi : \mathbb{R}^2 \to \mathbb{C} \ : \ \int_{\mathbb{R}^2} d^2\mathbf{r} \, |\phi(\mathbf{r})|^2 < \infty \right\} .$$

We will be interested in analyzing the action of linear and lossless optical elements on quantum states of light (single photon states). To do so, we will proceed in various steps. To a given system of linear and lossless optical elements (taking into account also the free propagation) we will associate a transformation matrix $\mathbf{T}$ (see Chap. 2 for more details). These optical elements act on the wave function, so they correspond to certain (unitary) operators defined on a Hilbert space $\mathscr{H}$.

At this point it is convenient to introduce the Dirac's bracket notation for the inner product of two state vectors (kets) $|\phi\rangle, |\varphi\rangle \in \mathscr{H}$

$$\langle \varphi | \phi \rangle = \int d^2\mathbf{r} \, \varphi^*(\mathbf{r}) \phi(\mathbf{r}) ,$$

where * denotes complex conjugation.

Let $\psi(\mathbf{r})$ denote the value of $\psi(\mathbf{r}; z)$ at $z = 0$, before the optical system. Then, the value $\psi(\mathbf{r}) = \langle \mathbf{r} | \psi \rangle$ at the point $\mathbf{r}$ is thus the component of the ket $|\psi\rangle$ on the basis vector $|\mathbf{r}\rangle$ of the $\{|\mathbf{r}\rangle\}$ (position) representation, with $\langle \mathbf{r} | \mathbf{r}' \rangle = \delta(\mathbf{r} - \mathbf{r}')$.

Lossless optical systems preserve the irradiance integrated over the transverse plane. For a given paraxial wave function ψ, intensity and irradiance (power per unit area) only differ by a multiplicative factor. The total power across a certain transverse plane at z is

$$I(z) = \int d^2\mathbf{r} \, |\psi(\mathbf{r}; z)|^2 . \tag{15.1}$$

For lossless propagation, $I(z)$ is independent of z. This fact can be expressed succinctly within the Hilbert-space representation. Let $|\psi\rangle, |\psi'\rangle \in \mathscr{H}$ and let $\langle \psi | \psi \rangle$ and $\langle \psi' | \psi' \rangle$ denote their norms. Suppose that $|\psi'\rangle = U(\mathbf{T}) |\psi\rangle$ where $U(\mathbf{T})$ is an operator corresponding to the transformation matrix of a (linear) lossless optical system. Since $\mathscr{I}(z) = \langle \psi | \psi \rangle = \langle \psi' | \psi' \rangle$, it follows that lossless systems should necessarily be represented by unitary operators $\hat{U} : \mathscr{H} \to \mathscr{H}$, that is $\hat{U}^\dagger \hat{U} = \hat{U} \hat{U}^\dagger = \hat{I}$, where $\hat{U}^\dagger$ is the adjoint of $\hat{U}$ and $\hat{I}$ the identity operator. Therefore, let $\hat{U}(\mathbf{T})$ be a unitary operator corresponding to $\mathbf{T}$, which acts on the input wave state $|\psi\rangle \in \mathscr{H}$. Its action is thus described as [57]

$$\begin{aligned} \hat{U}(\mathbf{T}) : \ |\psi\rangle \in \mathscr{H} &\xrightarrow{\mathbf{T}} |\psi'\rangle = \hat{U}(\mathbf{T}) |\psi\rangle, \quad \text{where } |\psi'\rangle \in \mathscr{H} , \\ \psi(\mathbf{r}) = \langle \mathbf{r} | \psi \rangle &\xrightarrow{\mathbf{T}} \psi'(\mathbf{r}) = \langle \mathbf{r} | \psi' \rangle = \int d^2\mathbf{r}' \, \langle \mathbf{r} | \hat{U}(\mathbf{T}) | \mathbf{r}' \rangle \, \psi(\mathbf{r}') , \end{aligned} \tag{15.2}$$

where $|\psi'\rangle$ is the output wave state and $\langle \mathbf{r}|\hat{U}(\mathbf{T})|\mathbf{r}'\rangle$ is the kernel of the unitary operator $\hat{U}(\mathbf{T})$. In Eq. (15.2) we have used the identity relation $\hat{I} = \int d^2\mathbf{r}'\ |\mathbf{r}'\rangle\langle\mathbf{r}'|$. This is nothing else but a restatement of Parseval's theorem for lossless transformations already encountered in Chap. 2. Moreover, from Eq. (15.2) it is clear that $\psi'(\mathbf{r})$ represents the output wave function of the input wave function $\psi(\mathbf{r})$ after traversing the optical system with transformation matrix $\mathbf{T}$. Hence, the $\hat{U}(\mathbf{T})$ represent linear canonical transformations acting on the wave state $|\psi\rangle \in \mathscr{H}$.

From a geometrical optics point of view, we have seen in previous chapters that a ray is defined by the phase-space vector $\boldsymbol{\xi} = (\mathbf{r}, \mathbf{q})$, and that the matrix $\mathbf{T}$ is a symplectic matrix. That is, $\mathbf{T} \in Sp(4,\mathbb{R})$ with $Sp(4,\mathbb{R})$ denoting the Lie group of 4×4 real matrices with $\det(\mathbf{T}) = 1$. Hence, matrix $\mathbf{T}$ transforms the phase-space vector as $\boldsymbol{\xi} \xrightarrow{\mathbf{T}} \boldsymbol{\xi}' = \mathbf{T}\boldsymbol{\xi}$. Within this framework the phase-space vector $\boldsymbol{\xi}$ becomes the canonical linear Hermitian operator $\hat{\boldsymbol{\xi}}$ acting on the Hilbert space $\mathscr{H}$ [27]

$$\boldsymbol{\xi} = \begin{pmatrix} x \\ y \\ q_x \\ q_y \end{pmatrix} \to \hat{\boldsymbol{\xi}} = \begin{pmatrix} \hat{x} \\ \hat{y} \\ \hat{q}_x \\ \hat{q}_y \end{pmatrix} = \begin{pmatrix} \hat{\mathbf{r}} \\ \hat{\mathbf{q}} \end{pmatrix}.$$

Just as the phase-space components of $\boldsymbol{\xi}$ satisfy the Poisson brackets $\{\xi_\alpha, \xi_\beta\} = \Omega_{\alpha\beta}$, with Ω being a 4×4 antisymmetric matrix[1]

$$\Omega = \begin{pmatrix} 0 & 0 & 1 & 0 \\ 0 & 0 & 0 & 1 \\ -1 & 0 & 0 & 0 \\ 0 & -1 & 0 & 0 \end{pmatrix}, \tag{15.3}$$

the four canonical operator components of $\hat{\boldsymbol{\xi}}$ obey the commutation relations [57]

$$[\hat{\xi}_\alpha, \hat{\xi}_\beta] = i\lambda\!\!\!^{-}\,\Omega_{\alpha\beta}\,, \tag{15.4}$$

where $\lambda\!\!\!^{-} = 1/k$ is the optical analog of Planck constant $\hbar$ in quantum mechanics. As in quantum mechanics, we find an ambiguous criterion when one wants to map a phase-space function $f(\boldsymbol{\xi})$ onto an operator function $f(\hat{\boldsymbol{\xi}})$. To do this, there are several routes, but the usual one is to resort to the Weyl or symmetric ordering [2, 3]. To connect the action of the optical system $\mathbf{T}$ on the paraxial waves with that found in geometrical optics [57], we need the average position and momentum (direction) of the paraxial wave

$$\bar{\xi}_\alpha \equiv \langle\hat{\xi}_\alpha\rangle = \langle\psi|\hat{\xi}_\alpha|\psi\rangle / I(z)\,,$$

[1]We use a slightly different symplectic metric matrix Ω to that used in Chap. 2. For every $\mathbf{T} \in Sp(4,\mathbb{R})$, the symplecticity condition reads as $\mathbf{T}\Omega\mathbf{T}^T = \Omega$.

where $\mathscr{I}$ is the total irradiance (15.1). Thus, after transformation $\mathbf{T}$, analogously to the results of rays optics, we impose

$$\bar{\xi}' = \langle\psi'|\hat{\xi}|\psi'\rangle/I(z) = \langle\psi|\,\hat{U}^{\dagger}(\mathbf{T})\,\hat{\xi}\,\hat{U}(\mathbf{T})\,|\psi\rangle/I(z) = \mathbf{T}\,\bar{\xi}\,,$$

which is satisfied for all $|\psi\rangle$ when

$$\hat{U}^{\dagger}(\mathbf{T})\,\hat{\xi}\,\hat{U}(\mathbf{T}) = \mathbf{T}\hat{\xi}\,. \tag{15.5}$$

Equation (15.5) is known as the Stone–von Neumann theorem and will be relevant for further considerations. It connects the symplectic formalism for geometrical optics with unitary operators acting on state vectors $|\psi\rangle$ whose position representation is provided by the paraxial wave $\psi(\mathbf{r})$. That is, each $\hat{U}(\mathbf{T})$ is in correspondence with the symplectic transformation matrix $\mathbf{T}$. The transformation $\{\mathbf{T}\hat{\xi}\}$ preserves the commutation relations (15.4).

The unitary transformation $\hat{U}(\mathbf{T})$ can be expressed in an exponential form $\hat{U}(\mathbf{T}) = e^{i\hat{\Theta}}$, where $\hat{\Theta}$ is an Hermitian operator ($\hat{\Theta} = \hat{\Theta}^{\dagger}$). The structure imposed by Eq. (15.5) indicates that the Hermitian operator $\hat{\Theta}$ is quadratic in the components of $\hat{\xi}$. Indeed, $\hat{\Theta}$ can be cast as a real linear combination of a quadratic-operator basis or generator $\hat{G}_{\alpha}$, i.e. $\hat{\Theta} = \sum_{\alpha} c_{\alpha}\hat{G}_{\alpha}$. By exploiting the Stone–von Neumann theorem (15.5) it is possible to construct the generators for $\hat{U}(\mathbf{T})$. From the generators $\mathbf{G}_{\alpha}$ for the symplectic group [6], i.e. $\mathbf{T} = e^{\mathbf{S}}$, where $\mathbf{S} = \sum c_{\alpha}\mathbf{G}_{\alpha}$, one can establish a one-to-one mapping between $\mathbf{G}_{\alpha}$ and $\hat{G}_{\alpha}$ to satisfy Eq. (15.5). Often, in the Mathematics literature, $\mathbf{G}$ is called an element of the Lie algebra of the symplectic group $\mathfrak{sp}(4)$. Using the property of the symplectic algebra that any $\mathbf{G}_{\alpha}$ is in one-to-one correspondence with a real symmetric 4×4 matrix $\mathbf{V}$; $\mathbf{G}_{\alpha} \to \mathbf{V}(\mathbf{G}_{\alpha}) = \Omega\mathbf{G}_{\alpha}$, we may cast the one-to-one mapping $\mathbf{G}_{\alpha} \leftrightarrow \hat{G}_{\alpha}$ as [57]

$$\mathbf{G}_{\alpha} \to \hat{G}_{\alpha} \equiv -\frac{1}{2}\hat{\xi}^{T}\mathbf{V}(\mathbf{G}_{\alpha})\hat{\xi} = -\frac{1}{2}\hat{\xi}^{T}\Omega\mathbf{G}_{\alpha}\,\hat{\xi}\,. \tag{15.6}$$

Using the well-known Campbell–Hausdorff formula [53] it is easy to check that Eq. (15.6) directly fullfils condition (15.5). Hence, Eq. (15.6) is satisfied by any element of the Lie algebra of the symplectic group, therefore $\mathbf{S} \to \hat{\Theta}(\mathbf{S})$.

In conclusion, any unitary transformation composed by a linear combination of Hilbert-space Hermitian operators that are quadratic in $\hat{\xi}_{\alpha}$ can be mapped, in general, to a symplectic matrix. We would like to emphasize the relevance of the previous statement because of its generality. Here, our Hilbert-space is applied to the transverse modes of light, but the same mathematical formalism can be applied to other physical systems. For example, imagine the 2D quantum harmonic oscillator [10]; the time evolution of any state is generated by its Hamiltonian that is quadratic in $\hat{\xi}_{\alpha}$. Therefore, one can map the time-unitary evolution onto a symplectic matrix.

The above unitary operators $\hat{U}(\mathbf{T})$ are elements of the, so-called, Metaplectic group $\hat{U}(\mathbf{T}) \in Mp(4)$, and the operator generators $\hat{G}_{\alpha}$ are elements of the algebra of the metaplectic group $\hat{G}_\alpha \in \mathfrak{mp}(4)$. For an exhaustive description of the metaplectic group, we refer the reader to [57]. Here, we underline the following properties:

1. The generators $\hat{G}_\alpha \in \mathfrak{mp}(4)$ obey the Lie algebra of $\mathfrak{sp}(4, \mathbb{R})$,

$$\left[-\frac{i}{\hbar}\hat{\Theta}(\mathbf{G}_\alpha), -\frac{i}{\hbar}\hat{\Theta}(\mathbf{G}_{\alpha'}) \right] = -\frac{i}{\hbar}\hat{\Theta}([\mathbf{G}_\alpha, \mathbf{G}_{\alpha'}]) \,. \tag{15.7}$$

2. The mapping $\hat{U}(\mathbf{T}) \in Mp(4) \rightarrow \mathbf{T} \in Sp(4, \mathbb{R})$ from the unitary operators to the symplectic matrices satisfies

$$\hat{U}(\mathbf{T}) \rightarrow \mathbf{T},\ \hat{U}(\mathbf{T}') \rightarrow \mathbf{T}' \quad \Rightarrow \quad \hat{U}(\mathbf{T})\hat{U}(\mathbf{T}') \rightarrow \mathbf{T}\mathbf{T}' \,. \tag{15.8}$$

3. The mapping introduced by the Stone–von Neumann theorem is defined up to a global phase, that is $e^{i\alpha}\hat{U}(\mathbf{T}) \rightarrow \mathbf{T}$. It can be proven [57] that the maximum possible reduction is to have two unitary operators, $\hat{U}(\mathbf{T})$ and $-\hat{U}(\mathbf{T})$, for each $\mathbf{T} \in Sp(4, \mathbb{R})$.[2]

To summarize, the linear canonical transformations acting on the wave state $|\psi\rangle \in \mathscr{H}$ are represented by $\hat{U}(\mathbf{T})$. The unitary operator $\hat{U}(\mathbf{T})$ belongs to the metaplectic group $Mp(4)$. It follows that these unitary operators can be satisfactorily parametrized, except for the sign ambiguity, by elements of the symplectic group $Sp(4, \mathbb{R})$. The unitary operators can be physically implemented by an arrange of cylindrical and spherical lenses [4, 58].

15.2.1 Canonical Operators on Paraxial Modes

In the previous section we established the connection of the unitary transformations $\hat{U}(\mathbf{T})$ with the geometrical optics picture. That allowed us to build up the general form of such unitary transformations. However, a detailed description of $\hat{U}(\mathbf{T})$ needs the introduction of the canonical operators which define a suitable framework for the next sections. The canonical operators have an identical structure to the well-known creation/annihilation operators of the 2D quantum harmonic oscillator. Here, we exploit this analogy to define the canonical operators acting on paraxial modes.

Any solution of the paraxial wave equation can be expressed in terms of a complete, orthogonal basis, such as the well-known Hermite–Gaussian mode basis or the Laguerre–Gaussian mode basis. In particular, we focus on the Hermite–Gaussian mode basis where the connection with the 2D quantum harmonic oscillator is straightforward. The Hermite–Gaussian modes read as [55]

[2]The metaplectic group is a double cover of the symplectic group.

$$\mathsf{HG}_{n_x,n_y}(x,y,z;k) = \mathsf{HG}_{n_x}(x,z;k)\mathsf{HG}_{n_y}(y,z;k)\,,$$
$$\mathsf{HG}_{n_j}(j,z;k) \equiv \sqrt{\frac{2^{1/2}}{2^{n_j}\,w(z)\,n_j!\sqrt{\pi}}}\,H_{n_j}\left(\frac{\sqrt{2}j}{w(z)}\right)\exp\left(-\frac{j^2}{w^2(z)}\right)$$
$$\exp\left[i\frac{kj^2}{2R(z)} - i\left(n_j + \frac{1}{2}\right)\arctan\left(\frac{z}{z_0}\right)\right], \tag{15.9}$$

where the indices n_x and n_y are nonnegative integers, k is the wave vector, $w(z) = w_0\sqrt{1+(z/z_0)^2}$, with w_0 being the width of the mode at $z = 0$, $R(z) = z\left[1+(z_0/z)^2\right]$ the phase-front radius, $z_0 = kw_0^2/2$ the Rayleigh range, $-(n_x+n_y+1)\arctan(z/z_0)$ the Gouy phase [30], and $H_{n_j}(\eta)$ the associated Hermite polynomials of order n_j

$$H_{n_j}(\eta) = \sum_{r=0}^{[n_j/2]}(-1)^r\frac{n!}{r!(n-2r)!}(2\eta)^{n_j-2r}\,.$$

The formal analogy between the Hermite–Gaussian modes (15.9) at $z = 0$ and the 2D harmonic-oscillator eigenstates offers the possibility of applying the operator algebra of the harmonic oscillator to paraxial waves [47]. We will proceed to analyze this connection with the harmonic oscillator Hamiltonian by defining, using the Dirac's bracket notation introduced in the previous section, $\langle\mathbf{r}|n_x,n_y\rangle = \mathsf{HG}_{n_x,n_y}(x,y,0;k)$. Note that in the brackets we only specify the indices n_x and n_y, but one should bear in mind that there is also a nonexplicit dependence on the width w_0 and wave vector $k = 1/\lambda$ of the Hermite–Gaussian mode at $z = 0$. The Hamiltonian of a 2D harmonic oscillator can be written as [52]

$$H_{HO} = \frac{w_0^2}{4\lambda^2}\left[\hat{q}_x^2+\hat{q}_y^2\right] + \frac{1}{w_0^2}\left[\hat{x}^2+\hat{y}^2\right] = \hat{a}_x^\dagger\hat{a}_x + \hat{a}_y^\dagger\hat{a}_y + 1\,, \tag{15.10}$$

where the annihilation $\hat{a}_x$ and creation $\hat{a}_x^\dagger$ (adjoint of $\hat{a}_x$) bosonic operators are defined in terms of the canonical operators $\hat{x}$ and $\hat{q}_x$ as [59]

$$\hat{a}_x = \frac{1}{w_0}\,\hat{x} + i\,\frac{w_0}{2\lambda}\,\hat{q}_x\,, \quad \text{and} \quad \hat{a}_x^\dagger = \frac{1}{w_0}\,\hat{x} - i\,\frac{w_0}{2\lambda}\,\hat{q}_x\,, \tag{15.11}$$

with analogous expressions for $\hat{a}_y$ and $\hat{a}_y^\dagger$. The only nonzero commutation relations among these operators are $[\hat{a}_x,\hat{a}_x^\dagger] = [\hat{a}_y,\hat{a}_y^\dagger] = 1$. Inversely,

$$\hat{x} = \frac{w_0}{2}(\hat{a}_x^\dagger + \hat{a}_x)\,, \quad \text{and} \quad \hat{q}_x = \frac{i\lambda}{w_0}(\hat{a}_x^\dagger - \hat{a}_x)\,, \tag{15.12}$$

and similarly for $\hat{y}$ and $\hat{q}_y$. Here, the ground state $|0,0\rangle$ represents the Gaussian mode whose spatial profile is given by Eq. (15.9) with $z = 0$ and $n_x = n_y = 0$.

As in quantum mechanics, any state can be constructed by the action of canonical operators on the fundamental state. In our case, the action of a canonical operator on a wave state (such as $\hat{a}_x^\dagger|0,0\rangle$) changes the mode profile of the wave ($\hat{a}_x^\dagger|0,0\rangle = |1,0\rangle$, which represents the first-order Hermite–Gaussian mode with transverse indices $n_x = 1$ and $n_y = 0$). In general, a Hermite–Gaussian mode can be written as

$$|n_x, n_y\rangle = \frac{(\hat{a}_x^\dagger)^{n_x}}{\sqrt{n_x!}} \frac{(\hat{a}_y^\dagger)^{n_y}}{\sqrt{n_y!}} |0,0\rangle \,, \tag{15.13}$$

whose spatial profile is given by Eq. (15.9) at $z = 0$. As a result, any transverse profile $|\psi\rangle = \sum_{n_x,n_y} c_{n_x,n_y}|n_x, n_y\rangle$ can be decomposed in a Hermite–Gaussian basis $|n_x, n_y\rangle$ or in canonical operators acting on the fundamental Gaussian mode. We emphasize that the Hamiltonian (15.10) has the harmonic oscillator form due to the structure of the Hermite–Gaussian mode (15.9) at $z = 0$. In general, if $z \neq 0$, $\mathsf{HG}_{n_x,n_y}(x, y, 0; k)$ does not satisfy the Hamiltonian (15.10). However, the free propagation of a Hermite–Gaussian mode from $z = 0$ up to a certain position z can be described by a linear canonical transformation.

In subsequent sections we will also use the Laguerre–Gaussian mode basis. The wave state for a Laguerre–Gaussian mode will be denoted as $|\ell, p\rangle$, where the indices $\ell = 0, \pm1, \pm2, \ldots$ and $p = 0, 1, 2, \ldots$ stand for the winding (or topological charge) and the number of nonaxial radial nodes of the mode, respectively. The spatial profile in cylindrical coordinates is [55]

$$\begin{aligned} \mathsf{LG}_{\ell,p}(r,\phi,z;k) = &\sqrt{\frac{2p!}{\pi(|\ell|+p)!}\frac{1}{w(z)}} \left(\frac{\sqrt{2}r}{w(z)}\right)^{|\ell|} L_p^{|\ell|}\left(\frac{2r^2}{w^2(z)}\right) \\ &\times \exp\left(-\frac{r^2}{w^2(z)}\right) \exp\left[i\ell\phi + i\frac{kr^2}{2R(z)} + i\Phi_G(z)\right], \end{aligned} \tag{15.14}$$

where $\Phi_G(z) = -(2p + |\ell| + 1)\arctan(z/z_0)$ is the Gouy phase, and $L_p^{|\ell|}(\eta)$ is the generalized Laguerre polynomial

$$L_p^{|\ell|}(\eta) = \sum_{m=0}^{p} (-1)^m \frac{(|\ell|+p)!}{(p-m)!\,(|\ell|+m)!\,m!}\,\eta^m. \tag{15.15}$$

In the position representation $\langle \mathbf{r}|\ell,p\rangle = \mathsf{LG}_{\ell,p}(\mathbf{r}, z = 0; k)$. Analogously to the Hermite–Gaussian basis, any paraxial wave can be expressed as $|\psi\rangle = \sum_{\ell,p} c_{\ell,p}|\ell,p\rangle$. The Hermite–Gaussian indices $\{n_x, n_y\}$ and Laguerre–Gaussian indices $\{\ell, p\}$ are related by the mode-order number $N = n_x + n_y = |\ell| + 2p$. The azimuthal structure of the $\mathsf{LG}_{\ell,p}$ modes is related to the OAM of light [5]; specifically, they carry an OAM $\hbar\,\ell$ per photon.

15.2.2 The Metaplectic Representation

In this subsection we make explicit the form of all possible unitary metaplectic representations acting on $\mathscr{H}$; $\hat{U}(\mathbf{T}) \in Mp(4)$ where $\mathbf{T} \in Sp(4,\mathbb{R})$. All these unitary operations $\hat{U}(\mathbf{T})$ are generated by ten Hermitian generators $\hat{G}_\alpha$, quadratic in $\hat{\boldsymbol{\xi}}$, in correspondence with the ten generators of $Sp(4,\mathbb{R})$, see Eq. (15.6). The ten Hermitian operators $\hat{G}_\alpha$ can be split into two sets [6]: *passive* and *active* generators. The passive set encompasses the maximal compact subgroup U(2):

$$\begin{aligned}
\hat{\mathscr{L}}_o &= \frac{1}{4}\left(\hat{a}_x^\dagger\hat{a}_x + \hat{a}_x\hat{a}_x^\dagger + \hat{a}_y^\dagger\hat{a}_y + \hat{a}_y\hat{a}_y^\dagger\right) = \frac{\hat{x}^2+\hat{y}^2}{2w_0^2} + \frac{(\hat{q}_x^2+\hat{q}_y^2)w_0^2}{8\lambda^2} - \frac{1}{2}, \\
\hat{\mathscr{L}}_x &= \frac{1}{2}\left(\hat{a}_x^\dagger\hat{a}_x - \hat{a}_y^\dagger\hat{a}_y\right) = \frac{\hat{x}^2-\hat{y}^2}{2w_0^2} + \frac{(\hat{q}_x^2-\hat{q}_y^2)w_0^2}{8\lambda^2}, \\
\hat{\mathscr{L}}_y &= \frac{1}{2}\left(\hat{a}_x^\dagger\hat{a}_y + \hat{a}_y^\dagger\hat{a}_x\right) = \frac{\hat{x}\hat{y}}{w_0^2} + \frac{\hat{q}_x\hat{q}_y w_0^2}{4\lambda^2}, \\
\hat{\mathscr{L}}_z &= -\frac{i}{2}\left(\hat{a}_x^\dagger\hat{a}_y - \hat{a}_y^\dagger\hat{a}_x\right) = \frac{\hat{x}\hat{q}_y - \hat{y}\hat{q}_x}{2\lambda}.
\end{aligned} \tag{15.16}$$

Here, $\hat{a}_x$ and $\hat{a}_y$ ($\hat{a}_x^\dagger$ and $\hat{a}_y^\dagger$) are the annihilation (creation) bosonic operators for all orthogonal transverse modes, see Eq. (15.11). The passive operators (15.16) have the form of the well-known Stokes operators. They obey the usual commutation relations $[\hat{\mathscr{L}}_i, \hat{\mathscr{L}}_j] = i\,\varepsilon_{ijk}\hat{\mathscr{L}}_k$ $(i,j,k = x,y,z)$, where ε_{ijk} is the totally antisymmetric Levi–Civita tensor. Of these U(2) generators, $\hat{\mathscr{L}}_o$ is the only commuting element while $\hat{\mathscr{L}}_z$ represents real spatial rotations on the transverse x–y plane. In contrast, $\hat{\mathscr{L}}_x$ and $\hat{\mathscr{L}}_y$ describe simultaneous rotations in the four-dimensional phase-space: $\hat{\mathscr{L}}_x$ produces rotations in the x–p_x and y–p_y planes by equal and opposite amounts, whereas $\hat{\mathscr{L}}_y$ gives rise to rotations in the x–p_y and y–p_x planes by equal amounts. Notice that $\hat{\mathscr{L}}_o$, when acting on paraxial modes, does not modify their transverse profile (it just gives rise to a global phase), whereas $\hat{\mathscr{L}}_z$ is proportional to the longitudinal component of the OAM operator for a light beam [20]. It is also important to mention that passive generators, due to their structure, preserve the mode-order number N for Laguerre–Gaussian, $N \equiv |\ell| + 2p$, and Hermite–Gaussian, $N = n_x + n_y$, modes.

Figure 15.1 depicts the action of the unitary operator $\hat{U} = e^{-is\hat{\mathscr{L}}_j}$, with $j = x, y, z$ and $s \in \mathbb{R}$, corresponding to the three passive generators $\hat{\mathscr{L}}_x$, $\hat{\mathscr{L}}_y$, and $\hat{\mathscr{L}}_z$ on two first order paraxial modes: the Laguerre–Gaussian mode $\mathsf{LG}_{\ell=1,p=0}$ and the Hermite–Gaussian mode $\mathsf{HG}_{n_x=1,n_y=0}$ for two different values of the parameter s. It can be observed that $\hat{\mathscr{L}}_z$ and $\hat{\mathscr{L}}_x$ leave invariant $\mathsf{LG}_{\ell=1,p=0}$ and $\mathsf{HG}_{n_x=1,n_y=0}$, respectively. More generally, the Laguerre–Gaussian mode state $|\ell, p\rangle$

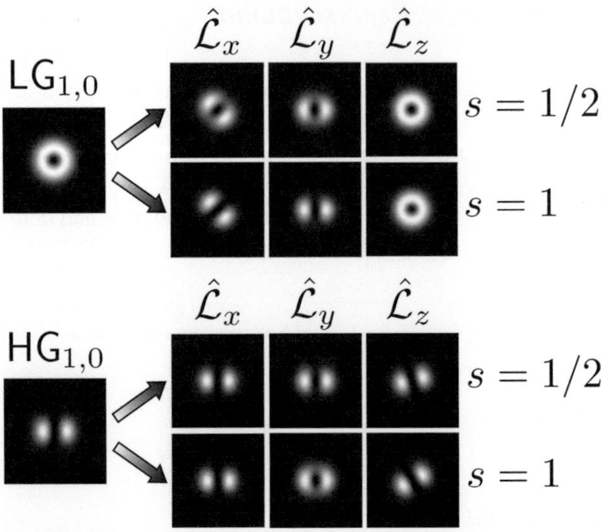

Fig. 15.1 Transformation properties of three of the passive generators (15.16) on a Laguerre–Gaussian mode $\mathsf{LG}_{\ell=1,p=0}$ and on a Hermite–Gaussian mode $\mathsf{HG}_{n_x=1,n_y=0}$. The transverse density plots are given for increasing squeezing parameter s

and the Hermite–Gaussian mode state $|n_x, n_y\rangle$ are eigenvectors of $\hat{\mathscr{L}}_z$ and $\hat{\mathscr{L}}_x$, respectively [20]. Also, Laguerre- and Hermite–Gaussian bases are unitarily related: $\mathsf{LG}_{\ell,p}$ transforms into HG_{n_x,n_y} via $e^{-i(\pi/2)\hat{\mathscr{L}}_y}$ (see also Chap. 2 for the connection between Hermite–Gaussian and Laguerre–Gaussian modes).

At variance with the passive set of generators, the active set is responsible for the noncompactness of $Sp(4,\mathbb{R})$:

$$\begin{aligned}
\hat{\mathscr{K}}_x &= -\frac{1}{2}(\hat{a}_x^\dagger \hat{a}_y^\dagger + \hat{a}_x \hat{a}_y)\,, \\
\hat{\mathscr{K}}_y &= \frac{1}{4}(\hat{a}_x^{\dagger\,2} + \hat{a}_x^2 - \hat{a}_y^{\dagger\,2} - \hat{a}_y^2)\,, \\
\hat{\mathscr{K}}_z &= -\frac{i}{4}(\hat{a}_x^{\dagger\,2} - \hat{a}_x^2 + \hat{a}_y^{\dagger\,2} - \hat{a}_y^2)\,, \\
\hat{\mathscr{M}}_x &= -\frac{i}{2}(\hat{a}_x^\dagger \hat{a}_y^\dagger - \hat{a}_x \hat{a}_y)\,, \\
\hat{\mathscr{M}}_y &= \frac{i}{4}(\hat{a}_x^{\dagger\,2} - \hat{a}_x^2 - \hat{a}_y^{\dagger\,2} + \hat{a}_y^2)\,, \\
\hat{\mathscr{M}}_z &= \frac{1}{4}(\hat{a}_x^{\dagger\,2} + \hat{a}_x^2 + \hat{a}_y^{\dagger\,2} + \hat{a}_y^2)\,.
\end{aligned} \tag{15.17}$$

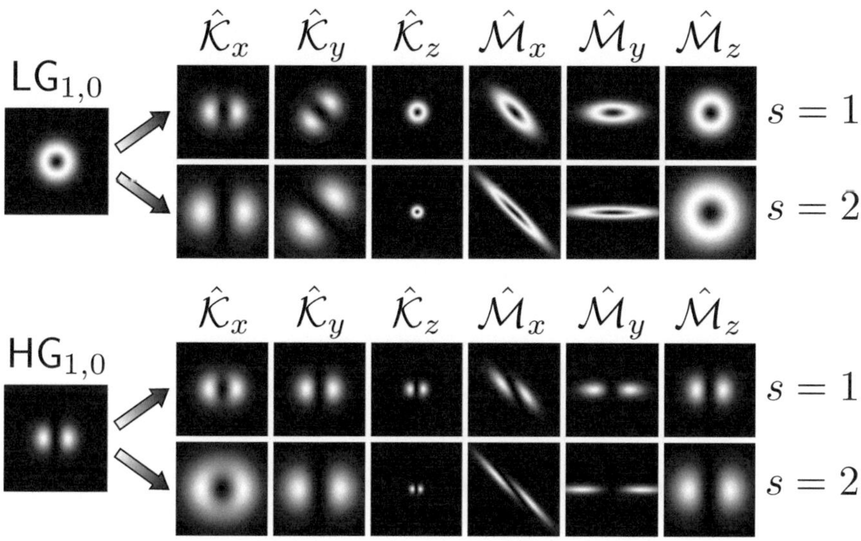

Fig. 15.2 Transformation properties of the six active generators (15.17) on a Laguerre–Gaussian mode $\mathsf{LG}_{\ell=1,p=0}$ and on a Hermite–Gaussian mode $\mathsf{HG}_{n_x=1,n_y=0}$. The transverse density plots are given for increasing squeezing parameter s

They satisfy the following commutators:

$$[\hat{\mathscr{L}}_i, \hat{\mathscr{K}}_j] = i\,\varepsilon_{ijk}\hat{\mathscr{K}}_k\,, \quad [\hat{\mathscr{L}}_i, \hat{\mathscr{M}}_j] = i\,\varepsilon_{ijk}\hat{\mathscr{M}}_k\,,$$
$$[\hat{\mathscr{L}}_o, \hat{\mathscr{K}}_j \pm i\hat{\mathscr{M}}_j] = \mp(\hat{\mathscr{K}}_j \pm i\hat{\mathscr{M}}_j)\,, \quad [\hat{\mathscr{K}}_i, \hat{\mathscr{M}}_j] = i\,\delta_{ij}\hat{\mathscr{L}}_o\,,$$
$$[\hat{\mathscr{K}}_i, \hat{\mathscr{K}}_j] = [\hat{\mathscr{M}}_i, \hat{\mathscr{M}}_j] = -i\,\varepsilon_{ijk}\hat{\mathscr{L}}_k\,.$$

Figure 15.2 displays the action of the unitary operators $\hat{U} = e^{-is\hat{\mathscr{K}}_j}$ and $\hat{U} = e^{-is\hat{\mathscr{M}}_j}$, with $j = x, y, z$ and $s \in \mathbb{R}$, corresponding to the six active generators (15.17) on two first order paraxial modes: the Laguerre–Gaussian mode $\mathsf{LG}_{\ell=1,p=0}$ and the Hermite–Gaussian mode $\mathsf{HG}_{n_x=1,n_y=0}$ for two different values of the parameter s. In contrast to passive generators, the active ones scale (squeeze) the spatial modes and change the order N, giving rise to infinite mode superpositions.

As any arbitrary sequence of symplectic transformations $\mathbf{T}_m$ is again another symplectic transformation $\mathbf{T} = \prod_m \mathbf{T}_m$, one concludes that the most general unitary metaplectic operator $\hat{U}(\mathbf{T})$ corresponding to $\mathbf{T}$ is given by a single exponential of i times real linear combinations of any of the above (15.16) and (15.17) generators [19]:

$$\hat{U}(\mathbf{T}) = \exp\left(-i\mathbf{s} \cdot \hat{\mathbf{G}}\right), \tag{15.18}$$

where $\hat{\mathbf{G}} \in \left\{\hat{\mathscr{L}}, \hat{\mathscr{K}}, \hat{\mathscr{M}}\right\}$ and $\mathbf{s}$ a ten-parameter real vector. Together with the four passive generators (15.16), only $\hat{\mathscr{K}}_z$ and $\hat{\mathscr{M}}_y$ suffice to describe the general metaplectic operator (15.18) by recourse to the following passive-active-passive (Euler) decomposition [6, 19]:

$$\hat{U}(\mathbf{T}) = \exp\left(-i\boldsymbol{\mu}\cdot\hat{\mathscr{L}}\right)\exp\left(-i\nu_y\hat{\mathscr{M}}_y - i\nu_z\hat{\mathscr{K}}_z\right)\exp\left(-i\boldsymbol{\eta}\cdot\hat{\mathscr{L}}\right), \quad (15.19)$$

still requiring ten real parameters via the two four-component vectors $\boldsymbol{\mu}$ and $\boldsymbol{\eta}$ plus constants ν_y and ν_z.

In the next sections we will elucidate how the action of the above generators can be easily understood within the Wigner function formalism.

15.2.3 The Wigner Function of Paraxial Modes

The phase space picture, which is a fundamental concept in classical mechanics, remains useful when passing to quantum mechanics. In a similar fashion with probability density distribution functions in classical systems governed by Liouville dynamics, quasiprobability distributions were already introduced in the early foundations of quantum mechanics. They provide a description of quantum systems at the level of density operators although not at the level of state vectors. Among them, the Wigner function stands out because it is real, nonsingular, yields correct quantum mechanical operator averages in terms of phase space integrals and possesses non-negative-definite marginal distributions [37, 43, 54]. It is, however, only positive for Gaussian pure states, according to the Hudson–Piquet theorem.

By exploiting the analogy between classical and quantum mechanics with geometrical and wave optics, Wigner distributions have been used in the context of classical wave optics of both coherent and partially coherent light fields [12, 27, 46, 59, 62, 63], where they are Fourier-related to the cross-spectral densities. Particularly outstanding has been the symplectic invariant approach by Simon and Mukunda [57], which has been applied to anisotropic Gaussian Schell-model beams via the relation between ray-transfer matrices of first-order optical systems (FOS) and metaplectic operators acting on wave amplitudes and cross-spectral densities.

The cross-spectral density $w_c(\mathbf{r}, \mathbf{r}')$ is defined by resorting to the density operator. The density operator for a paraxial wave state $|\psi\rangle$ (15.2) is $\hat{\rho} = |\psi\rangle\langle\psi|$. Here, we deal with coherent waves, but this formalism is more general, and can be extended to partially coherent waves. For a coherent wave, its amplitude $\psi(\mathbf{r}) = \langle\mathbf{r}|\psi\rangle$ describes the spatial form. In contrast, partially coherent waves do not have a deterministic amplitude. They are described by an ensemble of amplitudes with an associated probability. Within the density operator formalism, partially coherent waves can be cast as $\hat{\rho} = \sum_k p_k \hat{\rho}_k = \sum_k p_k |\psi_k\rangle\langle\psi_k|$, where $\hat{\rho}_k$ is a coherent wave with an associated probability p_k. Of course, $\sum_k p_k = 1$ must be satisfied, as it is expected for a probability distribution. When we have only an amplitude, $p_k = 1$, we return to the coherent wave. Therefore, for both coherent and partially coherent waves, the cross-spectral density is

$$w_c(\mathbf{r}, \mathbf{r}') = \langle \mathbf{r} | \hat{\rho} | \mathbf{r}' \rangle , \tag{15.20}$$

where $\hat{\rho}$ is the density operator of the wave. For coherent waves, the cross-spectral density reads as $w_c(\mathbf{r}, \mathbf{r}') = \psi(\mathbf{r})\psi^*(\mathbf{r}')$. As a density operator, we must impose hermiticity ($\hat{\rho} = \hat{\rho}^\dagger$) and nonnegativity ($\hat{\rho} \geq 0$). Notice that the total power (15.1) can be written in the new formalism as $I(z) = \int d^2\mathbf{r}\, w_c(\mathbf{r}, \mathbf{r}) = \mathrm{Tr}(\hat{\rho})$. On the other hand, when a symplectic transformation $\mathbf{T}$ acts on the state $|\psi\rangle$, we know that if $\hat{U}(\mathbf{T}) \in Mp(4)$, then $|\psi'\rangle = \hat{U}(\mathbf{T})|\psi\rangle$, $\hat{\rho} \xrightarrow{\mathbf{T}} \hat{\rho}' = \hat{U}(\mathbf{T})\, \hat{\rho}\, \hat{U}^\dagger(\mathbf{T})$ and

$$w_c(\mathbf{r}, \mathbf{r}') \xrightarrow{\mathbf{T}} w_c'(\mathbf{r}, \mathbf{r}') = \int d^2\mathbf{r}'' \int d^2\mathbf{r}''' \, \langle \mathbf{r} | \hat{U}(\mathbf{T}) | \mathbf{r}'' \rangle \, w_c(\mathbf{r}'', \mathbf{r}''') \langle \mathbf{r}''' | \hat{U}(\mathbf{T}) | \mathbf{r}' \rangle , \tag{15.21}$$

where $\langle \mathbf{r} | \hat{U}(\mathbf{T}) | \mathbf{r}' \rangle$ is the kernel of the symplectic transformation [22]. The description of the symplectic transformation with the cross-spectral density is not trivial, as one has to solve the integral in Eq. (15.21). Nevertheless, by resorting to the Wigner function formalism, the action of symplectic transformations is remarkably simple. The Wigner distribution $W(\mathbf{r}, \mathbf{q})$ is a faithful representation of the cross-spectral density

$$w_c(\mathbf{r}, \mathbf{r}') = \int d^2\mathbf{p}\, W\left(\frac{\mathbf{r} + \mathbf{r}'}{2}, \mathbf{q}\right) e^{i\mathbf{q}\cdot(\mathbf{r}-\mathbf{r}')} ,$$

or inversely

$$\begin{aligned} W(\mathbf{r}, \mathbf{q}) &= \frac{1}{(2\pi)^2} \int d^2\boldsymbol{\epsilon}\, w_c\left(\mathbf{r} + \frac{\boldsymbol{\epsilon}}{2}, \mathbf{r} - \frac{\boldsymbol{\epsilon}}{2}\right) e^{-i\mathbf{q}\cdot\boldsymbol{\epsilon}} \\ &= \frac{1}{(2\pi)^2} \int d^2\boldsymbol{\epsilon} \left\langle \mathbf{r} + \frac{\boldsymbol{\epsilon}}{2} \right| \hat{\rho} \left| \mathbf{r} - \frac{\boldsymbol{\epsilon}}{2} \right\rangle e^{-i\mathbf{q}\cdot\boldsymbol{\epsilon}} , \end{aligned} \tag{15.22}$$

where $\boldsymbol{\epsilon} = \mathbf{r} - \mathbf{r}'$. For fully coherent waves, $w_c(\mathbf{r} + \boldsymbol{\epsilon}/2, \mathbf{r} - \boldsymbol{\epsilon}/2) = \psi(\mathbf{r} + \boldsymbol{\epsilon}/2)\psi^*(\mathbf{r} - \boldsymbol{\epsilon}/2)$. We recall some of properties of the Wigner function:

- The Wigner function is real. This property is equivalent to the hermiticity of the density operator.
- The nonnegativity of the density operator translates into the Wigner function. The positivity condition $\hat{\rho} \geq 0$ is equivalent to saying that $\hat{\rho}$ is nonnegative if and only if $\mathrm{Tr}(\hat{\rho}\hat{\rho}') \geq 0 \;\; \forall\, \hat{\rho}'$. For Wigner functions it reads as

$$\mathrm{Tr}(\hat{\rho}\hat{\rho}') = \int d^2\mathbf{r} \int d^2\mathbf{r}' \; w_c(\mathbf{r}', \mathbf{r}) w_c'(\mathbf{r}', \mathbf{r})$$
$$= (2\pi)^2 \int d^2\mathbf{r} \int d^2\mathbf{q} \; W(\mathbf{r}, \mathbf{q}) W'(\mathbf{r}, \mathbf{q}) \geq 0 \,, \quad \forall \, W' \,.$$

- In general, the Wigner function is not pointwise nonnegative, but its marginals are always pointwise nonnegative

$$w_c(\mathbf{r}, \mathbf{r}) = \int d^2\mathbf{q} \; W(\mathbf{r}, \mathbf{q}) \geq 0 \,, \quad \forall \, W \,,$$
$$\tilde{w}_c(\mathbf{q}, \mathbf{q}) = \int d^2\mathbf{r} \; W(\mathbf{r}, \mathbf{q}) \geq 0 \,, \quad \forall \, W \,,$$

where $\tilde{w}_c(\mathbf{q}, \mathbf{q}')$ is the Fourier transform of $w_c(\mathbf{r}, \mathbf{r}')$. Here, $w_c(\mathbf{r}, \mathbf{r})$ and $\tilde{w}_c(\mathbf{q}, \mathbf{q})$ are the spatial and momentum probability distributions, respectively. For coherent waves $w_c(\mathbf{r}, \mathbf{r}) = |\psi(\mathbf{r})|^2$ and $\tilde{w}_c(\mathbf{q}, \mathbf{q}) = |\tilde{\psi}(\mathbf{q})|^2$, where $\tilde{\psi}(\mathbf{q})$ is the Fourier transform of the wave amplitude. Thus, the total irradiance

$$\mathscr{I} = \int d^2\mathbf{r} \, w_c(\mathbf{r}, \mathbf{r}) = \int d^2\mathbf{r} \int d^2\mathbf{q} \; W(\mathbf{r}, \mathbf{q}) \,.$$

- Mapping Hermitian operators of the Hilbert space $\mathscr{H}$ onto phase-space is not trivial, due to the noncommutativity of the operators. For example, suppose that one wishes to map the Hermitian operator $\hat{x}\hat{q}_x$ onto the phase-space. As $\hat{x}$ does not commute with $\hat{q}_x$, $\hat{q}_x\hat{x}$ is different from $\hat{x}\hat{q}_x$. In fact, $\hat{x}\hat{q}_x = \hat{q}_x\hat{x} + i\lambda$. If we choose, for example, the following map $\hat{x}\hat{q}_x \to xq_x$ and $\hat{q}_x\hat{x} \to q_x x$, naturally $xq_x = q_x x$, thus two different Hermitian operators are mapped onto the same phase-space dependence as a result of the chosen convention. As in quantum mechanics ordering problems, there is not a basic principle that imposes a unique convention. Some of the usual ordering conventions are the normal, antinormal, and symmetric or Weyl-ordering schemes. As pointed out previously, we choose here the Weyl-ordering. The Weyl-ordering can be set down by comparing the components of the monomials $\boldsymbol{\lambda}$ given by the correspondence

$$e^{i\boldsymbol{\lambda}\cdot\boldsymbol{\xi}} \leftrightarrow e^{i\boldsymbol{\lambda}\cdot\hat{\boldsymbol{\xi}}} \,,$$

where $\boldsymbol{\lambda}$ is a vector of the same dimension than $\boldsymbol{\xi}$. In the Wigner representation, the Weyl operator $\{\hat{x}^\alpha \hat{q}_x^\beta \hat{y}^\gamma \hat{q}_y^\delta\}_\mathrm{W}$ is in correspondence with the monomial $x^\alpha q_x^\beta y^\gamma q_y^\delta$, where α, β, γ, and δ are arbitrary powers, and obeys the following property

$$\frac{\mathrm{Tr}(\hat{\rho}\{\hat{x}^\alpha \hat{q}_x^\beta \hat{y}^\gamma \hat{q}_y^\delta\}_\mathrm{W})}{\mathrm{Tr}(\hat{\rho})} = \frac{\int d^2\mathbf{r} \int d^2\mathbf{q} \, x^\alpha q_x^\beta y^\gamma q_y^\delta \;\, W(\mathbf{r}, \mathbf{q})}{\int d^2\mathbf{r} \int d^2\mathbf{q} \, W(\mathbf{r}, \mathbf{q})} \,.$$

We previously showed how the density operator and the cross-spectral density are modified under symplectic transformations [see Eq. (15.21)]. However, the Wigner function takes an advantage over such representations; symplectic transformations are easily performed in the Wigner representation. Before going to the matter, it is worth reexpressing (15.22) in a more appropriate form,

$$W(\boldsymbol{\xi}) = \frac{1}{(2\pi)^4} \int d^4\boldsymbol{\xi}' \, e^{i\boldsymbol{\xi}^T \Omega \boldsymbol{\xi}'} \, \mathrm{Tr}(\hat{\rho}\, e^{i\boldsymbol{\xi}'^T \Omega \hat{\boldsymbol{\xi}}}) \,, \tag{15.23}$$

where $\boldsymbol{\xi}' = (\mathbf{r}', \mathbf{q}')$.

When a paraxial wave crosses a linear lossless optical system with transformation matrix $\mathbf{T}$, its density operator changes as $\hat{\rho} \xrightarrow{\mathbf{T}} \hat{U}(\mathbf{T})\hat{\rho}\hat{U}^{\dagger}(\mathbf{T})$, whereas the Wigner representation is transformed as a scalar field of the symplectic group,

$$W(\boldsymbol{\xi}) \to W(\mathbf{T}^{-1}\boldsymbol{\xi}) \,, \quad \text{where } \mathbf{T} \in Sp(4, \mathbb{R}) \,. \tag{15.24}$$

The covariant property (15.24) will be crucial to describe in subsequent sections the transformation of any paraxial wave along systems described by a symplectic matrix $\mathbf{T}$, since knowledge of both the Wigner function corresponding to the input wave and the symplectic matrix $\mathbf{T}$ of the optical system, the output Wigner function follows from Eq. (15.24). This property was firstly highlighted by Bastiaans in his pioneering papers introducing the Wigner function in Optics [11, 12].

15.3 Linear Canonical Transformations for Quantum Information

A controlled generation and manipulation of photon states encoded in their spatial degrees of freedom is a crucial ingredient in many quantum information tasks exploiting higher-than-two dimensional encoding. This section is mainly devoted to quantum information applications, in particular, we focus on the manipulation and control of quantum states encoded in the transverse spatial modes of light. Part of this section is based on the symplectic formalism presented in Sect. 15.2.

First of all, we show how photons with a well-defined spatial-transverse profile can be conceived in the quantum field theory framework. We also introduce the Poincaré Sphere to represent the quantum information bits encoded in the spatial-transverse profile. We will end this section with a discussion about quantum gates in a single photon by using linear canonical transformations.

15.3.1 The General Paraxial Single-Photon State

Before describing the unitary transformations generated by passive (15.16) and active (15.17) operators on the spatial mode of a photon, it is suitable at this stage to introduce in a nutshell the concept of a photon in the paraxial approximation.

In quantum electrodynamics the photon is interpreted as an excitation of the quantized mode field [44, 45]. In general, this excitation is associated with a plane wave: it has a well-defined linear momentum, energy, and polarization. However, a superposition of photons with precise linear momentum, energy and polarization, is still a photon [15]. Therefore, we can form a wavepacket with some spatial and temporal localization. As it was shown in [20], the electromagnetic field can be properly quantized accounting for the paraxial approximation and one can define photons in a certain superposition of spatial modes. Within this quantization, we can define a set of creation and annihilation operators: those that create a photon from the vacuum with a certain paraxial transverse profile, for example with a well-defined Laguerre–Gaussian mode or a Hermite–Gaussian mode.

The most general (paraxial) single-photon pure state can be described as [20]

$$|\psi\rangle = \sum_{\sigma,n_x,n_y} \int_0^\infty d\omega\, C_{\sigma,n_x,n_y}(\omega)\, \hat{b}^\dagger_{\sigma,n_x,n_y}(\omega)\, |\text{vac}\rangle \,. \tag{15.25}$$

Here, $\hat{b}^\dagger_{\sigma,n_x,n_y}(\omega)$ denotes the bosonic creation operator of a photon with a Hermite–Gaussian mode (n_x, n_y), a linear polarization σ, and a frequency ω acting on the vacuum state $|\text{vac}\rangle$. The commutation relations read as $[\hat{b}_{\sigma,n_x,n_y}(\omega), \hat{b}^\dagger_{\sigma',n'_x,n'_y}(\omega')] = \delta_{\sigma\sigma'}\delta_{n_x n'_x}\delta_{n_y n'_y}\delta(\omega-\omega')$. The complex coefficients $C_{\sigma,n_x,n_y}(\omega)$ can be interpreted as the probability amplitudes for finding a photon in the state $\hat{b}^\dagger_{\sigma,n_x,n_y}(\omega)\,|\text{vac}\rangle = |\sigma\rangle \otimes |n_x, n_y\rangle \otimes |\omega\rangle$. In the next sections we focus only on the spatial part of Eq. (15.25) and assume that it consists of a *finite* superposition of d orthogonal HG modes $|n_x, n_y\rangle \propto (\hat{a}^\dagger_x)^{n_x}(\hat{a}^\dagger_y)^{n_y}|0,0\rangle$ or d orthogonal LG modes $|\ell, p\rangle$ (for more details about the canonical operators acting on spatial modes, we refer the reader to Sect. 15.2.1, avoiding possible confusion between operators $\hat{a}$ with operators $\hat{b}$).

The superposition of d orthogonal modes can be interpreted as a qu*d*it. In the normalized qu*d*it form, we can either express

$$|\psi\rangle = \sum_{\ell,p} c_{\ell,p}\, |\ell, p\rangle \,, \tag{15.26}$$

in the Laguerre–Gaussian mode basis or

$$|\psi\rangle = \sum_{n_x,n_y} c_{n_x,n_y} |n_x, n_y\rangle \,. \tag{15.27}$$

in the Hermite–Gaussian mode basis. Any qu*d*it requires, at least, $2d$ independent real parameters, albeit normalization and invariance under a global phase reduce this number to $2(d-1)$. As expansions (15.26) and (15.27) suggest, one could choose other complete basis of transverse modes, for example the Ince–Gaussian (in elliptical coordinates) modes [8].

15.3.2 Passive Generators: The Poincaré Sphere

The basic unit of information for most quantum information and quantum computing schemes is the qu*d*it which, for $d = 2$, becomes a qubit. The qubit can be encoded in different degrees of freedom of a physical system, for example in the case of photons the usual encoding is by exploiting its polarization state. The Poincaré (also known as the Bloch) sphere representation is a convenient way to visualize the transformations on qubits. For polarization states, the north and south poles of the sphere correspond to left- $|\sigma = +1\rangle$ and right-handed $|\sigma = -1\rangle$ circularly polarized eigenstates, respectively. More generally, any polarized state can be described as a linear superposition of left- and right-handed circular polarization in the form (up to a global phase)

$$|\theta,\varphi\rangle = \cos\frac{\theta}{2}\,|\sigma = +1\rangle + e^{i\varphi}\sin\frac{\theta}{2}\,|\sigma = -1\rangle\,, \tag{15.28}$$

which, on the Poincaré sphere, corresponds to a point on the surface having polar and azimuthal angular coordinates θ and φ, respectively.

Analogously, Padgett and Courtial introduced the *orbital* Poincaré sphere for paraxial first-order-mode ($N = 1$) spatial beams carrying OAM [49]. Its underlying SU(2) symmetry was subsequently shown [1]. In this picture, the poles of the sphere correspond to Laguerre–Gaussian modes with radial-node number $p = 0$ and topological charge $\ell = \pm 1$ (plus and minus standing for the north and south poles, respectively). Figure 15.3a illustrates the first-order mode orbital Poincaré sphere. Hence, in complete analogy with Eq. (15.28), any state on the first-order-mode sphere can be written as

$$|\theta,\varphi\rangle_{N=1} = \cos\frac{\theta}{2}\,|\ell = 1, p = 0\rangle + e^{i\varphi}\sin\frac{\theta}{2}\,|\ell = -1, p = 0\rangle\,. \tag{15.29}$$

The Hermite–Gaussian modes with $N = 1$ are located in the equatorial plane of the sphere, for example the superposition at $\theta = \pi/2$ and $\varphi = 0$ ($\theta = \pi/2$, $\varphi = \pi$) gives rise to the Hermite–Gaussian mode with indexes $n_x = 1$ and $n_y = 0$ ($n_x = 0$ and $n_y = 1$).

One can obtain Eq. (15.29) by applying a unitary operator $\hat{U}(\mathbf{T})$ (induced by the passive generators (15.16)) on a paraxial mode with $N = 1$. In particular, we can express the state (15.29) as

$$|\theta,\varphi\rangle_{N=1} = \exp(-i\theta\,\hat{\mathscr{L}}\cdot\mathbf{u}_\varphi)|\ell = 1, p = 0\rangle \equiv \hat{U}(\theta,\varphi)|\ell = 1, p = 0\rangle\,. \tag{15.30}$$

Here, $\hat{\mathscr{L}} = (\hat{\mathscr{L}}_x, \hat{\mathscr{L}}_y, \hat{\mathscr{L}}_z)$ is a vector of the passive generators satisfying the SU(2) algebra, and $\mathbf{u}_\varphi = (-\sin\varphi, \cos\varphi, 0)$ is a unit vector in the equatorial plane of the sphere. The action of the unitary (metaplectic) operator $\hat{U}(\theta,\varphi)$ can be interpreted

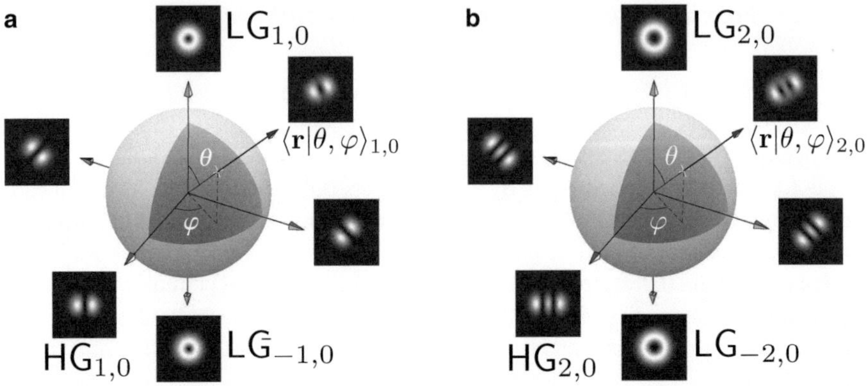

Fig. 15.3 Orbital Poincaré sphere of (**a**) first and (**b**) second order modes. The poles of the sphere correspond to Laguerre–Gaussian modes with (**a**) $\ell = \pm 1, p = 0$ and (**b**) $\ell = \pm 2, p = 0$ (positive and negative signs for north and south poles, respectively). The states in the equatorial plane ($\theta = \pi/2$) with $\varphi = 0$ and $\varphi = \pi$ yield the Hermite–Gaussian modes having indexes (**a**) $n_x = 1, n_y = 0$ and $n_x = 0, n_y = 1$ (not shown), respectively, and (**b**) $n_x = 2, n_y = 0$ and $n_x = 0, n_y = 2$ (not shown), respectively

as a rotation of θ following the unit vector $\mathbf{u}_r = (\cos\varphi \sin\theta, \sin\varphi \sin\theta, \cos\theta)$ for a given φ. Due to the structure of the passive generators, which preserve the mode-order number N, the operator $\hat{U}(\theta, \varphi)$ acting on the state $|\ell = 1, p = 0\rangle$ always yields states with $N = 1$.

15.3.3 The Poincaré Sphere for Higher-Order Modes

The first-order-mode orbital Poincaré sphere not only constitutes an elegant framework to represent states with mode-order number $N = 1$, but also all the transformations among them can be visualized as paths connecting points on the sphere. However, higher-order modes cannot be described by Eq. (15.29). That is, states on higher-order orbital Poincaré spheres involve more complex superpositions of Laguerre–Gaussian modes.

A generalization of the Poincaré sphere for higher-order modes was introduced in [18]. For each mode-order N an orbital Poincaré sphere $\mathscr{O}_{\ell,p}$ was constructed by applying the operator $\hat{U}(\theta, \varphi)$ (15.30) on a Laguerre–Gaussian mode state $|\ell, p\rangle$[3]

$$|\theta, \varphi\rangle_{\ell,p} = \exp(-i\theta\, \hat{\mathscr{L}} \cdot \mathbf{u}_\varphi)|\ell, p\rangle \,. \tag{15.31}$$

[3]For simplicity, we focus on the transverse profile of a single-photon, considering a certain polarization σ and a frequency ω, see Eq. (15.25). Also, without loss of generality, we consider all $\mathscr{O}_{\ell,p}$ having $\ell \geq 0$. Spheres with $\ell < 0$ exhibit a state configuration identical to those with $\ell > 0$, after inversion with respect to their centers.

Therefore, the state $|\theta,\varphi\rangle_{\ell,p}$, represented by the point (θ,φ) on the orbital Poincaré sphere $\mathscr{O}_{\ell,p}$, has mode-order $N = 2p + |\ell| \geq 0$. This generalization provides a suitable framework to visualize the geometric phases in paraxial optical beams under continuous mode transformations [33, 65]. An important observation is that such a rotation gives rise to a superposition of Laguerre–Gaussian mode states that can be cast as

$$|\theta,\varphi\rangle_{\ell,p} = \sum_{\ell'=-\ell}^{\ell} \sum_{p'=0}^{p} C_{\ell',p'}(\theta,\varphi;\ell,p)\,|\ell',p'\rangle\,, \tag{15.32}$$

where $2p' + |\ell'| = 2p + |\ell| = N$. The complex coefficients $C_{\ell',p'}(\theta,\varphi;\ell,p)$ depend on the point (θ,φ) on the Nth-order Poincaré sphere $\mathscr{O}_{\ell,p}$. For instance, in the particular case of Eq. (15.29) one obtains $C_{1,0}(\theta,\varphi;1,0) = \cos\theta/2$, $C_{0,0}(\theta,\varphi;1,0) = 0$, and $C_{-1,0}(\theta,\varphi;1,0) = e^{i\varphi}\sin\theta/2$. Figure 15.3b depicts representative modes associated with their angular orientations on the second-order-mode $\mathscr{O}_{\ell,p}$ sphere ($N = \ell = 2$).

The (θ,φ)-distribution of states on $\mathscr{O}_{\ell,p}$ can be computed by employing the Wigner function representation (15.22) of a photon in a pure state $\psi(\mathbf{q}) = \langle \mathbf{q}|\psi\rangle$

$$W(\boldsymbol{\xi}) = \frac{1}{(2\pi\lambda)^2}\int_{-\infty}^{\infty} d^2\boldsymbol{\xi}\,\exp(i\mathbf{r}\cdot\boldsymbol{\xi}/\lambda)\psi\left(\mathbf{q}+\frac{1}{2}\boldsymbol{\xi}\right)\psi^*\left(\mathbf{q}-\frac{1}{2}\boldsymbol{\xi}\right). \tag{15.33}$$

As shown in [18], it is possible to derive a closed-form expression for the Wigner representation of states $|\theta,\varphi\rangle_{\ell,p}$ without explicitly calculating the integrals in Eq. (15.33). By invoking the Stone–von Neumann theorem (15.5), we know that unitary operators whose generators are quadratic in $\hat{\boldsymbol{\xi}}$ [such as $\hat{U}(\theta,\varphi)$] induce linear canonical transformations, $\mathbf{T} : \hat{\boldsymbol{\xi}}' \to \mathbf{T}\hat{\boldsymbol{\xi}}$, in the optical phase space. Moreover, under the action of $\mathbf{T}$ the Wigner function (15.24) experiences a simple point transformation (15.24) $W(\boldsymbol{\xi}) \xrightarrow{\mathbf{T}} W'(\boldsymbol{\xi}) = W(\mathbf{T}^{-1}\boldsymbol{\xi})$. In our case, the linear canonical transformation generated by the quadratic operators (15.31) results from the relation $\hat{U}^{-1}(\theta,\varphi)\hat{\boldsymbol{\xi}}\hat{U}(\theta,\varphi) = \mathbf{T}(\theta,\varphi)\hat{\boldsymbol{\xi}}$, and reads as

$$\mathbf{T}(\theta,\varphi) = \begin{pmatrix} c_\theta & 0 & -\frac{w_0^2 s_\theta s_\varphi}{2\lambda} & \frac{w_0^2 s_\theta c_\varphi}{2\lambda} \\ 0 & c_\theta & \frac{w_0^2 s_\theta c_\varphi}{2\lambda} & \frac{w_0^2 s_\theta s_\varphi}{2\lambda} \\ \frac{2\lambda s_\theta s_\varphi}{w_0^2} & -\frac{2\lambda s_\theta c_\varphi}{w_0^2} & c_\theta & 0 \\ -\frac{2\lambda s_\theta c_\varphi}{w_0^2} & -\frac{2\lambda s_\theta s_\varphi}{w_0^2} & 0 & c_\theta \end{pmatrix}, \tag{15.34}$$

where $c_\theta = \cos(\theta/2)$, $s_\theta = \sin(\theta/2)$, $c_\varphi = \cos\varphi$, $s_\varphi = \sin\varphi$ (recall that w_0 is the mode width at $z = 0$). Also, the action of $\mathbf{T}(\theta,\phi)$ is independent of the chosen states at $\theta = \varphi = 0$, that is, besides $|\ell,p\rangle$, one could have chosen other state vectors, for instance, Hermite–Gaussian states $|n_x,n_y\rangle$ with equal sphere-order $N = n_x + n_y$.

The key point is thus to observe that owing to the unitary relation (15.31) between states belonging to the same sphere $\mathscr{O}_{\ell,p}$, knowledge of the Wigner function of any given state on $\mathscr{O}_{\ell,p}$ allows one to determine the Wigner function of all states on that same sphere. Laguerre–Gaussian states constitute the convenient choice here. Using their Wigner representation [56] and Eq. (15.34), the found normalized Wigner function is [18]

$$W_{\ell,p}(\boldsymbol{\xi};\theta,\varphi) = \frac{(-1)^N}{\pi^2 \lambda^2} e^{-Q_0} L_{\frac{N-\ell}{2}}\left(Q_0 - 4\mathbf{Q}\cdot\mathbf{u}_r\right) L_{\frac{N+\ell}{2}}\left(Q_0 + 4\mathbf{Q}\cdot\mathbf{u}_r\right), \quad (15.35)$$

where $Q_0 = 2(x^2+y^2)/w_0^2 + (p_x^2+p_y^2)w_0^2/2\lambda^2$, $L_m(\eta)$ are the mth order Laguerre polynomials, and the quadratic polynomials $\mathbf{Q}(\boldsymbol{\xi}) \equiv (Q_x, Q_y, Q_z)$ follow from $\hat{\mathscr{L}}_x$, $\hat{\mathscr{L}}_y$, and $\hat{\mathscr{L}}_z$ in Eq. (15.16) by replacing $\hat{\boldsymbol{\xi}} \to \boldsymbol{\xi}$. When $\theta = 0$ ($\theta = \pi$) one recovers from Eq. (15.35) the Wigner function of LG states $|\ell,p\rangle$ ($|-\ell,p\rangle$). If $\theta = \pi/2$ and $\varphi = 0$ ($\theta = \pi/2$ and $\varphi = \pi$) one obtains the Wigner function of Hermite–Gaussian states $|n_x, N-n_x\rangle$ ($|N-n_y, n_y\rangle$).

Equation (15.35) is a strictly positive and angle-independent Gaussian distribution only when $\ell = p = 0$ (in this case its associated Poincaré sphere becomes degenerated, i.e. all points (θ,φ) on the sphere represent the same Gaussian mode state). This is in accordance with the Hudson–Piquet theorem which states that the only non-negative Wigner function is a Gaussian distribution. Moreover, though $W_{\ell,p}(\boldsymbol{\xi};\theta,\varphi)$ does not explicitly contain the propagation variable z, its spatial evolution along z can be fully described by applying a Galilean boost $\mathbf{r} \to \mathbf{r} - z\mathbf{q}$.

To conclude, it is interesting to mention that: (1) any two states belonging to different spheres are mutually orthogonal; (2) if $\ell > 0$ and $p = 0$, only states corresponding to antipodal points on the orbital Poincaré sphere are mutually orthogonal. However, if $\ell > 0$ and $p > 0$, additional points exist on the sphere (apart from the antipodal) where their associated states are also orthogonal; (3) when $\ell = 0$, antipodal points no longer correspond to orthogonal states but to identical states.

15.3.4 Quantum Gates Using Linear Canonical Transformations

In the context of implementing quantum information protocols, a basic requirement is the construction of a complete set of quantum gates [9, 21, 26]. Single qu*d*it gates assure that any given qu*d*it state in the Hilbert space $\mathscr{H}$ can be transformed onto any other qu*d*it state belonging to the same Hilbert space $\mathscr{H}$. Hence, a necessary condition to perform unitary operators on (pure state) qu*d*its $|\psi\rangle = \sum_{j=1}^{d} \alpha_j |j\rangle$, consisting of a d-dimensional superposition of orthogonal eigenmodes $|j\rangle$, is to modify in a controlled way each of the complex coefficients α_j. This requires, in general, that the unitary operators obey the SU(d) algebra.

Here we consider unitary operators $\hat{U}(\mathbf{T})$ corresponding to the linear canonical transformations (see Sect. 15.2), acting on qu*d*its encoded in the transverse-spatial profile of a single photon, such as Eqs. (15.26) and (15.27). We will discuss the validity of these operators to construct single qu*d*it gates. In most of the experimental realizations oriented towards the use of spatial degrees of photons for high-dimensional encoding, phase holograms and reconfigurable spatial light modulators have been employed to approximately manipulate specific combinations of optical transverse modes [31, 64]. In practice, however, these elements do not strictly preserve paraxiality but, rather, behave as non-unitary transformations, thus constituting a source of mode noise that eventually destroys the desired large, but finite, multidimensionality of the quantum states to be exploited. The unitary operators $\hat{U}(\mathbf{T})$, implemented by lossless FOS [4, 13, 58], overcome these difficulties as they preserve unitarity and paraxiality in a controlled fashion.

In [19] it was proven that it is impossible to arbitrarily modify d-dimensional mode-superpositions of single-photon pure qu*d*it states (15.27) for $d > 2$, via unitary operations $\hat{U}(\mathbf{T})$ generated by symplectic transforms $\mathbf{T} \in \mathrm{Sp}(4,\mathbf{R})$. For $d = 2$ the unitary operators generated by the passive generators (15.16) satisfy the SU(2) algebra, therefore it is possible to construct single quantum gates on the Hilbert space spanned by first-order ($N = 1$) paraxial modes. However, for $d > 2$ the situation is different. For example, let us consider the Hilbert space with dimension $d = 3$ spanned by the paraxial modes $\{|0,0\rangle, |1,0\rangle, |0,1\rangle\}$. Here, passive unitary operators cannot connect paraxial modes with different mode-order number, hence we cannot carry out the conversion of the fundamental Gaussian mode $|0,0\rangle$ into a first-order Hermite–Gaussian mode. One may attempt to construct a qutrit (qu*d*it with $d = 3$) by using paraxial modes with the same mode-order number. For example, by considering the Hilbert space spanned by $\{|2,0\rangle, |1,1\rangle, |0,2\rangle\}$. In this case, one can perform some unitary transformations, but not all of them. In particular, it is impossible to convert the mode $|1,1\rangle$ into $|2,0\rangle$ by means of passive unitary operators. Furthermore, as unitary operators generated by active generators (15.17) involve transformations with an infinity number of paraxial modes, one cannot maintain mode conversion within a finite Hilbert space [19].

Hence, unitary operations $\hat{U}(\mathbf{T})$ generated by symplectic transforms $\mathbf{T} \in \mathrm{Sp}(4,\mathbf{R})$ do not allow a complete set of quantum gates for $d > 2$. Nevertheless, in [19, 66] a new set of non-Gaussian operations were introduced that truly enable the arbitrary manipulation (up to global phases) of qutrits. These transformations are related to aberrations [28, 70]. However, we will not describe these transformations here as they are beyond the scope of this chapter.

Unitary operations $\hat{U}(\mathbf{T})$ can also be used for quantum tomography problems, in which the spectrum of the spatial transverse profile can be reconstructed for a single photon. Such spatial spectrum analyzer can be implemented with a small number of elementary refractive elements embedded in a single Mach–Zehnder interferometer [21, 51].

15.4 Conclusions

The aim of this chapter has been to provide a condensed summary of how the symplectic invariant approach of lossless linear canonical transformations can be applied to manipulate the transverse spatial structure of single photons prepared in superpositions of paraxial modes. By resorting to the Stone–von Neumann theorem, we have made use of the well-known mapping (up to a global phase) between unitary operators belonging to the metaplectic group and the symplectic transformation matrices associated with linear-and-lossless optical elements. Particular emphasis has been put on the action of canonical unitary operations on two relevant paraxial mode bases: Hermite- and Laguerre–Gaussian modes. Within the metaplectic representation, the explicit form of all two-mode unitary operators has been split into two sets of passive and active generators. The transformation properties of the mode-order-preserving passive operators have been elucidated by means of the orbital Poincaré sphere, where the OAM of the paraxial modes displays an elegant geometric interpretation. We have also shown how the active operators differ from the mode-order-preserving passive operators, giving rise to a transverse squeezing and a change of order involving an infinite mode superposition. We have made explicit the action of these generators within the Wigner function formalism. We have ended the chapter by briefly discussing how linear canonical transformations on spatial paraxial modes can be employed for elementary quantum information tasks. It is noteworthy to remark that although the canonical operators introduced in Eq. (15.11) represent transverse-spatial modes of a single photon state, there is a vast literature about quantum information tasks with canonical operators, having an analogous formalism, representing different physical actions on photon states [7, 16, 40, 50, 60, 69].

Acknowledgements This work was supported by the U.S. Department of Energy, Basic Energy Sciences, Office of Science, under contract # DE-AC02-06CH11357. G.F.C. gratefully acknowledges the University of Castilla-La Mancha for financial support. G.F.C and A.P. acknowledge fruitful discussions with T. Alieva, J.A. Rodrigo, M.L. Calvo, M. VanValkenburgh, and S. Walborn.

References

1. G.S. Agarwal, SU(2) structure of the Poincaré sphere for light beams with orbital angular momentum. J. Opt. Soc. Am. A **16**, 2914–2916 (1999)
2. G.S. Agarwal, E. Wolf, Calculus for functions of noncommuting operators and general phase space methods in quantum mechanics: I. Mapping theorems and ordering of functions of noncommuting operators. Phys. Rev. D **2**, 2161–2186 (1970)
3. G.S. Agarwal, E. Wolf, Calculus for functions of noncommuting operators and general phase space methods in quantum mechanics: II. Quantum mechanics in phase space. Phys. Rev. D **2**, 2187–2205 (1970)
4. T. Alieva, M.J. Bastiaans, Mode mapping in paraxial lossless optics. Opt. Lett. **30**, 1461–1463 (2005)

5. L. Allen, M.W. Beijersbergen, R.J.C. Spreeuw, J.P. Woerdman, Orbital angular momentum of light and the transformation of Laguerre–Gaussian laser modes. Phys. Rev. A **45**, 8185–8189 (1992)
6. Arvind, B. Dutta, N. Mukunda, R. Simon, The real symplectic groups in quantum mechanics and optics. Pramana J. Phys. **45**, 471–497 (1995)
7. A. Aspuru-Guzik, P. Walther, Photonic quantum simulators. Nature Phys. **8**, 285–291 (2012)
8. M.A. Bandres, J.C. Gutiérrez-Vega, Ince Gaussian beams. Opt. Lett. **29**, 144–146 (2004)
9. A. Barenco, C.H. Bennett, R. Cleve, D.P. DiVincenzo, N. Margolus, P. Shor, T. Sleator, J.A. Smolin, H. Weinfurter, Elementary gates for quantum computation. Phys. Rev. A **52**, 3457–3467 (1995)
10. S.M. Barnett, *Methods in Theoretical Quantum Optics* (Oxford University Press, New York, 2003)
11. M.J. Bastiaans, The Wigner distribution function applied to optical signals and systems. Opt. Commun. **25**, 26–30 (1978)
12. M.J. Bastiaans, Wigner distribution function and its application to first-order optics. J. Opt. Soc. Am. A **69**, 1710–1716 (1979)
13. M.J. Bastiaans, T. Alieva, Synthesis of an arbitrary ABCD system with fixed lens positions. Opt. Lett. **31**, 2414–2416 (2006)
14. H. Bechmann-Pasquinucci, A. Peres, Quantum cryptography with 3-state systems. Phys. Rev. Lett. **85**, 3313–3316 (2000)
15. I. Bialynicki-Birula, Exponential Localization of Photons. Phys. Rev. Lett. **24**, 5247–5250 (1998)
16. S.L. Braunstein, P. van Loock, Quantum information with continuous variables. Rev. Mod. Phys. **77**, 513–577 (2005)
17. D. Bruss, C. Macchiavello, Optimal eavesdropping in cryptography with three-dimensional quantum states. Phys. Rev. Lett. **88**, 127901 (2002)
18. G.F. Calvo, Wigner representation and geometric transformations of optical orbital angular momentum spatial modes. Opt. Lett. **30**, 1207–1209 (2005)
19. G.F. Calvo, A. Picón, Manipulation of single-photon states encoded in transverse spatial modes: possible and impossible tasks. Phys. Rev. A **77**, 012302 (2008)
20. G.F. Calvo, A. Picón, E. Bagan, Quantum field theory of photons with orbital angular momentum. Phys. Rev. A **73**, 013805 (2006)
21. G.F. Calvo, A. Picón, R. Zambrini, Measuring the complete transverse spatial mode spectrum of a wave field. Phys. Rev. Lett. **100**, 173902 (2008)
22. O. Castaños, E. López-Moreno, K.B. Wolf, *Lie Methods in Optics* (Springer, Berlin, 1986)
23. N.J. Cerf, M. Bourennane, A. Karlsson, N. Gisin, Security of quantum key distribution using d-level systems. Phys. Rev. Lett. **88**, 127902 (2002)
24. D. Collins, N. Gisin, N. Linden, S. Massar, S. Popescu, Bell inequalities for arbitrarily high-dimensional systems. Phys. Rev. Lett. **88**, 040404 (2002)
25. D.-S. Ding, W. Zhang, Z.-Y. Zhou, S. Shi, G.-Y. Xiang, X.-S. Wang, Y.-K. Jiang, B.-S. Shi, G.-C. Guo, Quantum storage of orbital angular momentum entanglement in an atomic ensemble. Phys. Rev. Lett. **114**, 050502 (2015)
26. D.P. DiVincenzo, The physical implementation of quantum computation. Fortschr. Phys. **48**, 771–783 (2000)
27. D. Dragoman, Phase space correspondence between classical optics and quantum mechanics. Prog. Opt. **43**, 433–496 (2002)
28. A.J. Dragt, Lie algebraic theory of geometrical optics and optical aberrations. J. Opt. Soc. Am. **72**, 372–379 (1982)
29. L.-M. Duan, M.D. Lukin, J.I. Cirac, P. Zoller, Long-distance quantum communication with atomic ensembles and linear optics. Nature **414**, 413–418 (2001)
30. S. Feng, H.G. Winful, Physical origin of the Gouy phase shift. Opt. Lett. **26**, 485–487 (2001)
31. R. Fickler, R. Lapkiewicz, W.N. Plick, M. Krenn, C. Schaeff, S. Ramelow, A. Zeilinger, Quantum entanglement of high angular momenta. Science **338**, 640–643 (2012)

32. S. Franke-Arnold, L. Allen, M. Padgett, Advances in optical angular momentum. Laser Photonics Rev. **2**, 299–313 (2008)
33. E.J. Galvez, P.R. Crawford, H.I. Sztul, M.J. Pysher, P.J. Haglin, R.E. Williams, Geometric phase associated with mode transformations of optical beams bearing orbital angular momentum. Phys. Rev. Lett. **90**, 203901 (2003)
34. A.D. Greentree, S.G. Schirmer, F. Green, L.C.L. Hollenberg, A.R. Hamilton, R.G. Clark, Maximizing the Hilbert space for a finite number of distinguishable quantum states. Phys. Rev. Lett. **92**, 097901 (2004)
35. S. Gröblacher, T. Jennewein, A. Vaziri, G. Weihs, A. Zeilinger, Experimental quantum cryptography with qutrits. New J. Phys. **8**, 75 (2006)
36. N. Gisin, R. Thew, Quantum communication. Nat. Photonics **1**, 165–171 (2007)
37. M. Hillery, R.F. O'Connell, M.O. Scully, E.P. Wigner, Distribution functions in physics: fundamentals. Phys. Rep. **106**, 121–167 (1984)
38. D. Kaszlikowski, P. Gnaciński, M. Żukowski, W. Miklaszewski, A. Zeilinger, Violations of local realism by two entangled N-dimensional systems are stronger than for two qubits. Phys. Rev. Lett. **85**, 4418–4421 (2000)
39. H.J. Kimble, The quantum internet. Nature **453**, 1023–1030 (2008)
40. E. Knill, R. Laflamme, G.J. Milburn, A scheme for efficient quantum computation with linear optics. Nature **409**, 46–52 (2001)
41. B.P. Lanyon, M. Barbieri, M.P. Almeida, T. Jennewein, T.C. Ralph, K.J. Resch, G.J. Pryde, J.L. O'Brien, A. Gilchrist, A.G. White, Simplifying quantum logic using higher-dimensional Hilbert spaces. Nat. Phys. **5**, 134–140 (2009)
42. M. Lassen, G. Leuchs, U.L. Andersen, Continuous variable entanglement and squeezing of orbital angular momentum states. Phys. Rev. Lett. **102**, 163602 (2009)
43. H. Lee, Theory and application of the quantum phase-space distribution functions. Phys. Rep. **259**, 147–211 (1995)
44. R. Loudon, *The Quantum Theory of Light* (Clarendon Press, Oxford, 1983)
45. L. Mandel, E. Wolf, *Optical Coherence and Quantum Optics* (Cambridge University Press, New York, 1995)
46. W. Mecklenbraüker, F.F. Hlawatsch (eds.), *The Wigner Distribution-Theory and Applications in Signal Processing* (Elsevier, Amsterdam, 1997)
47. G. Nienhuis, L. Allen, Paraxial wave optics and harmonic oscillators. Phys. Rev. A **48**, 656–665 (1993)
48. J.L. O'Brien, A. Furusawa, J. Vučković, Photonic quantum technologies. Nat. Photonics **3**, 687–695 (2009)
49. M.J. Padgett, J. Courtial, Poincaré-sphere equivalent for light beams containing orbital angular momentum. Opt. Lett. **24**, 430–432 (1999)
50. T.C. Ralph, G.J. Pryde, Optical quantum computation. Prog. Opt. **54**, 209–269 (2010)
51. J. Řeháček, Z. Hradil, Z. Bouchal, R. Čelechovský, I. Rigas, L.L. Sánchez-Soto, Full tomography from compatible measurements. Phys. Rev. Lett. **103**, 250402 (2009)
52. J.J. Sakurai, *Modern Quantum Optics* (Addison-Wesley, Madrid, 1994)
53. F. Scheck, *Quantum Physics* (Springer, Heidelberg, 2007)
54. W.P. Schleich, *Quantum Optics in Phase Space* (Wiley-VCH, Berlin, 2001)
55. A.E. Siegman, *Lasers* (University Science Books, Sausalito, 1986)
56. R. Simon, G.S. Agarwal, Wigner representation of Laguerre–Gaussian beams. Opt. Lett. **25**, 1313–1315 (2000)
57. R. Simon, N. Mukunda, Optical phase space, Wigner representation, and invariant quality parameters. J. Opt. Soc. Am. A **17**, 2440–2463 (2000)
58. R. Simon, K.B. Wolf, Structure of the set of paraxial optical systems. J. Opt. Soc. Am. A **17**, 342–355 (2000)
59. K. Sundar, N. Mukunda, R. Simon, Coherent-mode decomposition of general anisotropic Gaussian Schell-model beams. J. Opt. Soc. Am. A **12**, 560–569 (1995)
60. S. Tanzilli, A. Martin, F. Kaiser, M.P. De Micheli, O. Alibart, D.B. Ostrowsky, On the genesis and evolution of integrated quantum optics. Laser Photonics Rev. **6**, 115–143 (2011)

61. D.S. Tasca, R.M. Gomes, F. Toscano, P.H. Souto Ribeiro, S.P. Walborn, Continuous-variable quantum computation with spatial degrees of freedom of photons. Phys. Rev. A **83**, 052325 (2011)
62. M.E. Testorf, B.M. Hennelly, J. Ojeda-Castañeda (eds.), *Phase-Space Optics: Fundamentals and Applications* (McGraw-Hill, New York, 2009)
63. A. Torre, *Linear Ray and Wave Optics in Phase Space* (Elsevier, Amsterdam, 2005)
64. J.P. Torres, L. Torner (eds.), *Twisted Photons* (Wiley-VCH, Weinheim, 2011)
65. S.J. van Enk, Geometric phase, transformations of gaussian light beams and angular momentum transfer. Opt. Commun. **102**, 59–64 (1993)
66. M. VanValkenburgh, Manipulation of semiclassical photon states. J. Math. Phys. **50**, 023501 (2009)
67. S.P. Walborn, D.S. Lemelle, M.P. Almeida, P.H. Souto Ribeiro, Quantum key distribution with higher-order alphabets using spatially encoded qudits. Phys. Rev. Lett. **96**, 090501 (2006)
68. X.-L. Wang, X.-D. Cai, Z.-E. Su, M.-C. Chen, D. Wu, L. Li, N.-L. Liu, C.-Y. Lu, J.-W. Pan, Quantum teleportation of multiple degrees of freedom of a single photon. Nature **518**, 516–519 (2015)
69. C. Weedbrook, S. Pirandola, R. García-Patrón, N.J. Cerf, T.C. Ralph, J.H. Shapiro, S. Lloyd, Gaussian quantum information. Rev. Mod. Phys. **84**, 621–669 (2012)
70. K.B. Wolf, Linear transformations and aberrations in continuous and finite systems. J. Phys. A Math. Theor. **41**, 304026 (2008)
71. Z.S. Yuan, X.H. Bao, C.Y. Lu, J. Zhang, C.Z. Peng, J.W. Pan, Entangled photons and quantum communication. Phys. Rep. **497**, 1–40 (2010)

MIX
Papier aus verantwortungsvollen Quellen
Paper from responsible sources
FSC® C105338

If you have any concerns about our products,
you can contact us on
ProductSafety@springernature.com

In case Publisher is established outside the EU,
the EU authorized representative is:
Springer Nature Customer Service Center GmbH
Europaplatz 3, 69115 Heidelberg, Germany

Printed by Libri Plureos GmbH
in Hamburg, Germany